P9-ELO-506

NATIONAL
GEOGRAPHIC

Field Guide to the
Birds
OF NORTH AMERICA

SIXTH EDITION

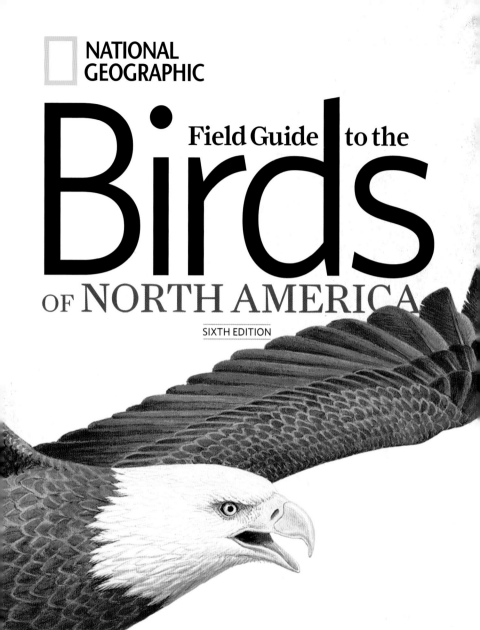

NATIONAL GEOGRAPHIC

Field Guide to the

Birds

OF NORTH AMERICA

SIXTH EDITION

Jon L. Dunn *and* **Jonathan Alderfer**
with maps by Paul Lehman

NATIONAL GEOGRAPHIC
WASHINGTON, D.C.

CONTENTS

INTRODUCTION

New illustration of
Wilson's Storm-Petrel
created for this edition

Recently recorded accidental
species from Florida:
Red-legged Thrush *plumbeus*

adult ♂

♀

Yellow Warbler *aestiva:*
a widespread
North American breeder

I n 2006, National Geographic released the fifth edition of *The Field Guide to the Birds of North America,* featuring 967 species. That was followed in 2008 by both *The Field Guide to the Birds of Eastern North America* and *The Field Guide to the Birds of Western North America,* compact versions of the main guide that are focused on those distinct faunal regions. Now, National Geographic releases this, the sixth edition of *The Field Guide to the Birds of North America,* in which we cover 990 species — quite an increase since the release of the first edition to the field guide in 1983, which included just over 800.

In addition to including more species, we have redesigned the guide so the illustrations are less crowded. This change has also made room for expanded species accounts that provide more in-depth coverage of vocalizations, range, and geographical variation (subspecies). We have added many new illustrations: Some are for the recently added species and others are replacements of previous figures.

The five artists who created the new art for this edition—Jonathan Alderfer, Killian Mullarney, David Quinn, John Schmitt, and Thomas Schultz—are all active birders with years of field experience. Their challenge was to render images that showed all the essential field marks of a species in lifelike postures, something a single photograph can rarely accomplish with such clarity. To this end they consulted many sources, including photographic references, museum specimens, and the most up-to-date bird identification literature. We are grateful to the institutions that have loaned specimens to the artists directly or photographed specimens (see Acknowledgments, page 562). These museums do more than just archive actual flat skins. They also house tissues with which important genetic work is done to further elucidate the relationships between species and the higher-level relationships at the genus, family, and order level. Such research has shown that some traditional treatments are no longer valid, which means each edition of this and other field guides are revised — often annoyingly so for the birding community. But no field of science is static.

Accordingly, we have also completely reworked and updated our maps. For the first time, we have included migration ranges and in many cases have specified subspecies' ranges, too. For more complex polytypic species (those having more than one subspecies), we have provided larger subspecies maps in the back that present much more detail (page 548). For a few species, we have included two maps: one for breeding range and one for winter. We have maintained the helpful thumb tabs and have introduced a quick-find visual index, first presented in the East and West guides.

At the end of the book is an illustrated list of 92 accidental species that have occurred along with specific details of their occurrence and at least brief details of their appearance (page 530). This section also includes the four species known to have gone extinct during the 19th and 20th centuries and two others that are also likely extinct. In a few cases we have chosen to include accidentals in the main section because we felt a direct comparison with similar species would be useful.

6

Species Selection

This guide includes all species that occurred in North America, defined here as the land extending northward from the northern border of Mexico as well as adjacent islands and seas within 200 nautical miles off the coast or offshore islands. In general, accidental visitors are included in the main text if they have been seen at least three times in the past two decades or five times in the past 100 years; those recorded fewer times are found in the appendix.

Innumerable exotic species from other continents have been introduced into North America as game, park, or cage birds and are seen in the wild frequently. Those with at least some degree of establishment, such as Nutmeg Mannikin, are included here, though it, Orange Bishop, and many species of Psittacidae (e.g., parrots and parakeets) have not yet been accepted by the American Birding Association (hereafter ABA). In general the ABA requires an introduced population to have been present and stable for at least 15 years. Some exotics (e.g., European Starling) have spread continentwide and are a nuisance, while others have maintained more stable populations or even have declined.

We strictly adhere to the American Ornithologists' Union (hereafter AOU) Committee on Classification and Nomenclature for issues of taxonomy and nomenclature — English and scientific names — as determined in the seventh edition of the *AOU Check-list* (1998) and the 11 subsequent supplements, published annually in the July issue of the AOU journal *The Auk*. We also follow either the ABA Checklist Committee or the above-mentioned AOU committee when deciding which species to include. Keep in mind that the AOU list includes the Hawaiian Islands, the West Indies, and all of Middle America, to the Colombian border, in addition to Bermuda, and now Greenland.

Families

Scientists organize animal species into family groups that share certain structural or molecular characteristics. Some bird families have more than a hundred members, others only one. Characteristics within a family are helpful in identifying birds in the field. With sturdy bills, strong claws, and short legs, members of the Picidae family, for example, are easily recognized as woodpeckers.

Brief family descriptions, with information applicable to all members of the family, can be found at the beginning of each group. Additionally, a description of a smaller group within a family (a genus) is sometimes provided, such as that given for the *Empidonax* flycatchers, which share some distinguishing traits.

Scientific Names

Each species has a unique two-part scientific name, derived from Greek or Latin (in italics). The first part, always capitalized, indicates the genus. For example, nine members of the family Picidae are placed in the genus *Picoides*. Together with the second part of the name (the specific epithet), which is not capitalized, this identifies the species.

Downy Woodpecker
Picoides pubescens

bacatus   dorsalis   fasciatus

Polytypic:
American Three-toed Woodpecker

Monotypic:
Cerulean Warbler

Lesser (top)
and **Greater Scaup** (bottom)
are best distinguished
by head shape

juvenile

Both **Short-billed** (above)
and **Long-billed Dowitchers**
move their bills up and down
like a sewing machine
needle when feeding

Alder Flycatcher (top) and
Willow Flycatcher (bottom) are
best separated by song

Picoides pubescens is the name of one specific kind of woodpecker, commonly known as Downy Woodpecker.

SUBSPECIES

Since the latter half of the 19th century, species have been further divided into subspecies, or *races,* when populations from different geographical regions show recognizable differences. Each subspecies bears a third scientific name, or trinomial (also in italics). For instance, of the three subspecies of American Three-toed Woodpecker, *Picoides dorsalis bacatus* (often abbreviated as *P. d. bacatus* or simply *bacatus*), identifies the dark-backed subspecies that inhabits the boreal forest of eastern North America. The Rockies subspecies is the paler-backed race, *dorsalis.* A third and intermediate subspecies, *fasciatus,* is found from Alaska to Oregon. If the third part of a scientific name is the same as the second, the subspecies in question is known as the *nominate subspecies,* the type for which the species was originally described. The subspecies *dorsalis* was named and described in the literature in 1858, earlier than *fasciatus* (1870) or *bacatus* (1900).

For some polytypic species, subspecies are divided into groups given their visual (now often genetic) differences, as with the "Sooty" group of dark Fox Sparrows from the Pacific region. Sometimes a group is known by a scientific name, as is the case with the *unalaschcensis* group of "Sooty" Fox Sparrows; these groups always use the oldest (or first published) name for the representative subspecies.

We rely on the fifth edition of the *AOU Check-list* (1957), the last edition that treated subspecies, for subspecies names; additional sources include Peter Pyle's *Identification Guide to North American Birds,* Allan R. Phillips's *The Known Birds of North and Middle America,* and the landmark *The Birds of North America* series. If the illustrations for a polytypic species show mainly or entirely one subspecies, that subspecies' name appears italicized under the English name, if it is known; illustrations of other subspecies are labeled as such.

Many species are monotypic (having no recognized subspecies), thus only the scientific binomial is used. For example, Cerulean Warbler is monotypic and is simply known by the scientific name *Setophaga cerulea.*

● How to Identify Birds

Field marks, a bird's physical features, are the clues by which birds are identified. They include plumage, or the bird's overall feathering; the shape of the body and its individual parts (see Parts of a Bird, page 10); and any actual markings such as bars, bands, spots, or rings. A field mark can be obvious, like a male Northern Cardinal's red plumage. Other field marks are more subtle, such as the difference in head shapes of Greater and Lesser Scaup. When a plumage in the species account is illustrated, it is noted in boldface (e.g., **first-winter male**).

Remember that the most important thing when birding is to look at the actual bird. There will be plenty of time to consult this field guide later.

● Behavior

Behavioral traits also provide many clues to species identity. Is the bird usually visible and approachable or shy and difficult to locate? Does it hop or walk? Does it flick its tail up or drop it down?

Voice

A bird's songs and calls reveal not only its presence, but also, in many cases, its identity. In fact, vocalizations are becoming increasingly useful to the taxonomist in determining species relationships. These are described under the heading **VOICE**.

Some species — particularly nocturnal or secretive birds such as owls, nightjars, and rails — are more often heard than seen. A few species are most reliably identified by voice even when they are seen well. When birds assemble or travel in flocks, they often keep in touch with a *contact* or *flight call* that may be markedly different from their other calls.

House Sparrows' appearance changes with wear, not molt

Molt and Plumage Sequence

The regular renewal of plumage, called *molt*, is essential to a bird's ability to fly and to its overall health. A molt produces a specific plumage, and not all birds go through the same sequence of plumages. Some species have a very simple sequence of plumages from juvenile to adult: For example, most raptors molt from juvenal plumage directly into adult plumage. Other species, such as many of the gulls, take more than three years to acquire adult plumage and go through a complicated series of interim plumages. Shown in this guide are all of the most distinctive plumages likely to be encountered in the field.

The first coat of true feathers, acquired before a bird leaves the nest, is called the *juvenal plumage;* birds in this plumage are referred to as *juveniles.* In many species, juvenal plumage is replaced in late summer or early fall by a *first-fall* or *first-winter* plumage that more closely resembles adult plumage. First-fall and any subsequent plumages that do not resemble the adult — known as *immature plumages* — may continue in a series that includes *first-spring* (when the bird is almost a year old), first-summer, and so on, until *adult plumage* is attained. When birds take several years to reach adult plumage, we label the interim plumages as *subadult* or note the specific year or season shown.

Western Tanagers undergo a complete molt from summer to late fall

In adult birds, the same annual sequence of molt and plumages is repeated throughout the bird's life. Most adults undergo a complete molt, replacing all their feathers, in late summer or early fall, after breeding. For some species this is the only molt of the year, so adults of these species have the same plumage year-round. Other species undergo a partial molt—usually involving the head, body, and some wing coverts—in late winter or early spring, resulting in a more colorful *breeding plumage* that is seen in spring and summer. For species with two annual molts, the plumage attained after breeding is referred to as the *nonbreeding* or *winter plumage*. Some changes are evident only during the brief period of courtship. In herons, for example, the colors of bill, lores, legs, and feet may change. When these colors are at their height, the birds are said to be in *high breeding plumage.* After breeding, most ducks molt into a briefly held *eclipse plumage* in which males acquire a femalelike plumage and females show little change, although some become paler and duller.

Wear and fading of a bird's feathers can also have a pronounced effect on its appearance. In species with a single annual molt in fall, the fresh colors exhibited right after molting are usually brighter than the faded colors seen at the end of the breeding season. For instance,

(continued on page 12)

Hammond's Flycatchers appear brightest in fall just after the complete molt

PARTS OF A BIRD

Because the size, the shape, and the configuration of birds' various parts differ among families and species, birders should become familiar with their names and locations—bird topography. Nonpasserines are quite variable, and we give three examples: hummingbird, shorebird, and gull. Notice how the feathers overlap, and remember that a bird's posture and activity level can affect which feathers are visible.

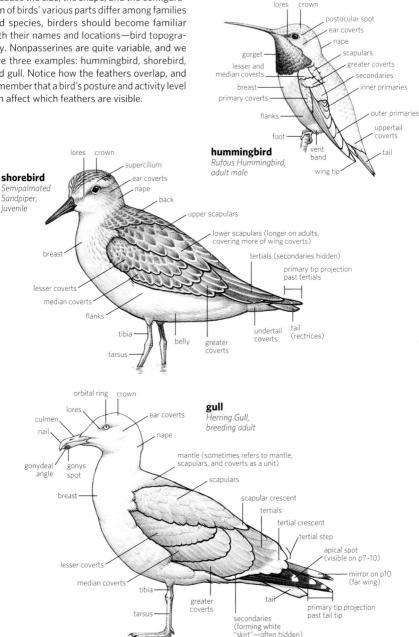

hummingbird
*Rufous Hummingbird,
adult male*

lores   crown
postocular spot
ear coverts
nape
gorget
scapulars
lesser and
median coverts
greater coverts
secondaries
breast
inner primaries
primary coverts
flanks
outer primaries
uppertail
coverts
foot
vent
band
tail
wing tip

shorebird
*Semipalmated
Sandpiper,
juvenile*

lores   crown
supercilium
ear coverts
nape
back
upper scapulars
lower scapulars (longer on adults,
covering more of wing coverts)
breast
tertials (secondaries hidden)
primary tip projection
past tertials
lesser coverts
median coverts
flanks
tibia
belly
greater
coverts
undertail
coverts
tail
(rectrices)
tarsus

gull
*Herring Gull,
breeding adult*

orbital ring   crown
lores
ear coverts
culmen
nail
nape
mantle (sometimes refers to mantle,
scapulars, and coverts as a unit)
gonydeal   gonys
angle   spot
scapulars
breast
scapular crescent
tertials
tertial crescent
tertial step
apical spot
(visible on p7-10)
lesser coverts
mirror on p10
(far wing)
median coverts
tibia
greater
coverts
tail
primary tip projection
past tail tip
tarsus
secondaries
(forming white
"skirt"—often hidden)

Passerines show less variation in body shape and feather organization. The Lark Sparrow example illustrates all the features of the head and wings. Many species show wing bars, formed by the pale tips of the greater and/or median coverts when the wing is folded.

upper wing
Lark Sparrow, adult

tertials

secondaries

scapulars

passerine
Lark Sparrow, adult

marginal coverts
lesser coverts
median coverts
greater coverts
bend of wing
carpal covert
alula

primary no. 1

p2

p3

p4

p5

p6

p7

p8

p9

median crown stripe
lores
lateral crown stripe
supraloral
supercilium
forehead
upper mandible (maxilla)
eye stripe
lower mandible
ear coverts
nape
chin
throat
back
moustachial stripe
submoustachial stripe
scapulars
malar stripe
lesser coverts
breast spot
breast
median coverts
side
greater coverts
primary coverts
flanks
belly
secondaries
primaries

primary coverts
primaries
emargination

tertials
undertail coverts

uppertail coverts

remiges:
overall term for the primaries, secondaries, and tertials

tail (rectrices)

tail spots

Some feather groups can be seen only when a bird is in flight. Other feather groups have special names when they are prominently marked, such as the carpal bar seen on a Common Tern (below). Most birds have twelve tail feathers. When you observe the folded tail from below, the two outermost tail feathers are visible.

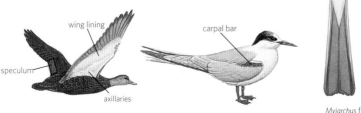

wing lining

carpal bar

speculum

axillaries

American Black Duck

Common Tern, first summer

Myiarchus flycatcher tails from below:
Great Crested, Ash-throated

Hammond's Flycatchers are brightest in fall; by the following summer, their plumage becomes worn and faded. Plumage patterns can also be affected. In certain songbirds, such as Snow Bunting or House Sparrow, the more patterned spring plumage appears as the dull tips of the fresh fall plumage gradually wear away, with no molt involved.

Barrow's x Common Goldeneye hybrid

Often the fall molt occurs before migration, so we see these birds in fresh fall plumage even if they spend winter outside of North America. Certain species, including many shorebirds, suspend their molt during fall migration and complete it on arrival at the wintering grounds. Some species molt over a considerably longer period than others. Birds of prey, for instance, must rely on their wing feathers for successful hunting and thus molt very gradually so that only a few wing feathers are missing at any one time.

● Plumage Variation

Where adult males (♂) and females (♀) of a species are similar, we show only one. When they differ, we usually show both. If spring and fall, or *breeding* and *nonbreeding,* plumages differ only slightly, or if only one of these plumages is seen in North America, we tend to show only one figure. Juveniles and immatures are illustrated when they hold a different-looking plumage after they are old enough to be seen away from their more easily recognizable parents.

Greater (top) and
Lesser Yellowlegs (bottom)
are distinguishable by size

A number of species have two or more *color morphs*—variations in plumage color occurring regionally or within the same population—as is the case with Swainson's Hawk. Some species (usually of the same genus) occasionally breed with each other, producing *hybrid* offspring that may appear intermediate. Different subspecies also interbreed, resulting in *intergrade* individuals or populations.

● Measurements

Knowing the size of a bird is important. When identifying a kinglet, for example, it helps to know that the bird in question is tiny. Relative size is also significant: In mixed flocks, Greater and Lesser Yellowlegs are easily distinguished from each other by size alone.

Average length (L) from tip of bill to tip of tail for each species is given. Where size varies greatly within a species, either because of sex or geographical variation among subspecies, a range of smallest to largest is provided. And for large birds often seen in flight, we give the wingspan (WS), measured from wing tip to wing tip.

Casual:
Eurasian Hobby

● Abundance and Habitat

Abundance must be considered in relation to habitat. Under the heading **RANGE,** the species accounts include supplemental information about habitat, abundance, and seasonal status that cannot be shown in a single map. Some species are highly local, found only in a very specialized habitat.

The following categories of abundance — given from most to least numerous — are used in this guide: *abundant, common, fairly common, uncommon, rare,* and *very rare.* Unlike rare species, *casual* species do not occur annually in North America or from the region detailed, but a pattern of their occurrence is apparent over decades. *Accidental* species have been seen only once or a few times in an area that is far out of their normal range. In fact, it may be decades before another one is

Accidental:
Eurasian Hoopoe

seen there again. Some other terms are used herein, such as *irruptive*, which describes species that are erratic in their movements: One year they may be numerous in a given region and the next year, or even the next decade, they may be totally absent.

● Range Maps

Maps are provided for all species with two general exceptions: (1) introduced species with very limited ranges that are described in the text or where the species is not believed established and (2) some species that do not breed in North America and are of rare, casual, or accidental occurrence here.

On each map, solid-colored ranges end approximately where the species ceases to be regularly seen. Keep in mind that nearly every species will be rare at the edges of its range. The map key on the back cover flap explains the colors and symbols used.

Irruptive:
Snowy Owl

Birds are not, however, bound by maps. Their ranges continually expand and contract. Irruptive species move southward in some years in large or small numbers and for great or small distances. Vagrant individuals of many species occur far outside their normal range. In some species, birds leave the nesting grounds in late summer and then move northward. These post-breeding wanderers, principally young birds, will subsequently migrate southward by winter. Range maps of pelagic species, which spend most of their time over the open sea, are somewhat conjectural. Migration is shown for many non-seabirds across open ocean when a substantial percentage of a species' population regularly makes such an overwater flight.

Range information is based on sightings and therefore depends on the number of knowledgeable and active birders in each area. There is much to learn about bird distribution in every part of North America. Birders assist by monitoring the expanding and contracting of ranges. Breeding-bird surveys and atlas projects make a vital contribution to the general fund of information about each species. The Cornell Laboratory of Ornithology has created eBird (www.ebird .org), a website where birders can post their observations of species.

Rapidly expanding species:
Eurasian Collared-Dove

In the accounts of birds presently on the U.S. federal list of threatened or endangered species, we have placed the symbols **E** (endangered) or **T** (threatened). Canada also has its own list of threatened species, as do many states and provinces. Extinct species are indicated with the symbol **EX.**

● How to Be a Better Birder

The time you spend studying your field guide at home will be repaid when you go out birding. In addition, a number of introductory books designed to help the beginning birder are available. One such book, *Birding Essentials*, published by National Geographic, covers all the basic skills and information.

Observing and studying birds can be an enriching, pleasurable experience with far-reaching rewards. Birders are becoming powerful advocates for the protection and stewardship of our natural resources. We urge you to get involved in conservation activities and to pass your knowledge and love of birds onto others.

Endangered species:
Red-cockaded Woodpecker

JON L. DUNN AND JONATHAN ALDERFER

DUCKS • GEESE • SWANS Family Anatidae

Worldwide family. Web-footed, gregarious birds, ranging from small ducks to large swans. Largely aquatic, but geese, swans, and some "puddle ducks" also graze on land.
SPECIES: 165 WORLD, 63 N.A.

Greater White-fronted Goose *Anser albifrons* L 28" (71 cm)

Grayish brown; irregular black barring on underparts; orange feet and legs. Bill pink or orangish with whitish tip. In flight, note grayish blue wash on wing coverts. Most **immatures** acquire white front above bill and white bill tip during first winter; acquire black belly markings by second fall. Distinguished from bean-geese by bill color; from Pink-footed Goose by bill and leg color; compare also with immature blue-morph Snow Goose (page 16). Color and size vary in **adults**: Pale, Arctic tundra birds have heavy barring; taiga-breeding (central AK) subspecies, *elgasi,* wintering in northern portion of Sacramento Valley, CA, is much larger and darker, with bigger bill, less barring on belly, and thicker, longer neck; some have yellowish eye ring. Greenland's intermediate-size tundra subspecies, *flavirostris,* rare in the Northeast, is darkest, with heaviest ventral barring and an orange bill.

VOICE: Call is a high-pitched, laughing *kah-lah-aluck;* calls of *elgasi* described as hoarser and lower pitched than other subspecies.

RANGE: Large populations are found on eastern Great Plains (fewer east to Mississippi Valley) and Pacific states. Rare winter visitor outside mapped range in all western states and north into southern Canada, on western Great Plains, and in eastern North America. Population of *elgasi* thought to be fewer than 10,000.

Tundra Bean-Goose *Anser serrirostris* L 28-33" (71-84 cm)

Eurasian species. Size increases across range from west to east; eastern nominate being largest. Shorter neck and stubbier bill than Taiga Bean-Goose; yellow-orange band before tip of bill.

VOICE: Calls said to be higher pitched than Taiga Bean-Goose.

RANGE: Breeds on tundra from Russian Northwest to Chukotski Peninsula, Anadyrland, Koryakland, Russian Far East; winter range similar to Taiga Bean-Goose. Specimens and photos from western AK and photo from YT. A specimen (reported as western *rossicus*) was obtained at Cap Tourmente, QC, in Oct. 1982.

Taiga Bean-Goose *Anser fabalis* L 30-35" (76-90 cm)

Eurasian species. Long, almost swanlike neck and long wedge-shaped bill; bill black, typically with yellow-orange band before tip. Size increases from west to east; easternmost *middendorffii* being largest.

VOICE: Calls loud and deep, very different from Greater White-fronted.

RANGE: Breeding on taiga from northern Scandinavia east to Anadyrland and wintering from northeastern Europe to China and Japan. Specimen from St. Paul Island, Pribilofs, on 19 Apr. 1946 and group of three photographed on Shemya Island, Aleutians, from Sept. 2007 into winter. Other photographed birds from western AK under evaluation. Other records from WA, NE, IA, and QC said to be this species, but a recent well-documented record from southeastern CA appears intermediate; Japanese authorities believe it may be from a more westerly population of Taiga Bean-Goose (perhaps an unnamed population).

orange
bill

dark neck

**Greater
White-fronted
Goose**
frontalis

gray on
upper wing

**Greenland
adult**
flavirostris

long, thick-
based bill

heavier
black
bars on
underparts

long, thick
dark neck

adult

orange
legs

white rump
band with
dark gray tail
base

high-pitched
laughing calls
heard in flight

bill color of
immature varies
from orangish
to pinkish

taiga adult
elgasi

immature

some white at
base of bill by end
of first winter

adult

juvenile

western AK *frontalis* is small
and pale; birds breeding from
northern AK (sometimes
recognized as *gambelli*) are
also pale, but larger

fewer black
bars (variable)

largely unmarked
belly

gray on upperwing
like Greater
White-fronted

darkish
head

**Tundra
Bean-Goose**
serrirostris

short black bill
with "grinning
patch" and
yellow-orange
subterminal ring

rather short
necked

adults

slightly larger than
Greater White-fronted

overall coloration and
marking like Tundra
Bean-Goose but much
larger and longer necked

long sloped bill
with little or no
"grinning patch"

**Salton Sea
record**

adults

**Taiga
Bean-Goose**
middendorffii

some Alaskan records and
one from Salton Sea, CA, have
appeared more intermediate
between Taiga and Tundra;
identification uncertain

adult

Pink-footed Goose *Anser brachyrhynchus* L 26" (66 cm)

Distinguished from plain-bellied immature Greater White-fronted Goose by pink legs and variable dark base and tip to stubbier pinkish bill; neck shorter than Greater White-fronted. Juvenile is browner, looks more scaly than barred; legs duller. In flight, shows an extensive area of bluish gray on mantle and all wing coverts; darker head and neck contrast with grayish body; some show very restricted white at bill base; tail base is grayer and paler than Greater White-fronted and bean-geese.

RANGE: East Coast records likely from breeding grounds in eastern Greenland. A scattering of records (almost annual in recent years) between eastern PA and Newfoundland; most sightings thought to be wild birds.

Snow Goose *Chen caerulescens* L 26-33" (66-84 cm)

Two color morphs. **Adults** distinguished from smaller Ross's Goose by larger, pinkish bill with black "grinning patch." Flies with slower wingbeat than Ross's; rusty stains often visible on face in summer. **White-morph juvenile** is grayish above, with dark bill. **Blue-morph adult** has mostly white head and neck, brown back, variable amount of white on underparts. **Blue-morph juvenile** has dark head and neck, overall slaty body coloring; distinguished from Greater White-fronted Goose (page 14) by dark legs and bill and lack of white on face. Intermediates between white and blue morphs have mainly white underparts and whitish wing coverts.

VOICE: Gives single high-pitched calls.

RANGE: Locally abundant. Occasionally hybridizes with Ross's Goose. A larger subspecies, the "Greater Snow Goose" (not shown), breeds around Baffin Bay, winters only along mid-Atlantic coast; blue morph almost unknown. Smaller subspecies, "Lesser Snow Goose" (morphs shown here), is rare in winter throughout interior U.S. and in southern Canada outside mapped range. Blue morph is abundant in central North America; rare west of the Great Plains; uncommon in East.

Ross's Goose *Chen rossii* L 23" (58 cm)

Stubby, triangular bill, mostly deep pinkish red with purplish blue-gray base (lacking in Snow Goose). More vertical demarcation between feathering and base of bill than Snow Goose. Neck shorter, head smaller and rounder than Snow Goose; white head generally lacks rusty stains. Ross's has two color morphs. **White-morph juvenile** may have very pale gray wash on head, back, and flanks, but far less than juvenile white-morph Snow Goose. Extremely rare **blue morph** is darker than blue-morph Snow Goose; face and belly are white.

VOICE: Calls are higher pitched and delivered more rapidly than Snow Goose.

RANGE: Seen during migration in grasslands and grainfields in eastern Great Plains and Mississippi Valley; rare but regular farther east to QC and western VT. Rare winter visitor to mid-Atlantic states; rare visitor also to much of West outside mapped range. Casual to YT and AK. Usually seen with Snow Geese, although in East often with Canada.

some have white rim at bill base

Pink-footed Goose

comparatively small head, dark neck

small, stubby, dark-based bill with pinkish band

warm cinnamon-buff tinge at base of neck

broad white tip to tail

pinkish legs and feet

light silvery or "frosty" gray upperparts on most adults; some are browner

more extensive blue-gray on upperwing than Greater White-fronted

broad white tail tip

adults

Snow Goose
caerulescens

black "grinning patch" on bill

blue-morph adults

rusty stains on face

dark bill

darker above than juvenile Ross's

white-morph juvenile

variably dark below

white-morph adult

blue-morph juvenile

mostly dark brown

adult

more rapid wingbeats than Snow

short neck

rounder head than Snow

Ross's Goose

very similar to adult

white-morph juvenile

white-morph adult

stubby bill

darker than blue-morph Snow

dark neck

blue-morph adult

Emperor Goose *Chen canagica* L 26" (66 cm)

Fairly stocky, small goose with short, thick neck. Head and back of neck white; chin and throat black; face often stained rusty in summer. Bill pinkish; lower mandible is sometimes black. Black-and-white edging to silvery gray plumage creates a scaled effect below; upperparts appear barred. **Juvenile** has dark head and bill. During first fall, acquires white flecking on head; resembles adult by first winter.

VOICE: Calls include a high-pitched, hoarse *kla-ha*. Less vocal than other geese.

RANGE: Breeds in tidewater marsh and tundra; winters on seashores, reefs. Casual south on Pacific coast to central CA and inland to Sacramento Valley.

Barnacle Goose *Branta leucopsis* L 27" (69 cm)

Part of world population breeds in northeastern Greenland; accidental vagrant in Maritime Provinces. Note distinctive head pattern and stubby bill. Bluish gray upperparts, barred with black; white, U-shaped rump band. Silver-gray wing linings show in flight.

VOICE: High-pitched cackling.

RANGE: Increasing Greenland population may be responsible for increase in reports from Maritime Provinces to mid-Atlantic region. Also fairly common in captivity. At least some sightings from eastern seaboard seem likely to be of wild birds; sightings from interior are of more problematic origin. Recorded west to NM, CA, and AK. More of a land goose than Brant, feeding in fields near the ocean.

See subspecies map, page 549

Brant *Branta bernicla* L 25" (64 cm)

A small, dark, stocky sea goose. Note whitish patch on side of neck. White uppertail coverts almost conceal black tail. White undertail coverts conspicuous in flight. Wings comparatively long and pointed, wingbeat rather rapid. **Immature** birds show bold white edging to wing coverts and secondaries, and fainter neck patches than **adults.** **Juveniles** sometimes lack neck patches entirely. In more easterly subspecies *hrota,* **"American Brant,"** pale belly contrasts with black chest, and neck patches do not meet in front. Western *nigricans,* **"Black Brant,"** has dark belly, and neck patches meet in front. **"Gray-bellied Brant,"** breeding on western islands of Canadian Arctic Archipelago, wintering mainly in western WA, has gray sides, paler belly than "Black Brant"; has intermediate necklace pattern.

VOICE: Call is a low, rolling, slightly upslurred *raunk-raunk.*

RANGE: Primarily a sea goose; flocks fly low in ragged formation and feed on aquatic plants of shallow bays and estuaries. Locally common. Western *nigricans* is regular at Salton Sea, CA, in spring and sometimes summers; casual elsewhere inland in West. Eastern *hrota* is casual inland in the East, except in the Great Lakes region, where generally rare but regular, with even small to large flocks in the more easterly portion of that region, particularly in fall. Actually, most *hrota* migrate overland southeast from James Bay to Atlantic coast, but most do so nonstop unless weather grounds them. Also casual on East Coast south of mapped winter range. Both subspecies are casual during migration and winter on opposite coasts.

Emperor Goose

appears stocky in flight

white tail

white head and neck

adult

extensive black throat

dusky head and neck

juvenile

scaly pattern overall

Barnacle Goose

barred upperparts

white face

sharp contrast between breast and belly

bluish gray upperparts

usually seen with Canada Geese

collar usually complete at bottom

"Gray-bellied Brant" adult

all Brant fly with rapid, shallow wingbeats

hrota

white neck collar broken in front

adults

juvenile lacks broken whitish collar

juvenile
hrota

underparts variable, usually much paler than *nigricans*; often like *hrota*

Brant

pale belly contrasts with black neck

"American Brant"
hrota

1st winter

white-fringed coverts on 1st winter and juvenile

complete white collar

white vent and black belly

"Black Brant"
nigricans

flanks like adult

blackish breast and belly

adult

dark border to rear flank

indistinct collar

juvenile

frosty white barring on flanks

grayish brown flanks

See subspecies map, page 548

Cackling Goose *Branta hutchinsii* L 23-33" (58-84 cm)

Formerly considered part of the Canada Goose complex, this species is smaller with a rounder head and a stubbier bill. Its shorter neck is most obvious in flight. Four subspecies vary from smallest and darkest *minima*, breeding in coastal southwestern AK, to largest *taverneri*, breeding in western and apparently northern AK. Most of both subspecies winter in OR and WA. Aleutian breeding *leucopareia*, with a prominent white neck ring (sometimes present on other subspecies), mostly winters in Central Valley, CA. Intermediately sized and palest *hutchinsii*, breeding in Canada's Arctic Archipelago and wintering from southern Great Plains to western Gulf Coast region, is generally otherwise rare to casual in eastern North America. Size of *parvipes* Canada and *taverneri* Cackling overlap; *taverneri* averages darker with stubbier bill, but identification criteria are still not adequately worked out.

VOICE: Calls are much higher pitched than Canada Goose.

RANGE: Aleutian *leucopareia* population has greatly increased after removal of arctic foxes from Aleutian Islands. A few now turn up annually east of the Sierra Nevada and in southern CA.

Canada Goose *Branta canadensis* L 30-43" (76-109 cm)

Our most common and familiar goose. Black head and neck marked with distinctive white "chin strap," stretching from ear to ear. In flight, shows large dark wings; white undertail coverts; white, U-shaped rump band; and long neck. The approximately seven recognized subspecies vary in overall color, generally darker in West; ranges from pale *canadensis* of eastern seaboard to dark *occidentalis,* which breeds on eastern Gulf of Alaska coast. Size decreases northward: The smallest subspecies, *parvipes* (**"Lesser Canada Goose"**) breeds from central AK to north-central Canada and winters in central portions of U.S. with some in the Pacific states. Many in East are of largest subspecies, *maxima*; once thought to be near extinction, this is now the breeding subspecies over much of the East.

VOICE: Call is a deep, musical *honk-a-lonk.*

RANGE: Flocks usually migrate in V-formations, stopping to feed in wetlands, grasslands, or agricultural areas. Breeding programs have produced expanding populations that are resident south of mapped range and along Pacific and Atlantic coasts. These involve mixes of multiple subspecies, but in East are mostly *maxima*, originally of the Midwest. Explosive increase has prompted control measures in East. In general, no longer winters as far south as in previous decades and accordingly now winters much farther north.

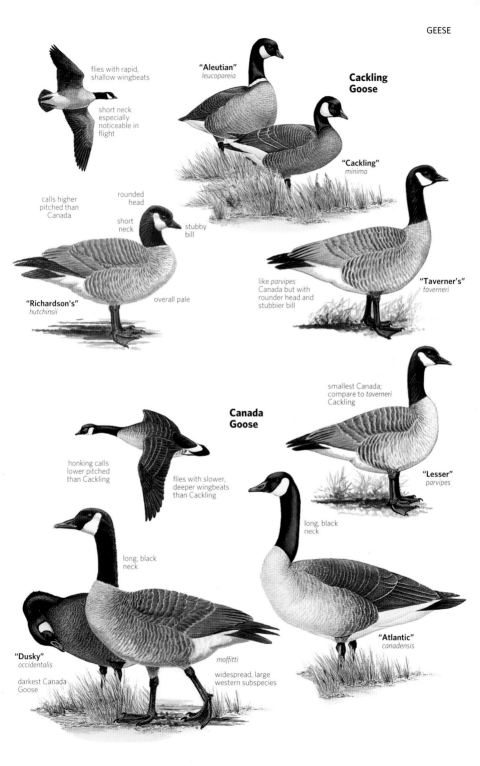

flies with rapid, shallow wingbeats

short neck especially noticeable in flight

"Aleutian"
leucopareia

Cackling Goose

"Cackling"
minima

calls higher pitched than Canada

rounded head

short neck

stubby bill

like *parvipes* Canada but with rounder head and stubbier bill

"Taverner's"
taverneri

"Richardson's"
hutchinsii

overall pale

smallest Canada; compare to *taverneri* Cackling

Canada Goose

honking calls lower pitched than Cackling

flies with slower, deeper wingbeats than Cackling

"Lesser"
parvipes

long, black neck

long, black neck

"Dusky"
occidentalis

darkest Canada Goose

moffitti

widespread, large western subspecies

"Atlantic"
canadensis

SWANS
Large and long-necked; North American species are white overall as adults; browner immatures are difficult to identify.

Tundra Swan *Cygnus columbianus* L 52" *(132 cm)*
In **adult,** black facial skin tapers to a point in front of eye and cuts straight across forehead; most birds have a yellow spot of variable size in front of eye. Head is rounded, bill slightly concave. In Eurasian subspecies, **"Bewick's Swan,"** seen very rarely in the Pacific states, facial skin and base of bill are yellow, but usually only behind the nostril; compare with Whooper Swan. **Juvenile** Tundras molt earlier than immature Trumpeter and Whooper Swans; appear much whiter by late winter. Immature "Bewick's" has whitish bill patch.
VOICE: Call is a noisy, high-pitched whooping or yodeling.
RANGE: Nests on tundra or sheltered marshes; winters in flocks in wetlands. Rare to uncommon in winter over parts of interior U.S.; casual south to the Gulf Coast and north to the Maritime Provinces.

Trumpeter Swan *Cygnus buccinator* L 60" *(152 cm)*
Adult's black facial skin tapers to broad point at the eye, dips down in a V on forehead. Forehead slopes evenly to straight bill. **Juveniles** retain gray-brown plumage through first spring.
VOICE: Common call is a single or double *honk* like an old car horn.
RANGE: Locally fairly common in its breeding areas. Reintroduced into parts of suspected former range and introduced elsewhere. Rare in winter south to CA and in western interior south of mapped range.

Whooper Swan *Cygnus cygnus* L 60" *(152 cm)*
Large yellow patch on lores and bill usually extends in a point to the nostrils; compare with "Bewick's" subspecies of Tundra Swan. Forehead slopes evenly to straight bill. **Juvenile** retains dusky plumage through first winter; by first fall, bill attains whitish patches in same shape as adult.
VOICE: Common call is a buglelike double note.
RANGE: Eurasian species. Regular winter visitor to outer and central Aleutians; has bred on Attu Island. Casual elsewhere in northwestern North America (south to northern CA). Records from eastern North America are all treated as escapes.

Mute Swan *Cygnus olor* L 60" *(152 cm)*
Prominent black knob at base of orange bill. **Juvenile** may be white or brownish; bill gray with black base. Darker juvenile begins to molt to white plumage by midwinter; bill becomes pinkish. Mute Swan often holds its long neck in an S-shaped curve, with bill pointed down. Often swims with wings arched over back.
VOICE: Gives a variety of hisses and snorts, but generally silent.
RANGE: An Old World species, introduced in the U.S. Seen in parks. East Coast populations are increasing. Now established on the southeastern portion of Vancouver Island, BC, and in parts of the Midwest. It is being systematically removed from the Great Lakes and other areas.

bill has more concave shape than Trumpeter

white on forehead forms a V

white on forehead cuts more straight across

eye stands out

Tundra Swan
columbianus

size of yellow lore spot variable, absent on some

extensive yellow squared off at base of bill

juvenile

adult

"Bewick's Swan" adult
bewickii

Trumpeter

Tundra

Whooper Swan

extensive yellow on bill extends forward in a point

pinkish gray on bill forms same shape as adult

adult

juvenile

Trumpeter Swan

white forms V-shape on forehead

retains darker plumage later in winter than Tundra

juvenile

adult

Mute Swan

sometimes arches wings

dark knob

dark border at base of bill

adult

juvenile

darker than other juvenile swans

WHISTLING DUCKS

Named for their whistling calls, these gooselike ducks have long legs and long necks. Wingbeats are slower than ducks, faster than geese.

Black-bellied Whistling-Duck *Dendrocygna autumnalis*
L 21" (53 cm) Gray face with white eye ring, red bill. Legs red or pink; belly, rump, and tail black. White wing patch shows as broad white stripe in flight. **Juvenile** is paler, with gray bill.
VOICE: Call is a high-pitched, four-note whistle.
RANGE: Increasing in southern Great Plains region; introduced populations expanding in FL, SC. Casual north to Canada and to southeastern CA, NM, west TX. Inhabits wetlands; nests in trees, nest boxes.

Fulvous Whistling-Duck *Dendrocygna bicolor* L 20" (51 cm)
Rich tawny color overall; back darker, edged with tawny. Dark stripe along hindneck is continuous in female, usually broken in male. Bill and legs dark. Whitish rump band conspicuous in flight.
VOICE: Call is a squealing *pe-chee.*
RANGE: Irregular summer wanderer north to dashed line on map; casual farther north. Nearly extirpated in the West. Forages in rice fields, marshes, shallow waters; often dives to feed. More active at night.

PERCHING DUCKS

These surface-feeding, woodland ducks have sharp claws for perching in trees. They nest in tree cavities or nest boxes.

Wood Duck *Aix sponsa* L 18½" (47 cm)
Male's glossy, colorful plumage and sleek crest are distinctive. Head pattern and bill colors are retained in drab eclipse plumage. **Female** identified by short crest and large, white, teardrop-shaped eye patch; compare with female Mandarin Duck (page 48). **Juvenile** resembles female but is spotted below. Flight profile is distinctive: large head with bill angled downward; long, squared-off tail.
VOICE: Male gives soft, upslurred whistle when swimming; female's squealing flight call is a rising *oo-eek.*
RANGE: Fairly common in open woodlands near water. Rare during winter throughout most of breeding range.

Muscovy Duck *Cairina moschata* L 26-33" (66-84 cm)
Large, blackish duck with green-and-purple gloss above and white patches on upper- and underwing. **Male** has blackish to dark reddish knob at base of bill, bare facial skin (usually brighter red in **domestic male,** whose color varies; can be all-white). Female is smaller, duller; lacks knob and bare facial skin. **Juvenile** even duller; acquires wing patches in first winter. Wild Muscovies are shy, usually silent; seen mostly in slow, gooselike flight at dawn and dusk.
RANGE: Tropical species; tame escaped birds found in parks across North America (now established in FL). A nest box program in northeastern Mexico helped spread of wild Muscovies to Rio Grande area, where they are now present near Falcon Dam, TX.

gooselike flight

adults

red bill

extensive white on wing

Fulvous Whistling-Duck

blackish rounded wings

white U-shaped rump band

white lines on sides and flanks

gray bill

Black-bellied juvenile

black belly

Black-bellied Whistling-Duck *autumnalis*

faint white streaks on flanks

white around and behind eye

♀

dark wings

♂

Wood Duck

juvenile ♂

longish tail

♀

♂

unique pattern

slow gooselike flight

female much smaller than male

juvenile ♀

knob

large white wing patch

adult ♂

extensive white underwing

adult ♂

Muscovy Duck

highly variable

domestic variety ♂

DABBLING DUCKS

Surface-feeding members of the genus *Anas:* the familiar "puddle ducks" of freshwater shallows and, chiefly in winter, salt marshes. Dabblers feed by tipping, tail up, to reach aquatic plants, seeds, and snails. They require no running start to take off but spring directly into flight. Most species show a distinguishing swatch of bright color, the speculum, on the secondaries. Many are known to hybridize.

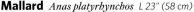

Mallard   *Anas platyrhynchos* L 23" (58 cm)

Male readily identified by metallic green head and neck, yellow bill, narrow white collar, chestnut breast. Black central tail feathers curl up. Both sexes have white tail, white underwing, bright blue speculum with both sides bordered in white. **Female's** mottled plumage resembles other *Anas* species; look for orange bill marked with black. Juvenile and **eclipse male** resemble female but bill is dull olive. Abundant and widespread. Mallard from the southwestern U.S. and south into Mexico, formerly treated as a separate species, **"Mexican Duck,"** darker, lacks distinctive male plumage; recent genetic studies show it is closer to American Black Duck and particularly Mottled Duck than Mallard. Extent of hybridization with Mallard has not been fully determined.
VOICE: Male, a *quack* and rasping *kreep;* female, a series of *quack* notes.
RANGE: Frequents a wide variety of shallow-water environments. Abundant and widespread. Feral and domestic birds are permanent residents, often found on park ponds, including south of mapped range; these are slightly discolored or misshapen, usually excessively tame.

Mottled Duck   *Anas fulvigula* L 22" (56 cm)

Both sexes closely resemble American Black Duck but body is slightly paler, throat and face are buffy and unstreaked; speculum bluish. Note dark spot at gape. Differs from female Mallard by darker plumage and absence of white in tail and black on bill. Western Gulf Coast subspecies, *maculosa*, is slightly darker than nominate FL subspecies. Some authorities consider Mottled Duck to be a subspecies of Mallard.
VOICE: Similar to Mallard.
RANGE: Has been introduced to coastal SC. Common year-round in coastal marshes. Casual north to ND, southern ON, and NC. Begins pairing in Jan. or Feb., earlier than migratory American Black Duck and Mallard.

American Black Duck   *Anas rubripes* L 23" (58 cm)

Blackish brown, paler on face and foreneck. In flight, white wing linings contrast more sharply with otherwise dark plumage than similar female Mallard. Violet speculum is bordered in black, may show a thin white trailing edge. **Male's** bill is yellow; **female's** dull green, may be flecked with black.
VOICE: Similar to Mallard.
RANGE: Nesting pairs favor woodland lakes and streams, freshwater or tidal marshes. Small introduced populations from western BC and WA appear to be extirpated. In many parts of range, especially deforested areas, Mallards are replacing American Black Ducks; the two species **hybridize** frequently.

Mallard

eclipse ♂

blue speculum bordered broadly with white ♂

underwing contrasts less with body than Black Duck ♀

mostly white tail

dark saddle on orange bill

male with yellow bill ♂

♀

yellow bill ♂

curled tail feathers

body darker than female Mallard

"Mexican Duck"

greenish yellow bill

unmarked buffy face and throat

black spot at gape ♂

Mottled Duck

mainly Florida
fulvigula

yellower bill

averages paler overall than *maculosa*

black spot at gape

♂

western Gulf Coast
maculosa

American Black Duck x Mallard hybrid ♂

American Black Duck

underwing contrasts strongly with dark body

purplish speculum lacks white forward border

♀

♂

streaked face

Eastern Spot-billed Duck *Anas zonorhyncha* L 22" (56 cm)

Pale tertials and sharply defined yellow tip of black bill, visible at a great distance, set this species apart from similar American Black Duck and female Mallard (page 26). Despite its name, this recently split (from *A. poecilorhyncha*) east Asian species lacks the red spots at bill base that the two related, more westerly taxa exhibit.

RANGE: Asian species, casual to Aleutians and Kodiak Island, AK.

Gadwall *Anas strepera* L 20" (51 cm)

Male is mostly gray, with white belly, black tail coverts, pale chestnut on wings. **Female's** mottled brown plumage resembles female Mallard (page 26), but belly is white, forehead steeper, upper mandible gray with orange sides. Both sexes have white inner secondaries that may show as small patch on swimming bird and identify the species in flight.

VOICE: Male's single-syllable call is loud and nasal; female's is a Mallard-like *quack*.

Falcated Duck *Anas falcata* L 19" (48 cm)

Named for **male's** long, falcated (sickle-shaped) tertials that overhang tail. Both sexes are chunky, with large head. **Female's** all-dark bill distinguishes her from female wigeons (page 30) and Gadwall; note slight bump on back of head. In flight, both sexes show a broad, dark speculum bordered in white.

RANGE: East Asian species, rare visitor to the western Aleutians; casual to Pribilofs and from West Coast region.

Green-winged Teal *Anas crecca* L 14½" (37 cm)

Our smallest dabbler. **Male's** chestnut head has dark green ear patch outlined in white. **Female** distinguished from other female teal (see also page 32) by smaller bill and by largely white undertail coverts that contrast with mottled flanks. A fast-flying, agile duck. In flight, shows green speculum bordered in buff on leading edge, white on trailing edge. In widespread North American *carolinensis,* male has vertical white bar on side. Eurasian subspecies, *crecca,* sometimes treated as a separate species, lacks the vertical bar, but has white stripe on scapulars and bolder buffy white facial stripes.

VOICE: Male, a liquid *preep* or *krick;* female, a weak, high-pitched *quack.*

RANGE: Eurasian *crecca* fairly common on the Aleutians and Pribilofs; rare to very rare elsewhere on West and East Coasts.

Baikal Teal *Anas formosa* L 17" (43 cm)

Adult male's intricately patterned head is distinctive. Long, dropping dark scapulars are edged in rufous and white. Gray sides are set off in front and rear by vertical white bars. **Female** similar to smaller female Green-winged Teal, but tail appears a bit longer; note distinct face pattern with a well-defined white spot at base of bill and white throat that angles up to rear of eye. Distinct eyebrow is bordered by darker crown. "Female" birds with "bridle" marking may be immature males. In flight, Baikal Teal's underwing is like Green-winged, but has more extensive and darker leading edge.

RANGE: Asian species, very rare, primarily fall, to western Aleutians and Pribilofs; casual elsewhere, including Pacific region.

Eastern Spot-billed Duck

dark facial stripes

distinct yellow tip to black bill

pale neck

mostly dark body

distinct white tertial edges

Gadwall

steep forehead

even line of orange on sides of bill

♀

mostly gray body

chestnut-edged scapulars

♂

black undertail coverts

small white speculum patch

♂

♀

white belly

♀

slight hint of crest

all-gray bill

distinctive head shape

dark speculum with faint pale borders

♂

Falcated Duck

♂

white throat with blackish band

♀

long sickle-shaped tertials

white horizontal bar

"Eurasian Teal"
crecca ♂

prominent buff or buffy white facial stripes

less white contrast to center of underwing than Blue-winged and Cinnamon

♂

♀

Green-winged Teal
carolinensis

♂

♀

small bill

dark line through eye with weaker dark area across cheek

distinctive pattern of white on throat

dark crown

prominent white spot bordered with dark

♀

Baikal Teal

distinctive head pattern

green on speculum usually hidden on swimming or standing bird

pale sides to undertail coverts

♂

long rufous-edged scapulars

more extensive dark on leading edge of underwing than Green-winged

American Wigeon *Anas americana* L 19" (48 cm)

Male's white forehead and cap are conspicuous in mixed flocks foraging in fields, marshes, and shallow waters; in flight, identified by mainly white wing linings and, in **adult male,** by large white patches on upperwing. Wing patches are grayish on **adult female** and immatures. Female lacks white on head, closely resembles gray-morph female Eurasian Wigeon; distinguishing field marks in flight are female American's white wing linings and contrast between gray throat and brown breast; Eurasian female's throat and breast are of uniform color.

VOICE: Male gives a piercing, three-note whistle; female call is a much lower *quack*.

RANGE: Common. A recently established breeder on the East Coast.

Eurasian Wigeon *Anas penelope* L 20" (51 cm)

Dark rufous head and gray back and sides make males conspicuous in flocks of American Wigeons; dusky wing linings are distinctive in all plumages. Hybridizes regularly with American Wigeon. **Adult male** has reddish brown head and neck with creamy crown; large white patches on upperwings. Many fall males retain some brown eclipse feathers but show distinctive reddish head. **Immature male** begins to acquire adult head and breast color but retains some brown juvenal plumage, particularly on forewing. **Gray-morph female** more closely resembles female American. **Rufous morph** has a more reddish head.

VOICE: Male gives a whistled *whee-ooo,* like American Wigeon but with two, not three, syllables, and even higher pitched.

RANGE: Rare but regular winter visitor along both coasts, more common in the West; rare in interior of North America. Regular migrant in western AK; a few winter in the Aleutians.

White-cheeked Pintail *Anas bahamensis* L 17" (43 cm)

White cheeks and throat contrast with dark forehead and cap; blue bill has a red spot near base. Long, pointed tail is buffy; tawny or reddish underparts are heavily spotted. Female is paler than **male;** tail slightly shorter. In flight, both sexes show green speculum broadly bordered on each side with buff.

RANGE: Casual vagrant from the West Indies to southern FL, but sightings even from FL are of uncertain origin; those away from FL are most likely birds escaped from captivity.

Northern Pintail *Anas acuta* ♂L 26" (66 cm) ♀L 20" (51 cm)

Male's chocolate brown head tops long, slender white neck. Black central tail feathers extend far beyond rest of long, wedge-shaped tail. **Female** is mottled brown, paler on head and neck; bill uniformly gray-ish. In both sexes, flight profile shows long neck; slender body; long, pointed wings; dark speculum bordered in white on trailing edge. In flight, female's mottled brown wing linings contrast with white belly; long, wedge-shaped tail lacks male's extended feathers.

VOICE: Male gives a whistled call, similar to Green-winged Teal; female gives a single or series of *quack*s.

RANGE: A common, widespread duck, found in marshes and open areas with ponds. Much more common in West than in East.

American Wigeon

white forewing

adult ♂

white underwing coverts and axillaries

white or buffy white crown

eclipse adult ♂

adult ♀

adult ♂

gray face contrasts with cinnamon-buff chest

bluish gray bill with dark tip

♀

head and chest uniform in color

rusty cast to head

rufous-morph ♀

Eurasian Wigeon

adult ♂

white forewing

dusky underwing and axillaries

cream crown

rufous head

gray-morph ♀

gray-morph ♀

adult ♂

red at bill base

sharply delineated white cheek and throat

White-cheeked Pintail

immature ♂

buffy pointed tail

♂

Northern Pintail

grayish bill

♂

white stripe extends up neck

♀

long pointed tail

long pointed tail

dark speculum with narrow white trailing edge

♂

♀

Northern Shoveler *Anas clypeata* L 19" (48 cm)

Large, spatulate bill, longer than head, identifies both sexes. **Male** distinguished by green head, white breast, brown sides; in early **fall** has a white crescent on each side of face, like Blue-winged Teal. **Female's** grayish bill is tinged with orange on cutting edges and lower mandible. In flight, both sexes show blue forewing patch.
VOICE: Male gives a two-syllable call; female gives a *quack*.
RANGE: Common to abundant in West; increasing in East. Found in marshes, ponds, and bays; often in large numbers on sewage ponds.

Blue-winged Teal *Anas discors* L 15½" (39 cm)

Violet-gray head with white crescent on each side identifies **male**. **Female** distinguished from smaller female Green-winged Teal (page 28) by larger bill, more heavily spotted undertail coverts, yellowish legs. Compare also with female Cinnamon Teal; note Blue-winged's grayer plumage, smaller bill, and bolder facial markings, including white at base of bill and more prominent, broken eye ring. Male in eclipse plumage resembles female. In flight, blue forewing, found in both sexes, is broadly bordered by white in male.
VOICE: Male gives a whistled *peeu*; female gives a nasal *quack*.
RANGE: Fairly common in marshes and on ponds and lakes in open country. Uncommon on the West Coast.

Cinnamon Teal *Anas cyanoptera* L 16" (41 cm)

Cinnamon coloration identifies **male. Female** like female Blue-winged but plumage is a richer brown; eye line and broken eye ring less distinct; bill longer, more spatulate. Compare also with Green-winged (page 28). Young birds and males in eclipse plumage resemble female. Older young males have red-orange eyes; Blue-winged's eyes are dark. Wing pattern is like Blue-winged.
VOICE: Male gives a chattering call; female gives a nasal *quack*.
RANGE: Common in marshes, ponds, and lakes. Regular from eastern Great Plains south to eastern TX. Casual to eastern Midwest and farther east. Some sightings may be escaped birds. Cinnamon Teal is known to interbreed with Blue-winged Teal.

Garganey *Anas querquedula* L 15½" (39 cm)

Prominent whitish edge to tertials is an important field mark in swimming birds; head shape is less rounded than Blue-winged and Cinnamon Teal. **Male's** bold white eyebrows separate dark crown, purple-brown face; in flight, shows gray-blue forewing and green speculum bordered fore and aft with white. Wing pattern is retained when male acquires femalelike supplemental plumage, held into winter. **Female** has strong facial pattern: dark crown, pale eyebrow, dark eye line, white lore spot bordered by a second dark line; note also dark bill and legs, dark undertail coverts. Larger and paler overall than female Green-winged Teal (page 28). Female in flight shows gray-brown forewing, white-bordered speculum lacks green. Note pale gray inner webs of primaries, visible in flight from above.
RANGE: Old World species. Regular migrant on western Aleutians; very rare on Pribilofs and in Pacific states; casual elsewhere in North America. Fewer records over past 20 years. Asian populations, at least in some areas, have plummeted recently.

Northern Shoveler

white facial crescent, compare to male Blue-winged Teal

fall ♂

blue forewing

orange line on bill ♀

large spatulate bill ♂

♀

like male Cinnamon, male Blue-winged has broad white border to forewing; much narrower in female

broken white eye ring

dark eye line

gray face

Blue-winged Teal

♂

♀

white face crescent

smaller bill than Cinnamon

extensive white underwing

♂

vertical white flank patch

♀

underwing of Cinnamon and Blue-winged Teal more extensively white than Green-winged Teal

all but youngest males have red-orange eyes

juvenile Cinnamon has slightly more evident eye line than adult Cinnamon

buffier face, less evident eye line and eye ring than female Blue-winged

♂

Cinnamon Teal

♂

larger spatulate bill than Blue-winged

♀

pale gray inner webs to primaries

lacks green in speculum; thin light edge to each feather

fall ♂

pale blue forewing

Garganey

adult ♀

♂

purple-brown head with long white, curved supercilium

two dark lines on face

less rounded head than Blue-winged or Green-winged

long silvery scapulars

♀

rather long tail

unstreaked creamy throat and foreneck

♂

POCHARDS

Diving ducks of the genus *Aythya* have legs set far back and far apart, which makes walking awkward. Heavy bodies require a running start on water for takeoff. Various species hybridize. Always carefully check potential vagrants to make sure they are not hybrids. All are mostly silent, except during display.

Canvasback *Aythya valisineria* L 21" (53 cm)
Forehead slopes to long, black bill. **Male's** head and neck are chestnut, back and sides whitish. **Female** and eclipse male have pale brown head and neck, pale brownish gray back and sides. In flight, whitish belly contrasts with dark breast, dark undertail coverts. Wings lack contrasting pale gray stripe of smaller Common Pochard and Redhead.
RANGE: Locally common in marshes, lakes, and bays; feeds in large flocks; has decreased significantly but decline has stabilized. Migrating flocks fly in irregular V-formations or in lines.

Common Pochard *Aythya ferina* L 18" (46 cm)
Resembles Canvasback in plumage and head shape. Bill similar to Redhead but dark at base and tip, gray in center. Gray on upperparts immediately distinguishes **female** from female Redhead and Ring-necked Duck. In flight, wings show gray stripe along trailing edge.
RANGE: Eurasian species, rare migrant to Pribilofs and to western and central Aleutians; accidental to south coastal AK and southern CA.

Redhead *Aythya americana* L 19" (48 cm)
Rounded head and shorter, tricolored bill separate this species from Canvasback. Bill is mostly pale blue (male) or slate (female), with narrow white ring bordering black tip. **Male's** back and sides are smoky gray. Eye is paler, more yellow, than male Canvasback. **Female** and eclipse male are tawny brown, with slightly darker crown, pale patch bordering black bill tip; may show buffy eye ring; compare to female Greater and Lesser Scaup (page 36) and female Ring-necked Duck. Redheads in flight show gray stripe on trailing edge of wings.
RANGE: Locally common in marshes, ponds, and lakes; is declining in the East.

Ring-necked Duck *Aythya collaris* L 17" (43 cm)
Peaked head; bold white ring near tip of bill. **Male** has second white ring at base of bill; white crescent separates black breast from gray sides. Narrow cinnamon collar is often hard to see in the field. **Female** has dark crown, white eye ring; may have a pale line extending back from eye; face is mainly gray. In flight, all plumages show a gray stripe on secondaries.
RANGE: Fairly common in freshwater marshes and on woodland ponds, lakes; during winter, found also in southern coastal marshes. Range is variable; may breed south or winter north of mapped range.

Canvasback

smoothly sloped forehead

grayish wing stripe

♀

♂

all-dark bill

pale silvery back contrasts with rufous head and neck

♂

♀

pale gray center to bill in both sexes

♀

head shape and coloration of both sexes resemble Canvasback

gray wing stripe

♂

Common Pochard

♂

♀

more rounded and uniform buffy head than female Ring-necked

head and back uniform in color

♀

Redhead

gray wing stripe

♂

♀

rounded head

darker gray back than Canvasback ♂

pale blended band on bill

Ring-necked Duck

adult ♂

grayish stripe

peaked head

white eye ring and grayish face ♀

♀

black back

adult ♂

vertical white bar on side

cinnamon collar is hard to see

prominent white ring near tip of bill

Tufted Duck *Aythya fuligula* L 17" (43 cm)

Head is rounded; crest distinct in **male,** smaller in **female** and immatures; may be absent in eclipse male. Gleaming white sides further distinguish male from male Ring-necked Duck (page 34). **First-winter male** has gray sides but lacks the white crescent conspicuous in male Ring-necked. Female is blackish brown above; lacks white eye ring and white bill ring of female Ring-necked. Bills of both male and female Tufted have a wide black tip. Some females also have a small white area at base of bill. In flight, all plumages show a broad white stripe on secondaries and extending onto primaries. May hybridize with scaup, especially Greater Scaup.

RANGE: Small numbers found, mostly in migration in western and central Aleutians and on the Pribilofs. Very rare elsewhere in AK and down the West Coast, where mostly found in winter; perhaps fewer in recent years. Also rare to very rare along East Coast as far south as MD; very rare in Great Lakes region. Found on ponds, rivers, bays, often with Ring-necked Ducks and especially scaup.

Greater Scaup *Aythya marila* L 18" (46 cm)

Larger size and smoothly rounded head help distinguish this species from Lesser Scaup. In close view, note Greater Scaup's slightly larger bill with wider black tip. **Male** averages paler on back and flanks; in good light, head may show a green gloss. In both species, **female** has bold white patch at base of bill. Some female Greater Scaup, especially in spring and summer, have a paler head with a distinct whitish ear patch. In flight, Greater Scaup typically shows a bold white stripe on secondaries and well out onto primaries, unlike Lesser Scaup.

RANGE: Locally common; found on large, open lakes and bays. Migrates and winters in small or large flocks, often with Lesser Scaup. Rare to uncommon winter visitor throughout the Gulf Coast states and on many larger inland lakes and reservoirs.

Lesser Scaup *Aythya affinis* L 16½" (42 cm)

Smaller size and peaked crown help distinguish from Greater Scaup. In close view, note Lesser Scaup's slightly smaller bill with smaller black tip. In good light, **male's** head may show a purple gloss, sometimes mixed with green. **Female** is brown overall, with bold white patch at base of bill. In some females, especially in spring and summer, head is paler, with whitish ear patch less distinct than female Greater Scaup. In flight, Lesser Scaup shows bold white stripe on secondaries only.

RANGE: Common; breeds in marshes, small lakes, and ponds. In winter, found in large flocks on sheltered bays, inlets, and lakes.

some females whitish around bill

Tufted Duck

extensive and bold white wing stripe

adult ♂

adult ♂ (in flight)

crest length variable on males and females

crest

♀

♀

dark brown back

broad black bill tip

1st winter ♂

black back

adult ♂

pure white sides

many females with prominent whitish ear patches

♀

larger, rounder head than Lesser

extensive white stripe

adult ♂

♀

♀

Greater Scaup
nearctica

1st winter ♂

more black on bill tip than Lesser

white restricted to secondaries

adult ♂

♀

adult ♂

Lesser Scaup

smaller, more peaked head than Greater

♀

less black on bill tip than Greater

adult ♂

EIDERS

These large, bulky diving sea ducks have dense down feathers that help insulate them from the cold northern waters. Females pluck their own down to line nests. Mostly silent, except during breeding season.

See subspecies map, page 549

Common Eider *Somateria mollissima* L 24" (61 cm)

Female distinguished from female King Eider by larger size, sloping forehead, and evenly barred sides and scapulars. Feathering extends along sides of bill to or beyond nostril, with minimal feathering on top of bill. Females range in overall color from rust to gray. Eastern *dresseri* is reddish brown. Western *v-nigrum* is duller brown. **Male's** head pattern is distinctive. Most western and a few eastern males show a thin black V on throat; *v-nigrum* male has narrow and tapered frontal lobes, orange-yellow bill; *borealis*, which winters in numbers to Atlantic Canada, is intermediate; frontal lobe shape closer to *v-nigrum*. Eclipse and **first-winter males** are dark; first-winter has white on breast; full adult plumage is attained by fourth winter. In flight, adult male shows solid white back and wing coverts.

RANGE: Locally abundant on shallow bays, rocky shores; casual on Great Lakes. Rare in winter on East Coast south to SC; casual to FL and on Great Lakes; accidental on West Coast.

King Eider *Somateria spectabilis* L 22" (56 cm)

Female distinguished from female Common Eider by smaller size, more rounded head, and crescent or V-shaped markings on sides and scapulars. Feathering extends only slightly along sides of bill but extensively down the top, making bill look stubby. **Male's** head pattern is distinctive. In flight, shows partly black back, black wings with white patches. **First-winter male** has brown head, pinkish or buffy bill, buffy eye line; lacks white wing patches. Full adult plumage attained by third winter.

RANGE: Common on tundra and coastal waters in northern part of range; generally very rare on Great Lakes except on Lake Ontario, where a few occur regularly. Rare and possibly declining in winter on East Coast to VA; casual to FL and on West Coast; accidental on Gulf Coast.

Spectacled Eider *Somateria fischeri* L 21" (53 cm) **T**

Male has green head with white, black-bordered eye patches and orange bill. In flight, black breast separates adult male from other eiders, smaller size from Common Eider. White on scapulars, present on even drabber immature males, separate Spectacled from slightly larger King Eider. Drab **female** has fainter spectacle pattern; bill is gray-blue; feathering extends far down upper mandible.

RANGE: Uncommon and declining; found on coastal tundra near lakes and ponds. Flocks winter in openings in ice pack on Bering Sea. Accidental on Aleutians and in BC.

broadly rounded frontal lobe

greenish bill

"Atlantic Eider"
♂ *dresseri*

frontal lobe very similar to *dresseri*

"Hudson Bay Eider"
♂ *sedentaria*

more tapered frontal lobe

dull yellow bill

"Northern Eider"
♂ *borealis*

frontal lobe tapers to a fine point

orange bill

"Pacific Eider"
♂ *v-nigrum*

adult ♂

extensive white on back
v-nigrum

Common Eider

feathering extends to nostril

♀ *v-nigrum*

overall color variable
♀ *dresseri*

♀

evenly barred on sides and flanks

flat forehead

white back

1st winter ♂
dresseri

adult ♂

more rounded head than Common

buff-pink bill

1st winter ♂

King Eider

adult ♂

more restricted white on upperwing than Common

isolated white vent patch

adult ♂

small scapular points

feathering falls short of nostril

♀

crescent-shaped marks on sides and flanks

green head with white "goggles"

adult ♂

feathering on all birds extends well out on bill

pattern of white on upperparts similar to Common Eider

adult ♂

black extends into breast

female has pale "goggles," but pattern is subdued

♀

Spectacled Eider

♀

Steller's Eider *Polysticta stelleri* L 17" *(43 cm)* **T**

Greenish head tufts, black eye patch, chin, and collar identify **male.** **Female** is dark cinnamon-brown with pale eye ring, unfeathered dark bill. In flight, adults, immature males, and some immature females show blue speculum bordered fore and aft in white.

RANGE: Found along rocky coasts; nests on inland grassy areas or tundra. Winters casually south to BC (six records), WA (two records, including one inland at Walla Walla River Delta, 9 to 13 Sept. 1995), OR (once), and northern CA coast (three records). Accidental to Northeast (ME, MA) and NU. Numbers reduced over recent decades.

Harlequin Duck *Histrionicus histrionicus* L 16½" *(42 cm)*

Small duck, with rounded head, stubby bill. **Male's** colorful plumage appears dark at a distance. **Female** has three white spots on each side of head. Juvenile resembles adult female. Flight is rapid, low. Compare female in flight with female Bufflehead (page 45).

VOICE: Male's call is a high-pitched nasal squeaking, but mostly silent, except during breeding season.

RANGE: Locally common on rocky coasts; moves inland along swift and turbulent streams for nesting. Rare on Great Lakes in migration and winter on East Coast south to Carolinas and on West Coast to southern CA; casual to accidental elsewhere in the interior. Formerly bred in Sierra Nevada through 1920s. Some that appear well south of normal range can remain for long periods of time, even years.

Long-tailed Duck *Clangula hyemalis*

♂L 22" *(56 cm)* ♀16" *(41 cm)* Formerly known in North America as Oldsquaw. **Adult male's** long tail is conspicuous in flight, may be submerged in swimming bird. Male in winter and spring is largely white; breast and back dark brown, scapulars pearl gray; stubby bill shows pink band. By late spring, male becomes mostly dark, with pale facial patch, bicolored scapulars; in later, supplemental molt, acquires paler crown and shorter, buff-edged scapulars. Molt into full eclipse plumage continues until early fall. **Female** lacks long tail; bill is dark; plumage whiter in winter, darker in summer. **First-fall birds** are even darker. Long-tailed is identifiable at some distance by its swift, careening flight and its calls. Females and immatures best sexed by bill color. Both sexes show uniformly dark underwing. At sea when flying away, white flanks and dark underwings could lead to confusion with murres or Razorbills.

VOICE: Loud, yodeling, three-part calls, heard all year.

RANGE: Nests on tundra ponds. Winters along and a short distance off coast; also on bays. Away from Great Lakes, where status varies between lakes, most common on Lake Ontario; rare in the interior and south to Gulf Coast. On West Coast generally rare south of WA. Like scoters and loons, individuals sometimes summer within their winter range.

Steller's Eider

white forewing

adult ♂

greenish head tufts

1st winter ♂

blue speculum bordered by white

♀

rather small unfeathered, dark bill

adult ♂

cinnamon-buff underparts

♀

dark cinnamon-brown color overall

Harlequin Duck

unique face pattern

round head with small bill

adult ♂

♀

chestnut sides

adult ♂

adult ♂

white belly

head spots develop over winter

1st winter ♂

1st winter ♂

males have pink band on bill

Long-tailed Duck

winter adult ♂

dark wings

small bill

early summer adult ♂

long pointed tail

1st fall ♀

winter adult ♂

females have grayish bills

winter ♀

variably white on face

winter ♀

SEA DUCKS
Stocky, short-necked diving ducks, most species breed in the far north and migrate in large, compact flocks to and from their coastal wintering grounds.

Surf Scoter *Melanitta perspicillata* L 20" (51 cm)
Male's black plumage sets off colorful bill, white eye, white patch on forehead and nape; forehead is sloping, not rounded. **Female** is brown, with dark crown; usually has two white patches on each side of face; feathering extends down top of bill only. Adult female and **first-winter male** may have whitish nape patch. All juveniles have whitish belly, usually white face patches. In flight, more uniform color of underwing helps distinguish Surf from Black Scoter; also orangish, not dark, legs and feet.
VOICE: Mostly silent, except during display.
RANGE: Common; nests on tundra and in wooded areas near water. Rare inland migrant. A few winter on southern Great Lakes; most in coastal waters. Small numbers of nonbreeders oversummer in winter range on West Coast.

Black Scoter *Melanitta americana* L 19" (48 cm)
Male is black, with orange-yellow knob at base of dark bill. **Female's** dark crown and nape contrast with pale face and throat; feathering does not extend onto bill. In both, forehead is rounded. In flight, adult male's blackish wing linings contrast with paler flight feathers. Juveniles resemble females but are whitish on belly; **first-winter male** has some yellow at base of bill by winter. Adult male **Common Scoter,** *M. nigra,* with smaller area of yellow at base of bill, casual to Greenland.
VOICE: Most vocal of the scoters. Drakes call even in winter; most frequent call is a prolonged and plaintive whistle.
RANGE: Nests on tundra. Small numbers seen in fall on Great Lakes (common on Lake Ontario, where some winter); rare elsewhere in eastern interior; casual in western interior.

White-winged Scoter *Melanitta fusca* L 21" (53 cm)
White secondaries, conspicuous in flight, may show as a small white patch on swimming bird. Forehead slightly rounded. Feathering extends almost to nostrils on top and sides of bill. **Female** and juveniles lack contrasting dark crown and paler face of other scoters; white facial patches are distinct on juveniles, often indistinct on adult female. Juveniles and immatures are whitish below. **Adult male** has black knob at base of colorful bill; crescent-shaped white patch below white eye, brownish flanks. Black-flanked adult male Asian *stejnegeri,* casual to western AK, has more obvious nasal hook and different bill color pattern. Western Palearctic adult male *fusca,* casual to Greenland, also with black flanks and largely lacks bill knob; note bill color. Females of all subspecies very similar to one another.
VOICE: Mostly silent, except during display.
RANGE: Fairly common on inland lakes in breeding season, coastal areas in winter. Uncommon inland migrant. Large numbers winter on Lake Ontario since introduction of zebra mussels. Rarest scoter in the South.

feathering out culmen

white head patches

forward whitish patch vertically shaped

white nape on adult female

Surf Scoter

DUCKS

adult ♂

more uniformly dark wings than Black

adult ♂

immature ♀

1st winter ♂

whitish belly

forward whitish patch vertically shaped

1st winter ♀

bill shows color by Dec.

immatures of all scoters have pale bellies

black knob

female like Black Scoter

yellow-orange knob at base of bill

round head

no feathering out on bill

pale face contrasts sharply with dark cap

small yellow area on bill

thinner neck than Black Scoter

dull yellow at bill base

adult ♂

adult ♀

Common Scoter
adult ♂

1st winter ♂

recorded from Greenland

secondaries and primaries paler than rest of wing, especially on adult males

adult ♂

Black Scoter
americana

adult ♀

White-winged Scoter
deglandi

adult ♂

1st winter ♀

juveniles of both sexes show whitish patches; forward patch horizontally shaped

1st winter ♂

only scoter with white in wings

adult ♀

adult ♂

adult male has white slash under eye

brownish flanks

feathering extends out bill

adult ♀

adult ♂
stejnegeri

white not always visible on folded wing

"Velvet Scoter"
fusca

small knob

recorded from Greenland

bill color and white behind eye differs from *deglandi*; note bulbous nostrils

black flanks

yellow on sides of bill

adult ♂

black flanks like *stejnegeri*

Bufflehead *Bucephala albeola* L 13½" (34 cm)

A small duck with a large, puffy head, steep forehead, short bill. **Male** is glossy black above, white below, with large white patch on head. **Female** is duller, with small, elongated white patch on each side of head. **First-winter male** and male in eclipse resemble female. In flight, males show white patch across entire wing; female has white patch only on inner secondaries.

VOICE: Mostly silent. Display sounds feeble in comparison to goldeneyes.

RANGE: Generally common; nests in woodlands near small lakes and ponds. During migration and winter, found also on sheltered bays, rivers, lakes. Casual breeder south of mapped range. Rare western AK.

Common Goldeneye *Bucephala clangula* L 18½" (47 cm)

Male has round white spot on each side of face; scapulars are mostly white. **Female** and eclipse male closely resemble Barrow's. Head of Common is more triangular; forehead more sloped; bill longer. Female's head is slightly paler than female Barrow's; bill generally all-dark or with yellow near tip only; rarely completely dull yellow. In all plumages, subtle differences between the two species in white wing patches visible in flight.

VOICE: Wings in flight produce a whistling sound in both goldeneye species. Drakes give a raspy, high-pitched, double-note call when throwing head back in elaborate display; females of both species give a grunting note.

Barrow's Goldeneye *Bucephala islandica* L 18" (46 cm)

Male has white crescent on each side of face; white patches on scapulars show on swimming bird as a row of spots; dark color of back extends forward in a bar partially separating white breast from white sides. **Female** and male in eclipse plumage closely resemble Common. Puffy, oval-shaped head, steep forehead, and stubby, triangular bill help identify Barrow's. Adult female's head is slightly darker than female Common; bill mostly yellow, except in young females, which may have only a yellow band near tip of bill. In all plumages, white wing patches visible in flight differ subtly between the two species. **Hybrids** between the two goldeneye species are regularly noted.

VOICE: In display (less elaborate than Common), drake gives an *e-eng* call.

RANGE: Both species summer on open lakes and small ponds; winter in sheltered coastal areas, inland lakes, and rivers. Overall, Barrow's is much less common; rare to casual outside mapped winter range.

MERGANSERS

Long, thin, serrated bills help these divers catch fish, crustaceans, and aquatic insects. In flight, they show pointed wings.

Smew *Mergellus albellus* L 16" (41 cm)

Dark, relatively short bill. In **female,** white throat and lower face contrast sharply with reddish head and nape. **Adult male** white with black markings; black-and-white wings conspicuous in flight.

RANGE: Eurasian species, rare visitor to Aleutians; casual on Pribilofs and Pacific states; accidental in East, including three records from ON.

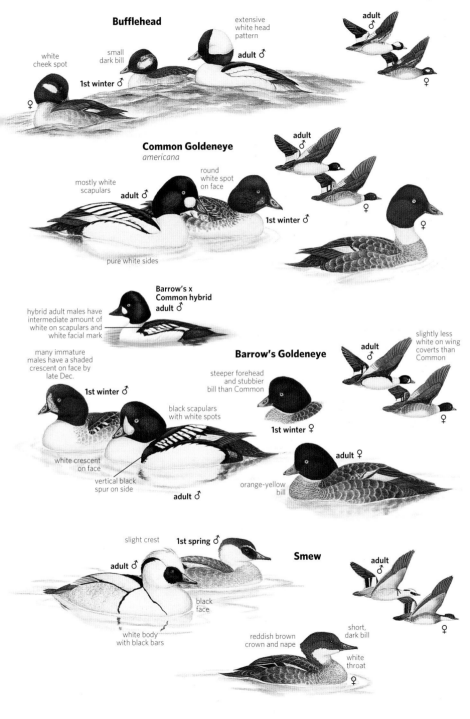

Bufflehead

white cheek spot

small dark bill

1st winter ♂

extensive white head pattern

adult ♂

adult ♂

♀

♀

Common Goldeneye
americana

mostly white scapulars

adult ♂

round white spot on face

1st winter ♂

pure white sides

adult ♂

♀

♀

Barrow's x Common hybrid
adult ♂

hybrid adult males have intermediate amount of white on scapulars and white facial mark

many immature males have a shaded crescent on face by late Dec.

1st winter ♂

Barrow's Goldeneye

steeper forehead and stubbier bill than Common

black scapulars with white spots

1st winter ♀

slightly less white on wing coverts than Common

adult ♂

adult ♀

white crescent on face

vertical black spur on side

adult ♂

orange-yellow bill

♀

slight crest

1st spring ♂

adult ♂

black face

Smew

white body with black bars

reddish brown crown and nape

short, dark bill

white throat

adult ♂

♀

♀

Hooded Merganser *Lophodytes cucullatus* L 18" (46 cm)

Puffy, rounded crest; thin bill. **Male's** bill is dark; white head patches are fan-shaped and conspicuous when crest is raised. Compare with male Bufflehead (page 44). **Female** brownish overall; upper mandible dark, lower yellowish. Rapid wingbeats in flight; both sexes show black-and-white inner secondaries. Crest is flattened in flight; male's head patch shows only as a white line.

VOICE: Generally silent, except in display, when drake gives a rolling froglike note; female gives a single harsh note.

RANGE: Uncommon in West; common over much of East. In breeding season, found on woodland ponds, rivers, and backwaters. Winters chiefly on fresh or brackish water. Casual to central AK.

Common Merganser *Mergus merganser* L 25" (64 cm)

Large duck with long, slim neck and thick-based, hooked, red bill. White breast and sides, often tinged with pink, and lack of crest distinguish **male** from Red-breasted Merganser. **Female's** bright chestnut, crested head and neck contrast sharply with white chin, white upper breast. Adult male in flight shows white patch on upper surface of entire inner wing, partially crossed by a single black bar. Eclipse male resembles female but retains wing pattern. Female's white inner secondaries and greater coverts are partially crossed by a black bar. Old World **"Goosander"** (nominate *merganser*), recorded from western Aleutians, lacks dark bar on wing. Note different bill shape. As in all species on this page, young male resembles adult female but may show some darkening on face by spring.

VOICE: Generally silent, except in display, when drake gives single, bell-like calls. Both sexes give single harsh calls.

RANGE: Nests in woodlands near lakes and rivers; in winter, sometimes also found on brackish water. Casual to Gulf Coast.

Red-breasted Merganser *Mergus serrator* L 23" (58 cm)

Shaggy double crest, white collar, and streaked breast distinguish **male** from male Common Merganser. **Female's** head and neck are paler than female Common; chin and foreneck whitish, with no sharp contrasts as in Common. Adult male in flight shows white patch on upper surface of inner wing, partly crossed by two black bars. Eclipse male resembles female but retains male wing pattern. Female's white inner secondaries and greater coverts are crossed by a single black bar. Smaller size than Common and thinner bill also help distinguish Red-breasted Merganser in mixed flocks.

VOICE: Generally silent, except in display, when drake gives various notes, including mewing calls; female gives various harsh calls.

RANGE: Nests in woodlands near fresh water or in sheltered coastal areas; prefers brackish or salt water in winter. Abundant migrant on Great Lakes, where moderate numbers winter; elsewhere, fairly common to common migrant in interior.

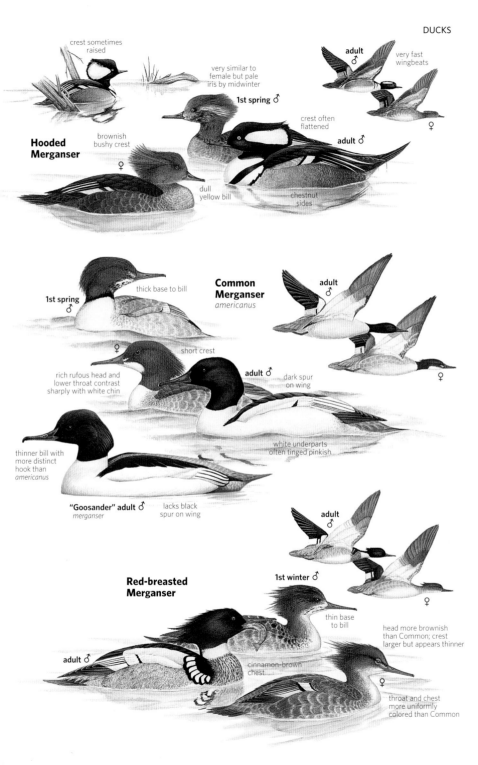

crest sometimes raised

adult ♂

very fast wingbeats

very similar to female but pale iris by midwinter

1st spring ♂

♀

Hooded Merganser

brownish bushy crest

crest often flattened

adult ♂

♀

dull yellow bill

chestnut sides

1st spring ♂

thick base to bill

Common Merganser
americanus

adult ♂

♀

short crest

rich rufous head and lower throat contrast sharply with white chin

adult ♂

dark spur on wing

♀

thinner bill with more distinct hook than *americanus*

white underparts often tinged pinkish

"Goosander" adult ♂
merganser

lacks black spur on wing

adult ♂

1st winter ♂

thin base to bill

Red-breasted Merganser

♀

adult ♂

cinnamon-brown chest

head more brownish than Common; crest larger but appears thinner

♀

throat and chest more uniformly colored than Common

STIFF-TAILED DUCKS
Long, stiff tail feathers serve as a rudder for these diving ducks. In both species, male's bill is blue in breeding season.

Ruddy Duck *Oxyura jamaicensis* L 15" (38 cm)
Chunky, with large head, broad bill, long tail, often cocked up. **Male's** white cheeks are conspicuous both in **breeding** plumage and in dull **winter** plumage. In **female,** single dark line crosses cheek. Young resemble female through first winter.
VOICE: Mostly silent. During display, male produces soft ticking and popping sounds.
RANGE: Common; nests in dense vegetation of freshwater wetlands. During migration and winter, found on lakes, bays, and salt marshes.

Masked Duck *Nomonyx dominicus* L 13½" (34 cm)
Generally shy. **Male's** black face on reddish brown head is distinctive. In **female, winter male,** and **juvenile,** two dark stripes cross face and barred back. White wing patches show in flight.
RANGE: Tropical species. Found on densely vegetated ponds. Rare and irregular visitor to southern and southeastern TX; casual in LA and FL. Accidental in East, north to WI and New England.

EXOTIC WATERFOWL
Many waterfowl species are brought into North America from other continents for zoos and private collections. Escapes are frequent. The species shown here are among those seen most frequently.

Ruddy Shelduck *Tadorna ferruginea* L 26" (66 cm)
Afro-Eurasian species often kept in captivity. A record of a flock of six on 23 July 2000 at Southampton Island, NU, may have been vagrants from Old World.

Common Shelduck *Tadorna tadorna* L 25" (64 cm)
Eurasian species. Female smaller, lacks knob on bill. Recent records from Newfoundland and MA of uncertain origin.

Egyptian Goose *Alopochen aegyptiacus* L 27" (68 cm)
African species. Widespread escape. Note white wing patches.

Mandarin Duck *Aix galericulata* L 16" (41 cm)
Asian species. Compare female to female Wood Duck (page 24).

Bar-headed Goose *Anser indicus* L 30" (76 cm)
Asian species. Fairly common in zoos and private collections.

Graylag Goose *Anser anser* L 32" (81 cm)
Eurasian species, progenitor of most domestic geese. A recent record of a wild bird off Newfoundland (page 530). Compare carefully to Greater White-fronted Goose (page 14).

Ruddy Duck

dark cap contrasts sharply with white face

winter ♂

long pointed tail often raised

winter ♀

breeding ♂

raised tail is spread in display

breeding ♀

bright blue bill

rufous body

single blurry line on face

breeding ♂

flies low with rapid wingbeats

dark wings

♀

breeding ♂

long pointed tail

white wing patches, but species seldom seen in flight

♀

Masked Duck

winter ♂

all plumages but breeding male have two dark facial bars

smaller, stubbier bill than Ruddy

breeding ♀

juvenile

blackish head

winter ♀

body more richly colored than Ruddy

breeding ♂

Ruddy Shelduck

adult ♂

Common Shelduck

adult ♂

Egyptian Goose

adults

Mandarin Duck

compare to female Wood Duck

♀

♂

Graylag Goose

wild type

domestic type

Bar-headed Goose

adult

Perching Ducks

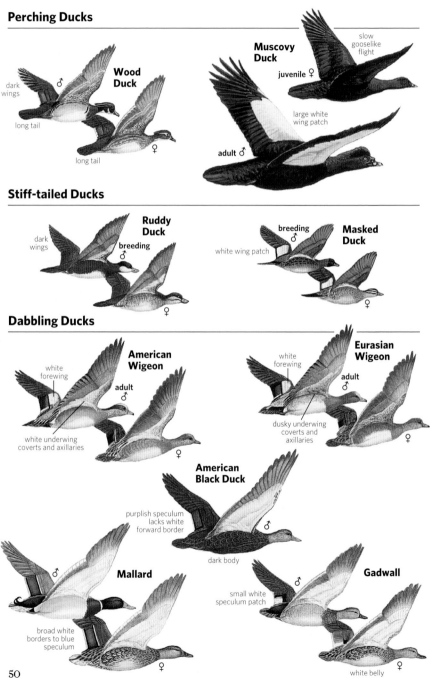

Wood Duck

dark wings

♂

long tail

long tail

♀

Muscovy Duck

slow gooselike flight

juvenile ♀

large white wing patch

adult ♂

Stiff-tailed Ducks

Ruddy Duck

dark wings

breeding ♂

♀

Masked Duck

breeding ♂

white wing patch

♀

Dabbling Ducks

American Wigeon

white forewing

adult ♂

white underwing coverts and axillaries

♀

Eurasian Wigeon

white forewing

adult ♂

dusky underwing coverts and axillaries

♀

American Black Duck

purplish speculum lacks white forward border

♂

dark body

small white speculum patch

Mallard

♂

broad white borders to blue speculum

♀

Gadwall

♂

white belly

♀

DUCKS IN FLIGHT

Northern Shoveler
blue forewing

Northern Pintail
long pointed tail
white edge to secondaries
long neck

Blue-winged Teal
male with prominent white bar
extensive white underwing

Green-winged Teal *carolinensis*
pale borders to green speculum
white confined to center of underwing

Garganey
pale gray inner webs to primaries and no color in speculum
eclipse ♂
pale blue forewing

Cinnamon Teal
male with prominent white bar
extensive white underwing

Baikal Teal
more extensive dark on leading edge on underwing than Green-winged

Falcated Duck
dark speculum with faint pale borders
white belly

Bay Ducks *(Aythya)*

Redhead
gray wing stripe

all three species lack conspicuous white on primaries and/or secondaries as in the two scaup species and Tufted Duck (page 52)

Canvasback
grayish wing stripe
long neck

Common Pochard
gray wing stripe

51

Bay Ducks *(Aythya)* continued

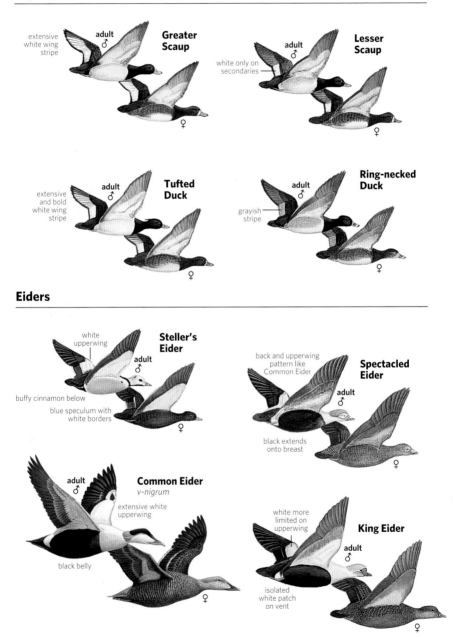

extensive
white wing
stripe

adult
♂

**Greater
Scaup**

♀

adult
♂

white only on
secondaries

**Lesser
Scaup**

♀

extensive
and bold
white wing
stripe

adult
♂

**Tufted
Duck**

♀

adult
♂

grayish
stripe

**Ring-necked
Duck**

♀

Eiders

white
upperwing

adult
♂

**Steller's
Eider**

buffy cinnamon below

blue speculum with
white borders

♀

back and upperwing
pattern like
Common Eider

adult
♂

**Spectacled
Eider**

black extends
onto breast

♀

adult
♂

Common Eider
v-nigrum

extensive white
upperwing

black belly

♀

white more
limited on
upperwing

adult
♂

King Eider

isolated
white patch
on vent

♀

Sea Ducks

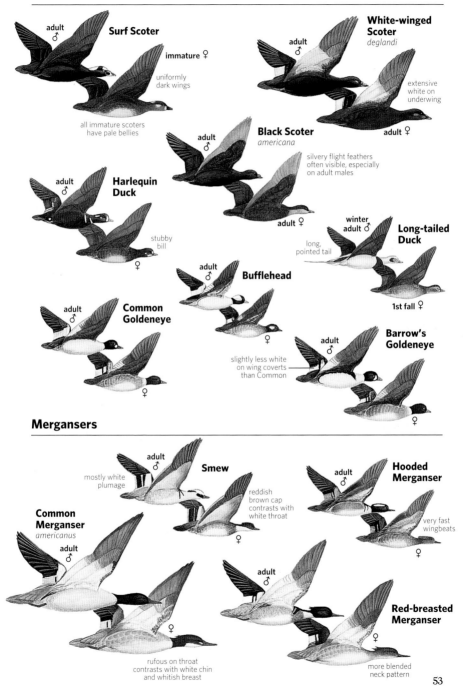

Surf Scoter

adult ♂

immature ♀

uniformly
dark wings

all immature scoters
have pale bellies

**White-winged
Scoter**
deglandi

adult ♂

extensive
white on
underwing

adult ♀

Black Scoter
americana

adult ♂

silvery flight feathers
often visible, especially
on adult males

adult ♀

**Harlequin
Duck**

adult ♂

stubby
bill

♀

winter
adult ♂

**Long-tailed
Duck**

long,
pointed tail

1st fall ♀

adult ♂

Bufflehead

♀

**Common
Goldeneye**

adult ♂

**Barrow's
Goldeneye**

adult ♂

slightly less white
on wing coverts
than Common

♀

♀

Mergansers

adult ♂

Smew

mostly white
plumage

reddish
brown cap
contrasts with
white throat

adult ♂

**Hooded
Merganser**

very fast
wingbeats

♀

**Common
Merganser**
americanus

adult ♂

♀

adult ♂

**Red-breasted
Merganser**

♀

rufous on throat
contrasts with white chin
and whitish breast

more blended
neck pattern

53

NEW WORLD QUAIL Family Odontophoridae

Scientific evidence has recently placed the New World Quail in their own family. All have chunky bodies and crests or head plumes. In North America, most live in the West.
SPECIES: 31 WORLD, 6 N.A.

Gambel's Quail *Callipepla gambelii* L 11" (28 cm)

Grayish above, with prominent teardrop-shaped plume or double plume. Chestnut sides and crown, and lack of scaling on underparts, distinguish Gambel's from California. **Male** has dark forehead, black throat, black patch on belly. Smaller **juvenile** is tan and gray with pale mottling and streaking. Shows less scaling and streaking than darker California juvenile; nape and throat are grayer. Sometimes **hybridizes** with Scaled (page 56) and California where ranges overlap.

VOICE: Calls include varied grunts and cackles and a plaintive *qua-el;* loud, querulous *chi-ca-go-go* call is similar to California but higher pitched and usually has four notes.

RANGE: Common in desert scrublands and thickets, usually near permanent water source. Gregarious; in fall and winter, forms large coveys. Introduced populations exist in ID and on San Clemente Island, CA.

California Quail *Callipepla californica* L 10" (25 cm)

Gray and brown above, with prominent teardrop-shaped plume or double plume. Scaled underparts, and brown sides and crown separate California from Gambel's. Body color varies from grayish, seen over most of range, to brown in coastal mountains of CA; extremes are shown here in **females. Male** has pale forehead, black throat, and chestnut patch on belly. **Juvenile** is smaller; resembles Gambel's juvenile, but is darker, with traces of scaling on underparts.

VOICE: Calls include varied grunts and cackles; loud, emphatic *chi-ca-go* call is similar to Gambel's but lower pitched and usually has three notes rather than four.

RANGE: Common in open woodlands, brushy foothills, stream valleys, suburbs, usually near permanent water source. Gregarious; in fall and winter, assembles in large coveys. Populations in northeastern portion of range and UT are probably introduced.

Mountain Quail *Oreortyx pictus* L 11" (28 cm)

Gray and brown above, with two long, thin head plumes; often appears as one plume. Gray breast; chestnut sides boldly barred with white; chestnut throat outlined in white. **Male** and **female** are alike; female has shorter head plumes. Amount of brown and gray in upperparts varies in different subspecies; birds of humid coastal Northwest are browner than three gray interior subspecies. **Juvenile** told from young Gambel's and California by head plume shape and grayer breast.

VOICE: Male's mating call is a loud, clear, descending *quee-ark;* both sexes give whistled notes.

RANGE: Uncommon to fairly common, but declining in parts of range; in chaparral, brushy ravines, mountain slopes, at altitudes up to 10,000 feet. Nonmigratory but descends to lower altitudes in winter. Gregarious, forming small coveys in fall and winter. Secretive; best seen in late summer in family groups along roadsides.

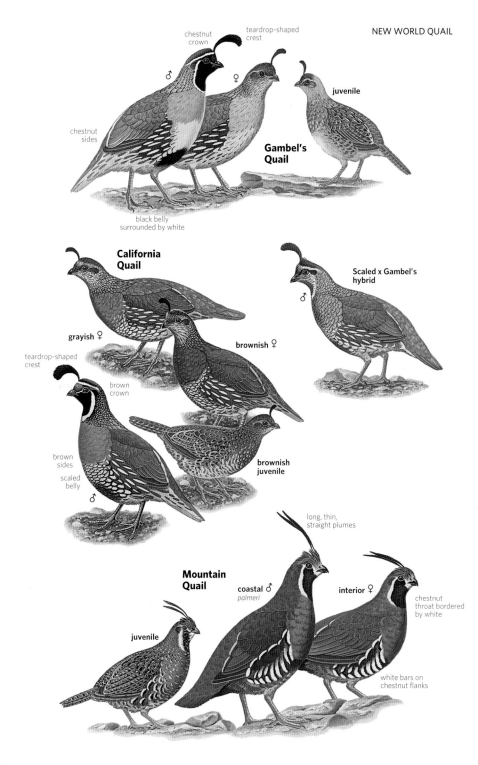

chestnut crown

teardrop-shaped crest

♂

♀

juvenile

chestnut sides

Gambel's Quail

black belly surrounded by white

California Quail

grayish ♀

brownish ♀

Scaled x Gambel's hybrid

♂

teardrop-shaped crest

brown crown

brownish juvenile

brown sides

scaled belly

♂

long, thin, straight plumes

Mountain Quail

coastal ♂
palmeri

interior ♀

chestnut throat bordered by white

juvenile

white bars on chestnut flanks

Northern Bobwhite *Colinus virginianus* L 9¾" (25 cm)

Mottled, reddish brown quail with short gray tail. Flanks are striped with reddish brown. Throat and eye stripe are white in **male,** buffy in **female. Juvenile** is smaller and duller. Birds from peninsular FL (Gainesville and south), *floridanus,* are smaller and darker than nominate subspecies, while *taylori* and more southerly *texanus* (not shown) on Great Plains from western part of native range are paler. Male **"Masked Bobwhite,"** *ridgwayi* (**E**), from south-central AZ, where historically found in Altar Valley and upper Santa Cruz Valley, and northern Sonora, Mexico, has black throat and cinnamon underparts. **VOICE:** Male's call is a rising, whistled *bob-white,* heard chiefly in late spring and summer; whistled *hoy* call is heard year-round. Also gives soft clucking notes. **RANGE:** Uncommon to common in brushlands and open woodlands; feeds and roosts in coveys except during nesting season. The population in the Northwest is introduced. At northern edge of range, numbers have greatly declined over the last couple of decades. Southwestern *ridgwayi* extirpated from U.S. part of range by about 1900. Small populations remain due to conservation efforts at several ranches in northern Sonora, especially south-southwest of Benjamin Hill, about 100 miles south of the AZ border. Reintroductions began in AZ prior to 1950 but were unsuccessful. They were released into the Altar Valley from wild populations in Sonora in the 1980s. The population estimate is some 300 to 500 birds but is greatly augmented each year by new releases of some 100 birds raised on site.

Montezuma Quail *Cyrtonyx montezumae* L 8¾" (22 cm)

Plump, short-tailed, round-winged quail. **Male** has distinctive facial pattern and rounded pale brown crest on back of head. Back and wings mottled black, brown, and tan; breast dark chestnut; sides and flanks dark gray with white spots. **Female** is mottled pinkish brown below with less distinct head markings. **Juvenile** is smaller, paler, with dark spotting on underparts. **VOICE:** Call given by male in breeding season is a loud, quavering, descending whistle. Female gives multisyllabic whistles on one pitch. **RANGE:** Uncommon, secretive, and local in grassy undergrowth of open juniper-oak or pine-oak woodlands on semiarid mountain slopes. Recently rediscovered in Chisos Mountains of southwest TX.

Scaled Quail *Callipepla squamata* L 10" (25 cm)

Grayish quail with conspicuous white-tipped crest. Bluish gray breast and mantle feathers have dark edges, creating a shingled or scaly effect. Female's crest is buffy and smaller. **Males** in southernmost Texas (*castanogastris*) tend to show a dark chestnut patch on belly, unlike the common subspecies, *pallida,* found over much of the U.S. range. **Juvenile** resembles adult but is more mottled above, with less conspicuous scaling. **VOICE:** During breeding season, both sexes give a location call when separated, a low, nasal *chip-churr,* accented on the second syllable. **RANGE:** Fairly common; found on barren mesas and plateaus, semi-desert scrublands, and grasslands with mixed scrub; often frequents roadsides. In fall, forms large coveys. Recent records of unknown origin in southern UT and southwestern CO.

slight crest

white supercilium

buffy supercilium

white throat

buffy throat

♂

♀

juvenile

Northern Bobwhite
virginianus

Florida ♂
floridanus

smaller and darker

♂ *taylori*

paler overall

black throat

cinnamon underparts

"Masked Bobwhite" ♂
ridgwayi

Montezuma Quail

exotic head pattern

dark buffy crest laid over nape

juvenile

slightly crested look

♀

♂

solid blackish brown chest

round white spots on sides and flanks

pale buffy top to crest

Scaled Quail

juvenile

♂ *pallida*

south Texas ♂
castanogastris

scaly breast

dark chestnut belly patch

CURRASOWS • GUANS Family Cracidae

These tropical-forest birds have short, rounded wings and long tails. Generally secretive but highly vocal. One species of this family is found in the United States. SPECIES: 50 WORLD, 1 N.A.

Plain Chachalaca *Ortalis vetula* L 22" (56 cm)

Gray to brownish olive above, with small head, slight crest; long and rounded, lustrous, dark green tail tipped with white. Patch of bare skin on throat, usually grayish, is carmine-pink in **male,** duller in female. Juvenile is duller. Can be found on the ground both foraging and hopping; also hops from branch to branch in trees. Usually found in small flocks.

VOICE: Male's call is a deep, ringing *cha-cha-lac,* often given in a loud chorus with other birds; female's voice is higher pitched.

RANGE: Inhabits tall chaparral thickets along the Rio Grande; feeds in trees, chiefly on leaves and buds; often best seen at feeding stations, which many habituate. Introduced to Sapelo Island, GA.

PARTRIDGES • GROUSE • TURKEYS • OLD WORLD QUAIL
Family Phasianidae

Ground dwellers with feathered nostrils, short, strong bills, and short, rounded wings. Flight is brief but strong. Males perform elaborate courting displays. In some species, birds gather at the same strutting grounds, known as leks, every year. SPECIES: 177 WORLD, 17 N.A.

Chukar *Alectoris chukar* L 14" (36 cm)

Old World species, introduced in North America as a game bird in the 1930s. Gray-brown above; flanks boldly barred black and white; buffy face and throat outlined in black; breast gray; belly buff; outer tail feathers chestnut, best seen in flight just prior to landing. Bill and legs are red. Sexes are similar, but males are slightly larger and have small leg spurs. **Juvenile** is smaller and mottled; lacks bold black markings of **adults.**

VOICE: Calls include a series of loud, rapid *chuck chuck chuck* notes and a shrill *whitoo* alarm note.

RANGE: Has become established in rocky, arid, mountainous areas of the West. Game farm Chukars or hybrids with Rock Partridge (*A. graeca*) are released for hunting in the East. In fall and winter, Chukars feed in coveys of 5 to 40 birds.

Gray Partridge *Perdix perdix* L 12½" (32 cm)

Grayish brown bird with rusty face and throat, paler in **female. Male** has dark chestnut patch on belly; patch is smaller or absent in females. Flanks are barred with reddish brown; outer tail feathers rusty.

VOICE: Calls include a hoarse *kee-uck,* likened to a rusty gate.

RANGE: Widely introduced from Europe in early 1900s. Uncommon in most areas, has declined over parts of North American range. Inhabits open farmlands, grassy fields. In fall, forms coveys of 12 to 15 birds. Easiest to find when there is snow cover.

carmine-pink wattle, more pinkish in female, is often hidden

♂

Plain Chachalaca

long blackish tail

white tail tips

rufous face and throat

♀

Gray Partridge

extensive rufous on outer tail feathers

rufous outer tail feathers

gray breast

adults

dark belly

♂

Chukar

buffy throat outlined by black

black-and-white barred flanks

juvenile

Ring-necked Pheasant *Phasianus colchicus*

♂ L 33" (84 cm) ♀ L 21" (53 cm) Introduced from Asia, this large, flashy bird has a long, pointed tail and short, rounded wings. **Male** is iridescent bronze overall, mottled with brown, black, and green; head varies from dark, glossy green to purplish, with fleshy red eye patches and iridescent ear tufts. Often shows a broad white neck ring. **Female** is buffy overall, much smaller and duller than male. Distinguished from female Sharp-tailed Grouse (page 66) by larger size, longer tail that lacks white and barring below.

VOICE: Male's territorial call is a loud, penetrating *kok-cack*. Both sexes give hoarse, croaking alarm notes. When flushed, rises almost vertically with a loud whirring of wings.

RANGE: Locally common; declining in parts of the East. Found in open country, farmlands, brushy areas, and edges of woodlands and marshes. Local hunting releases help maintain some populations and account for presence of some individuals outside normal range. A group of subspecies with white wing coverts (not shown) has become established in parts of the West. **Green Pheasant,** *P. versicolor,* introduced in Tidewater region of VA and southern DE, is apparently gone; considered a subspecies of Ring-necked.

Himalayan Snowcock *Tetraogallus himalayensis*

L 28" (71 cm) Large, gray-brown overall, with tan streaking above. Whitish face and throat, outlined with chestnut stripes; undertail coverts white. Note white in wing in flight. **Male** almost identical to female, except female is slightly smaller, lacks spurs, has buff forehead and grayer area around eye. Inhabits mountainous terrain; flies downhill in the morning, then walks back up, feeding.

VOICE: Calls include various clucks and cackles while feeding. One advertising call suggestive of Long-billed Curlew, male's call rises, female's descends.

RANGE: Asian species, introduced 1963, successfully established only at high elevations in the Ruby Mountains of northeastern NV.

Wild Turkey *Meleagris gallopavo*

♂ L 46" (117 cm) ♀ L 37" (94 cm) Largest game bird in North America; slightly smaller, more slender than the domesticated bird. **Male** has dark, iridescent body, flight feathers barred with white, red wattles, blackish breast tuft, spurred legs; bare-skinned head is blue and pink. Tail, uppertail coverts, and lower rump feathers are tipped with chestnut on eastern birds, buffy white on western birds. **Female** and immature are smaller and duller than male, often lack breast tuft. Of the subspecies seen in North America, *silvestris* predominates in the East, *merriami* in the West. Birds from KS to Mexico *(intermedia)* are intermediate, with buffy tips to the uppertail coverts and a glossy black rump. Birds from peninsular FL *(osceola)* are like *silvestris,* but smaller. These birds of the open forest, forest openings, and field edges forage mostly on the ground for seeds, nuts, acorns, and insects. At night they roost in trees.

VOICE: In spring a male's gobbling call may be heard a mile away.

RANGE: Restocked in much of its former range and introduced in other areas, often involving multiple subspecies.

white neck ring

♂

**Ring-necked
Pheasant**

♀

long tail; compare
carefully with female
Sharp-tailed Grouse

Green
Pheasant

♂

♀

♀

**Himalayan
Snowcock**

♂

white areas
bordered by
chestnut bands

dark belly

unfeathered
reddish head

Wild Turkey
eastern *silvestris*

breast tuft

displaying ♂

western
merriami

♀

unfeathered
gray head

rufous tips to
tail feathers

Ruffed Grouse *Bonasa umbellus* L 17" (43 cm)
Small crest; black ruff on sides of neck, usually inconspicuous; tail with wide dark band near tip, incomplete in **female.** Two color **morphs, red** and **gray;** note tail color difference. Red morphs predominate in the humid Pacific Northwest and Appalachian region; gray morphs in the North and West outside the Pacific Northwest. Ten subspecies.
VOICE: In spring, **male** displays by raising ruff and crest, fanning tail, and beating wings to make a hollow, accelerating, drumming noise. Both sexes give soft clucking notes.
RANGE: Uncommon to fairly common in deciduous and mixed woodlands. Numbers fluctuate, but declining in East.

Spruce Grouse *Falcipennis canadensis* L 16" (41 cm)
Male has dark throat and breast, edged with white; red eye combs. Over most of range, both sexes have black tail with chestnut tip. Birds of the northern Rockies and Cascades, **"Franklin's Grouse,"** *franklinii,* have white spots on uppertail coverts; **male**'s tail is all-dark. In all subspecies, **females** have two color **morphs, red** and **gray;** resemble female Sooty and Dusky Grouse but are smaller and have black barring and white spots below. Juveniles resemble red-morph female.
VOICE: Both sexes give soft clucking notes. In courtship display, male spreads tail, erects red eye combs, rapidly beats wings. In territorial flight display, male flutters upward on shallow wingstrokes; "Franklin's" ends this performance by beating wings together, making a clapping sound. Female's high-pitched call is thought to be territorial.
RANGE: Inhabits open coniferous forests with dense undergrowth. Frequents roadsides, especially in fall.

Sooty Grouse *Dendragapus fuliginosus* L 20" (51 cm)
Formerly (with Dusky Grouse) known as Blue Grouse. **Male**'s sooty gray plumage sets off yellow-orange eye comb. On neck, white-based feathers cover an inflatable bare yellow sac. Female is mottled brown above, with plain gray belly. On both sexes, the 18 tail feathers are round and tipped with a gray terminal band. Chicks are yellowish.
VOICE: Male's display call is a series of loud low hoots audible at considerable distance, usually delivered from perch in tree.
RANGE: Inhabits coniferous forest but will forage at meadow edges. Believed extirpated from mountains of southern CA.

Dusky Grouse *Dendragapus obscurus* L 20" (51 cm)
All plumages similar to Sooty Grouse, but paler overall; closed tail squarer, less graduated; the 20 tail feathers are more square tipped. Northern subspecies, *richardsonii* and *pallidus,* lack or virtually lack the gray terminal band. **Male** neck sac is purplish and smoother, with broader white-feathered border than Sooty. Male's display, usually from ground, often involves low fluttering or making short circular flights, then strutting with tail fanned, body tipped forward, head drawn in, wings dragging. Chicks are grayish.
VOICE: Male's display call, usually given from ground, softer and lower pitched than Sooty and audible only at close range.
RANGE: Often prefers more open forest than Sooty; sometimes found in sagebrush. Range almost entirely separate from Sooty; hybrids recorded from interior of BC.

male with solid
dark tail band

crest

red-morph ♂

**Ruffed
Grouse**

red-morph ♀

displaying
gray-morph ♂

female with
broken tail band

red-morph ♀

gray-
morph ♀

red comb

**Spruce
Grouse**

rufous tail
band

white tips on
uppertail coverts

displaying ♂

"Franklin's
Grouse"
franklinii

Dusky Grouse

northern
subspecies lack
gray tail band

broad gray
tail band

Sooty Grouse

purple
air sac

displaying
northern
Rockies ♂
richardsonii

yellow
air sac

displaying
coastal ♂
fuliginosus

southern Rockies ♀
obscurus

White-tailed Ptarmigan *Lagopus leucura* L 12½" *(32 cm)*

As with all ptarmigans, legs and feet are feathered and plumage is molted three times a year, matching seasonal changes in habitat. Distinguished from other ptarmigans in all seasons by white tail. **Winter** bird is white except for small dark bill and eyes and red eye combs. In **summer,** body is mottled blackish or brown with white belly, wings, and tail. Spring and **fall molts** give a patchy appearance. May form flocks in fall and winter.

VOICE: Calls include a henlike clucking and soft, low hoots.

RANGE: Locally common on rocky alpine slopes, high meadows. Small numbers have been successfully introduced in the central Sierra Nevada, Wallowa Mountains in OR, Unita Mountains in UT, and Pike's Peak in CO. Reintroduced into northern NM. Extirpated from WY. Moves to slightly lower elevations during severe weather.

Rock Ptarmigan *Lagopus muta* L 14" *(36 cm)*

Mottled **summer** plumage is black, dark brown, or grayish brown; **male** generally lacks the reddish tones of male Willow Ptarmigan. There are many recognized subspecies, with color variations according to geography. In **winter** plumage, male has a black line from bill through eye, lacking in male Willow. Acquires breeding plumage later in spring than Willow. In both sexes, bill and overall size are slightly smaller than Willow. **Females** are otherwise difficult to distinguish from Willow. Plumage is patchy white during spring and fall molts. Both species retain white wings and black tail year-round.

VOICE: Calls include low growls and croaks and noisy cackles.

RANGE: Common on high, rocky slopes and tundra. In breeding season, generally prefers higher and more barren habitat than Willow. Irregular fall and winter movements slightly south of normal range. Accidental in northern MN and Queen Charlotte Islands, BC, in spring.

Willow Ptarmigan *Lagopus lagopus* L 15" *(38 cm)*

Largest ptarmigan. Mottled plumage of **summer male** is generally redder than Rock Ptarmigan. White **winter** plumage lacks the black eye line of male Rock; bill and overall size are slightly larger in Willow Ptarmigan. **Female** is otherwise difficult to distinguish from Rock. Both species retain white wings and black tail year-round. Plumage is patchy white during **spring** and fall **molts.** Ptarmigans' red eye combs can be concealed or raised during courtship and aggression.

VOICE: Calls include low growls and croaks, noisy cackles. In courtship and territorial displays, male utters a raucous *go-back go-back go-backa go-backa go-backa.*

RANGE: May form flocks in fall and winter. Common on tundra, especially in thickets of willow and alder. In breeding season, generally prefers wetter, brushier habitat than Rock Ptarmigan. Irregular fall and winter movements slightly south of normal range. Casual in spring and winter to northern tier of U.S. states.

winter

White-tailed Ptarmigan

slighter bill than Willow

molting fall ♂

summer ♀

all plumages have all-white tail

summer ♂

bold black eye line

winter ♀

Rock Ptarmigan

slighter bill than Willow

summer ♀

summer ♂

winter ♂

black tail

fall ♂

Willow Ptarmigan

dark rufous head and neck

thick bill

winter

molting spring ♂

summer ♀

summer ♂

summer ♂

black tail

Greater Prairie-Chicken *Tympanuchus cupido* L 17" (43 cm)
Heavily barred above and below. Short, rounded tail all-dark in **male,** barred in **female.** Male has fleshy yellow-orange eye combs. Both sexes have elongated dark neck feathers, longer in males and erected during courtship to display inflated golden orange neck sacs.
VOICE: Courting males make a deep *oo-loo-woo* sound known as "booming," like blowing over top of an empty bottle.
RANGE: Uncommon, local, and declining. Found in areas of natural tall-grass prairie interspersed with cropland. A smaller, darker subspecies, endangered "Attwater's Prairie-Chicken," *attwateri* (**E**) of southeastern TX, is nearly extinct. The "Heath Hen" (nominate *cupido*), formerly resident along the Atlantic seaboard from MA to VA, is now extinct—last record on Martha's Vineyard in 1932.

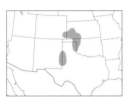

Lesser Prairie-Chicken *Tympanuchus pallidicinctus*
L 16" (41 cm) Resembles Greater Prairie-Chicken, but slightly smaller, paler, less heavily barred below; **male's** neck sacs dull orange-red.
VOICE: Male's courtship notes are higher pitched than Greater.
RANGE: Uncommon, local, and declining; found in sagebrush and shortgrass prairie country, especially where shinnery oak grows. Will forage in cropland.

Sharp-tailed Grouse *Tympanuchus phasianellus* L 17" (43 cm)
Similar to prairie-chickens, but underparts scaled and spotted; tail mostly white and pointed; yellowish eye combs less prominent. Compare with female Ring-necked Pheasant (page 60). Birds darkest in AK and northern Canada (standing figure), palest in the Plains (flying figure). **Male's** purplish neck sacs are inflated during courtship display.
VOICE: Male's courting notes include cackling and a single, low *coo-oo* call accompanied by the rattling of wing quills.
RANGE: Inhabits grasslands, sagebrush, woodland edges, and river canyons. Fairly common over much of range but rare in western U.S. and extirpated from most of southern range. Where ranges overlap, can hybridize with Greater Prairie-Chicken and Dusky Grouse.

Gunnison Sage-Grouse *Centrocercus minimus*
♂ L 22" (56 cm) ♀ L 18" (46 cm) Smaller than Greater Sage-Grouse, with more strongly white-banded tail. Longer, denser filoplumes are erected to form a distinct, recurved crest on **displaying male.**
VOICE: Male's display call lower pitched, more uniform than Greater Sage-Grouse.
RANGE: Small, declining population in south-central CO and southeastern UT is geographically isolated from Greater Sage-Grouse.

Greater Sage-Grouse *Centrocercus urophasianus*
♂ L 28" (71 cm) ♀ L 22" (56 cm) Blackish belly, long pointed tail feathers, and large size distinctive. **Male** larger than **female,** has yellow eye combs, black throat and bib, large white ruff on breast. In flight, dark belly, absence of white outer tail feathers, and larger size distinguish it from Sharp-tailed Grouse. **Displaying male** fans tail.
VOICE: In display, male rapidly inflates and deflates air sacs, emitting a loud, bubbling popping.
RANGE: Uncommon and local; declining. Found in sagebrush.

short, square black tail

displaying ♂

golden orange air sacs

Greater Prairie-Chicken

♀

Lesser Prairie-Chicken

displaying ♂

reddish air sacs

♀

weaker barring below than Greater Prairie-Chicken

white-banded tail

larger crest than Greater Sage-Grouse

displaying ♂

Sharp-tailed Grouse

displaying ♂

pointed tail shorter than female Ring-necked Pheasant

purplish air sacs

♀

grayish plumage overall, heavily mottled and barred with dark

Gunnison Sage-Grouse

rather uniform-colored tail feathers

♂

♀

Greater Sage-Grouse

graduated, pointed tail

black belly patch

displaying ♂

displaying males on lek

LOONS Family Gaviidae
In all species, juvenal-like plumage held through the first summer. SPECIES: 5 WORLD, 5 N.A.

Red-throated Loon *Gavia stellata* L 25" (64 cm)
Tends to hold head tilted up; thin bill often appears slightly upturned. **Breeding adult** has gray head with brick red throat patch that appears dark in flight; dark brown upperparts lack contrasting white patches on scapulars found in all other loons in breeding plumage. **Winter adult** has sharply defined white on face and extensive white spotting on back. **Juvenile's** head is grayish brown; throat may have dull red markings. In all plumages, white on flanks extends upward a bit on sides of rump, which may cause confusion with Arctic Loon. In flight, shows smaller head and feet than Common and Yellow-billed Loon (page 70); wingbeat is quicker; often flies with drooping neck, unlike other loons.
VOICE: Flight call, heard on breeding range, is a rapid, gooselike *kak-kak-kak*.
RANGE: Migrates coastally; also overland in the East, where most numerous on northern and eastern Great Lakes. Always casual inland in western North America, and in the eastern interior during winter.

Pacific Loon *Gavia pacifica* L 26" (66 cm)
In all plumages, has dark flanks, with no white extending upward on sides of rump. Bill is slim and straight; head smoothly rounded and held level. **Breeding adult's** head and nape are pale gray; white stripes on sides of neck show only moderate contrast; throat's iridescent purple patch, sometimes washed with green, usually appears black unless seen clearly on swimming bird. **Juvenile's** crown and nape are slightly paler than back, unlike Common Loon (page 70); in juveniles and **winter adults,** dark cap extends to eye. Winter adults and most juveniles have a thin, brown "chin strap," though it may be faint in juveniles. In flight, resembles Common, but head and feet are smaller.
VOICE: On breeding grounds, gives various yodeling calls.
RANGE: A coastal and offshore migrant; unlike other loons, often migrates in small to moderate-size flocks. Rare inland throughout the West; very rare in Midwest including Great Lakes region; casual on East Coast.

Arctic Loon *Gavia arctica* L 28" (73 cm)
Larger than Pacific Loon, with less-rounded head; best distinguished in all plumages from Pacific by more extensive white on flanks, coming up over sides of rump. Visibility of white area depends on how buoyantly the bird is swimming. When diving, often only a small white rump patch is evident. At rest, Arctic Loon shows much more white; note Pacific can also show some white. Nape in **breeding adult** is darker, and black-and-white stripes are bolder, than Pacific; white stripes on face connect more to sides of neck; greenish on throat very hard to see.
VOICE: On breeding grounds, yodeling calls are deeper than Pacific.
RANGE: Old World species. To date, only the larger Asian subspecies, *viridigularis,* has been recorded in North America. Breeds in northwestern AK. Seen in migration in coastal western AK, especially at St. Lawrence Island. Casual elsewhere on West Coast.

Red-throated

in flight, neck often droops

winter adults

Arctic

Pacific

more dark on sides of neck than Red-throated

Red-throated

no white on scapulars

grayish brown stripe on side

breeding adults

darker gray nape with bolder white neck stripes

white scapulars

even black-and-white line down flanks

black stripe on side

pale head and nape

Pacific

Arctic

white extends up to sides of rump

Red-throated Loon

bill slightly upturned and often holds head up

face and neck washed with dusky

juvenile

1st spring

breeding adult

reddish throat often looks dark

extensive pure white face

winter adult

white speckling on upperparts

Pacific Loon

pale gray nape

winter adult

faint white vertical streaks on sides of neck

some young birds lack "chin straps"

juveniles

white scaling on scapulars

dark "chin strap"

winter adult

dark upperparts

breeding adult

sharp and even division between front and rear of neck

often holds head up, especially in display

Arctic Loon
viridigularis

darker gray nape

winter adult

broader and more contrasty white stripes across throat and on sides of neck

breeding adults

white flank patch in all plumages

juvenile

never has a "chin strap"

Common Loon *Gavia immer* L 32" (81 cm)

Large, thick-billed loon with slightly curved culmen. Bill is black in **breeding** plumage, blue-gray in **winter adults** and **juveniles,** but the culmen remains dark. In winter plumage, crown and nape are darker than back; dark on nape extends around sides of neck, but note the white indentation above this. In winter adults the white extends up and around the eye; the face pattern is more blended in juveniles. Forehead is steep, crown is peaked at front. Holds head level. Juvenile Common and Yellow-billed Loons have whitish scalloping on their scapulars, distinguishing them from the plainer-backed winter adults. Full juvenal plumage is kept through most of the winter, with a partial molt in spring. Most winter adults retain at least a few spotted coverts, often visible on swimming birds. Full adult breeding plumage is not acquired until nearly three years of age. Under most conditions, Common and Yellow-billed fly high above the water when migrating whereas other loon species fly lower. Note that Common and Yellow-billed have slower wingbeats and paddle-shaped feet that are usually visible beyond the tail. Also note that in flight, large head and feet help distinguish Common from Arctic, Pacific, and Red-throated Loons (flight figures on page 69).

VOICE: Loud yodeling calls delivered on water and in flight are heard all year, but most often on breeding grounds.

RANGE: Fairly common; nests on large lakes. Migrates overland as well as coastally. Winters mainly in coastal waters or on large, ice-free inland bodies of water. Small numbers of nonbreeders oversummer in winter range. Generally rare in Southwest due to lack of appropriate habitat.

Yellow-billed Loon *Gavia adamsii* L 34" (86 cm)

Breeding adult has straw yellow bill, usually longer than Common Loon; culmen is straight, giving bill a slightly uptilted look; head often tilted back, which enhances this effect. Crown is peaked at front and rear, giving a subtle double-bump effect. Bill is duskier at the base in **winter adults** and **juveniles,** but always shows strong yellow cast toward the tip. Note also pale face and distinct dark mark behind eye; eye is smaller, back and crown are paler and browner than Common. As with Common, full adult **breeding** plumage is not acquired until three years of age.

VOICE: Calls are similar to Common.

RANGE: Breeds on tundra lakes and rivers. Migrates coastally; rare south of Canada on West Coast, where it is recorded annually south to northern CA, casually to southern CA. Very rare inland in West; casual east to Great Lakes region and south to TX and GA. Accidental on East Coast.

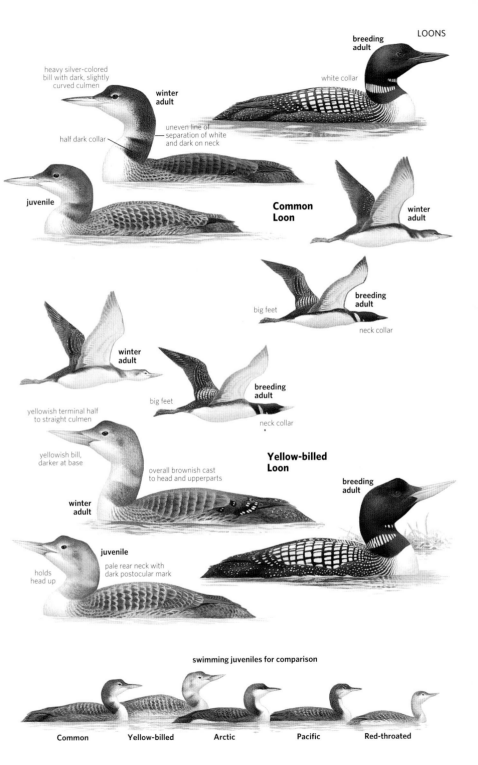

LOONS

Common Loon

breeding adult

white collar

heavy silver-colored bill with dark, slightly curved culmen

winter adult

uneven line of separation of white and dark on neck

half dark collar

juvenile

winter adult

big feet

breeding adult

neck collar

winter adult

big feet

breeding adult

neck collar

yellowish terminal half to straight culmen

Yellow-billed Loon

breeding adult

yellowish bill, darker at base

overall brownish cast to head and upperparts

winter adult

holds head up

juvenile

pale rear neck with dark postocular mark

swimming juveniles for comparison

Common   Yellow-billed   Arctic   Pacific   Red-throated

GREBES Family Podicipedidae
A worldwide family of aquatic diving birds. Lobed toes make them strong swimmers. Grebes are infrequently seen on land or in flight. SPECIES: 22 WORLD, 7 N.A.

Least Grebe *Tachybaptus dominicus* L 9¾" *(25 cm)*
A small, grebe with golden yellow eyes, a slim, dark bill, and purplish gray face and foreneck. **Breeding adult** has blackish crown, hindneck, throat, and back. **Winter** birds have white throat, paler bill, less black on crown. In flight, shows large white wing patch.
VOICE: Gives nasal *beep*; in display, a descending, rapid, buzzy trill.
RANGE: Rather uncommon and local; may hide in vegetation near shores of ponds, sloughs, and ditches. May nest at any season on any quiet, inland water. Casual straggler to southern AZ, southeastern CA, south FL, and upper TX coast.

Pied-billed Grebe *Podilymbus podiceps* L 13½" *(34 cm)*
Breeding adult is brown overall, with black ring around stout, whitish bill; black chin and throat; pale belly. **Winter** birds lose bill ring; chin is white, throat tinged with pale rufous. **Juvenile** resembles winter adult but throat is much redder, eye ring absent, head streaked with brown and white. In flight, shows almost no white on wing.
VOICE: On breeding grounds, delivers a loud series of gulping notes.
RANGE: Nests around marshy ponds and sloughs; sometimes hides from intruders by sinking until only its head shows. Common but not gregarious. Winters on fresh or salt water. Casual to AK.

Horned Grebe *Podiceps auritus* L 13½" *(34 cm)*
Breeding adult has chestnut foreneck and golden "horns." In **winter** plumage, white cheeks and throat contrast with dark crown and nape; some are dusky on lower foreneck. Black on nape narrows to a thin stripe. All birds show a pale spot in front of eye. In flight (page 74), white secondaries show as patch on trailing edge of wing. Bill is short and straight, thicker than Eared Grebe; neck is thicker too, crown flatter. Smaller size and shorter, dark bill most readily separate winter Horned from Red-necked Grebe (page 74).
VOICE: Mostly silent, except on nesting grounds.
RANGE: Breeds on lakes and ponds. Winters on salt water but also on ice-free lakes of eastern North America; a few winter inland in West.

Eared Grebe *Podiceps nigricollis* L 12½" *(32 cm)*
Breeding adult has blackish neck, golden "ears" fan out behind eye. In **winter** plumage, throat is variably dusky, cheek dark; whitish on chin extends up as a crescent behind eye; compare with Horned Grebe. Note also Eared Grebe's longer, thinner bill; thinner neck; more peaked crown. Lacks pale spot in front of eye. Generally rides higher in the water than Horned Grebe, exposing fluffy white undertail coverts. In flight, white secondaries show as white patch on trailing edge of wing.
VOICE: Most vocal on breeding grounds; most frequent call is a rising, whistled note.
RANGE: Usually nests in large colonies on freshwater lakes. Rare in eastern North America. Casual to AK.

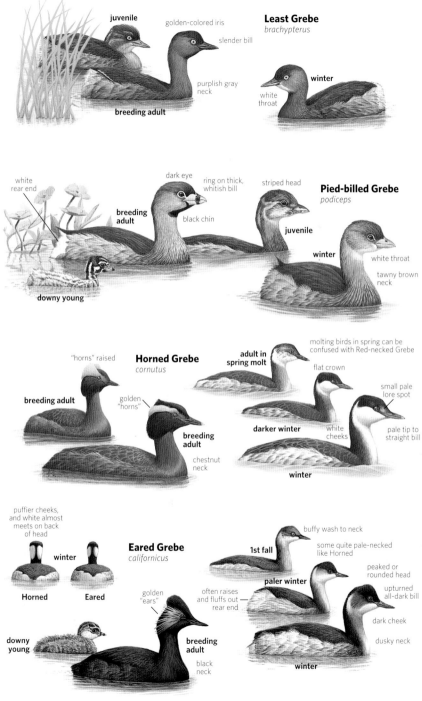

Least Grebe
brachypterus

juvenile
golden-colored iris
slender bill
purplish gray neck
breeding adult
winter
white throat

Pied-billed Grebe
podiceps

white rear end
dark eye
ring on thick, whitish bill
striped head
breeding adult
black chin
juvenile
winter
white throat
tawny brown neck
downy young

Horned Grebe
cornutus

"horns" raised
adult in spring molt
molting birds in spring can be confused with Red-necked Grebe
flat crown
breeding adult
golden "horns"
small pale lore spot
breeding adult
chestnut neck
darker winter
white cheeks
pale tip to straight bill
winter

Eared Grebe
californicus

puffier cheeks, and white almost meets on back of head
winter
Horned
Eared
golden "ears"
often raises and fluffs out rear end
breeding adult
black neck
downy young
1st fall
buffy wash to neck
some quite pale-necked like Horned
paler winter
peaked or rounded head
upturned all-dark bill
dark cheek
dusky neck
winter

Red-necked Grebe *Podiceps grisegena* L 20" (51 cm)

Large grebe with heavy, tapered, yellowish bill almost as long as the head. **Breeding adult's** whitish throat and cheeks contrast with reddish foreneck. In **winter** plumage, throat is dusky, white of chin extends onto rear of face in a crescent. **First-winter** bird has rounder head, darker bill, paler eye; lacks strong facial crescent. **Juvenile** has striped head. In flight, Red-necked Grebe shows a white leading and trailing edge on inner wing; thick neck is often held slouched down. Generally solitary.

VOICE: Calls, usually heard only on breeding grounds, include a *crick-crick* note and drawn-out braying calls.

RANGE: Breeds on shallow lakes; winters mostly along coasts. Rare in interior south of northern tier of states; occasionally, moderate numbers winter in mid-Atlantic region, especially in years when Great Lakes freeze. Rare to very rare in interior West and coastal southern CA. Casual south to Southwest and Gulf Coast states.

Clark's Grebe *Aechmophorus clarkii* L 25" (64 cm)

Resembles Western Grebe but bill is orange; back and flanks are paler; black cap does not extend to eye in **breeding** plumage; **downy young** are paler. In **winter adult,** lore region acquires more dark color, pattern looks more like Western; best distinction then is bill color. In flight, Clark's Grebe's white wing stripe is more extensive than Western. Both species have an elaborate courtship that includes both sexes rising out of the water and rushing forward in almost perfect synchronization. Formerly considered one species with Western; hybrids sometimes noted and are undoubtedly more frequent than reported.

VOICE: Call is a single, two-syllable, upslurred *kree-eek* note.

RANGE: Limits of range in both species are not well known; Clark's occupies same general area and habitat as Western but is much less common in northern and eastern part of range. Accidental to eastern North America.

Western Grebe *Aechmophorus occidentalis* L 25" (64 cm)

Large grebe, strikingly black and white, with a long, thin neck and long bill. Resembles Clark's Grebe but bill is yellow-green; black cap extends to include eyes; back and flanks are darker; **downy young** are darker. In **winter adult,** lore region acquires more whitish color, and pattern can closely resemble winter Clark's. In flight, Western Grebe's white wing stripe is less extensive than Clark's. Like Clark's, gregarious. The two species often occur together.

VOICE: Call is a loud, two-note *crick-kreek.*

RANGE: Nests in reeds along broad, freshwater lakes. Winters on seacoasts and sheltered bays and large inland bodies of water. Occupies same general range and habitat as Clark's but greatly predominates in northern and eastern part of range. Casual during migration and winter to eastern North America and to YT.

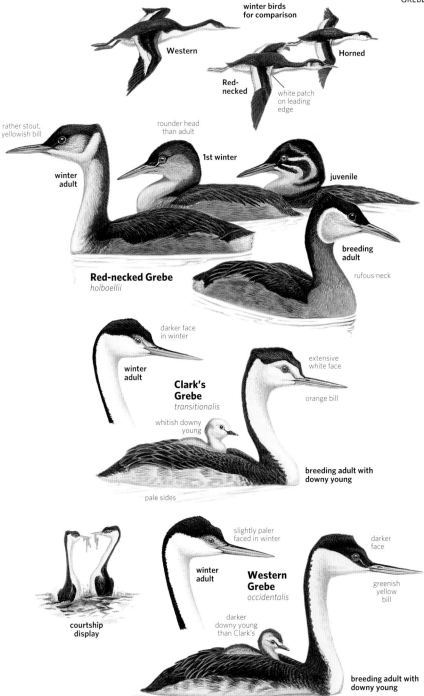

winter birds
for comparison

Western

Horned

Red-
necked

white patch
on leading
edge

rather stout,
yellowish bill

rounder head
than adult

1st winter

juvenile

winter
adult

breeding
adult

rufous neck

Red-necked Grebe
holboellii

darker face
in winter

winter
adult

extensive
white face

orange bill

**Clark's
Grebe**
transitionalis

whitish downy
young

breeding adult with
downy young

pale sides

courtship
display

slightly paler
faced in winter

winter
adult

**Western
Grebe**
occidentalis

darker
face

greenish
yellow
bill

darker
downy young
than Clark's

breeding adult with
downy young

slightly darker sides than Clark's on average

ALBATROSSES Family Diomedeidae

Gliding on extremely long narrow wings, these largest of seabirds spend most of their lives at sea, alighting on the water when becalmed or when feeding on squid, fish, and refuse. Pelagic; most species nest in colonies on oceanic islands; pairs mate for life. A number of species, especially those in Southern Hemisphere, are threatened by long-line fishing. Largely silent at sea. SPECIES: 15 WORLD, 8 N.A.

Short-tailed Albatross *Phoebastria albatrus*

L 36" (91 cm) WS 85-91" (215-230 cm) **E** Large size, but best field mark is the long and massive pink (initially dark on very young juveniles) bill with pale bluish tip. Dark humerals and pale feet distinctive in post-juvenal plumages. **Adult** is mostly white, with golden wash on head. **Juvenile** is blackish brown, except for traces of white below and behind eye and on chin. Older juvenile has more white around bill; compare with Black-footed Albatross. **Subadult** shows white forehead and face, dark cap; acquires white patches on scapulars and inner secondary coverts; with age becomes progressively white, but retains dark hindneck. Full adult plumage takes more than a decade to acquire, but can breed in subadult plumages.

RANGE: Common until end of 19th century, then decimated and was on verge of extinction by 1930s. Very small numbers of breeders reappeared on Torishima Island, beginning in 1951. Now protected and population slowly recovering—global population believed to number about 2,000. Presently breeds on Torishima (most) and Minami-kojima Islands off southern Japan. Still very rare to rare, but sightings are increasing in North Pacific off North America from Aleutians to central CA. The great majority of these sightings are of juveniles and subadults.

Laysan Albatross *Phoebastria immutabilis*

L 32" (81 cm) WS 77-80" (195-203 cm) Back, upperwing blackish brown, except for white flash in primaries; underwing with black margins and variable internal markings. Note blurry dark mark surrounding eye; pinkish bill. Adults and juveniles similar. Occasionally hybridizes with Black-footed Albatross.

RANGE: Most numerous spring through summer off AK; rare to uncommon off West Coast from late fall through spring; casual inland in winter and spring; most records from southeastern CA in spring; accidental southwestern AZ. Breeds mainly on islands of HI; small colonies recently established in the Revillagigedo Archipelago and Isla Guadalupe, Mexico.

Black-footed Albatross *Phoebastria nigripes*

L 32" (81 cm) WS 80" (203 cm) Mostly dark in all plumages. White area around bill is more extensive on **old** birds; reduced or absent on immatures. Most birds of all ages have dark undertail coverts. Some **adults** have white undertail coverts, and white may extend onto belly; these birds can be confused with subadult Short-tailed Albatross, but lack white upperwing patches and have thinner, shorter, darker bills.

RANGE: Seen year-round off West Coast; most common in spring and summer. Breeds mainly on islands of HI; overall population declining.

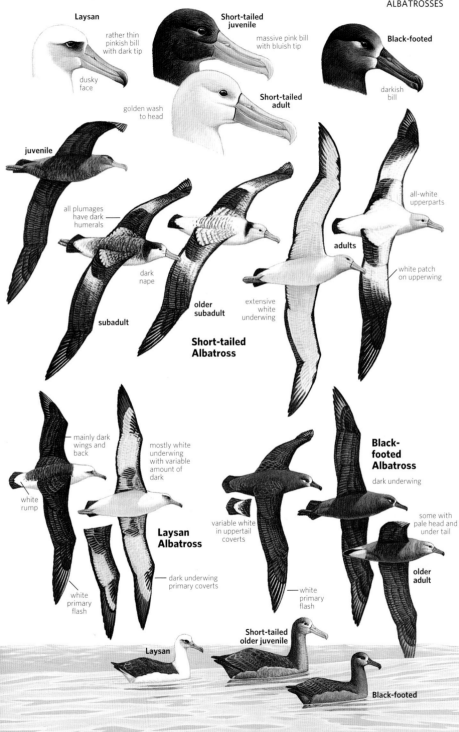

ALBATROSSES

Laysan

rather thin
pinkish bill
with dark tip

dusky
face

Short-tailed
juvenile

massive pink bill
with bluish tip

golden wash
to head

Short-tailed
adult

Black-footed

darkish
bill

juvenile

all plumages
have dark
humerals

dark
nape

subadult

older
subadult

extensive
white
underwing

adults

all-white
upperparts

white patch
on upperwing

**Short-tailed
Albatross**

mainly dark
wings and
back

white
rump

mostly white
underwing
with variable
amount of
dark

**Laysan
Albatross**

white
primary
flash

dark underwing
primary coverts

variable white
in uppertail
coverts

white
primary
flash

**Black-
footed
Albatross**

dark underwing

some with
pale head and
under tail

older
adult

Laysan

Short-tailed
older juvenile

Black-footed

Shy Albatross *Thalassarche cauta*

L 35-39" (90-99 cm) WS 87-101" (220-256 cm) **Adults** of *cauta/steadi* are white-headed, unlike all ages of *salvini* (**"Salvin's"**), which have a grayish wash on head; younger *cauta/steadi* head also grayish. Distinctive in all ages is underwing pattern: primaries more extensively dark in *salvini* (and *eremita*); larger than Laysan, which has smaller, thinner, mostly pinkish bill. Note Shy's paler back, longer and grayer tail, more extensive white on the rump, and more languid flight style; all ages show extensively white underwing and characteristic dark "thumb mark" at beginning of leading edge. Yellow-tipped bill in *cauta* adults (*steadi* very similar); grayer and darker-tipped bill in all younger birds. Bill is dusky with a dull yellow ridge and dark lower tip in adult *salvini*.

RANGE: Casual off Pacific coast from WA to northern CA; once off Kasatochi Island, Aleutians. Only specimen (nominate *cauta*) was off WA in 1951; nine other records since 1996. Taxonomic opinions differ, but presently divided into four subspecies breeding in the following locations: *cauta* breeds on islands off Tasmania; nearly identical *steadi* breeds on Auckland, Antipodes, and Chatham Islands, New Zealand; *salvini* breeds on Snares and Bounty Islands off New Zealand, and on Îles Crozet in southwestern Indian Ocean; and unrecorded *eremita* group (dark gray head) breeds on Chatham Island. Remaining nine North American records are divided between those of *cauta/steadi* and those believed to be of *salvini* (including the Aleutians record).

Yellow-nosed Albatross *Thalassarche chlororhynchos*

L 32" (81 cm) WS 80" (203 cm) Confused with Black-browed, but Yellow-nosed is smaller and slimmer, with longer neck. **Adult's** bill appears black; at close range, yellow ridge on top and reddish tip are visible. Note light grayish wash on head and blackish triangular patch in front of eye of nominate subspecies. Underwing extensively white with a narrow dark border; some **juveniles** show more dark on leading edge, which can cause confusion with adult Black-browed. Otherwise juvenile resembles adult, except for all-dark bill and reduced eye patch.

RANGE: Casual off Atlantic and Gulf coasts; a few inland sightings in East. Most of nominate subspecies breed on Tristan de Cunha and Gough Islands in South Atlantic, the probable source of North American records. The other subspecies, *bassi,* breeding on southern Indian Ocean islands, has purer white head and reduced dark smudge in front of eye.

Black-browed Albatross *Thalassarche melanophris*

L 35" (89 cm) WS 88" (224 cm) From similar Yellow-nosed, note larger size with thicker neck and chunkier body. **Adult** has broad dark leading edge to underwing and heavier orange bill with redder tip; black eyebrow. **Juveniles** have darker bills and gray shading about head and neck forming collar; underwing mainly dark. **Subadults** have more yellowish bill with dark tip and some white in underwing.

RANGE: A circumpolar Southern Hemisphere species. Casual in North Atlantic, most recorded in northeastern portion. Several recent records documented with photographs; about 20 other reports (some possibly correct).

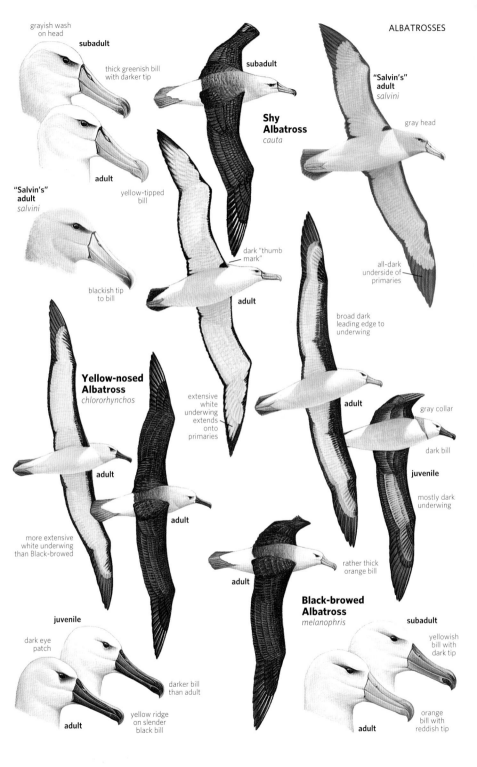

ALBATROSSES

grayish wash on head

subadult

thick greenish bill with darker tip

subadult

"Salvin's" **adult**
salvini

gray head

adult

yellow-tipped bill

"Salvin's" **adult**
salvini

blackish tip to bill

Shy Albatross
cauta

all-dark underside of primaries

dark "thumb mark"

adult

broad dark leading edge to underwing

Yellow-nosed Albatross
chlororhynchos

extensive white underwing extends onto primaries

adult

adult

gray collar

dark bill

juvenile

mostly dark underwing

adult

adult

more extensive white underwing than Black-browed

juvenile

dark eye patch

adult

rather thick orange bill

Black-browed Albatross
melanophris

subadult

yellowish bill with dark tip

darker bill than adult

yellow ridge on slender black bill

orange bill with reddish tip

adult

PETRELS • SHEARWATERS Family Procellariidae

Pelagic seabirds rarely seen from shore; bills have nostril tubes. Fly with rapid wingbeats, stiff-winged glides. Most species generally silent at sea. SPECIES: 82 WORLD, 29 N.A.

Northern Fulmar *Fulmarus glacialis*

L 19" (48 cm) WS 42" (107 cm) **Light morphs** predominate over much of North Atlantic. In Pacific light morphs predominate in Bering Sea; **dark morphs** farther south. Pacific subspecies (*rodgersii*) more slender billed than Atlantic *glacialis* and more southerly and even longer billed *auduboni; rodgersii* also shows greater variation in color morphs (darker dark morphs and paler lights), and darker tail contrasts with rump. **Intermediates** of all shades are frequent. Distinguished from gulls by short, thick bill with nostril tubes, and shearwater-like flight; from shearwaters by thick, yellow bill, stockier shape, and rounder wings. **RANGE:** Common and increasing. Within winter range, numbers fluctuate annually; some summer south to ME and southern CA. Casual to Hudson and James Bay in fall.

White-chinned Petrel *Procellaria aequinoctialis*

L 21½" (55 cm) WS 55½" (140 cm) Almost wholly blackish petrel; rather thick yellowish bill, lined with black; feet and legs dark. White chin (variable) very difficult to see in field. Note usually entirely pale bill, including tip; a few with very limited black on tip. **RANGE:** Widespread in southern oceans. Three scattered records, all well photographed: TX (late Apr.), ME (Aug.), and CA (mid-Oct.). An additional and more poorly photographed bird was noted east-northeast of Oregon Inlet, NC, on 12 and 17 Oct. 1996.

Parkinson's Petrel *Procellaria parkinsoni*

L 18" (46 cm) WS 45" (115 cm) Like larger White-chinned in body and bill coloration, except bill tip mostly dark. Compare also to similarly colored Flesh-footed Shearwater (page 88); note bill shape and color. Westland Petrel (*P. westlandica*) from South Pacific is similar but is even larger, blockier headed, and thicker billed; a few have paler-tipped bills approaching White-chinned; unrecorded north of Equator. **RANGE:** Breeds on islands off New Zealand; ranges north in austral winter to east-central Pacific, north to southern Mexico. One certain record (photos): 1 Oct. 2005, about 18 miles off Point Reyes, CA.

GADFLY PETRELS

Fast-flying petrels with arcing, acrobatic flight. Unlike shearwaters, they typically hold their wings slightly forward from the shoulder and bent sharply back at the "wrist."

Great-winged Petrel *Pterodroma macroptera*

L 16" (41 cm) WS 38" (97 cm) Similar to Murphy's Petrel (page 82) but browner (less gray); more uniform underwing; white more evenly distributed around bill; dark legs and feet. **RANGE:** Southern oceans species; three records (late summer, Oct.) off central CA of presumed *gouldi* subspecies with whiter face.

intermediate
morph

light
morph

white flash on
inner primaries

Pacific birds' plumage
more variable than
Atlantic birds

Pacific birds
have dark tails
that contrast
with paler rump

variable, some
individuals
even paler

light
morph

dark
morph

uniformly
dark

**Northern
Fulmar**
Pacific *rodgersii*

heavier bill than
Pacific birds

Atlantic dark morph

heavy
yellowish
bill

light
morph

Atlantic
light morph

tail blends with
rump in both
Atlantic morphs

large and
heavily built

**White-chinned
Petrel**

**Parkinson's
Petrel**

smaller and more
lightly built than
White-chinned

primary flash
fainter than
Murphy's

uniform dark
underwing

usually
all-pale
bill tip

thick yellowish
bill with
blackish tip

light
morph

overall color
browner, less
gray, than
Murphy's

white evenly
distributed
around bill

variable white
chin usually
difficult to see

White-chinned

dark tip to bill

Parkinson's

**Great-winged
Petrel**
gouldi

Murphy's Petrel *Pterodroma ultima*
L 16" (41 cm) WS 38" (97 cm) Dark brownish gray with faint dark M-pattern on back, wedge-shaped tail, and white underwing flash; white most conspicuous below bill; legs and feet pink. The similar but larger Solander's Petrel (*P. solandri*), reported off CA, WA, and BC (photo) but no accepted records yet, has heavier bill with equal or more white above bill than below; prominent dark primary covert tips. Compare with dark-morph Northern Fulmar.
RANGE: Breeds on remote central South Pacific islands. Uncommon spring visitor far off West Coast to southern WA; casual in summer and fall off Pacific Northwest coast.

Hawaiian Petrel *Pterodroma sandwichensis*
L 17" (43 cm) WS 39" (98 cm) **E** Recently split from Galapagos Petrel (*P. phaeopygia*), which together were formerly known as Dark-rumped Petrel. Both have a mostly black crown that extends down sides of neck, dark upperparts with white on uppertail coverts, a prominent black bar on underwing, and a long tail. Hawaiian shows less of a "shawl" and is whiter below, appears more capped with a white insertion into the shawl. All well-photographed records show characters of Hawaiian, not Galapagos.
RANGE: Nests only on HI. Casual, recorded off CA and OR from late April to Oct.; over 20 records, but no specimens.

Mottled Petrel *Pterodroma inexpectata*
L 14" (36 cm) WS 32" (81 cm) White throat, breast, and vent contrast with rest of mostly gray underparts. Shows prominent black bar on otherwise white underwing; dark M across upperwings.
RANGE: Breeds on islands off New Zealand. Probably regular well off southern AK and Aleutians in summer and fall; rare and irregular well off the West Coast, chiefly in late fall. Accidental in western NY (Livingston County, early Apr. 1880).

Stejneger's Petrel *Pterodroma longirostris*
L 11" (28 cm) WS 23" (58 cm) Resembles Cook's Petrel but distinct dark half hood contrasts with grayish back and white forehead; tail is longer and more uniformly colored, with less white in outer tail feathers.
RANGE: Breeds on Juan Fernandez Islands off Chile. Casual well off the CA coast, chiefly in fall. Accidental to coastal TX (Port Aransas, 15 Sept. 1995, tideline corpse).

Cook's Petrel *Pterodroma cookii* *L 10¾" (27 cm) WS 26" (66 cm)*
This and Stejneger's considered part of the *Cookilaria* petrel subgenus. They are smaller and more acrobatic than the larger *Pterodroma* petrels. Cook's is small, with long wings and rather short tail. Crown and back uniformly gray; blackish eye patch. Note mainly white underwing and distinct dark M across upperwings. White on outer tail feathers and dark tip to central tail feathers can be hard to see.
RANGE: Breeds on islands off New Zealand. Found well off CA coast from spring through late fall, where most numerous (and increasing) *Pterodroma*. Casual off Aleutians and on the Salton Sea in summer.

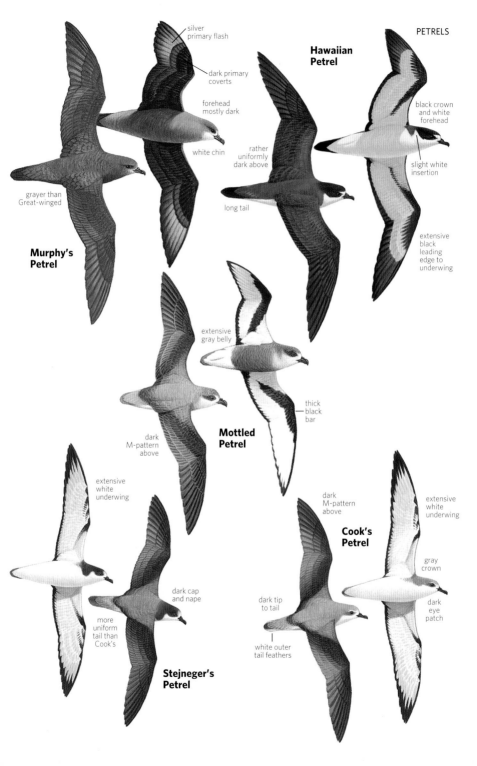

PETRELS

Hawaiian Petrel

black crown and white forehead

slight white insertion

extensive black leading edge to underwing

silver primary flash

dark primary coverts

forehead mostly dark

white chin

rather uniformly dark above

long tail

grayer than Great-winged

Murphy's Petrel

extensive gray belly

thick black bar

dark M-pattern above

Mottled Petrel

extensive white underwing

dark M-pattern above

Cook's Petrel

extensive white underwing

gray crown

dark eye patch

dark tip to tail

white outer tail feathers

dark cap and nape

more uniform tail than Cook's

Stejneger's Petrel

Black-capped Petrel *Pterodroma hasitata*

L 16" (41 cm) WS 37" (94 cm) Distinct dark cap; white collar; broad white band on uppertail coverts and base of tail. White wing lining, with variable dark diagonal bar on leading edge. Wing and bill shape, white forehead, broader band on tail, and languid, arcing flight distinguish this species from Great Shearwater (page 86). Some birds have less white at base of tail and duskier collar.

RANGE: Breeds on Hispaniola and Cuba. Common in Gulf Stream off NC from late May to mid-Oct.; uncommon in winter; casual north to NS and in Gulf of Mexico. Recorded inland in the East after hurricanes north to the eastern Great Lakes.

Fea's Petrel *Pterodroma feae* *L 14" (36 cm) WS 37" (94 cm)*

Brownish gray above with dark M-pattern, pale uppertail coverts, and long, pale tail. White below, partial breast band, and mostly dark underwing. Formerly, this and other taxa all considered part of one species, the Soft-plumaged Petrel. Recent taxonomic revisions now restrict the Soft-plumaged Petrel (*P. mollis*) to cold water in the southern oceans. Fea's comprises two subspecies: nominate *feae* in the Cape Verde Islands and slightly larger-billed *deserta* on Bugio Island, Desertas, Madeira. The smaller, more slender-billed Zino's Petrel (*P. madeira*) breeds only on the summit of Madeira. All North Atlantic taxa are migratory. Distinct differences in genetics, timing of breeding, and vocalizations have led some to argue for species recognition between *feae* and *deserta*. All of these North Atlantic taxa are very similar in appearance, although the small-billed extremes of Zino's and the larger-billed Fea's (*feae* or *deserta*) should be distinct. But with all, males are larger billed than females and immatures are smaller billed than adults. A minority of Zino's show moderate to rather extensive white in the underwing coverts (greater primary and secondary), unlike any Fea's.

RANGE: Rare visitor off NC in late May and early June, casual into the fall; accidental to NS; sightings (supported by excellent photos but no specimens) believed to be Fea's rather than Zino's, based on likelihood and particularly for those with very thick-appearing bills. To date, American Ornithologists' Union and American Birding Association committees have accepted North American records only as Fea's/Zino's. Recent studies using geolocators might elucidate the origin of birds of this complex seen in North American waters.

Bermuda Petrel *Pterodroma cahow*

L 15" (38 cm) WS 35" (89 cm) **E** Note that larger Black-capped Petrel has heavier bill and disproportionately shorter wings. Bermuda Petrel's whitish rump, sometimes lacking, is restricted to base of uppertail coverts. Without Black-capped's white collar, dark on head is more like cowl than cap; Bermuda more buoyant in flight, with darker underwing than Black-capped.

RANGE: Endangered species. Believed extinct, but rediscovered in 1951; population increasing, now estimated at about 200. Nests only on islets off Bermuda; since mid-1990s, more than two dozen well-documented records, in late spring (mostly) and summer off Outer Banks, NC; once well off MA.

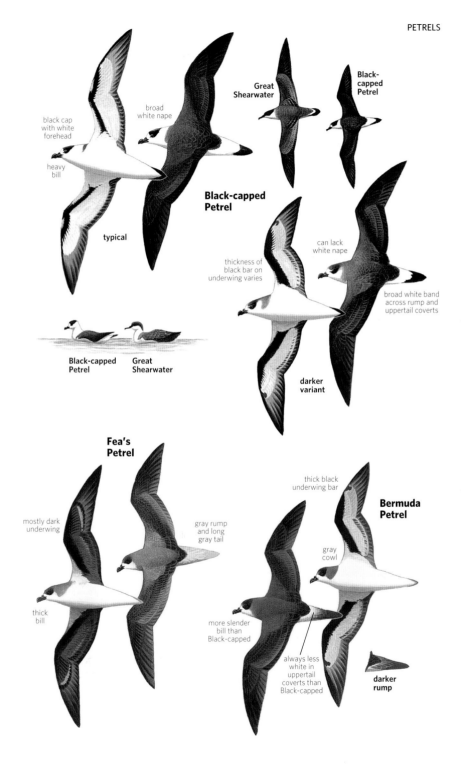

black cap with white forehead

broad white nape

heavy bill

typical

Great Shearwater

Black-capped Petrel

Black-capped Petrel

can lack white nape

thickness of black bar on underwing varies

broad white band across rump and uppertail coverts

Black-capped Petrel

Great Shearwater

darker variant

Fea's Petrel

mostly dark underwing

gray rump and long gray tail

thick bill

thick black underwing bar

Bermuda Petrel

gray cowl

more slender bill than Black-capped

always less white in uppertail coverts than Black-capped

darker rump

Herald Petrel *Pterodroma arminjoniana*

L 15½" (39 cm) WS 37½" (95 cm) Long-winged, slender-bodied species with languid wingbeats. Occurs in three morphs; **dark morph,**
predominant in U.S., has pale-based flight feathers and greater primary coverts on underwing. Compare to chunkier Sooty Shearwater
with its shorter tail, thinner bill, paler underwing, and faster wingbeats.
Light morph has brownish gray head and chest, variably whitish
throat, white belly, pale underwing. **Intermediate morphs** are variably mottled below.

RANGE: Tropical Southern Hemisphere species. South Atlantic birds
of the nominate subspecies ("Trindade Petrel") nest on islands off
Brazil; mid-Atlantic region sightings likely from these populations.
Rare visitor late May to late Sept. in Gulf Stream off NC. Accidental
inland after hurricanes.

SHEARWATERS
**Flies with quick flaps and glides. Thinner-billed with on aver-
age less arcing flight than gadfly petrels.**

Cory's Shearwater *Calonectris diomedea*

L 18" (46 cm) WS 46" (117 cm) Grayish brown upperparts merge into
white underparts without sharp contrast; bill is yellowish. Flight is
more languid than most other shearwaters.

RANGE: Prefers warmer waters. Nominate *diomedea* breeds in the
Mediterranean; slightly larger *borealis* breeds on the Azores, Canary,
and Salvage Islands. Underside of primaries darker in *borealis*. Both
occur (*borealis* much more common) off East Coast, mainly late
spring through fall. Uncommon in Gulf of Mexico. Rare well off coast
of Atlantic provinces. Accidental off CA.

Cape Verde Shearwater *Calonectris edwardsii*

L 15½" (39 cm) WS 39½" (101 cm) Much smaller, slightly longer tailed
than Cory's with darker gray upperparts and more contrasting face;
much slimmer bill is olive-gray, not yellowish.

RANGE: Breeds Cape Verde archipelago, ranges to West African coast.
Accidental off Hatteras Inlet, NC (15 Aug. 2004) and possibly off
Ocean City, MD (21 Oct. 2006).

Great Shearwater *Puffinus gravis*

L 18" (46 cm) WS 44" (112 cm) Dark brown cap contrasts with grayish
brown upperparts and white cheeks. Rump usually shows a narrow,
white, U-shaped band. Bill is dark; underparts white with indistinct
dusky patch on belly. Many have a white nape. Similar to thicker-billed
Black-capped Petrel (page 84), but lacks white forehead and wide
black bar on underwing; also has shorter, less wedge-shaped tail with
less extensive white at base.

RANGE: Breeds in South Atlantic. Fairly common off the East Coast
during migration, chiefly in spring. Summers in large numbers from
the Gulf of Maine north. Casual in Gulf of Mexico and off West Coast.

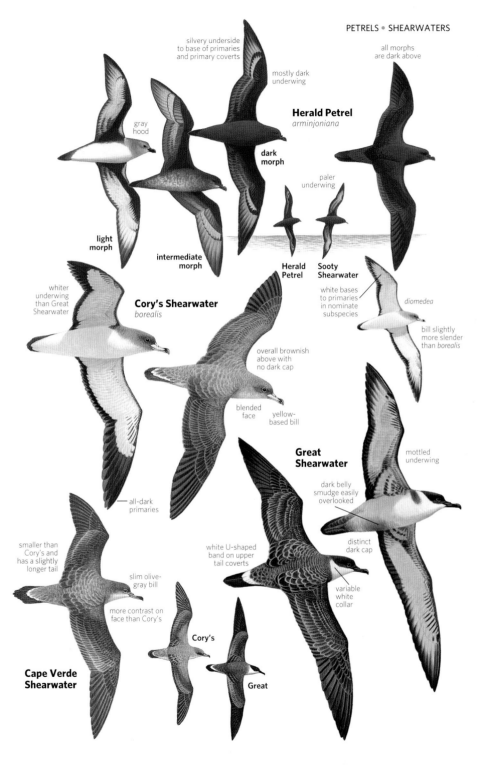

silvery underside to base of primaries and primary coverts

mostly dark underwing

all morphs are dark above

Herald Petrel
arminjoniana

gray hood

dark morph

paler underwing

light morph

intermediate morph

Herald Petrel

Sooty Shearwater

white bases to primaries in nominate subspecies

diomedea

bill slightly more slender than *borealis*

whiter underwing than Great Shearwater

Cory's Shearwater
borealis

overall brownish above with no dark cap

blended face

yellow-based bill

Great Shearwater

mottled underwing

dark belly smudge easily overlooked

distinct dark cap

all-dark primaries

smaller than Cory's and has a slightly longer tail

slim olive-gray bill

more contrast on face than Cory's

white U-shaped band on upper tail coverts

Cory's

Great

variable white collar

Cape Verde Shearwater

Bulwer's Petrel *Bulweria bulwerii* L 10" (26 cm) WS 26" (66 cm)
Sooty brown overall with pale diagonal bar across secondary coverts.
Long tail usually held in a point; wedge shape visible only when
fanned. Flight is buoyant and erratic, long wings slightly bowed and
held forward. Flies within a few feet of the water; flight, when it is
windy, is more like gadfly petrels. A larger (L 12½" [32 cm], WS 31"
[79 cm]), thicker-billed species of *Bulweria*, Jouanin's Petrel, from the
northwestern Indian Ocean, has been well documented from Lisianski
Island, HI, in the central Pacific (4 Sept. 1967), so could occur in the
eastern Pacific as well.
RANGE: Bird of tropical and subtropical oceans; accidental summer
visitor off Monterey, CA (photo 26 July 1998), and Outer Banks, NC
(sight record 1 July 1992 and photo 8 Aug. 1998).

Flesh-footed Shearwater *Puffinus carneipes*
L 17" (43 cm) WS 41" (104 cm) Dark above and below except for pale
flight feathers, distinctive pale pink base of bill. Compare especially
with Sooty Shearwater, which has whitish (not dark) wing linings
and all-dark bill. Also, languid flight suggests Pink-footed Shearwater
(page 90), not Sooty.
RANGE: Breeds on islands off Australia and New Zealand. Winters
(our summer) in North Pacific; rare off West Coast; recorded most
frequently in fall; casual in winter.

Short-tailed Shearwater *Puffinus tenuirostris*
L 17" (43 cm) WS 39" (99 cm) Plumage variable. Separated with difficulty from slightly larger Sooty Shearwater. Usually dark overall,
but often with pale wing linings like Sooty Shearwater; white is more
evenly distributed, when present, forming a panel on the inner wing
(not on primary coverts). Best field mark is shorter, more slender
bill, with thinner tip (nail). Also note slightly steeper forehead and
more rounded crown. Some birds, unlike Sooty, have pale throat
and dark-capped appearance. Often follows boats where it can be
photographed for a more certain identification.
RANGE: Breeds off Australia. Winters (our summer) in North Pacific to
AK, when regularly noted in the tens of thousands off the Aleutians,
by late summer north in the Bering Sea to St. Lawrence Island. Seen
along West Coast from BC to CA during southward migration in late
fall and winter. Accidental to FL; a sight record off VA.

Sooty Shearwater *Puffinus griseus*
L 18" (46 cm) WS 40" (101 cm) Whitish underwing coverts contrast
with overall dark plumage. Flies with fast wingbeats and, except when
it is windy, short glides. Almost identical to Short-tailed Shearwater.
White on underwing usually most prominent on primary coverts. A
few have darker underwing.
RANGE: Breeds in Southern Hemisphere. Fairly common off East Coast
in spring, and in summer off New England and Canada. Abundant off
West Coast north to Kodiak region of AK, often seen from shore. Very
rare in Gulf of Mexico.

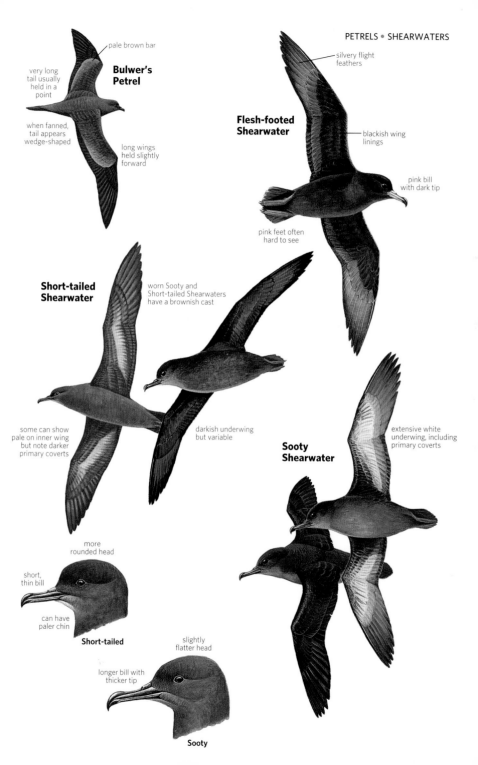

pale brown bar

Bulwer's Petrel

very long tail usually held in a point

when fanned, tail appears wedge-shaped

long wings held slightly forward

silvery flight feathers

Flesh-footed Shearwater

blackish wing linings

pink bill with dark tip

pink feet often hard to see

Short-tailed Shearwater

worn Sooty and Short-tailed Shearwaters have a brownish cast

some can show pale on inner wing but note darker primary coverts

darkish underwing but variable

Sooty Shearwater

extensive white underwing, including primary coverts

more rounded head

short, thin bill

can have paler chin

Short-tailed

slightly flatter head

longer bill with thicker tip

Sooty

Wedge-tailed Shearwater *Puffinus pacificus*

L 18" (46 cm) WS 40" (101 cm) Long tail held in a point; wedge shape visible only when fanned. Slender, grayish bill has darker tip. Head and upperparts of **light morph** grayish brown; wing linings and under-parts white. Most sightings are of wholly brown **dark morph,** which has paler base to flight feathers. Very languid flight with prolonged soaring on bowed wings angled forward to "wrist," then swept back.
RANGE: Polymorphic species from warm waters of Pacific and Indian Oceans. Casual in summer and fall to waters off central CA; accidental on Salton Sea.

Streaked Shearwater *Calonectris leucomelas*

L 19" (48 cm) WS 48" (122 cm) Head is variable, largely white to rather heavily streaked; often looks white at a distance. Pale fringes give upperparts a scaly look. White uppertail coverts on some forms a pale "horseshoe." Bill base color varies from pale gray to pale pink, tip dark. Note the white axillaries and dark underwing primary coverts. Languid, soaring flight is typical of *Calonectris* shearwaters (see also Cory's, page 86).
RANGE: Asian species, casual off CA (most records from Monterey Bay) in the fall, accidental inland in upper Central Valley, CA, and WY.

Pink-footed Shearwater *Puffinus creatopus*

L 19" (48 cm) WS 43" (109 cm) Uniformly blackish brown above; white wing linings and underparts are variably mottled (sight records of much darker birds than illustrated); pink bill and feet distinctive at close range. Flies with slower wingbeats and more soaring than Sooty Shearwater (page 88). Spring birds of this and other Southern Hemisphere species may be in heavy **molt,** often resulting in whitish wing bars and odd wing shape.
RANGE: Breeds on islands off Chile; winters (our summer) in the northern Pacific. Common from spring through fall; rare in midwinter. Rare off southeastern AK.

Buller's Shearwater *Puffinus bulleri*

L 16" (41 cm) WS 40" (102 cm) Gleaming white below, including wing linings. Gray above, with a darker cap and a long, dark, wedge-shaped tail. Dark bar across leading edge of upperwing extends across back, forming a distinct M. Flight is graceful, buoyant, with long periods of gliding.
RANGE: Breeds on islands off New Zealand. Irregular off West Coast during southward migration. Most common from WA to central CA; rarer north (regular off southern AK) and south along West Coast. Accidental off NJ and on Salton Sea.

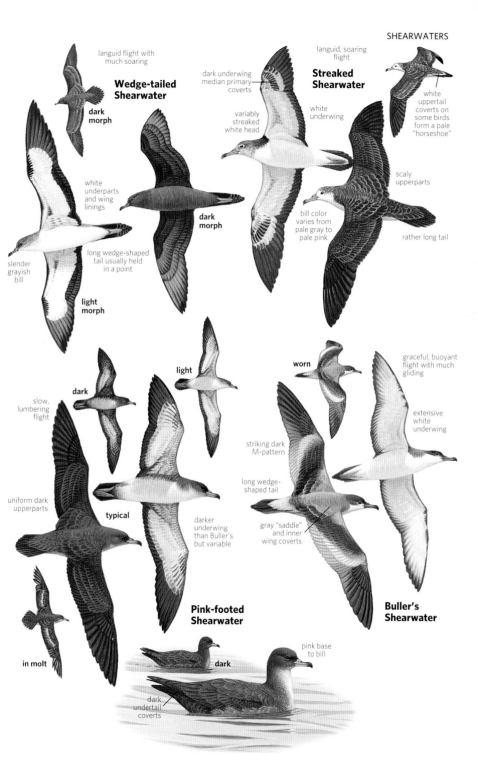

languid flight with much soaring

Wedge-tailed Shearwater

dark morph

white underparts and wing linings

slender grayish bill

long wedge-shaped tail usually held in a point

light morph

dark morph

dark underwing median primary coverts

Streaked Shearwater

variably streaked white head

white underwing

languid, soaring flight

white uppertail coverts on some birds form a pale "horseshoe"

scaly upperparts

bill color varies from pale gray to pale pink

rather long tail

slow, lumbering flight

dark

light

uniform dark upperparts

typical

darker underwing than Buller's but variable

in molt

worn

graceful, buoyant flight with much gliding

extensive white underwing

striking dark M-pattern

long wedge-shaped tail

gray "saddle" and inner wing coverts

Pink-footed Shearwater

Buller's Shearwater

pink base to bill

dark

dark undertail coverts

Manx Shearwater *Puffinus puffinus*

L 13½" (34 cm) WS 33" (84 cm) Blackish above, white below with white wing linings. Pure white undertail coverts extend to end of short tail. White wraps around dark ear coverts.

RANGE: Most breed on islands around United Kingdom; one small colony in Newfoundland. Winters off eastern South America. Fairly common off the northern Atlantic coast in summer. Rare in winter from MD south. Rare in summer and fall off West Coast to AK.

Black-vented Shearwater *Puffinus opisthomelas*

L 14" (36 cm) WS 34" (86 cm) Dark brown above, white below, with dark undertail coverts. Variable dusky mottling on sides of breast, often extending across entire breast.

RANGE: Seen off CA from Aug. to May; often visible from shore. Numbers vary from year to year; scarce in some. Nests chiefly on islands off Mexico's Baja peninsula. Strictly casual north of central CA to southwestern BC. Formerly considered a subspecies of the Atlantic species Manx Shearwater, which is rare in fall off central CA. Manx is much darker above and much cleaner white below, including pure white undertail coverts; it has the pale wrapping around the dark ear coverts; and its flight is more buoyant.

Audubon's Shearwater *Puffinus lherminieri*

L 12" (31 cm) WS 27" (69 cm) Dark brown above, white below, with long tail, dark undertail coverts (a few with pale ones). Small white area in front of eye. Flight fairly rapid and usually close to water.

RANGE: Prefers warmer waters. Breeds on Caribbean islands. Common off the southern Atlantic coast, uncommon to fairly common in Gulf of Mexico, chiefly from May through Oct. Rare in winter. Small numbers found most years in late summer north to waters off southern New England. Accidental inland (KY). Rarely seen from shore.

Little Shearwater *Puffinus assimilis*

L 11" (28 cm) WS 25" (64 cm) Similar to larger Audubon's Shearwater, but has shorter tail, whiter underwing, and grayer two-tone upperwing with white tips to median and greater secondary coverts; white undertail coverts; face whiter. Flies with rapid, stiff, shallow wingbeats. Records believed to be *baroli* ("Barolo's Shearwater") breeding on the Azores, Madeira, and the Canary Islands; *baroli* and the darker-faced *boydi* from the Cape Verde Islands are distinct at the molecular level (the two combined as one species known as Macaronesian Shearwater, *P. baroli*) from Southern Hemisphere taxa.

RANGE: Casual, two fall specimens from NS and SC (*baroli*, breeds Azores); photo record from off MA and sight records from off NC and NS; photo record from Monterey County, CA (29 Oct. 2003) controversial, but if a Little Shearwater, then likely Southern Hemisphere subspecies.

blackish
brown above

soars more than
Black-vented

white often
comes up
side of rump

light

dark

extent of dark
on underwing
variable

pure white
undertail
coverts

extensive white
underwing

**Manx
Shearwater**

**Black-vented
Shearwater**

flies with
rapid flaps and
brief glides

white wraps
around
ear coverts

underwing
darker than
Manx

white
undertail
coverts

typical

dark
undertail
coverts

face pattern
more blended
than Manx

dark

dark
undertail
coverts

dark
bill

short
wings

dark
undertail
coverts

flies with stiff,
shallow wingbeats

white
undertail
coverts

white extends
prominently over
eye in *baroli*

longer tail
than Manx

extensive white
underwing

**Audubon's
Shearwater**
lherminieri

pale gray creates
two-tone upperwing

**Little
Shearwater**
baroli

STORM-PETRELS Family Hydrobatidae

These small seabirds hover close to the water, pattering or hopping across the waves to pluck up small fish and plankton. Some species follow ships. Identification is often difficult. Flight behavior helps to distinguish the various species, but can vary deceptively depending on weather, especially wind speed. Silent away from nesting colonies.
SPECIES: 24 WORLD, 14 N.A.

White-faced Storm-Petrel *Pelagodroma marina*

L 7½" (19 cm) WS 17" (43 cm) Flies with stiff, shallow wingbeats and short glides. Often angles toward water surface with a slightly sideways slant; it then bounces off the surface with its long legs and flies a short distance before repeating the process. Distinctive white underparts, wing linings, and face. Dark eye stripe, crown, and upperparts; paler rump; black tail often spread and tilted slightly to side. Distant white-plumaged phalaropes (page 206) and even nonbreeding Black Tern have been briefly mistaken for White-faced Storm-Petrel.

RANGE: Widespread in southern oceans. In eastern North America, breeds on Cape Verde Islands (*eadesi*) and on the Salvages (similar *hypoleuca*). Rare off the Atlantic coast from NC to MA in late summer.

European Storm-Petrel *Hydrobates pelagicus*

L 5½-6½" (14-17 cm) WS 15" (38 cm) Small, blackish storm-petrel with prominent white rump and underwing bar. Smaller than, and overall color darker than, both Wilson's and especially even larger Leach's (page 96). Also much darker on upperwing; lacks broad pale brown secondary coverts found on Wilson's; instead has narrow pale brown tips to the greater secondary coverts. Has narrower and longer inner wing (arm) than Wilson's; feet do not project beyond tail. Flight is direct and rapid with little gliding.

RANGE: Nests on islands in northeastern Atlantic and Mediterranean. (Latter are larger billed than Atlantic breeders.) Specimen record from Sable Island, NS (Aug. 1970), and more than a dozen recent photo records in late spring off the Outer Banks, NC.

Wilson's Storm-Petrel *Oceanites oceanicus*

L 7¼" (18 cm) WS 16" (41 cm) Flies with shallow, fluttery wingbeats. The inner wing is short, giving an overall short-winged, somewhat triangular appearance. Long legs; in flight, feet trail behind tip of squarish or rounded tail. Often hovers to feed, pattering its long yellow-webbed feet (color of webbing difficult to see in flight) on the water. Bold white, U-shaped rump band extends onto undertail coverts; visible even on sitting bird. Smaller than Leach's and Band-rumped Storm-Petrels (page 96). Most spring adults seen in our waters are molting primaries, unlike Leach's.

RANGE: Breeds in Antarctic waters. Common off Atlantic coast; rare off eastern Gulf Coast and CA (fall).

broad wings

**White-faced
Storm-Petrel**

pushes off water and
changes direction

bold white
supercilium

pale gray
rump

long bill

very long legs

white underparts
and wing linings

lacks obvious
carpal bar

smaller and darker
overall than Wilson's

**European
Storm-Petrel**

long "arm"

white
underwing bar

European

short legs

dropped
greater
coverts

old (faded)
outer primaries

**Wilson's
different stages of wing molt
from spring to summer**

new inner
primaries

feet
project
beyond
tail

molt gap

typical
foot-pattering
behavior

short "arm"

**Wilson's
Storm-Petrel**

white extends
well below tail

long legs

feet with
yellow webs

Band-rumped Storm-Petrel *Oceanodroma castro*

L 9" (23 cm) WS 17" (43 cm) Rather shallow wingstrokes followed by stiff-winged glides, like the flight of a shearwater but unlike the erratic flight of Leach's Storm-Petrel or the fluttery flight of Wilson's (page 94). White rump patch narrower and less extensive on undertail coverts than Wilson's. Tail squarish or very slightly notched; no foot projection in flight. Larger, longer winged than Wilson's; fainter covert bar and thicker bill than Leach's.

RANGE: Breeds widely on tropical and subtropical islands. In eastern North Atlantic, breeds on Azores, Berlengas, Canary Islands, Madeira Archipelago, Salvages, and Cape Verde Islands. Recent studies have shown distinct difference in vocalizations and timing of breeding (up to two populations on the same island, but at different seasons). Up to four cryptic species proposed just from the North Atlantic breeders. North American birds are believed to mainly involve widespread **"Grant's"** from above-named island groups, except Cape Verde Islands. Some with slightly broader rump bands may be **"Cape Verde"** from Cape Verde Islands. Fairly common from late May to late Aug. in Gulf Stream, especially off NC; rare or casual farther north to waters off MA; uncommon well off Gulf Coast. Accidental onshore and inland in East following hurricanes. No confirmed records off the West Coast.

See subspecies map, page 549

Leach's Storm-Petrel *Oceanodroma leucorhoa*

L 8" (20 cm) WS 18" (46 cm) Distinctive erratic flight, with deep strokes of long, pointed wings. In close view, note dusky line dividing white rump band on most birds. Little white visible on flanks of sitting bird. Brown overall, with pale wing stripes and forked tail. Nominate subspecies in North Atlantic and North Pacific shows extensive white on rump, typically with dusky down center; off Baja California, birds on Coronado and San Benito Islands to the south, **"Chapman's"** (*chapmani*), are mostly dark rumped to entirely dark rumped (San Benito Islands). Two additional subspecies, breeding in different seasons, described from islets off Guadalupe Island: "Ainley's" (*cheimomnestes*) is white rumped; **"Townsend's"** (*socorroensis*) is polymorphic, rump bright white on some, mostly dark on others. Off southern CA in summer, at times *chapmani* is dominant, but a few small, dark individuals with a bold white rump, shallower tail fork, and fainter covert bar are believed to be "Townsend's." Some reports off southern CA of Band-rumped, Wedge-rumped, and even Wilson's (page 94) may be due to confusion with "Townsend's."

RANGE: Fairly common well off Pacific coast and in Atlantic from ME northward; also off NC; rare and remains well offshore elsewhere. Very rare in Gulf of Mexico; casual onshore and inland following hurricanes and nor'easters; also on Salton Sea, CA.

Wedge-rumped Storm-Petrel *Oceanodroma tethys*

L 6" (15 cm) WS 13¼" (34 cm) Distinctive bold white triangular patch of uppertail coverts gives the appearance of a white tail with dark corners. Compare with the rounded rump band and white flanks of Wilson's Storm-Petrel (page 94). Wedge-rumped is almost as small as Least Storm-Petrel, with similar deep wingbeats.

RANGE: Breeds on the Galápagos Islands (nominate *tethys*) and on islets off Peru (much smaller *kelsalli*). Casual off the CA coast from Aug. to Jan.; single specimen (Jan.) is of *kelsalli*.

Band-rumped Storm-Petrel

old (faded) outer primaries

possible **"Grant's"**

new (blackish) inner primaries

type with broader white rump band

possible **"Grant's"**

white extends noticeably onto sides of rump, unlike Leach's

short legs do not project beyond tail

possible **"Cape Verde"**

type with narrow white rump band, and narrower wings

primary molt in spring adults, unlike Leach's

rump often divided in center about the length of tail

bold carpal bar extends to leading edge of wing

"Chapman's"
chapmani

this type has the deepest tail fork

smaller and more blackish with a more compact build

compare to Wedge-rumped

leucorhoa

usually dark rumped or with white at sides

short tail with shallow fork

little white visible below level of tail on sides of rump

Leach's Storm-Petrel

white rump patch larger and more solid, including center, and longer than tail

"Townsend's"
socorroensis

Wedge-rumped Storm-Petrel
kelsalli

triangular white rump almost reaches end of tail

Leach's
leucorhoa

rather long thin bill

little visible white on sides of undertail

Band-rumped

rather heavy bill

Wedge-rumped

white on sides of undertail often hidden by wings

small size and relatively dark plumage

Wilson's

extensive white on sides of undertail

Fork-tailed Storm-Petrel *Oceanodroma furcata*

L 8½" (22 cm) WS 18" (46 cm) Wingbeats shallow, rapid, often followed by glides. Looks fairly long tailed in flight. Occasionally makes shallow dives for food. Distinctively bluish gray above, pearl gray below. Note also dark gray forehead and eye patch, dark wing linings. Nominate *furcata* from coastal northeast Asia and Aleutians is larger and paler than *plumbea* from off coastal southern AK to northern CA.
RANGE: Found regularly south to central CA; rare off southern CA coast and in northern Bering Sea.

Least Storm-Petrel *Oceanodroma microsoma*

L 5¾" (15 cm) WS 15" (38 cm) Our smallest storm-petrel. Swift, indirect flight, low over the water, with deep wingbeats like the much larger Black Storm-Petrel. Blackish brown overall. Short tailed; appears almost tailless in flight; often confused with molting Ashy Storm-Petrels. Unique wedge-shape of tail is sometimes visible on a close bird.
RANGE: Irregular; rare to common off the coast of southern CA in late summer and fall; in peak years, small numbers occur to central CA; after tropical storms, may be seen in southeastern CA and southwestern AZ, exceptionally in large numbers (e.g., Sept. 1976 after Hurricane Kathleen).

Black Storm-Petrel *Oceanodroma melania*

L 9" (23 cm) WS 19" (48 cm) Deep, languid wingstrokes, graceful flight. Largest of the all-dark storm-petrels. Blackish brown overall with pale bar on upper surface of wing. Tail forked and fairly long. Slow, deep wingbeats and larger size distinguish Black Storm-Petrel from dark-rumped individuals of Leach's Storm-Petrel (page 96).
RANGE: Breeds from May to Dec. off Baja California and in Gulf of California; small colony in vicinity of Santa Barbara Island, CA. Fairly common off southern CA coast from late spring; by late summer north to Monterey Bay, a few to off Marin County. Casual in interior of southern CA and in southwestern AZ after tropical storms.

Ashy Storm-Petrel *Oceanodroma homochroa*

L 8" (20 cm) WS 17" (43 cm) Fluttery wingbeats, but flight fairly direct; not as swallowlike as Wilson's Storm-Petrel (page 94). Gray-brown overall; pale mottling on underwing coverts may be visible at close range. Viewed from the side, Ashy appears to have long tail, unless in molt. Distinguished from larger Black Storm-Petrel by rapid, shallow wingbeats and overall paler, grayer appearance, along with a disproportionately longer tail.
RANGE: Fairly common most of the year; rare in winter. Breeds on islands off central and southern CA.

Fork-tailed Storm-Petrel

rounded wing tip

dark mask can suggest a phalarope

prominent pale carpal bar whitens near body

pearly gray with blackish wing lining

distinctive fork to tail

slow wing flaps suggest Black Storm-Petrel

tiny size and dark coloration

Least Storm-Petrel

large and dark

flies with languid wingbeats like Black Tern

wedge-shaped tail sometimes apparent

short tail

Black Storm-Petrel

Ashy Storm-Petrel

rather long neck and small head

molting tail looks more like Least

molting fall adult

long winged

distinctly long tailed with deep tail fork

dark wing linings

slightly paler gray on uppertail coverts

some pale visible in center of underwing, but can be hard to see

paler than Black Storm-Petrel with ashy-gray plumage

TROPICBIRDS Family Phaethontidae

Long central tail feathers identify adults. They are usually seen far out at sea, where they are mostly silent. Here, each of these three species is often first spotted right over the highest point of the boat; they circle a few times and then fly off. SPECIES: 3 WORLD, 3 N.A.

White-tailed Tropicbird *Phaethon lepturus*

L 30" (76 cm) WS 37" (94 cm) Smaller than Red-billed Tropicbird; distinctive black stripe on upperwing coverts; primaries show less black than Red-billed. Bill orange in Atlantic subspecies (*catesbyi*) **adults. Juvenile** lacks tail streamers; upperparts are boldly barred, bill more yellowish. Pacific *dorotheae* has a greenish bill and can have a reddish cast to its long central tail feathers.

RANGE: Tropical species, now most regular, but still rare, in Gulf Stream off NC in mid- to late summer (nests on Bermuda); casual in spring around Fort Jefferson on Garden Key, Dry Tortugas, FL, where a few decades ago it was thought they might nest; many fewer in recent years. Casual elsewhere off East Coast where recorded north to NS. Note that a tropicbird seen in the Atlantic away from the Dry Tortugas and the Gulf Stream off NC is as likely, perhaps more likely, to be Red-billed than White-tailed. Even off NC, late spring birds are more likely Red-billed. Accidental to Upper Newport Bay, Orange County, CA, from 24 May to 23 June 1964 (photographs, *dorotheae*), and 22 Aug. 1980 in Scottsdale, AZ (specimen, subspecies identification controversial).

Red-billed Tropicbird *Phaethon aethereus*

L 40" (102 cm) WS 44" (112 cm) Flies with rapid, stiff, shallow wing-beats, not unlike a large falcon species, unlike other tropicbirds, whose flight is more ternlike. **Adult** has red bill, black primaries, barring on back and wings, white tail streamers. **Juvenile** has black collar; lacks streamers; tail is tipped with black; barring on upperparts is finer than other young tropicbirds. Also bill is yellowish, but soon becomes orange-red. Interestingly, very few, if any, of the sightings off CA involve younger juveniles.

RANGE: Tropical species, rare well off southern CA coast chiefly in late summer and fall; exceptionally a concentration of 15 to 20 off San Clemente Island on 29 July 2001; casual off central CA; accidental off Gray's Harbor, WA, on 18 June 1941 (specimen); casual to southeastern CA (two records) and southern AZ (five records). Very rare in Gulf of Mexico and off Atlantic coast to NC; casual north to ME.

Red-tailed Tropicbird *Phaethon rubricauda*

L 37" (94 cm) WS 44" (112 cm) Broadest-winged tropicbird; flies with languid wingbeats. Flight feathers mostly white. **Adult** has red bill; red tail streamers, narrower than other tropicbirds. **Juvenile's** all-white tail lacks streamers; upperparts barred; bill black, gradually changing to yellow and then red. Note also lack of black collar on nape.

RANGE: Species from tropical and subtropical Pacific and Indian Oceans. Very rare, usually well off CA coast (over 20 records and others just outside of our waters, including one 261 miles off Curry County, OR, on 26 Aug. 2005), but a few records from Farallones and on coast of southern CA. Accidental from Vancouver Island, BC (partial remains found on shore).

White-tailed Tropicbird
catesbyi

coarse bars on back and wing coverts

blackish diagonal bar on upperwing coverts

greenish yellow bill

juvenile

orange bill

Pacific adult
dorotheae

adult

black cut off by white primary coverts

slower, more ternlike flight

Red-billed Tropicbird

fine bars on back and wing coverts

flies with rapid, shallow wingflaps, almost like Peregrine Falcon

black collar

juvenile

duller bill

adult

red bill

extensive black in primaries and primary coverts

extensive black

narrow, red central tail feathers; Pacific *dorotheae* White-tailed Tropicbird can also have reddish tail feathers

no collar

barred upperparts

no collar

adult

red bill

juvenile

broader winged than other tropicbirds and flies with languid wingbeats

black bill on very young birds gradually turns to yellow and then red

Red-tailed Tropicbird

white flight feathers

FRIGATEBIRDS Family Fregatidae

These large, dark seabirds have the longest wingspan, in proportion to weight, of all birds. SPECIES: 5 WORLD, 3 N.A.

Magnificent Frigatebird *Fregata magnificens*

L 40" (102 cm) WS 90" (229 cm) Long, forked tail; long, narrow wings. **Male** is glossy black; orange throat pouch becomes bright red when inflated in courtship display. **Female** is blackish brown, with white at center of underparts. **Juveniles** show varying amount of white on head and underparts; require four to six years to reach adult plumage. Frigatebirds skim the sea, snatching up food from surface; also harass other birds in flight, forcing them to disgorge food.

RANGE: Generally seen along coast, but also casual inland, especially after storms. Breeds on Dry Tortugas off FL. Rare on East Coast north to NC, casual farther north. Casual in interior West (most frequent at Salton Sea, CA) and on CA coast; even fewer records farther north.

BOOBIES • GANNETS Family Sulidae

High-diving seabirds that plunge into water. Gregarious, nesting in colonies on small islands. The rest of the year, gannets roost at sea, boobies primarily on land. Mostly silent at sea. SPECIES: 10 WORLD, 5 N.A.

Masked Booby *Sula dactylatra* L 32" (81 cm) WS 62" (158 cm)

Proportionately, the shortest tailed booby. **Adult** distinguished from Northern Gannet by yellow bill and extensive black facial skin; black tail; and solid black trailing edge to wing. On **juvenile,** note more white on underwing, with contrasting dark median primary coverts and pale collar. **Subadult** has paler head and broader collar; note yellow on bill.

RANGE: Breeds on Dry Tortugas off FL; uncommon in Gulf of Mexico in summer. Rare in Gulf Stream north to Outer Banks, NC. Casual to coastal CA, where Nazca Booby (page 532) must be considered.

Northern Gannet *Morus bassanus*

L 37" (94 cm) WS 72" (183 cm) Large, white seabird with long, black-tipped wings, pointed white tail. **Juvenile** is dark gray above, with pale speckling; grayish below. **First-summer** birds are whiter below; distinguished from juvenile and immature Masked Booby by more uniformly dark underwing and, at close range, by different feathering pattern around the bill. Full **adult** plumage is acquired in three to four years.

RANGE: Common; breeds in large colonies on rocky cliffs; winters at sea. Often seen from shore during migration and winter. Casual in Great Lakes region in late fall.

white head

juvenile

Magnificent Frigatebird

long, thin, angled wings

all-dark head

white chest

displaying adult ♂

adult ♂

long forked tail, usually closed

adult ♀

all-white wing linings

adult

dark extends to body

adult

black mask

subadult

adult

dull yellow-green bill

white collar on most young birds

subadult

mostly white underwing

juvenile

juvenile

Masked Booby
dactylatra

2nd year

3rd year

juvenile

no pale collar

Northern Gannet

tawny head

1st summer

juvenile

dark bill

adult

fine white speckles

black primaries

adult

adult courtship display

Blue-footed Booby *Sula nebouxii*

L 32" (81 cm) WS 62" (158 cm) Feet bright blue in adults, darker in young; long, attenuated bill is dark bluish gray. **Adult** has streaked head, whitish patches on upper back, lower back, and upper tail coverts; note also white-fringed scapulars and pale iris. **Juvenile** has darker head and neck; compare with immature Masked Booby (page 102), which has an even darker neck and usually a contrasting pale collar; note also pale dorsal patches, all-dark underwing primary coverts and differently shaped bill.

RANGE: Breeds on islands in Gulf of California. Rare and irregular to inland CA and southwestern AZ in late summer and fall (most to Salton Sea). Casual to CA coast; accidental to WA, NM, and TX. Absent in U.S. most years; formerly occurred somewhat more regularly.

Brown Booby *Sula leucogaster* L 30" (76 cm) WS 57" (145 cm)

Adults of nominate subspecies and female *brewsteri* from western Mexico have dark brown heads and necks with sharply contrasting white bellies and underwing coverts; **adult male** *brewsteri* has white on head and neck. **Adult female's** bill, facial skin, legs, and feet are bright yellow; male's soft parts washed with grayish green, throat bluish. **Juveniles** are dark brown, with little or no contrast between breast and belly; underwing muted. **Subadults** show white on belly and sharp line of contrast with darker neck.

RANGE: A few found on Dry Tortugas and vicinity; rare north off both FL coasts. Very rare through Gulf of Mexico; casual up Atlantic coast to NS. Widespread tropical species, breeds as close to the East as Caribbean. In West (*brewsteri*), a small colony now breeds on Islas Coronados in northern Baja California Norte, just south of San Diego, CA. Noted regularly to San Diego County. Rare to very rare farther north on CA coast, casually to OR. Also casual in interior Southwest, most frequently recorded at the Salton Sea, CA.

Red-footed Booby *Sula sula* L 28" (71 cm) WS 60" (152 cm)

Smallest booby, with rounder head and gentler look to face than Brown Booby, which has a flatter head shape. All **adults** show bright coral red feet, and blue and pink at base of bill. Four principal morphs occur: **brown morph, white-tailed brown morph, white morph,** and **black-tailed white morph;** note that white morphs have black primaries, secondaries, and underwing median primary coverts. Note that the adult Masked Booby (page 102) has black lower scapulars, a black mask, white underwing primary coverts, and a yellow bill. All **juveniles** and **subadults** are brownish overall, sometimes with grayish cast, with darker chest band, with mainly dark underwing and flesh pink legs and feet. Immature also told from darker brown immature Brown Booby by presence of pinkish at bill base.

RANGE: Widespread tropical species. In Caribbean, nominate subspecies breeds mostly on remote islands; large colony of 18,000 on Little Cayman. In Pacific, breeds on HI (*rubipes*) and off western Mexico (*websteri*) on Revillagigedo Islands (nearly all white morphs) and Clipperton Island (mostly brown morphs), fewer on Islas Tres Marias and Isla Isabel. Casual to FL, most frequently on the Dry Tortugas, where recently recorded in slightly more than half the years. Accidental to SC and TX.

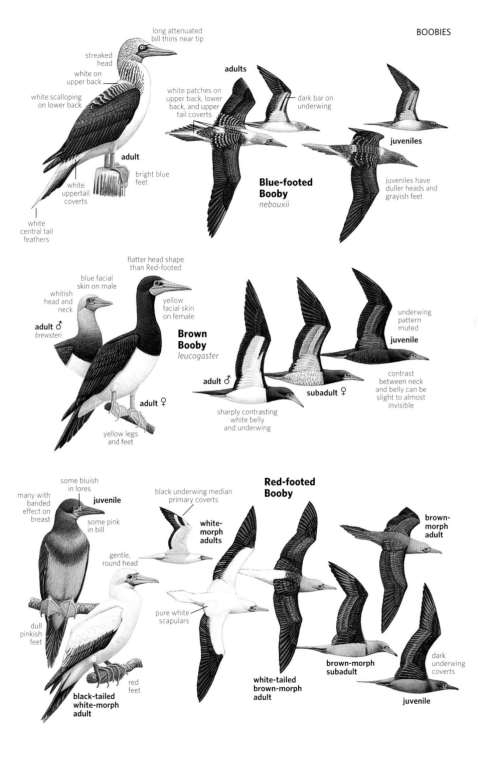

long attenuated
bill thins near tip

streaked
head

white on
upper back

white scalloping
on lower back

adult

white
uppertail
coverts

bright blue
feet

white
central tail
feathers

adults

white patches on
upper back, lower
back, and upper
tail coverts

dark bar on
underwing

**Blue-footed
Booby**
nebouxii

juveniles

juveniles have
duller heads and
grayish feet

flatter head shape
than Red-footed

blue facial
skin on male

whitish
head and
neck

adult ♂
brewsteri

yellow facial
skin on female

**Brown
Booby**
leucogaster

adult ♂

adult ♀

yellow legs
and feet

sharply contrasting
white belly
and underwing

subadult ♀

underwing
pattern
muted

juvenile

contrast
between neck
and belly can be
slight to almost
invisible

many with
banded
effect on
breast

some bluish
in lores

juvenile

some pink
in bill

gentle,
round head

dull
pinkish
feet

black-tailed
white-morph
adult

red
feet

**Red-footed
Booby**

black underwing median
primary coverts

white-
morph
adults

pure white
scapulars

white-tailed
brown-morph
adult

brown-
morph
adult

brown-morph
subadult

dark
underwing
coverts

juvenile

CORMORANTS Family Phalacrocoracidae

Dark birds with set-back legs; long, hooked bill; and colorful bare facial skin and throat pouch. Dive from the surface for fish. May briefly soar; may swim partially submerged. Mostly silent, except around nesting colonies. SPECIES: 40 WORLD, 6 N.A.

Great Cormorant *Phalacrocorax carbo*

L 36" (91 cm) WS 63" (160 cm) Large, short-tailed cormorant with small, lemon yellow throat pouch broadly bordered with white feathering. **Breeding adult** shows white flank patches and wispy white plumes on head. Smaller Double-crested Cormorant has orange throat pouch; lacks flank patches; note also Great Cormorant's larger, blockier head and heavier bill. **Juvenile** birds are brown above; white belly contrasts with streaked brown neck, breast, and flanks. Second-year immatures resemble nonbreeding adults more closely but have a brown tinge above; compare with young Double-crested, which has a slimmer bill, deep orange facial skin, and, often, a darker belly. **RANGE:** Winters in small numbers regularly south to NC, very rarely as far south as FL. Casual on Lake Ontario. Accidental elsewhere in the eastern interior.

Neotropic Cormorant *Phalacrocorax brasilianus*

L 26" (66 cm) WS 40" (102 cm) Small, long-tailed cormorant with white-bordered, yellow-brown or dull yellow throat pouch that tapers to a sharp point behind bill. In **breeding** plumage, **adult** acquires short white plumes on sides of neck. Distinguished from Double-crested Cormorant by smaller size, longer tail, and smaller, angled throat pouch that does not extend around eye. Neotropic **juveniles** are overall browner than adults, particularly on underparts. **RANGE:** Fairly common; found at marshy ponds or shallow inlets near perching stumps and snags. Regular to AZ; casual to southeastern CA, UT, CO, and the western Midwest; accidental to mid-Atlantic region and Great Lakes. Formerly called Olivaceous Cormorant.

Double-crested Cormorant *Phalacrocorax auritus*

L 32" (81 cm) WS 52" (132 cm) Large, rounded throat pouch is yellow-orange year-round. **Breeding adult** has a tuft curving back on both sides of its head from behind eyes. Tufts are largely white in western birds, black and less conspicuous in eastern birds. **Juvenile** is brown above, variably pale below, but usually palest on upper breast and neck. Immatures sometimes have pouch edged with white, which can cause confusion with Neotropic Cormorant but note more extensive yellow-orange on face. Among West Coast cormorants, Double-crested's kinked neck is distinctive in flight; its wings are also longer and more pointed than Brandt's and Pelagic Cormorants (page 108). **RANGE:** Common and widespread; found along coasts, inland lakes, and rivers. Breeding populations in the interior have greatly increased in the last three decades and some control measures have been instituted.

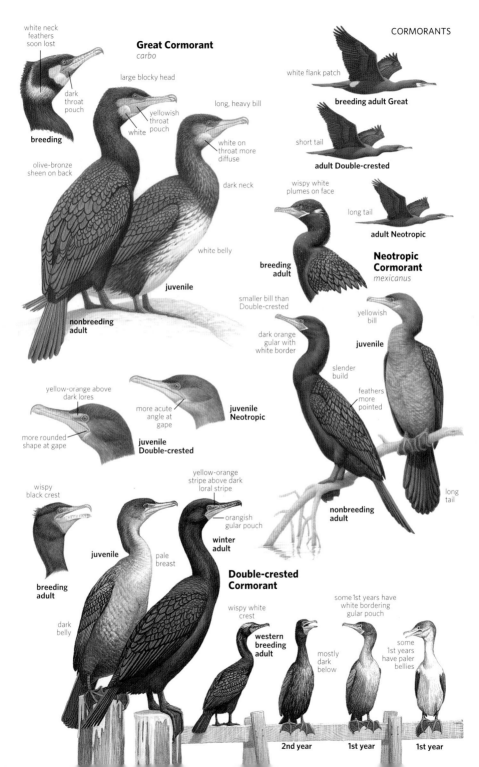

Great Cormorant
carbo

white neck feathers soon lost

dark throat pouch

breeding

large blocky head

yellowish throat pouch

white

long, heavy bill

white on throat more diffuse

olive-bronze sheen on back

dark neck

white belly

juvenile

nonbreeding adult

white flank patch

breeding adult Great

short tail

adult Double-crested

wispy white plumes on face

long tail

adult Neotropic

Neotropic Cormorant
mexicanus

breeding adult

smaller bill than Double-crested

dark orange gular with white border

slender build

yellowish bill

juvenile

feathers more pointed

nonbreeding adult

long tail

yellow-orange above dark lores

more rounded shape at gape

more acute angle at gape

juvenile Neotropic

juvenile Double-crested

yellow-orange stripe above dark loral stripe

orangish gular pouch

winter adult

wispy black crest

wispy white crest

western breeding adult

juvenile

pale breast

breeding adult

dark belly

Double-crested Cormorant

mostly dark below

some 1st years have white bordering gular pouch

some 1st years have paler bellies

2nd year

1st year

1st year

Brandt's Cormorant *Phalacrocorax penicillatus*

L 35" (89 cm) WS 48" (122 cm) A band of pale buffy feathers bordering the throat pouch identifies all ages. Throat pouch becomes bright blue in **breeding** plumage; head, neck, and scapulars acquire fine, white plumes. **Juvenile** is dark brown above, slightly paler below. In all ages, appears more uniformly dark above than Double-crested Cormorant (page 106); wings and tail are shorter. Head and bill are larger than Pelagic Cormorant. Both Brandt's and Pelagic fly low over the ocean with their necks held straight out, while the larger-headed, longer-winged, and thicker-necked Double-crested has a distinct kink to the neck and also often flies at a higher altitude.

RANGE: Common and gregarious; often fishes in large flocks; flies in long lines between feeding and roosting grounds. Much more likely to be seen over open ocean than Pelagic Cormorant. A tiny breeding colony may still exist on Seal Rocks near Prince William Sound, AK. Accidental to inland CA. Rare to southeastern AK.

Pelagic Cormorant *Phalacrocorax pelagicus*

L 26" (66 cm) WS 39" (99 cm) **Adults** are dark and glossy overall; bill dark. Smaller and slenderer than other western cormorants. Birds from AK and northern BC (nominate *pelagicus*) average larger than birds from southern BC south *(resplendens)*. **Breeding adult** has tufts on crown and nape; fine white plumes on sides of neck; white patches on flanks. Distinguished from Red-faced Cormorant by darker and less extensive red facial skin and lack of yellow in bill. **Juvenile** is uniformly dark brown; closely resembles young Red-faced, but note dark bill and smaller size. Pelagic Cormorant is distinguished in flight from Brandt's Cormorant by smaller head, slimmer neck, smaller overall size, and disproportionately longer tail. Its bill is thinner and different in color than Red-faced Cormorant.

RANGE: Despite name, rare out over open ocean. Less gregarious than other species; breeds in smaller colonies. Accidental to inland CA (Mono County).

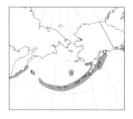

Red-faced Cormorant *Phalacrocorax urile*

L 31" (79 cm) WS 46" (117 cm) Heavier and paler bill distinguishes all ages from Pelagic Cormorant. In **adult,** dull brown wings contrast with glossy upperparts; more uniform in Pelagic. Extensive yellow on bill with bluish at base. Facial skin is red and becomes enlarged and brighter in the **breeding** season. **Juvenile** is uniformly dark brown, with pale yellowish gray bill and narrow ring of pinkish around edge. More gregarious than Pelagic.

RANGE: Nests in colonies on the ledges of steep coastal cliffs and on rocky sea islands, alongside gulls, murres, and auklets.

Brandt's Cormorant

buffy band under bill in all plumages

juvenile

blue skin

breeding adult

winter adult

appears shorter tailed than Pelagic

Brandt's

breeding adults

thick kinked neck

shorter tail than Neotropic

Double-crested

Pelagic

thin neck

Red-faced

browner wings

Pelagic Cormorant

winter adult

all have thicker two-tone bill than Pelagics

all have slender, dark bill

juvenile

uniformly dark brown, including chin

limited dark red facial skin

breeding adult

white lower flank patch

Red-faced Cormorant

winter adult

narrow pinkish eye ring

extensive bright red on face

breeding adult

juvenile

white lower flank patch

DARTERS Family Anhingidae

Long, slim neck helps to distinguish anhingas from cormorants. Anhingas often swim submerged to the neck. Sharply pointed bill is used to spear fish. SPECIES: 4 WORLD, 1 N.A.

Anhinga *Anhinga anhinga* L 35" (89 cm) WS 45" (114 cm)

Black above, with green gloss; silvery white spots and streaks on wings and upper back. During breeding season, **male** acquires pale, wispy plumes on upper neck; bill and bare facial skin become brightly colored. **Female** has buffy neck and breast. Immatures resemble adult female but are browner overall. In flight, profile looks headless. Flies with slow, regular wingbeats and circles like raptor on thermals; note that soaring Double-crested Cormorants (page 106) with slightly splayed tails may be mistaken for Anhingas. Anhingas often seen perched on branches or stumps with wings spread.

VOICE: Mostly silent; may give occasional croaks and rattles.

RANGE: Prefers freshwater habitats. Casual wanderer north of breeding range to ON, Southwest, and CA.

PELICANS Family Pelecanidae

These large, heavy waterbirds have massive bills and huge throat pouches used as dip nets to catch fish. In flight, pelicans hold their heads drawn back. Mostly silent away from breeding colonies. SPECIES: 8 WORLD, 2 N.A.

American White Pelican *Pelecanus erythrorhynchos*

L 62" (158 cm) WS 108" (274 cm) White, with black primaries and outer secondaries. **Breeding adult** has pale yellow crest; bill is bright orange, usually with a fibrous plate on upper mandible. Plate is shed after eggs are laid; crown and nape become grayish. Juvenile is white with brownish wash on head, neck, and lesser coverts; soft parts more dully colored. Does not dive for food but dips bill into the water while swimming. Usually found in flocks.

RANGE: Nonbreeding birds are seen in summer throughout area enclosed by dashed line on map. In fall, vagrants may appear almost anywhere, increasingly in Northeast.

Brown Pelican *Pelecanus occidentalis*

L 48" (122 cm) WS 84" (213 cm) **E** **Nonbreeding adult** has white head and neck, often washed with yellow; grayish brown body; blackish belly. In **breeding** bird, hindneck is dark chestnut; yellow patch appears at base of neck. On eastern *carolinensis,* gular pouch is grayish; breeding *californicus* from the West Coast has a bright red gular pouch. Molt during incubation and **chick feeding** produces speckled head and foreneck; light eye darkened during chick feeding. Juvenile is grayish brown above; underparts whitish. **Immatures** are browner; acquire adult plumage by third year. Dives from the air after prey.

RANGE: Large numbers move north to Salton Sea, CA, after breeding, many winter; in some years, small numbers to elsewhere in Southwest, where normally very rare; casual elsewhere in U.S. interior and southern ON. A few wander on Atlantic coast, mainly spring and summer, to limit of dashed line on map; casual to NS. On Pacific coast, rare to BC, casual to southeastern AK.

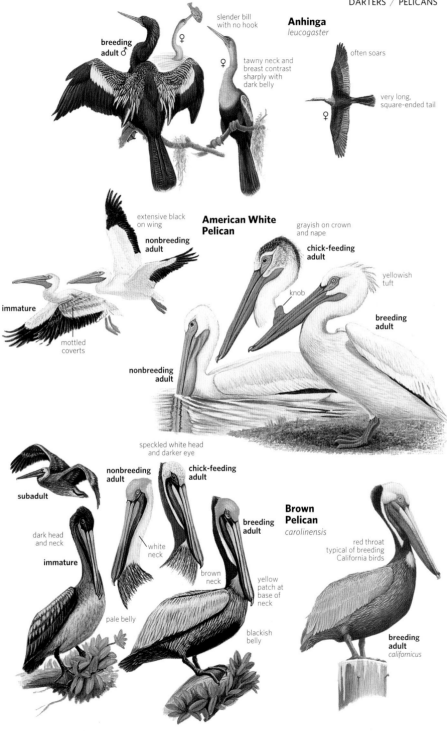

Anhinga
leucogaster

breeding adult ♂

slender bill with no hook ♀

♀ tawny neck and breast contrast sharply with dark belly

often soars

very long, square-ended tail

♀

American White Pelican

extensive black on wing

nonbreeding adult

grayish on crown and nape

chick-feeding adult

yellowish tuft

immature

knob

breeding adult

mottled coverts

nonbreeding adult

speckled white head and darker eye

subadult

nonbreeding adult

chick-feeding adult

Brown Pelican
carolinensis

dark head and neck

immature

white neck

breeding adult

brown neck

red throat typical of breeding California birds

pale belly

yellow patch at base of neck

blackish belly

breeding adult
californicus

HERONS • BITTERNS • ALLIES Family Ardeidae

Wading birds; most have long legs, neck, and bill for stalking food in shallow water. Graceful crests and plumes adorn some species in breeding season. SPECIES: 64 WORLD, 20 N.A.

American Bittern *Botaurus lentiginosus*

L 28" (71 cm) WS 42" (107 cm) Mottled brown upperparts and brownish neck streaks. Contrasting dark flight feathers are conspicuous in flight; note also that wings are longer, narrower, and more pointed, not rounded as in night-herons. **Juvenile** lacks neck patches. When alarmed, freezes with bill pointing up, or flushes with rapid wingbeats.
VOICE: Distinctive spring and early summer song, *oonk-a-lunk,* is most often heard at dusk in dense marsh reeds.
RANGE: Uncommon and declining; casual breeder south of usual range.

Least Bittern *Ixobrychus exilis*

L 13" (33 cm) WS 17" (43 cm) Buffy inner wing patches identify this small, rather secretive heron as it flushes briefly from dense marsh cover. When alarmed, it may freeze with bill pointing up. In **male** back and crown are black; in **female** they are browner. **Juvenile** resembles female but has more prominent streaking on back and breast. Rare dark morph, **"Cory's Least Bittern"** of eastern North America not documented since mid-20th century, is dark chestnut where typical plumage is pale. It was best known from Ashbridge Bay, Toronto, ON, which has been buried under a landfill for decades.
VOICE: Calls include a series of harsh *kek* notes delivered year-round, which, when learned, is often the best means of detection; song, a softer series of *ku* notes, is heard only on the breeding ground.
RANGE: Rare to fairly common. May breed sporadically beyond mapped range in West.

Great Blue Heron *Ardea herodias*

L 46" (117 cm) WS 72" (183 cm) Large, gray-blue heron; black stripe extends above eye; white foreneck is streaked with black. **Breeding adult** has yellowish bill and ornate plumes on head, neck, and back. Nonbreeding adult lacks plumes; bill is yellower. **Juvenile** has black crown, no plumes. All-white morph found in southern FL formerly considered a separate species, **"Great White Heron." "Wurdemann's Heron"** morph, found chiefly in FL Keys, has all-white head.
VOICE: Occasional deep croaks.
RANGE: Common. A few winter far north into breeding range; also may wander at others season north of mapped breeding range. "Great White Heron" has occurred casually north of FL and west to coastal TX, mostly in summer and early fall.

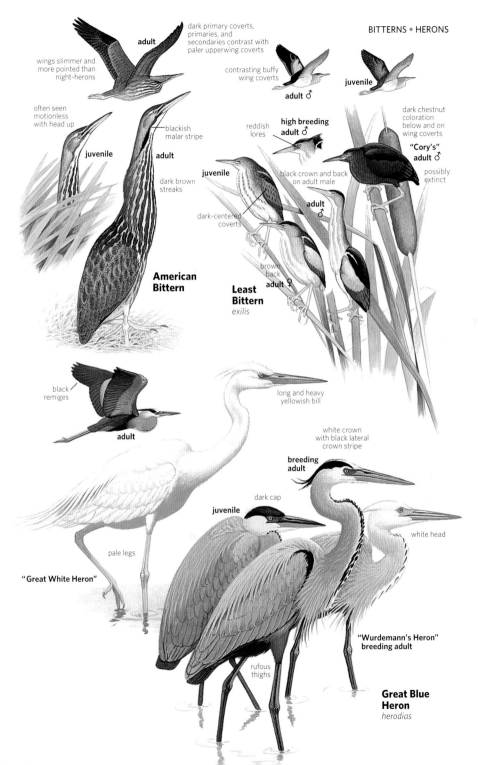

American Bittern

wings slimmer and more pointed than night-herons

dark primary coverts, primaries, and secondaries contrast with paler upperwing coverts

adult

often seen motionless with head up

juvenile

adult

blackish malar stripe

dark brown streaks

Least Bittern
exilis

contrasting buffy wing coverts

adult ♂

juvenile

reddish lores

high breeding
adult ♂

dark-centered coverts

black crown and back on adult male

adult ♂

brown back
adult ♀

dark chestnut coloration below and on wing coverts

"Cory's"
adult ♂

possibly extinct

Great Blue Heron
herodias

black remiges

adult

long and heavy yellowish bill

white crown with black lateral crown stripe

breeding adult

dark cap

juvenile

white head

"Great White Heron"

pale legs

rufous thighs

"Wurdemann's Heron"
breeding adult

Cattle Egret *Bubulcus ibis* L 20" *(51 cm)* WS 36" *(91 cm)*

Small, stocky white heron with rounded head; throat feathering extends on bill. **Breeding adult** with orange-buff plumes on crown, back, and foreneck. In high breeding season, bill is red-orange, lores purplish, legs dusky red. **Nonbreeding adult** has short yellow bill, darkish legs. Juvenile's bill is black; begins to turn yellow in late summer. In flight, resembles Snowy Egret but is smaller; bill and legs shorter; wingbeats faster. Larger and longer-necked Asian *coromandus,* with cinnamon on head and neck in breeding plumage, recorded on Agattu, Aleutians (19 June 1988), flies with slower wingbeats than nominate *ibis.*
VOICE: Mostly silent.
RANGE: Old World species. Came to South America from Africa, spread to FL in the early 1950s, reached CA by the mid-1960s. Prefers fields, often with livestock. In spring, summer, and especially fall, wanders well north of breeding range.

Snowy Egret *Egretta thula* L 24" *(61 cm)* WS 41" *(104 cm)*

Note black legs and bright golden yellow feet. Graceful plumes on head, neck, and back (where they curve upward) are striking in **breeding adult.** In **high breeding** plumage, lores turn red, feet orange. In nonbreeding, similar but plumes shorter; yellow on backs of legs. Compare to Little Egret. **Juvenile** lacks plumes and shows some bluish gray at base of lower mandible. Distinguished from immature Little Blue Heron (page 116) by its slimmer, mostly black bill; yellow lores; predominantly dark legs; and white wing tips. Active feeder.
VOICE: Low, raspy note, mostly at nest site.
RANGE: Common in various wetland habitats. Disperses north of mapped range in spring and after breeding season.

Little Egret *Egretta garzetta* L 24" *(60 cm)* WS 36" *(91 cm)*

Closely resembles Snowy Egret, but often appears larger, with longer neck; longer, thicker bill and legs, the latter always entirely black; mostly grayish lores; and more extensive throat feathering out on lower mandible; crown is flatter; feet are yellow, like Snowy, but average slightly duller. In **breeding** plumage, lore color is variable, but can be yellow. Note the two or three long, tapering plumes on back of head, rather than Snowy's many curved plumes. Often feeds less frenetically than Snowy, with long neck bent over in a posture like Little Blue Heron.
RANGE: Old World species. Casual spring and summer visitor to East Coast, from Newfoundland to mid-Atlantic states.

Great Egret *Ardea alba* L 39" *(99 cm)* WS 51" *(130 cm)*

Large white heron with heavy yellow bill, blackish legs and feet. In **breeding** plumage, long plumes trail from back, extending beyond tail. In immature and **nonbreeding adult,** bill and leg colors are duller, plumes absent. Distinguished from most other white herons by large size; from white morph of the larger Great Blue Heron by black legs and feet. Slightly smaller Asian *modesta* casual to western and central Aleutians; accidental VA; bill is black during breeding season.
VOICE: Occasional deep croaks.
RANGE: Common in wetlands, damp fields. Partial to open habitats for feeding; stalks prey slowly. Rarely wanders and sometimes breeds well north of mapped range.

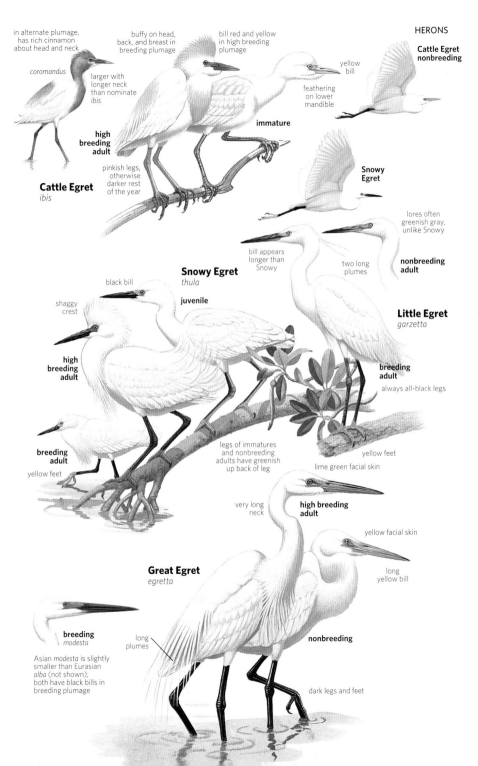

in alternate plumage, has rich cinnamon about head and neck

coromandus

larger with longer neck than nominate *ibis*

buffy on head, back, and breast in breeding plumage

bill red and yellow in high breeding plumage

Cattle Egret nonbreeding

yellow bill

feathering on lower mandible

immature

high breeding adult

pinkish legs, otherwise darker rest of the year

Cattle Egret
ibis

Snowy Egret

lores often greenish gray, unlike Snowy

nonbreeding adult

bill appears longer than Snowy

two long plumes

black bill

juvenile

Snowy Egret
thula

shaggy crest

Little Egret
garzetta

high breeding adult

breeding adult

always all-black legs

breeding adult

yellow feet

legs of immatures and nonbreeding adults have greenish up back of leg

yellow feet

lime green facial skin

very long neck

high breeding adult

yellow facial skin

long yellow bill

Great Egret
egretta

breeding
modesta

Asian *modesta* is slightly smaller than Eurasian *alba* (not shown); both have black bills in breeding plumage

long plumes

nonbreeding

dark legs and feet

Tricolored Heron *Egretta tricolor*

L 26" (66 cm) WS 36" (91 cm) White belly and foreneck contrast with mainly dark blue upperparts; bill long and slender. **Juvenile** has chestnut hindneck and wing coverts.

VOICE: Mostly silent; some low croaking at nest site.

RANGE: Common inhabitant of salt marshes and mangrove swamps of the East and Gulf Coasts. Rare inland, but has bred in ND and KS. Rare but regular on southern CA coast, chiefly in winter; casual in the Southwest and elsewhere in the interior.

Little Blue Heron *Egretta caerulea*

L 24" (61 cm) WS 40" (102 cm) Slate blue overall. During most of the year, plumage, head, and neck are dark purple; legs and feet dull green. In high **breeding** plumage, head and neck become reddish purple, legs and feet black. **Juvenile** is easily confused with immature Snowy Egret (page 114); note Little Blue Heron's dull yellow legs and feet; two-tone bill with thicker, gray base and dark tip; mostly grayish lores; and, often, narrow, dusky primary tips. During first spring, juvenile's white plumage begins gradual **molt** to adult plumage, during which time the mottled birds are said to be in their "calico phase." Slow, methodical feeders.

VOICE: Mostly silent; some low croaking and squawking, mostly at nest site.

RANGE: Fairly common at freshwater ponds, lakes, and marshes and coastal saltwater wetlands. Disperses north in spring and during post-breeding dispersal. Rare north of mapped range in spring (especially in West) and during post-breeding dispersal. Casual to BC and NL.

Reddish Egret *Egretta rufescens* *L 30" (76 cm) WS 46" (117 cm)*

While feeding, this heron lurches, dashing about with wings spread in a canopy. **Dark-morph breeding adult** has shaggy plumes on rufous head, neck. Bill is pink with black tip; legs cobalt blue. Nonbreeding plumage varies, but in general duller, with shorter plumes, darker bill. **Dark-morph juvenile** is gray; some pale cinnamon on head, neck, inner wing; bill is dark. **White-morph adults** in breeding season, with pink-based bills, are distinctive but nonbreeding adults and immatures, with all-dark bills, are more cryptic and resemble immature Little Blue Heron or Snowy Egret (page 114); note larger size, longer bill, dark legs and feet; also note behavior. A few dark-morph birds have much white on wings and resemble molting immature Little Blue Heron.

VOICE: Mostly silent; some low grunts at nest site.

RANGE: Uncommon. Inhabits shallow, open salt pans, salt marshes. Wanders along Gulf Coast in post-breeding dispersal; casual inland to Midwest, Southwest, up the Atlantic coast to New England; rare to coastal southern CA and Salton Sea; casual elsewhere in West.

Tricolored Heron
ruficollis

rufous on neck
and wing coverts

juvenile

long, slim
neck

long bill

breeding
adult

white belly

mottled plumage seen
on one-year-old birds
in spring and summer

bicolored bill
with grayish lores

molting
immature

juvenile

dusky
on wing
tips visible
on some
immatures

maroon head
and neck

**Little Blue
Heron**

juvenile

blue base
to bill

greenish
yellow legs

breeding
adult

nonbreeding
adult

very deliberate
feeding behavior

thick-based, slightly
downcurved bill

some juveniles show
dusky wing tips;
best viewed in flight

dark-morph
breeding adult

shaggy rufous
feathers on
head and neck

overall gray with
some pale cinnamon

dark-morph
breeding adult

**Reddish
Egret**
rufescens

dark-
morph
juvenile

when feeding,
rushes about
with wings
open

gray body

white-morph
winter adult

distinctly larger than
Little Blue Heron

dark legs
and feet

only birds in breeding
plumage have pink-
based bills; bill otherwise
blackish in all birds in
fall and winter

Green Heron *Butorides virescens* L 18" (46 cm) WS 26" (66 cm)

Small, chunky heron with short legs. Back and sides of **adult's** neck are deep chestnut; green on upperparts is mixed with blue-gray; center of throat and neck white. Greenish black crown feathers. Legs are usually dull yellow but in male turn bright orange in high breeding plumage. **Juvenile** is browner above; white throat and underparts heavily streaked with brown. Compare with Least Bittern (page 112). When alarmed, raises crest and flicks tail.

VOICE: Common call is a loud, sharp *kyowk*.

RANGE: Usually solitary; found in a variety of habitats, but prefers streams, ponds, and marshes with woodland cover; often perches in trees. Generally common; a few winter north of resident limit. Very rare visitor to much of southern Canada.

Black-crowned Night-Heron *Nycticorax nycticorax*

L 25" (64 cm) WS 44" (112 cm) Stocky heron with short neck and legs. **Adult** has black crown and back; white hindneck plumes are longest in breeding season. In **high breeding** plumage, legs turn bright pink. **Juvenile** distinguished from young Yellow-crowned Night-Heron by browner upperparts with bolder white spotting; thicker neck and stockier overall body shape; paler, less contrasting face with smaller eyes; and longer, thinner bill with mostly pale lower mandible. In flight, feet barely extend beyond tail. Full adult plumage is not acquired until third year. Mainly nocturnal feeder.

VOICE: Calls include a low, harsh *woc,* which is more guttural than Yellow-crowned.

RANGE: Very local in much of northern part of range. Declining in some regions. Occurs very rarely north of mapped range. Casual to AK, including to the western and once central Aleutians; five of these are specimens of nominate *nycticorax* from Eurasia, but sight records are almost certainly this subspecies too. Often roosts in trees and bushes, often in groups. Compare immature in flight also to American Bittern (page 112).

Yellow-crowned Night-Heron *Nyctanassa violacea*

L 24" (61 cm) WS 42" (107 cm) **Adult** has buffy white crown, black face with white cheeks; acquires head plumes in breeding season. **Juvenile** told from young Black-crowned Night-Heron by grayer upperparts with less conspicuous white spotting; longer neck; stouter, mostly dark bill, although recently fledged juveniles have some yellow at bill base ; and larger eyes. In flight, its feet extend well beyond its tail and it shows darker flight feathers and trailing edge on wings. Overall less stocky than Black-crowned with thinner, more-pointed wings. Full adult plumage is acquired in third year.

VOICE: Calls include a short *woc,* which is higher and less harsh than Black-crowned.

RANGE: Uncommon to fairly common; roosts in trees in wet woods and swamps. Single pairs may be found nesting well away from heron colonies. A few nest in coastal southern CA (San Diego and Ventura Counties). A few hybrids with Black-crowned have also been produced. Elsewhere casual in CA and the Southwest; also casual to NL and north to dashed line on map, mostly as a spring overshoot and during post-breeding dispersal.

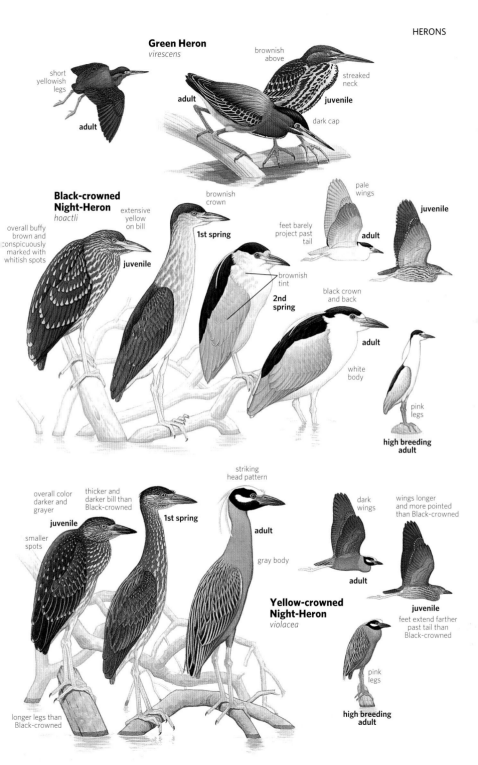

Green Heron
virescens

short
yellowish
legs

adult

brownish
above

streaked
neck

adult

juvenile

dark cap

**Black-crowned
Night-Heron**
hoactli

brownish
crown

pale
wings

juvenile

extensive
yellow
on bill

1st spring

feet barely
project past
tail

adult

overall buffy
brown and
conspicuously
marked with
whitish spots

juvenile

brownish
tint

2nd
spring

black crown
and back

adult

white
body

pink
legs

high breeding
adult

overall color
darker and
grayer

thicker and
darker bill than
Black-crowned

striking
head pattern

smaller
spots

juvenile

1st spring

adult

dark
wings

wings longer
and more pointed
than Black-crowned

gray body

adult

**Yellow-crowned
Night-Heron**
violacea

juvenile

feet extend farther
past tail than
Black-crowned

pink
legs

longer legs than
Black-crowned

high breeding
adult

IBISES • SPOONBILLS Family Threskiornithidae

Gregarious, heronlike birds, these long-legged waders feed with long, specialized bills: slender and curved downward in ibises, wide and spatulate in spoonbills. SPECIES: 32 WORLD, 5 N.A.

Glossy Ibis *Plegadis falcinellus* L 23" (58 cm) WS 36" (91 cm)

Breeding adult's chestnut plumage is glossed with green or purple; looks all-dark at a distance. Distinguished from White-faced Ibis by brown eye, gray-green legs with red joints, and lack of distinct white border to bare facial skin. Blue edge to gray facial skin does not extend behind eye or under chin. **Winter adult** closely resembles winter White-faced; look for gray facial skin partially bordered by blue line. **Juvenile** closely resembles juvenile White-faced Ibis, but note gray facial skin and at least trace of blue line on most. Adult breeding plumage is acquired in second spring. Identification should be made cautiously as hybrids with White-faced are frequent.

VOICE: Low grunts.

RANGE: Inhabits freshwater and saltwater marshes. Fairly common but local. Expanding north along the East Coast. Rare but annual in TX. Rare inland wanderer, chiefly in spring. Rather recent substantial increase in records from TX and Great Plains, chiefly in spring and now through much of the West (e.g., annual in CA). No winter records yet.

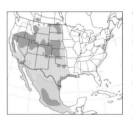

White-faced Ibis *Plegadis chihi* L 23" (58 cm) WS 36" (91 cm)

Breeding adult distinguished from Glossy Ibis, with which it sometimes hybridizes, by bronzer tones in chestnut plumage, reddish bill, red eye, mostly red legs, and white feathered border around red facial skin; border extends behind eye and under chin. **Winter adult** plumage is like Glossy, but lacks pale blue line from eye to bill; facial skin is pale pink. **Juvenile** closely resembles juvenile Glossy until winter, when facial skin turns pinkish; look for lack of blue line (or a hint of white border) and reddish tinge to eye.

VOICE: Like Glossy Ibis.

RANGE: Breeds in freshwater marshes. Rare but regular north to WA, casually farther. Very rare in Midwest; casual on East Coast north to New England in spring and summer.

White Ibis *Eudocimus albus* L 25" (64 cm) WS 38" (97 cm)

Adult's white plumage and pink facial skin are distinctive. In **breeding adult,** facial skin, bill, and legs turn scarlet. Dark tips of primaries most easily seen in flight. **Immatures** have white underparts and wing linings, pinkish bill; gradually molt into adult plumage by second fall. Closely related **Scarlet Ibis** (*E. ruber*), a South American species introduced or escaped in FL, hybridizes with White Ibis; offspring are various shades of pink or scarlet.

VOICE: Occasional low grunts.

RANGE: Locally common to abundant in coastal salt marshes, swamps, mangroves. Expanding north, now breeds to VA. Casual north to NJ, Midwest, Southwest; accidental to OK.

Glossy Ibis

all ages have dark eyes

breeding adult

juvenile

juvenile

juvenile

winter adult

most juveniles show bluish lines above and below lores

powder blue lines don't extend around eye

reddish joints

breeding adult

red iris

pinkish skin on lores

winter adult

often with faint whitish supraloral line; iris red by late winter

facial skin grayish, sometimes with some pinkish by fall

juvenile

White-faced Ibis

breeding adult

extensively reddish legs

red facial skin bordered by white feathered margin

Scarlet Ibis adult

juvenile

breeding adult

black wing tips

red face and white eye

White Ibis

dark eye

pinkish red bill

white belly

breeding adult

red legs

juvenile

1st spring

Roseate Spoonbill *Platalea ajaja* *L 32" (81 cm) WS 50" (127 cm)*
Adult has pink body with red highlights; long, spatulate bill; unfeathered greenish head. The head may become buffy during courtship.
Juvenile has white feathering on head; body is mostly pale pink.
Spoonbills feed in shallow waters, swinging their bills from side to side.
VOICE: Mostly silent; occasional low grunts.
RANGE: Fairly common in swamps and marshes along the Gulf Coast; a few wander north in summer to OK; casual north to mid-Atlantic, Midwest, Southwest, and CA coast.

STORKS Family Ciconiidae
Large, long-legged birds that fly with slow beats of their long, broad wings, soaring and circling like hawks. SPECIES: 19 WORLD, 2 N.A.

Wood Stork *Mycteria americana*
L 40" (102 cm) WS 61" (155 cm) **E** Black flight feathers and tail contrast with white body. **Adult** has bald, blackish gray head; thick, dusky, downcurved bill. **Juvenile's** head is feathered; bill is yellow.
VOICE: Mostly silent; some bill clacking at nest site.
RANGE: Uncommon. Inhabits wet meadows, swamps, ponds, and coastal shallows. A few wander north in summer and fall, including flocks from Mexican grounds to TX and lower Mississippi Valley. Casual north to Dakotas, ON and ME. In Southwest now uncommon and declining post-breeding visitor to south end of Salton Sea; formerly common. Casual elsewhere in Southwest. Casual to southern CA coast, where formerly regular; accidental north to BC.

Jabiru *Jabiru mycteria* *L 52" (132 cm) WS 90" (229 cm)*
Distinguished from Wood Stork by larger size; large bill, slightly upturned; and all-white wings and tail. Red throat pouch brightens and inflates during breeding season. **Juvenile** is patchy brown-gray; head is blackish brown. Usually seen with flocks of Wood Storks.
RANGE: Huge stork of Central and South America, casual straggler in south TX (eight records); recorded once in OK, LA, and MI.

FLAMINGOS Family Phoenicopteridae
Large waders with big, bent bills, used to strain food from the waters of shallow lakes and lagoons. SPECIES: 6 WORLD, 1 N.A.

American Flamingo *Phoenicopterus ruber*
L 46" (117 cm) WS 60" (152 cm) Note pink legs, black flight feathers, tricolored bill. **Immature** is grayer, with pink wash below; paler bill.
VOICE: Mostly silent; occasional honking notes.
RANGE: Breeds as close to FL as southern Bahamas, Cuba, and the Yucatán Peninsula. A few seen most years in fall in winter in Florida Bay; casual on TX coast (at least one involved a banded bird from the Yucatán). Others, especially away from these regions, more likely escapes. In addition, escapes include Greater Flamingo (*P. roseus*), with pink-and-white plumage; Chilean Flamingo (*P. chilensis*), with grayish legs with pink joints; and Lesser Flamingo (*P. minor*), with dark red bill and blotchy red wing coverts and axillaries.

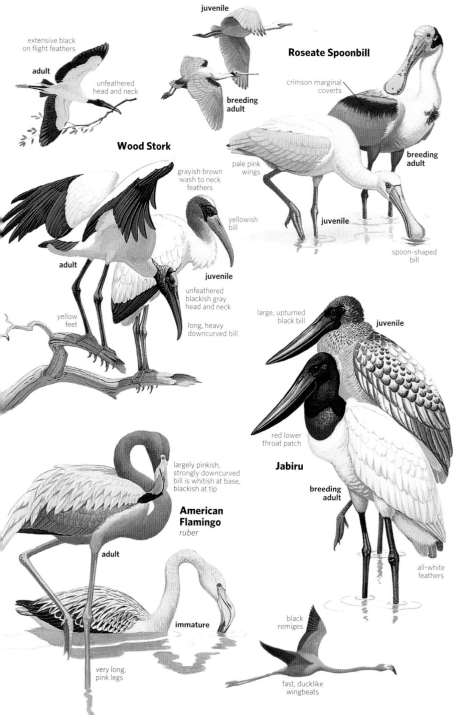

juvenile

Roseate Spoonbill

extensive black
on flight feathers

adult

unfeathered
head and neck

crimson marginal
coverts

breeding
adult

**breeding
adult**

Wood Stork

grayish brown
wash to neck
feathers

pale pink
wings

yellowish
bill

juvenile

adult

unfeathered
blackish gray
head and neck

juvenile

spoon-shaped
bill

yellow
feet

long, heavy
downcurved bill

large, upturned
black bill

juvenile

red lower
throat patch

Jabiru

largely pinkish,
strongly downcurved
bill is whitish at base,
blackish at tip

**American
Flamingo**
ruber

breeding
adult

adult

all-white
feathers

immature

black
remiges

very long,
pink legs

fast, ducklike
wingbeats

NEW WORLD VULTURES Family Cathartidae

Small, unfeathered head and hooked bill aid in consuming carrion. Generally silent away from nesting site. Latest research indicates that these species are more closely related to hawks than storks; placement here restores an earlier treatment. SPECIES: 7 WORLD, 3 N.A.

Turkey Vulture *Cathartes aura* L 27" (69 cm) WS 69" (175 cm)

In flight, rocks side to side with little flapping and wings held upward in a shallow V; dark wing linings contrast with silvery flight feathers. Rather long tailed. **Adult** has red head, white bill, brown legs; **juvenile**'s head and bill are dark, legs are paler. Feeds chiefly on carrion and refuse. Always check every Turkey Vulture for a Zone-tailed Hawk (page 136) within the hawk's range, or even outside of it for a stray. The two look remarkably similar at a distance.

RANGE: Common in mapped range. Often seen in spiraling flocks, especially in migration. Has expanded both summer and winter ranges northward. Casual to AK.

Black Vulture *Coragyps atratus* L 25" (64 cm) WS 57" (145 cm)

In flight, shows large white patches at base of primaries. Tail is shorter than Turkey Vulture; wings shorter and broader; legs pale gray; feet usually extend to edge of tail or beyond. Flight includes rapid flapping and short glides, usually with wings flat. Gregarious and aggressive, but less efficient at spotting carrion than Turkey Vulture, which, unlike Black Vulture, has a well-developed sense of smell. The two species are often seen together, either in flight or at carrion.

RANGE: Common in open country and near human settlements, often scavenges in garbage dumps. Range expanding in the Northeast; rare north to ON; casual to Maritime Provinces and CA; accidental to southwestern NM, BC, and southern YT.

California Condor *Gymnogyps californianus*

L 47" (119 cm) WS 108" (274 cm) **E** Huge size distinctive. **Adult** has white wing linings, orange head; **juvenile**'s wing linings mottled, head dusky. Soars on flat wings without flapping, in search of carrion. Population in wild now over 200; over 150 more in captivity.

RANGE: Last two wild birds captured 19 Apr. 1987 in Kern County, CA. Historically found north to Columbia River (OR and WA) and south to San Pedro Mártir, Baja California Norte, Mexico. Last reported in WA in 1897, OR in 1904, San Diego County, CA, in 1910, and Baja California in about 1930. By 1940, found mainly in hills and mountains fringing San Joaquin Valley, southern and central CA. Decline to near extinction caused mostly by lead poisoning and illegal shooting. Release program began in 1992 and now populations found in CA, northern AZ, and Baja California, with some successful breeding outside captivity in CA and AZ.

Turkey Vulture

reddish head

adult

grayish head

juvenile

adults

soars on slight dihedral; wingflaps are slow and deep

long tail

two-toned underwing

soars on flat wings; wingflaps are rapid and shallow

adult

extensive white base to outer primaries

pale grayish legs

short tail

Black Vulture

wings are long, broad, and of rather uniform width

unlike Turkey Vulture soars with no distinct dihedral

whitish wing linings

juvenile

orange head

California Condor

adult

huge size

white wing linings contrast with black flight feathers

adult

short tail

white tips to greater coverts and white on edges of secondaries visible from above

adult

OSPREYS Family Pandionidae
Large, eagle-like raptor with reversible outer toes. Feeds almost exclusively on fish, normally caught live and then transported head first and belly down. SPECIES: 1 WORLD, 1 N.A.

Osprey *Pandion haliaetus*
L 22-25" (56-64 cm) WS 58-72" (147-183 cm) Dark brown above, white below, with white head, prominent dark eye stripe. Females average darker streaking on neck; **juvenal** plumage is fringed with pale buff above. In flight, long, narrow wings are bent back at "wrist," dark carpal patches conspicuous; wings slightly arched in soaring. Eats mostly fish. Hovering over water, dives down, then plunges feetfirst to snatch prey.
VOICE: Call is a series of loud, whistled *kyew* notes.
RANGE: Nests near fresh or salt water. Bulky nests are built in trees, on sheds, poles, docks, and special platforms. Conservation programs successful and the species now fairly common.

KITES • EAGLES • HAWKS Family Accipitridae
Worldwide family of diurnal birds of prey, with hooked bills and strong talons. SPECIES: 236 WORLD, 27 N.A.

Snail Kite *Rostrhamus sociabilis* L 17" (43 cm) WS 46" (117 cm) **E**
Wings are paddle-shaped; bill thin and deeply hooked. **Male** is gray-black above and below, with white uppertail and undertail coverts; square, white tail with broad, dark band and paler terminal band; legs orange-red; eyes and facial skin reddish. **Female** is dark brown, with distinctive head pattern. **Juvenile** has dark brown eyes, duller facial skin and legs, streaked crown and underparts. Hunting flight is slow, with considerable flapping of wings, and head held down as the kite searches for apple snails, its chief and perhaps only food.
VOICE: Mostly silent; occasionally a nasal grating sound.
RANGE: A tropical species. Endangered; uncommon and local resident in south FL. Accidental in TX and SC.

Hook-billed Kite *Chondrohierax uncinatus*
L 18" (46 cm) WS 36" (91 cm) Plumage varies, but look for large, heavy bill with long hook, white eyes, banded tail, and heavily barred underparts, including underwings. **Males** are generally gray overall. **Females** are brown, with a rufous collar and rufous, barred underparts and wing linings. **Juveniles** have brown eyes, white collar, and whitish underparts with variable dark brown barring. In the **black morph,** rarely seen in the U.S., **adult** is all-black except for a single white or grayish tail band and whitish tail tip. Flies with deep, languid wingbeats, its "wrists" slightly cocked upward and "hands" angled down. Wings are paddle-shaped, slightly tapered in at the base. Eats insects and small amphibians, but prefers snails of various kinds.
VOICE: Loud rattling notes given near nest when disturbed and in courtship.
RANGE: Tropical species, uncommon over most of its range. Found in dense woodlands from which it thermals upward in midmorning. Rare resident in Rio Grande Valley from Falcon Dam to Santa Ana.

dark "wrist"

barred flight feathers

pale wing linings

gull-like flight

uniformly dark above

bold dark eye stripe

Osprey
carolinensis

prominent pale tips

juvenile

long angled wings

adult

adult ♂

whitish supercilium and chin

long, thin curved bill

strong eye line

white uppertail coverts

adult ♀

heavy streaks on underparts

adult ♂

white undertail

gray tail tip

Snail Kite
plumbeus

juvenile

black-morph adult

black-morph juvenile

paddle-shaped wings

hooked bill

rufous wing linings

adult ♀

adult ♂

juvenile

barred grayish underparts

white collar

variable dark barring below

whitish underparts

adult ♀

rufous collar

barred rufous underparts

Hook-billed Kite
uncinatus

juvenile

Mississippi Kite *Ictinia mississippiensis*

L 14½" (37 cm) WS 35" (89 cm) Long, pointed wings with first primary distinctly shorter; long, flared tail. Dark gray above, paler below, with pale gray head, averaging paler on **male. Female** with white shaft on outer tail feather and often whitish in vent region. White secondaries show in flight as white wing patch. Black tail readily distinguishes Mississippi from White-tailed Kite. Compare also with male Northern Harrier (page 132); note Mississippi never hovers. **Juvenile** is heavily streaked and spotted, with pale bands on tail, but pattern and overall darkness highly variable on underparts, underwings, and tail. **First-summer** bird (page 152) more like adult but retains juvenal flight feathers. At all ages, may be confused with Peregrine Falcon (page 152); compare wing and tail shapes. Captures and eats prey, mainly insects, on the wing. Gregarious; often hunts in groups, nests in loose colonies.

VOICE: Downward whistle, given mainly on breeding grounds.

RANGE: Found in woodlands, swamps, rangelands. Regular straggler (chiefly immatures in spring) to mid-Atlantic states. Casual north to Great Lakes region, west to NV and CA. Winters in South America.

White-tailed Kite *Elanus leucurus*

L 16" (41 cm) WS 42" (107 cm) Long, pointed wings; long tail. White underparts and mostly white tail distinguish **adults** from similar Mississippi Kite. Compare also with male Northern Harrier (page 132). **Juvenile's** underparts and head are lightly streaked with rufous, which rapidly fades. In all ages, black shoulders show in flight as black leading edge of inner wings from above, small black patches from below. Hovers while hunting, unlike any other North American kite. Eats mainly rodents, insects.

VOICE: Calls include various whistled notes.

RANGE: Populations fluctuate. Fairly common in grasslands, farmlands, even highway median strips. Some visit Channel Islands off southern CA coast. Casual well north of mapped range to BC, northern Great Plains, Midwest, and Northeast. Often forms winter roosts of more than a hundred birds.

Swallow-tailed Kite *Elanoides forficatus*

L 23" (58 cm) WS 48" (122 cm) Seen in flight, deeply forked tail and sharply defined pattern of black and white are like no other large bird except the young Magnificent Frigatebird (page 102). Perched, coloring more closely resembles White-tailed and Mississippi Kites; again, note long, forked tail. Juvenile is similar to **adult,** but tail is shorter, flight feathers and tail narrowly tipped with white. Agile and graceful, Swallow-tailed snatches flying insects; also drops down upon snakes, lizards, young birds; does not hover. Often eats prey in flight; also drinks in flight, skimming the water like a swallow.

VOICE: Mostly silent.

RANGE: Found in open woods, bottomlands, and wetlands. Nests in the tops of tall trees. Somewhat social; several may hunt in the same territory. Casual in spring and summer as far north as ON and NS; accidental as far west as AZ and CA. Most winter in South America.

Mississippi Kite

never hovers

adult ♂

whitish secondary patch

pale gray head

adult ♀

dark wings

black tail

juvenile

banded tail held through 1st summer

adult ♂

usually seen in flight; white body and wing linings contrast with black flight feathers

long forked black tail

adult

Swallow-tailed Kite

habitually hovers when foraging

dark spot near "wrist"

adults

long whitish tail

juvenile

buffy on chest

black shoulders

adult

adult

White-tailed Kite
majusculus

EAGLES
Large raptors with broad, long wings. Three of the species recorded *(Haliaeetus)* are sea-eagles and are found around water; two of those are casual visitors from the Old World. Golden Eagle is our only *Aquila*. Ten other species are found in the Old World.

Golden Eagle *Aquila chrysaetos*
L 30-40" (76-102 cm) WS 80-88" (203-224 cm) Brown, with variable yellow to tawny brown wash over back of head and neck; bill mostly horn colored; tail faintly banded. Tawny greater upperwing coverts form a bar. **Juveniles,** seen in flight from below, show well-defined white patches at base of primaries, white tail with distinct dark terminal band. Compare with juvenile Bald Eagle's larger head, shorter tail, blotchier tail and underwing pattern. **Adult** plumage is acquired in four years. Often soars with wings slightly uplifted.
VOICE: A rather faint and thin *kee-yep* or *yep,* sometimes in a series; usually silent away from nesting area.
RANGE: Nests on cliffs or in trees. Inhabits mountainous or hilly terrain, hunting over open country for small mammals, snakes, birds, and carrion. Also found in valleys and western plains, especially in migration and winter. Uncommon to rare in the East; uncommon to fairly common in the West.

Bald Eagle *Haliaeetus leucocephalus*
L 31-37" (79-94 cm) WS 70-90" (178-229 cm) **T Adults** readily identified by white head and tail, large yellow bill. **Juveniles** are mostly dark, may be confused with juvenile Golden Eagle; compare blotchy white on underwing coverts, axillaries, and tail with Golden Eagle's more sharply defined pattern; note also Bald Eagle's disproportionately larger head, shorter tail. Neck is shorter and tail longer than White-tailed Eagle; Steller's Sea-Eagle has longer, wedge-shaped tail. Flat-winged soar distinguishes young Bald Eagle from Turkey Vulture (page 124). Bald Eagles require four or five years to reach full adult plumage. The various interim subadult plumages may be highly variable; some **second-** and **third-year** birds show an Osprey-like dark patch through the eye. Feeds mainly on fish in breeding season, regularly on carrion, and on roadkill in winter, particularly in the Southwest.
VOICE: A variety of calls including a series of high-pitched twitterings or whistles, often delivered in a staggered rhythm.
RANGE: Nests in tall trees or on cliffs. Seen most often on seacoasts or near rivers and lakes. Most abundant in AK; common in winter along Mississippi and Missouri Rivers and at large lakes and reservoirs, fairly common in the Northwest. Birds raised in FL may wander north as far as southern Canada. Banning of pesticides and intense recovery programs have increased populations that had been seriously diminished in the East.

Golden Eagle *canadensis*

juvenile
whitish wing patch
short head projection
whitish tail base

adult

golden nape

adult

adult

dark or faintly barred tail

juvenile
whitish underwing coverts and axillaries

2nd year
longer head projection than Golden

Bald Eagle

note Osprey-like face pattern

larger bill than Golden

juvenile

white head

white tail

adults

3rd year

tail shorter than Golden

White-tailed Eagle *Haliaeetus albicilla*

L 26-35" (66-89 cm) WS 72-94" (183-239 cm) Note short, wedge-shaped white tail. Plumage mottled; head may be very pale and appear white at a distance; undertail coverts are dark, unlike subadult and adult Bald Eagles. **Juvenile**'s tail has variable dark mottling and tip is less wedge shaped, underwing darker, than Bald Eagle.

RANGE: Flies over northern Eurasia and Greenland in diminishing numbers. Very rare visitor to western Aleutians, especially Attu Island, where it nested from at least the late 1970s until 1996. Recorded east in Aleutians to Kiska Island; casual (spring) to St. Lawrence Island, AK. Accidental to MA.

Steller's Sea-Eagle *Haliaeetus pelagicus*

L 33-41" (84-104 cm) WS 87-96" (221-244 cm) In flight, white shoulders show as white leading edge of wings; trailing edge of wings more curved than White-tailed or Bald Eagles. Immense yellow-orange bill; long, white, wedge-shaped tail; white thighs. **Juvenile** lacks white shoulders; end of tail is dark.

RANGE: Nests in northeast Asia; casual in AK; recorded on Aleutians; Unimak Island (eastern Aleutians), Alaska Peninsula, and Simeonof Island, Shumagin Islands (all involving one bird); Nushagak River in western AK, Kodiak Island; and near Juneau in southeastern AK; some have involved multiple individuals that have returned multiple years.

Northern Harrier *Circus cyaneus*

L 16-20" (41-51 cm) WS 38-48" (97-122 cm) White uppertail coverts and owl-like facial disk distinctive in all ages and both sexes. Body slim; wings long and narrow with somewhat rounded tips; tail long. **Adult male** is grayish above; mostly white below with variable chestnut spotting; has black wing tips and black tips to secondaries. **Female** is brown above, whitish below with heavy brown streaking on breast and flanks, lighter streaking and spotting on belly. **Juveniles** resemble adult female but are cinnamon below, fading to creamy buff by spring; streaked only on the breast; wing linings are cinnamon, distinctly darker on inner half. Generally perches low and flies close to the ground, wings upraised, while searching for birds, mice, frogs, and other prey. Seldom soars high except during migration and in exuberant, acrobatic courtship display.

VOICE: Occasionally gives a high, downslurred call; also a series of *keee* notes when agitated.

RANGE: Fairly common in wetlands and open fields. Very local as breeder in southern part of range. Adult males migrate later in fall and earlier in spring than females and immatures. In winter, forms communal ground roosts, sometimes with Short-eared Owls. New World birds represent the subspecies (split as a full species by some authorities) *hudsonius*, which has been documented several times from northwest Europe and at least once from the Azores. Nominate *cyaneus,* known widely as the Hen Harrier, is widespread in the Old World. Sightings in the western Aleutians may be of this subspecies; a partial specimen (wing) salvaged on Attu Island in June 1999 tentatively has been identified as *cyaneus.* Adult males are paler gray overall with reduced chestnut markings below and more black in the wing tips. Juveniles are more streaked below and the underparts average less cinnamon.

broader winged than Bald Eagle, giving paddle-shaped appearance

pale head

adult

long dark undertail coverts

underwing darker than immature Bald Eagle

short, whitish wedge-shaped tail

juvenile

white spikes into dark-tipped tail create sawtooth pattern

White-tailed Eagle

palish head

overall scaly appearance

short tail

adult

massive size

paddle-shaped appearance

huge orange bill

white leading edge to wing

adult

long white acutely wedge-shaped tail

juvenile with mostly orange bill

white axillaries

juvenile

white primary patch

long wedge-shaped tail is white in adult, dark tipped in juvenile

Steller's Sea-Eagle

huge bright orange bill

white shoulders

adult

white leggings

gray above

adult ♂

roundish owl-like head

white underwing with dark wing tip and trailing edge

adult ♂

juvenile

darker secondaries

unstreaked belly

barred remiges

adult ♀

streaked belly

juveniles

whitish rump

Northern Harrier
hudsonicus

long tail

adult ♂

long, thin wings with rounded tips

ACCIPITERS
Comparatively long tails and short, rounded wings give these woodland hawks great agility. Flight is several quick wing-beats and a glide. Females are noticeably larger than males.

Sharp-shinned Hawk *Accipiter striatus*
L 10-14" (25-36 cm) WS 20-28" (51-71 cm) Distinguished from Cooper's Hawk by shorter, squared tail, often appearing notched when folded, thinner legs, and by smaller head and neck. **Adult** lacks Cooper's strong contrast between crown and back. **Juveniles** are whitish below, some streaked with brown (like Cooper's), others spotted with reddish brown. Note also the pale eyebrow, narrow white tip on tail, entirely white undertail coverts, less tawny head than other accipiters. In flight, wingbeats quick and choppy, slower on Cooper's.
VOICE: Mostly silent, except around nest. Adults give a single, sharp passerine-like note; juveniles give a high-pitched call.
RANGE: Fairly common, found in mixed woodlands. Preys chiefly on small birds, often at feeders.

Cooper's Hawk *Accipiter cooperii*
L 14-20" (36-51 cm) WS 29-37" (74-94 cm) Distinguished from Sharp-shinned Hawk by longer, rounded tail, larger head, and, in **adult,** stronger contrast between back and crown. **Juvenile** has whitish or buffy underparts with fine streaks on breast, streaking reduced or absent on belly; tawny rufous color on head much richer, white tip on tail broader, than Sharp-shinned; undertail coverts entirely white. Some juveniles may have a pale eyebrow like Sharp-shinned. In flight, again compare larger head and longer tail. Preys largely on songbirds, some small mammals. Often perches on telephone poles, unlike Sharp-shinned.
VOICE: Most notes with a nasal quality, include a series of *kek* notes. Juveniles give a squeaky whistle.
RANGE: Increasing. Nests in variety of wooded habitats, even in towns. Rare, mainly in fall, in the Maritime Provinces.

Northern Goshawk *Accipiter gentilis*
L 21-26" (53-66 cm) WS 40-46" (102-117 cm) Conspicuous eyebrow, flaring behind eye, separates **adult's** dark crown from blue-gray back. Underparts are white with dense gray barring; appear gray at a distance; has wedge-shaped tail with fluffy undertail coverts. Note disproportionately shorter tail, longer wings, than Cooper's Hawk. **Juvenile** is brown above, buffy below, with thick, blackish brown streaks, heaviest on flanks; tail has wavy dark bands bordered with white and a thin white tip; undertail coverts usually have dark streaks. In flight, note tawny bar on upperwing on greater secondary coverts. Juvenile also can be confused with Gyrfalcon (page 150) and Red-shouldered Hawk (page 138). Preys on birds and mammals as large as hares. Often most easily found in late summer, when juveniles have fledged and incessantly call.
VOICE: Calls include a loud, wailing *kee-ah*, more plaintive in juvenile, and a loud series of single notes.
RANGE: Inhabits deep, conifer-dominated, mixed woodlands. Uncommon; winters irregularly south of mapped range in the East. Southward irruptions occur in some winters.

small head

Sharp-shinned Hawk
velox

larger head,
longer neck,
and tawny nape

Cooper's Hawk

juvenile ♀

juvenile ♀

adult ♂

thin legs

adult ♂

curved
leading
edge

head
projects

juvenile

long
rounded tail

straight
leading
edge

shorter
square tail

juvenile

long wings
for an accipiter

prominent pale
supercilium

juvenile ♀

juvenile

long, pale gray supercilium
contrasts sharply
with dark auriculars

adult ♂

streaked
undertail
coverts

Northern Goshawk
atricapillus

gray barred
underparts

thin, pale, wavy bands
border dark bands

Common Black-Hawk *Buteogallus anthracinus*

L 21" (53 cm) WS 50" (127 cm) Wings broad and rounded; tail short, broad. **Adult** blackish overall; tail has broad white band. Legs and cere orange-yellow. Distinguished from Zone-tailed Hawk by broader wings; shorter, broader, less banded tail; larger bill; legs that are longer and appear slightly thicker; and more orange-yellow in lore region. Also note blended whitish area at base of outer primaries and brownish tinge to upper side of secondaries. **Juvenile** has strong face pattern; heavily streaked underparts; many bands on tail; buffy wing panel visible from above and below.

VOICE: Call is a series of loud whistles, falling in intensity near end.

RANGE: Found along waterways. Rare, local, and declining; very rare in southwestern UT, NV, and south TX; casual in CA and CO.

Harris's Hawk *Parabuteo unicinctus*

L 21" (53 cm) WS 46" (117 cm) Chocolate brown overall, with chestnut shoulder patches, leggings, and wing linings; white at base and tip of long tail; rounded wing tips. **Juvenile** is heavily streaked below; chestnut shoulder patches are less distinct. Gregarious; sometimes hunts in small, cooperative groups.

VOICE: Call is a long, harsh, grating *eeaarr.*

RANGE: Inhabits semiarid woodland, and brushland. From mapped range, may straggle north to Great Plains and west to CA (formerly nested in southeastern CA until early 1960s), but many may be escapes.

BUTEOS

These high-soaring hawks use rising thermals for energy-efficient flight; they are among the easiest of our birds of prey to spot. All have broad wings and relatively short tails. Most species also hunt from perches. Many species were persecuted by humans until the mid-1900s.

Zone-tailed Hawk *Buteo albonotatus*

L 20" (51 cm) WS 51" (130 cm) Grayish black overall, with barred flight feathers. Legs and cere yellow. Much slimmer winged than Common Black-Hawk; longer tail, variably banded according to age and sex: male with one broad white mid-tail band and one narrower white inner band; female also with a broad white mid-tail band but three to four narrower white inner bands. Flies remarkably like Turkey Vulture (page 124); this similarity may keep prey from recognizing it. Compare Zone-tailed's banded tail and barred primaries and secondaries; smaller bill; yellow cere; larger, feathered head. **Juvenile** has grayish tail, some white flecking on breast.

VOICE: Gives a loud screaming call, not too different from Red-tailed, but more of a whistle and less harsh, especially at end.

RANGE: Uncommon; found in mesa and mountain country, often near watercourses; drops from low glide onto small birds, rodents, lizards, and fish. Rare in southern CA, where it has nested, and south TX (mostly in winter). Accidental to northern CA, CO, NS, LA, and FL.

Common Black-Hawk
anthracinus

extensive orange-yellow in lores

adult

juvenile

stocky body shape with short tail

longer tail than adult

finely banded remiges

juvenile

banded tail

short tail with single, broad, white band

whitish patch

adult

short broad wings

short rounded wings

juvenile

heavily streaked below with blackish brown

adult

white undertail base and tip

chestnut wing linings

Harris's Hawk
harrisi

Turkey Vulture for comparison

long, narrow wings

flies with dihedral very similar to Turkey Vulture

juvenile

finely banded tail

adult

chestnut wing coverts

chestnut thighs

smaller bill than Common Black-Hawk

white undertail coverts

long tail with white tip

Zone-tailed Hawk

barred remiges differ from Turkey Vulture

adult ♂

long tail with one broad white mid-tail band and one white band at tail base

adult

Broad-winged Hawk *Buteo platypterus*

L 16" (41 cm) WS 34" (86 cm) Pointed wing tips; white underwing has dark border; tail has broad black and white bands, with last white band broader than the others. Wings broad but more pointed than Red-shouldered Hawk; wing linings buffy or white; tail shorter, broader. Wingbeats are slower than *elegans* subspecies of Red-shouldered. **Juveniles** typically have black moustachial streak; dark-bordered underwing, indistinct bands on tail; very similar to juvenile eastern Red-shouldered but paler below; may have a pale area at base of primaries but lack the distinct pale crescent. Rare **dark morph** breeds in western and central Canada and is only casually seen east of the Great Plains.

VOICE: Call, heard on breeding and winter grounds, is a thin, shrill, slightly descending whistle: *pee-teee.*

RANGE: A woodland species; may be seen perched on poles and power lines near forest edges. Preys primarily on small mammals, amphibians, reptiles, birds, and large insects. Often migrates in very large flocks. Rare migrant in the West, when most often seen at favored hawk-watching spots. Most winter in South America; a few winter in southern FL and very rarely in south TX and coastal CA.

See subspecies map, page 549

Red-shouldered Hawk *Buteo lineatus*

L 15-19" (38-48 cm) WS 37-42" (94-107 cm) Relatively long tailed and long legged. In flight, shows pale crescent at base of primaries. **Adult** has reddish shoulders and wing linings and extensive pale spotting above. Widespread eastern nominate subspecies *lineatus* shows dark streaks on reddish chest. Southeastern *alleni* (not shown) is smaller, with grayish cast to head and back; usually lacks breast streaking. South FL *extimus* is the smallest and palest subspecies. CA *elegans* is decidedly more rufous below; *elegans* is often solidly rufous across the chest and has broader white tail bands. Central TX *texanus* is merged into *alleni* by some; adults may average slightly redder below, juveniles are identical. **Juveniles** show extensive variations; *lineatus* shows more finely streaked breast and more closely resembles juvenile Broad-winged Hawk; other eastern subspecies show more coarsely marked underparts; *elegans* is quite dark and has more adultlike features, including some rufous on shoulders and wing linings. Flight of all ages of *elegans* is accipiter-like, with several quick wingbeats and a glide, while *lineatus* flies with slower wingbeats, more like Broad-winged.

VOICE: Call is an evenly spaced series of clear, high *kee-ah* or *kah* notes.

RANGE: Found in moist, mixed woodlands, including woodlots bordering residential areas; often seen near water. Preys primarily on small mammals, amphibians, reptiles, and crawfish. CA *elegans* and especially FL *extimus* more apt to be seen perched in the open. Migratory *lineatus* is an early spring and late fall migrant. Very rare in Maritime Provinces.

Broad-winged Hawk

juvenile

black moustachial streak

adult

rufous-brown barred underparts

black wing tips, more pointed than Red-shouldered

black trailing edge

adult

one broad white band shows on blackish tail

variable whitish underwing

juveniles

juvenile

dark-morph adult (rare)

juvenile *elegans*

adult *elegans*

rufous underparts

juvenile *elegans*

pale crescent

rufous wing lining

heavily marked underparts

banded tail

adult *elegans*

pale crescent

juvenile *elegans*

long banded tail

rufous wing lining

reddish shoulders

white crescent

Red-shouldered Hawk

juvenile *lineatus*

pale gray head

Florida adult *extimus*

Eastern adult *lineatus*

pale crescent

rufous wing lining

banded tail

adult *lineatus*

longer tail

juvenile *lineatus*

pale crescent

juvenile *lineatus*

streaked underparts

pale crescent

Roadside Hawk *Buteo magnirostris*

L 14" (36 cm) WS 30" (75 cm) A small, slim, long-legged raptor with banded tail. **Adult** with brown bib, barred belly. **Juvenile** with some streaking on chest. Flies with stiff, rapid wingbeats; wing tips rounded with rufous patch on inner primaries.

VOICE: In U.S., birds have been silent, but in Mexico, where common in many places, one of the Roadside's characteristic sounds is a drawn-out, complaining scream, delivered from a perch.

RANGE: Tropical species, casual (about ten records) in winter in lower Rio Grande Valley, TX.

Gray Hawk *Buteo nitidus* L 17" (43 cm) WS 35" (89 cm)

Gray upperparts, gray-barred underparts and wing linings, and rounded wing tips distinguish Gray from Broad-winged Hawk (see also page 138). Flight is accipiter-like: several rapid, shallow wing-beats and a glide. **Juvenile** resembles juvenile Broad-winged, but has much longer tail projection, stronger face pattern with outlined white cheek, and white, U-shaped rump band; dark trailing edge on wings is smaller or absent.

VOICE: Calls include a loud, descending whistle.

RANGE: Inhabits deciduous growth along streams. Tropical species; local nester in southeastern AZ. Rare in lower Rio Grande Valley year-round. Rare in summer upriver to Big Bend and in southwestern and southeastern NM.

Short-tailed Hawk *Buteo brachyurus*

L 15½" (39 cm) WS 35" (89 cm) Small hawk with two color morphs. Secondaries seen from below are darker than primaries. **Dark morph** more numerous in FL. **Light morph** has dark helmet and underwing resembling Swainson's Hawk (page 142), wings and tail are shorter, broader; lacks chest band. Dark morph has whitish area at base of outer primaries; adult has broad black border to trailing edge of wing. **Adults** have wide, dark subterminal tail band; more of equal width on juveniles. Faint streaks on sides of light morph; some dark-morph juveniles have all-dark wing linings, others are variably spotted with white. Exclusively an aerial hunter, most often seen in flight.

VOICE: Generally silent, especially in nonbreeding season. Has a high-pitched, drawn-out, slightly descending, two-syllable call.

RANGE: Found in woodland, savanna, and swamps; in winter even some suburbs. In recent decades has appeared in summer in the mountains of southeastern AZ (especially Chiricahuas, where nesting recently proven, and the Huachucas) and southwestern NM (Animas Mountains). Has wintered in Tucson. Casual to south TX in spring and summer; accidental northern MI (Whitefish Point).

short rounded wings

adult

barred rufous patch on inner primaries

pale eye

adult

brown bib

Roadside Hawk

short, pale supercilium

long legs

long banded tail

juvenile
streaked breast

distinct head pattern with dark eye line, dark malar, and pale cheek

juvenile

Gray Hawk
plagiata

fine gray bars on underparts

adult

longer tail than Broad-winged

juveniles

whitish uppertail coverts

adult

banded black-and-white tail

dark helmet

rarely seen perched

light-morph adult

Short-tailed Hawk
fuliginosus

uniform white underparts

light-morph adult

white wing linings

whitish patch at base of outer primaries

dark-morph adult

black border to trailing edge

Swainson's Hawk *Buteo swainsoni*

L 21" (53 cm) WS 52" (132 cm) Distinguished from most other buteos by long, narrow, pointed wings; plumage is extremely variable. Lacks Red-tailed Hawk's pale mottling on scapulars; bill is smaller. All but darkest birds show contrast between paler wing linings and dark flight feathers; most show pale uppertail coverts. In **light morph,** whitish or buffy white wing linings contrast with darkly barred brown flight feathers (see also page 154); dark bib; underparts otherwise whitish to pale buff. **Dark-morph** bird is dark brown with white undertail coverts; shows less sharp contrast between wing linings and flight feathers; darkest birds show none. Compare with first-year White-tailed Hawk. **Intermediate** colorations between light and dark morphs include a rufous morph. Intermediate and **light-morph juveniles** have dark moustachial stripe and conspicuous whitish eyebrows that meet on the forehead; variable streaking below, very heavy on dark morphs. Show less contrast between wing linings and flight feathers than adult birds. Swainson's soars over open plains and prairie with uptilted wings in teetering, vulturelike flight. Gregarious; usually migrates in large flocks, often with Broad-winged Hawks (page 138).
VOICE: A drawn-out scream, usually heard near nest site.
RANGE: Very rare spring and fall migrant in eastern North America and to AK. Winters chiefly in South America; rarely in south FL, south TX, and Central Valley of CA.

White-tailed Hawk *Buteo albicaudatus*

L 20" (51 cm) WS 51" (130 cm) Wings fairly long and pointed; at rest, appears long-legged and **adult's** wing tips project well beyond end of short tail; tail is white with single black band and other finer bands. Rusty shoulders contrast with gray upperparts. Underparts and wing linings vary from white on most to lightly barred. Females are darker above, more barred below. **Juveniles** brown above, variable below from mostly blackish to paler; most show a white patch on breast; tail is pale gray; undertail and uppertail coverts whitish, the latter forming a pale U at tail base. Compare with dark morphs of Swainson's and Ferruginous Hawks. Identifiable second-year plumage is intermediate.
VOICE: Rarely heard except when disturbed at nest site.
RANGE: Resident in open coastal grasslands and semiarid brush country. Casual to southwestern LA.

Ferruginous Hawk *Buteo regalis* L 23" (58 cm) WS 56" (142 cm)

Pale head; extended "gape line" going back under eye; tail is a mixture of pale rust, white, and gray. Wings are long, broad, and pointed; note large, white, crescent-shaped patches on upperwing. Seen from below, flight feathers lack barring. **Adults** show rusty above; rusty leggings form a conspicuous V. Much scarcer **dark morph** is rare; varies from dark rufous to dark brown, with dark undertail coverts. Lacks dark tail bands of dark-morph Rough-legged Hawk (page 144). **Juvenile** lacks rusty leggings and is less rufous above; resembles "Krider's" type of Red-tailed Hawk (page 144), but wings longer and more pointed. Often hovers when hunting or soars in a dihedral. Often sits on ground.
VOICE: Gives harsh alarm calls, *kree-a,* chiefly in breeding season.
RANGE: Casual east to WI, IL, AR, LA, and FL in migration and winter. Very rare migrant to MN; casual in summer. Accidental to BC and VA.

Swainson's Hawk

dark-morph adult

light-morph adult

whitish underwing coverts contrast with dark flight feathers

broad dark subterminal band

intermediate-morph adult

smaller bill than Red-tailed

often with darker lateral breast patches

light-morph juvenile

brown breast band

light-morph adult

White-tailed Hawk
hypospodius

long pointed wings

adult

white below

short white tail with black subterminal tail band

grayish head and back

rufous scapulars and marginal coverts

adult

whitish tail base

smudgy white cheek patches

size of white breast patch and undertail coverts color variable

dark wing linings contrast with pale flight feathers

juvenile

juvenile

dark juvenile

some juveniles all-dark below except for undertail coverts

pale tail

Ferruginous Hawk

variable rufous feathering in underwing coverts

adults

all show broad whitish crescents

long gape extends back under rear of eye

yellow cere and gape

juvenile

mostly white below

feathered legs

basal third of tail is white

palish head with whitish supercilium and dark postocular

pale head

extended gape

adult

rufous underparts

rufous leggings

juvenile

juveniles whiter below than adults

dark-morph adult

whitish tail lacks bands

Rough-legged Hawk *Buteo lagopus*

L 21" (53 cm) WS 53" (135 cm) White tail with dark band or bands helps to identify this hawk in all plumages; bill small. Thin legs are feathered to the toes, the feathering barred in adults, unbarred in juveniles. **Adult male** has multibanded tail with a broad blackish subterminal band. **Adult female's** tail is brown toward tip with a thin, black subterminal band. **Juveniles** have a single broad, brown tail band. Wings are long, fairly narrow. Seen in flight from above, white at base of tail is conspicuous; note also the small white patches at base of primaries on upperwing. In the common light morph, pale head contrasts with darker back and dark belly band, especially in females and immatures. Adult male has darker breast markings that may create a bib effect; belly is paler. Observe the square, black carpal patches at the "wrists" of the wings. **Dark morph** is less numerous. Often hovers while hunting. With delicate legs, often perches on thin branches, unlike other buteos.

VOICE: During breeding season gives a soft, plaintive courting whistle. Alarm call is a loud screech or squeal.

RANGE: Numbers from more southerly part of regular winter range vary from year to year, but in general, they are declining; this may indicate that Rough-leggeds are now wintering farther north. Casual to the Southeast and to coastal southern CA. A bird of the open country, also seen in marshes in winter.

See subspecies map, page 550

Red-tailed Hawk *Buteo jamaicensis*

L 22" (56 cm) WS 50" (127 cm) Our most common buteo; wings broad and fairly rounded; plumage extremely variable. Looks heavy billed, unlike Swainson's Hawk (page 142) and Rough-legged Hawk. Variable pale mottling on scapulars contrasts with dark mantle, often forming a broad-sided V on perched birds. Most **adults,** especially in the East, show a belly band of dark streaks on whitish underparts; dark bar on leading edge of underwing, contrasting with paler wing linings (see also page 154). Some pale breeding birds on the Great Plains designated as *krideri,* known as **"Krider's Red-tailed,"** have paler upperparts and whitish tail with pale reddish wash; in flight, shows pale rectangular patches at base of primaries on upperwing. Many southwestern birds of the *fuertesi* subspecies lack belly band and have entirely light underparts. Widespread **dark morph** and rufous morph of western subspecies, *calurus,* have dark wing linings and underparts, obscuring the bar on leading edge and belly band; tail is dark reddish above. In *harlani,* **"Harlan's Hawk,"** formerly considered a separate species, dark morph has dusky white tail, diffuse blackish terminal band; shows some white streaking on its dark breast; may lack scapular mottling; rare *harlani* light morph has typical tail pattern, but plumage resembles *krideri;* a few in the West. Dark and rufous morphs of *calurus* may be found east to Mississippi River Valley, very rarely farther east. **Juveniles** of all morphs except *harlani* have gray-brown tails with many blackish bands; otherwise heavily streaked and spotted with brown below.

VOICE: Distinctive call is a harsh, descending *keeeeer.*

RANGE: Habitat highly variable: woods with nearby open land; also plains, prairie groves, agricultural areas, and desert. Preys primarily on rodents; also on reptiles, amphibians, and birds. "Harlan's Hawk" breeds in AK and east to northwestern Canada; winters primarily in central U.S.

Rough-legged Hawk
sanctijohannis

squarish carpal patch

small bill

dark-morph adult ♂

adult ♂

adult ♀

juvenile with pale head and blackish belly

white wing linings with sparse, black markings

white tail base with broad, black subterminal band

adult light-morph "Harlan's"

mostly white below

adult males with multiple blackish bands at tail base

feathered tarsus

juvenile

white head with variable gray

tail variable with smudgy dark terminal band

adult dark-morph "Harlan's"

black wing linings with variable spotting

light rufous tail often with white base

adult "Krider's"

pale underparts and wing linings

moderate to faint patagial bar

pale head

adult *fuertesi*

dark patagial bar

uniform pale underparts

variable white on forehead and around eye

slightly paler scapulars

adult dark-morph "Harlan's"

extensive pale above

pale head

dark-morph adult *calurus*

wings more slender than adult

juvenile *borealis*

streaked belly

adult "Krider's"

pale underparts

white marks on black underparts

Red-tailed Hawk

heavy bill

juvenile *calurus*

adult *calurus*

belly band

white chest with heavily mottled belly

warm buffy wash to underparts

rufous tail

pale breast contrasts with darkish head and variably marked belly band

rather rounded wing tip

dark patagial bar

adult *borealis*

rufous tail

pale rufous tail with narrow or no dark subterminal band

adult *borealis*

CARACARAS • FALCONS Family Falconidae

These powerful hunters are distinguished from hawks by their long wings, which are bent back at the "wrist" and, except in the Crested Caracara, narrow and pointed. Females are larger than males. Birds of the genus *Falco* use their notched bills to kill prey by severing its spinal column at the neck. SPECIES: 65 WORLD, 11 N.A.

Eurasian Hobby *Falco subbuteo*

L 12¼" (31 cm) WS 30¼" (77 cm) Small, short-tailed falcon with long, slender wings; in folded wing, wing tips extend well past tip of tail. Graceful and powerful flier. White cheeks; thin, pale eyebrow; thin, dark moustachial stripe; heavily streaked below. **Adult** has rufous-red undertail coverts; is dark gray above. **Juvenile** is blackish brown above with buffy feather fringes; lacks rufous below. By following spring some look like adults, others intermediate in appearance. Compare all ages carefully to Merlin (page 148) and Peregrine Falcon (page 150).
RANGE: Old World species. Casual in late spring and summer in Bering Sea region and on western Aleutians. Record of a bird on a ship off Newfoundland and an Oct. record from Seattle, WA.

Aplomado Falcon *Falco femoralis*

L 15-16½" (38-42 cm) WS 40-48" (102-122 cm) **E** In flight, often hovers; long, pointed wings and long, banded tail; underwing is dark, with pale trailing edge. Note slate gray crown, boldly marked head. Pale eyebrows join at back of head. Dark patches on sides sometimes extend across breast. **Juvenile** is cinnamon below with a streaked breast, and browner above.
VOICE: Call is a series of *kek* notes.
RANGE: Once found in open grasslands and deserts from south TX to southeastern AZ. Disappeared by the early 20th century: last recorded in AZ in 1940 and in NM in 1952 (a nesting pair); birds seen more recently in NM and west TX from 1990s probably from a small extant population in northern Chihuahua, Mexico; recent nesting records southern NM; however, recent releases to NM and west TX will complicate determining origin of future sightings. A reintroduction project started in 1995 is ongoing in coastal south TX.

Crested Caracara *Caracara cheriway*

L 23" (58 cm) WS 50" (127 cm) **T** Large head, long neck, and long legs. Blackish brown overall, with white throat and neck and red-orange to yellow bare facial skin; underparts barred with black. **Juvenile** is browner; upperparts are edged and spotted with buff; underparts streaked with buff, unlike **adult** barring; second-year plumage closer to adult. In flight, shows whitish patches near ends of rounded wings. Flapping, ravenlike flight; soars with flat wings. Often seen on the ground in company with vultures. Feeds chiefly on carrion; also hunts insects and small animals.
VOICE: Calls include a low rattle and a single *wuck* note.
RANGE: Inhabits open brushlands. Fairly common in TX part of range. Rare in LA and southern AZ. Casual to southern NM and west TX. Records from well outside known range are questioned by some in terms of origin, but as population increases in TX, more records well to the north seem likely. Numerous recent CA records.

thin white supercilium, white cheeks, and dark mustache

Eurasian Hobby
subbuteo

long wings

adult

rufous red undertail

black streaked underparts

adult

juvenile

juvenile

long tail

juvenile

adult ♀

dark belly band and underwing coverts

buffy streaked breast

juvenile ♀

bold white supercilium and breast

adult ♂

Aplomado Falcon
septentrionalis

blackish side patches

blackish cap

slight crest

adult

pink facial skin

barred white chest

Crested Caracara

black belly

juvenile

streaked chest

conspicuous white wing patches

adults

long white-based tail barred with black

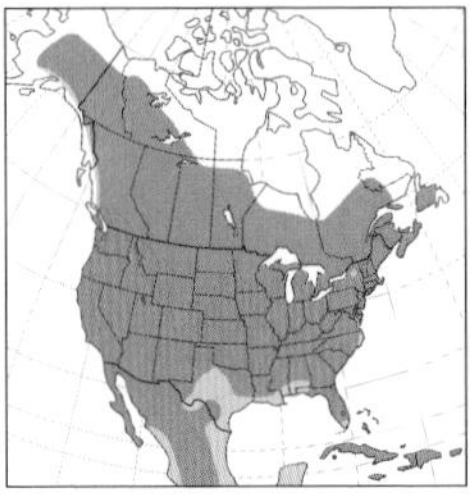

American Kestrel *Falco sparverius*

L 10½" (27 cm) WS 23" (58 cm) Smallest and most common of our falcons. Identified by russet back and tail, double black stripes on white face. Seen in flight from below, **adult** shows pale underwing, and **male** shows a distinctive row of white, circular spots on trailing edge of wings. Male also has blue-gray wing coverts; compare with Merlin. **Juvenile** male is like adult male, but breast heavily streaked, back completely barred; by first fall looks more like adult, but some dark markings remain. Feeds on insects, reptiles, and small mammals, hovering over prey before plunging. Also eats small birds, chiefly in winter. Often perches on telephone wires; frequently bobs its tail.
VOICE: Call is a shrill *killy killy killy.*
RANGE: Found in open country and in cities. Declining in East.

Eurasian Kestrel *Falco tinnunculus*

L 13½" (34 cm) WS 29" (74 cm) Resembles American Kestrel, but note larger size and single, not double, dark facial stripe. In flight, distinguished by wedge-shaped tail and two-tone upperwing, with back and inner wing paler. Hovers as it hunts. **Adult male** has russet wings, gray tail; **female** duller, often with gray rump. **Juvenile** similar to adult female, but dark barring heavier on upperparts and tail.
RANGE: Casual on western Aleutians and in Bering Sea region; accidental in fall, winter, and spring on the East Coast from NB and NS to NJ and FL; and on the West Coast to WA and CA.

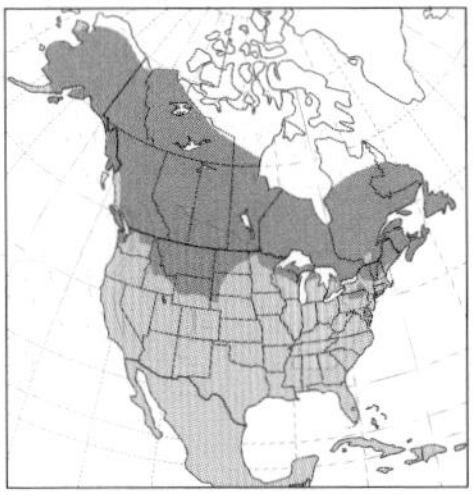

See subspecies map, page 550

Merlin *Falco columbarius* *L 12" (31 cm) WS 25" (64 cm)*

Adult male is gray-blue above; **female** and juvenile usually dark brown. Lacks the strong facial markings and russet upperparts of kestrels, and has broader wings than American Kestrel. Plumage varies geographically from the very dark subspecies, *suckleyi,* of the Pacific Northwest to the pale *richardsonii* that breeds on northern Great Plains from southern Canada to northern U.S. A few *suckleyi* winter to southern CA; this subspecies has dark cheeks and narrow, incomplete tail bands. All *richardsonii* have pale cheeks; male is paler blue-gray above; female and juvenile are pale brown, the latter with wide, pale tail bands. Winters to southern Great Plains, a few to the Great Basin and Pacific states. The widespread nominate subspecies, *columbarius,* which breeds in the taiga region, is intermediate in plumage; western *columbarius* averages slightly paler than eastern. In flight, strongly barred tail distinguishes Merlin from the much larger Peregrine and Prairie Falcons (page 150). Underparts and underwings darker than kestrels, particularly in *suckleyi* and *columbarius,* and head larger. Powerful flyer; does not hover. Catches birds in flight by a sudden burst of speed rather than by diving. Also eats large insects and small rodents.
VOICE: Silent except near nest site, where it gives shrill notes.
RANGE: Nests in open woods or wooded prairies; otherwise found in a variety of habitats. Fairly common to common on East Coast during fall migration. Many individuals in the Prairie Provinces do not migrate but winter in or near cities. Generally uncommon throughout U.S. in winter.

uniformly
rufous-brown
back and tail
adult ♀

all with two
dark facial
stripes

frequently
hovers
adult ♂

**American
Kestrel**

adult ♂

two-tone
upperwing
adult ♂

wedge-shaped
tail

**Eurasian
Kestrel**

juvenile

rufous tail

pale gray
head

single dark
moustache

juvenile
♂

adult ♀

male has
bluish gray
wings
adult ♂

adult ♂

gray
tail
base

rapid, powerful
flight with
no hovering

♀

adult ♂

Merlin
columbarius

faint
moustache

dark
bluish
above

♀

dark
brown
above

"Black Merlin"

♀ *suckleyi*

darker than *columbarius*
with fainter tail bars and
darker head

adult ♂

white flank
spots

adult ♂
suckleyi

♀
richardsonii

overall, color
of Prairie
Falcon

pale bluish
gray above

adult ♂
richardsonii

Prairie Falcon *Falco mexicanus*

L 15½-19½" (39-50 cm) WS 35-43" (89-109 cm) Pale brown above; creamy white and heavily spotted below. Brown crown, dark moustachial stripe, and broad pale area below and behind eye; facial markings narrower and plumage paler overall than Peregrine Falcon. Compare also with female and juvenile male Merlin (page 148), especially subspecies *richardsonii*. In flight, all ages show distinctive dark axillaries and wing bar on wing lining, broader on females. Juvenile is streaked below (not spotted) and darker above; bluish cere. Preys chiefly on birds and small mammals.

VOICE: Calls are higher than Peregrine. Largely silent away from nest.

RANGE: Inhabits dry, open country. Uncommon. Rare migrant and winter visitor in western Midwest. Casual elsewhere in Midwest and Southeast. Small numbers winter throughout breeding range.

Peregrine Falcon *Falco peregrinus*

L 16-20" (41-51 cm) WS 36-44" (91-112 cm) Crown and nape black; black wedge extends below eye, forms a distinctive helmet. Tail is shorter than Prairie; wing tips almost reach end of tail; also lacks dark bar and axillaries on underwing. Plumage varies from pale in highly migratory subspecies *tundrius* of the North to very dark in *pealei* from northwest Olympic Peninsula to Aleutians. In *pealei,* the largest subspecies, adult has heavy spotting on whitish breast, underparts very dark. Rather sedentary intermediate *anatum* subspecies has thickest moustachial stripe; **adult** shows rufous wash below; **juvenile** is dark brownish above, underparts are heavily streaked. Juvenile *tundrius* has a pale eyebrow and larger pale area on side of face; underparts more finely streaked.

VOICE: Gives harsh *cack* notes when agitated at nest site.

RANGE: Inhabits open wetlands; preys chiefly on birds. Nests on cliffs but now established also in cities where seen year-round; nests on bridges, tall buildings. Use of pesticides helped eliminate eastern *anatum* breeding populations; banning of these toxins and reintroduction programs led to rebounding populations. Most East Coast sightings in the fall are of *tundrius;* this subspecies appears much scarcer in West. Uncommon to rare in winter in U.S.

Gyrfalcon *Falco rusticolus*

L 20-25" (51-64 cm) WS 50-64" (127-163 cm) Heavily built; wings broader based than other falcons. **Adult** has yellow-orange eye ring, cere, and legs (bluish gray in juveniles). Tail broad and tapered; may be barred or unbarred; in perched bird, tail extends far beyond wing tips, unlike Peregrine. Compare also with Northern Goshawk (page 134). Plumages vary from **white morph** to **gray morph,** to very **dark morph;** paler gray morphs intermediate between typical gray and white. Facial markings range from none on white morph to all-dark cheeks on dark morph. **Juveniles** of gray and dark morphs show darker wing linings, paler flight feathers; juveniles of white and gray morphs much browner above. Flies with slow, powerful wingbeats. Preys chiefly on birds.

VOICE: Gives harsh *cack* notes when agitated at nest site.

RANGE: Inhabits open tundra near rocky outcrops, cliffs. Uncommon; winters irregularly south to dashed line on map. Casual to central CA, southern Great Plains, southern Great Lakes, mid-Atlantic regions.

blackish axillaries and underwing coverts

head pattern differs from Peregrine Falcon

pale brown above

adults

Prairie Falcon

adult *anatum*

uniform underwing

juvenile *pealei*

adult *pealei*

heavily spotted breast

heavily streaked underparts

juvenile *anatum*

broad, dark moustachial stripe

thinner dark moustache than *anatum*

juvenile *tundrius*

Peregrine Falcon

long wing tips extend nearly to tail tip

adult *anatum*

adult *tundrius*

faint moustache

gray-morph adult

dark-morph juvenile

Gyrfalcon

very dark brown above

darker wing coverts contrast with slightly paler remiges

gray-morph juvenile

long tail extends well beyond wing tips

white-morph adult

Kites

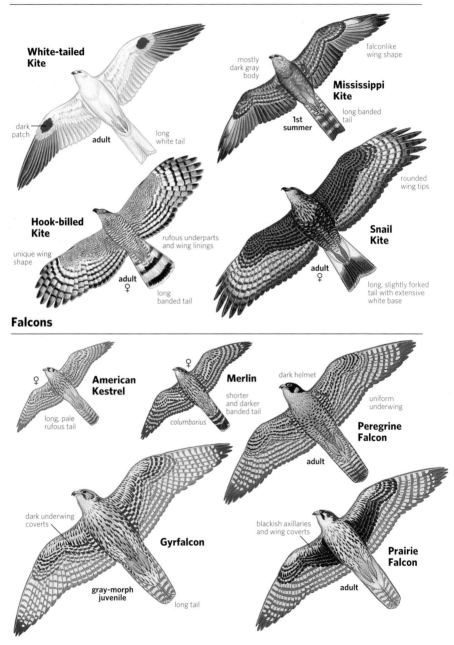

White-tailed Kite

dark patch

adult

long white tail

Mississippi Kite

falconlike wing shape

mostly dark gray body

1st summer

long banded tail

Hook-billed Kite

unique wing shape

rufous underparts and wing linings

adult ♀

long banded tail

Snail Kite

rounded wing tips

adult ♀

long, slightly forked tail with extensive white base

Falcons

♀

American Kestrel

long, pale rufous tail

♀

Merlin

shorter and darker banded tail

columbarius

dark helmet

uniform underwing

Peregrine Falcon

adult

dark underwing coverts

Gyrfalcon

gray-morph juvenile

long tail

blackish axillaries and wing coverts

Prairie Falcon

adult

Accipiters, Harrier, Smaller Buteos

Sharp-shinned Hawk

short head projection

rather short, square-ended tail

juvenile

Cooper's Hawk

longer head projection than Sharp-shinned

unstreaked whitish undertail

longer, rounder, more white-tipped tail

juvenile

Northern Goshawk

long winged for an accipiter

streaked undertail

juvenile

Roadside Hawk

short rounded wings

adult

barred rufous patch on inner primaries

Gray Hawk

gray underparts

banded black-and-white tail

adult

Short-tailed Hawk
fuliginosus

white wing linings

light-morph adult

dark-morph adult

Northern Harrier

long, thin wings with rounded tips

barred remiges

adult ♀

streaked belly

Red-shouldered Hawk

juvenile
elegans

whitish crescent

rufous underparts and wing linings

long banded tail

Broad-winged Hawk

black wing tips, more pointed than Red-shouldered

black trailing edge

adult

one broad white band shows on blackish tail

153

Larger Buteos, Black-Hawk

rather rounded
wing tip

**Red-tailed
Hawk**

dark
patagial
bar

rufous
tail

adult
borealis

whitish underwing with
variable rufous markings
on underwing coverts

**Ferruginous
Hawk**

rufous leg
feathering

adult

whitish
tail

pale
head

**Rough-legged
Hawk**

dark carpal
patch

dark
belly

**adult
♀**

extensive white tail
base with broad dark
subterminal band

**Swainson's
Hawk**

dark chest
band

thin
pointed
wings

pale wing linings
contrast with
darker flight
feathers

**light-
morph
adult**

short rounded
wings

**Harris's
Hawk**

adult

chestnut
wing
linings

long tail with
white base
and tip

whitish
patch

**Common
Black-Hawk**

broad
short
wings

short tail with
single, broad,
white band

adult

barred
remiges
differ from
Turkey
Vulture

long pointed
wings

**White-tailed
Hawk**

white
below

short white tail with
black subterminal
tail band

adult

**Zone-tailed
Hawk**

long tail
with broad
white band

adult ♂

154

Osprey, Eagles, Caracara

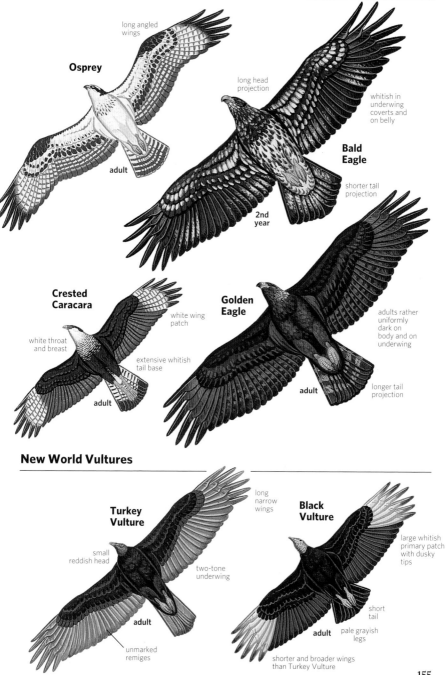

Osprey

long angled wings

adult

long head projection

whitish in underwing coverts and on belly

Bald Eagle

shorter tail projection

2nd year

Crested Caracara

white throat and breast

white wing patch

extensive whitish tail base

adult

Golden Eagle

adults rather uniformly dark on body and on underwing

longer tail projection

adult

New World Vultures

Turkey Vulture

small reddish head

long narrow wings

two-tone underwing

adult

unmarked remiges

Black Vulture

large whitish primary patch with dusky tips

short tail

pale grayish legs

adult

shorter and broader wings than Turkey Vulture

155

LIMPKINS Family Aramidae

Large, long-necked wading bird, named for its limping gait. SPECIES: 1 WORLD, 1 N.A.

Limpkin *Aramus guarauna* L 26" (66 cm)

Chocolate brown overall, densely streaked with white above. Long bill, slightly downcurved. Long legs and large, webless feet are dull grayish green. Juvenile is paler than **adult.**

VOICE: Call, heard chiefly at night, is a wailing *krr-oww*.

RANGE: Uncommon in swamps and wetlands, where it wades or swims in search of snails, frogs, and insects. Rare to fairly common in FL; casual in southern GA; accidental north to MD and NS.

RAILS • GALLINULES • COOTS Family Rallidae

These marsh birds have short tails and short, rounded wings. Most species are local and secretive. Some, especially the rails, are identified chiefly by call and habitat. SPECIES: 145 WORLD, 13 N.A.

Yellow Rail *Coturnicops noveboracensis* L 7¼" (18 cm)

A small, dark rail, deep tawny yellow above with wide dark stripes crossed by white bars. In flight, shows a large white patch on trailing edges of wings. Bill is short and thick; color varies from yellowish to greenish gray. **Juvenile** is darker than adult.

VOICE: Distinctive call, heard chiefly in breeding season, is a four- or five-note *tick-tick, tick-tick-tick* in alternate twos or twos and threes, sounds like tapping two pebbles together.

RANGE: Uncommon and local; secretive. Breeds in grassy marshes, boggy swales; not in deepwater marshes. Rare in the West. Winters in freshwater, brackish, or salt marshes, rice fields, dry fields.

Sora *Porzana carolina* L 8¾" (22 cm)

Short, thick bill, yellow or greenish yellow. **Breeding adult** is coarsely streaked above. Face and center of throat and breast are black. **Juvenile** lacks black on face and throat; underparts are paler. Compare with Yellow Rail; juvenile Sora is paler above; upperparts streaked, not barred, with white.

VOICE: Calls heard year-round are a descending whinny and a sharp, high-pitched *keek;* a whistled *ker-wheer* is heard on breeding grounds.

RANGE: Common in freshwater and brackish marshes, grain fields; also found in saltwater marshes during winter.

Black Rail *Laterallus jamaicensis* L 6" (15 cm)

Very small, extremely secretive. Blackish above, with white speckling; chestnut nape. Bill short and black. Underparts grayish black, with narrow white barring on flanks. Newly hatched juveniles of other rails resemble Black Rail.

VOICE: Most vocal in the middle of the night. Distinctive call, heard chiefly in breeding season, is a repeated *kik-kee-do* or *kik-kee-derr;* sometimes four notes: *kik-kik-kee-do*.

RANGE: Uncommon and local; inhabits marshes, wet meadows. Irregular inland; range speculative; declining in some coastal areas.

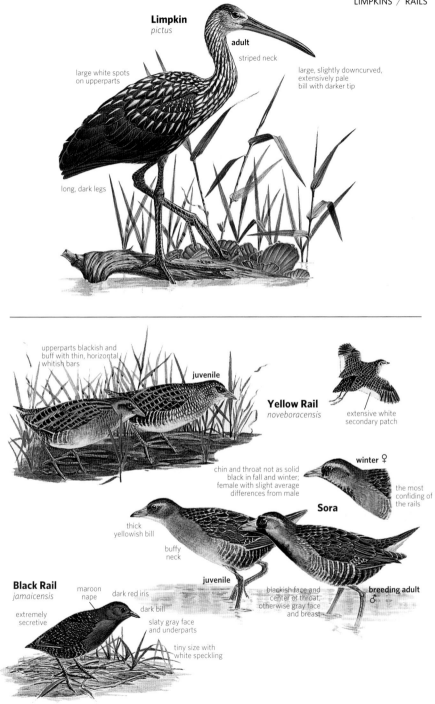

Limpkin
pictus

adult

striped neck

large white spots
on upperparts

large, slightly downcurved,
extensively pale
bill with darker tip

long, dark legs

upperparts blackish and
buff with thin, horizontal
whitish bars

juvenile

Yellow Rail
noveboracensis

extensive white
secondary patch

chin and throat not as solid
black in fall and winter;
female with slight average
differences from male

winter ♀

the most
confiding of
the rails

Sora

thick
yellowish bill

buffy
neck

blackish face and
center of throat;
otherwise gray face
and breast

breeding adult
♂

juvenile

Black Rail
jamaicensis

maroon
nape

dark red iris

dark bill

extremely
secretive

slaty gray face
and underparts

tiny size with
white speckling

Corn Crake *Crex crex* L 10½" (27 cm)

Dull buffy yellow overall, with short, thick, brownish bill; distinctive large chestnut wing patch. Extremely secretive.

RANGE: European species, formerly a very rare vagrant in fall along the East Coast; recently only three fall records: St.-Pierre, NS, and Avalon Peninsula, NL. Western European populations have seriously declined over last decades but have recently started to rebound. Found in damp, grassy fields, croplands, not in marshes.

Clapper Rail *Rallus longirostris* L 14½" (37 cm)

Much larger than Virginia Rail. Plumage variable but always has gray-ish edges on brown-centered back feathers, olive wing coverts. East Coast subspecies, such as *waynei* from extreme southern NC to Merritt Island, FL (very similar to *crepitans* from farther north), are much duller than King Rail: buffy below; cheeks gray; flanks less strongly barred than King. Peninsular FL and Gulf Coast subspecies such as *scottii* are brighter cinnamon below. West Coast subspecies, such as cinnamon-breasted *levipes* (**E**) from southern CA, similar but more buffy-breasted *obsoletus* (**E**) from San Francisco Bay Area, and thinner billed and duller inland *yumanensis* (**E**) are brighter below than East Coast birds but have brownish gray cheeks.

See subspecies map, page 550

VOICE: Call is a series of ten or more dry *kek kek kek* notes, like King but accelerating and then slowing. Some calls are nearly identical to King.

RANGE: Common in coastal salt marshes except on West Coast, where declined since introduction of the red fox; also along lower Colorado River and at Salton Sea, CA. In East, casual north to Maritime Provinces.

King Rail *Rallus elegans* L 15" (38 cm)

Large freshwater rail with long, slightly downcurved bill. Much larger than similar Virginia Rail. Adult distinguished from Clapper Rail by tawny edges on black-centered back feathers, tawny wing coverts. Head slate, with brown or grayish cheeks, buffy eyebrow; underparts cinnamon; flanks strongly barred black and white. **Juvenile** is darker above, paler below. Secretive.

VOICE: As with Clapper Rail, most often heard at dusk and dawn. Usually distinctive call is a series of fewer than ten *kek kek kek* notes, fairly evenly spaced. Also a series of grunting notes.

RANGE: Favors freshwater and brackish marshes. Fairly common to common near Gulf Coast; generally rather rare and local well inland in East. Rare in west TX, where it may breed. Casual west to CO and NM. Some birds winter in coastal marshes with Clapper Rails. Hybridizes with Clapper Rail in narrow zone of overlap.

Virginia Rail *Rallus limicola* L 9½" (24 cm)

Similar to King Rail but smaller; cheeks grayer; wings richer chestnut; legs and bill often redder. **Juvenile** is blackish brown above, mottled black or gray below.

VOICE: Song is a series of *kid kid kidick kidick* phrases, also a *tic tic turrr;* heard chiefly in breeding season; common call, heard year-round, is a descending series of *oink* notes.

RANGE: Fairly common but a bit secretive; found in freshwater and brackish marshes and wetlands; also in coastal salt marshes. Casual north to southeastern AK.

Corn Crake

broad gray eyebrow

broadly edged with gray or buff above

thick stubby bill

gray throat and breast

reddish chestnut wing patch

coastal southern California
levipes

dark and rich rusty neck

obsoletus from San Francisco Bay Area is similar but is slightly less richly colored below

slightly paler than the two California coastal subspecies

interior Southwest
yumanensis

Clapper Rail

rusty neck much like King

south Atlantic coast
waynei

flank bars weaker, more disorganized than King

Florida
scottii

all subspecies have brown centered feathers on upperparts with grayish edges and grayish faces

buffy face

feathers on upperparts with black centers and rich buff edges

rusty neck

King Rail
elegans

strong and aligned black and white flank bars

juvenile

gray cheek

long, thin, mostly reddish bill

deep rusty neck

juvenile

Virginia Rail

extensively blackish overall

Purple Gallinule *Porphyrio martinica* L 13" *(33 cm)*

Mostly purplish blue. Legs and long toes bright yellow. **Juvenile** is buffy brown overall with olive bill. Molts into winter plumage after fall migration but may retain traces of juvenal plumage into first spring. In all ages, all-white undertail coverts are conspicuous.

VOICE: Call is a sharp *kek;* also a series of grunting notes.

RANGE: Fairly common in overgrown swamps, lagoons, and marshes. Highly migratory; winters from southern FL to Argentina. Wanderers are seen in all seasons far north of mapped range; frequently breeds north of area shown. Casual in West to CO, UT, and CA.

Purple Swamphen *Porphyrio porphyrio* L 18-20" *(45-50 cm)*

Resembles a huge Purple Gallinule with reddish bill, frontal shield, iris, and legs. Various subspecies groups are recognized. FL birds appear to belong to the *poliocephalus* group subspecies (three included) from south Asia with grayish blue neck, or possibly closely allied *viridis* (mainland Southeast Asia), but some bluer-headed swamphens likely represent other subspecies.

VOICE: Give a wide variety of calls.

RANGE: Found from southern Europe to island groups in tropical South Pacific. Introduced into south FL in 1996 and has spread.

Common Gallinule *Gallinula galeata* L 14" *(36 cm)*

Black head and neck, with red forehead shield, red bill with yellow tip. Back brownish olive; white along flanks is diagnostic. Outer undertail coverts white. Legs and feet yellow. **Juvenile** is paler, browner; throat whitish; bill and legs dusky. **Winter adult** has brownish facial shield and usually a brownish bill with dusky yellow tip. **Eurasian Moorhen** *(G. chloropus)* recently split from our New World species; a juvenile specimen from Shemya Island, Aleutians, likely Eurasian Moorhen on probability.

VOICE: A high *keek;* also a descending series of nasal clucking notes. Calls are quite distinct from Old World species.

RANGE: Common in freshwater marshes, ponds, placid rivers; now uncommon to rare and declining from much of interior range.

American Coot *Fulica americana* L 15½" *(39 cm)*

Overall blackish; outer undertail coverts white. Whitish bill has dark subterminal band; reddish brown forehead shield. Leg color yellow in **adult,** duller in **juvenile.** Toes lobed, unlike gallinules. Juvenile paler; similar to adult by first winter. In flight, distinctive white trailing edge on wing. Often dives to feed. A few have extensively white facial shields like Caribbean Coot, *F. caribaea*, of the West Indies. Regarded by some as a subspecies of American, Caribbean not yet verified in FL.

VOICE: Variety of grunting and clucking calls; also a sharp *krrp,* slightly lower than Common Moorhen.

RANGE: Common to abundant in most regions. Nests in freshwater habitats; winters in both fresh and salt water, usually in flocks; frequently grazes on golf courses, lawns.

Eurasian Coot *Fulica atra* L 15¾" *(40 cm)*

Slightly larger and darker than American Coot; undertail coverts all-black. Forehead shield and bill entirely white.

RANGE: Old World species. Accidental to NL, QC, and the Pribilofs.

green above, bright purplish blue on head and below

light violet shield

greenish wings

juvenile

Purple Gallinule

overall buffy color

duller legs

long yellow legs and toes

Eurasian Moorhen adult breeding *chloropus*

shorter bill and more rounded top to frontal shield

record from western Aleutians, a juvenile, presumably this species

darker bill

winter

thick red bill

light violet-gray neck

red bill with yellow tip

breeding

bronze-brown back

juvenile

Purple Swamphen *poliocephalus*

large size

Common Gallinule *cachinnans*

thin whitish stripe

American Coot *americana*

whitish bill with subterminal band

adult

juvenile

some lack dark top to shield

variant

Eurasian Coot *atra*

all-white frontal shield and bill

lobed toes

swims like a duck most of the time

black undertail

CRANES Family Gruidae

Tall birds with long necks and legs. Tertials droop over the rump in a "bustle" that distinguishes cranes from herons. Cranes fly with their necks fully extended and circle in thermals like raptors. Courtship includes a frenzied, leaping dance. SPECIES: 15 WORLD, 3 N.A.

See subspecies map, page 550

Sandhill Crane *Grus canadensis*

L 41-48" (104-122 cm) WS 73-84" (185-213 cm) Subspecies vary in size: northern nominate subspecies smallest; more southerly *tabida* largest. Resident FL subspecies, *pratensis,* and endangered Gulf Coast subspecies, *pulla* (**E**), and migratory *"rowani,"* are intermediate. **Adult** is gray, with dull red skin on crown and lores; whitish chin, cheek, and upper throat; and slaty primaries. **Juvenile** lacks red patch; head and neck vary from pale to tawny; gray body is irregularly mottled with brownish red; full adult plumage reached after two and a half years. Great Blue Heron (page 112), sometimes confused with Sandhill Crane, lacks bustle. Preening with muddy bills, cranes may stain feathers of upper back, lower neck, and breast with ferrous solution in mud.
VOICE: Common call is a trumpeting, *gar-oo-oo,* audible for more than a mile. Young birds give a wholly different, cricket-like call.
RANGE: Locally common; breeds on tundra and in marshes. In winter, feeds in dry fields, returning to water at night. Resident near parts of the Gulf Coast, FL, and Cuba; other North American subspecies migratory. Rare during fall and winter on East Coast from MA south. Migrating flocks fly at great heights.

Common Crane *Grus grus*

L 44-51" (112-130 cm) WS 79-91" (202-231 cm) **Adult** distinguished from Sandhill Crane by blackish head and neck marked by broad white stripe. **Juvenile** like juvenile Sandhill; may show trace of white head stripe by spring. In flight, in all ages, black primaries and secondaries show as a broad black trailing edge on gray wings.
RANGE: Eurasian species, casual vagrant on the Great Plains, accidental farther east; almost always with migrating flocks of Sandhill Cranes. Some records involve a mixed Common-Sandhill pair with accompanying hybrid young.

Whooping Crane *Grus americana*

L 52" (132 cm) WS 87" (221 cm) **E Adult** is white, with red facial skin; black primaries show in flight. **Juvenile** is whitish, with pale reddish brown head and neck and scattered reddish brown feathers over the rest of body; begins to acquire adult plumage after first summer. A few abnormally colored Sandhill Cranes of *tabida* subspecies ("Greater Sandhill Crane") have been called Whooping Cranes; check wing-tip pattern. Endangered: Noncaptive population is now over 400, including introductions. Intensive management and protection seem to be succeeding.
VOICE: Call is a shrill, trumpeting *ker-loo ker-lee-loo.*
RANGE: Sparse wild population breeds in freshwater marshes of Wood Buffalo National Park, NT, and winters in Aransas National Wildlife Refuge on Gulf Coast of TX. A small population has been introduced in FL; some migrate to WI.

all cranes have tertial "bustles"

juvenile

red crown

adult

Sandhill Crane
"rowani"

stained adult

intermediate-size *"rowani"* breeds from northern BC to James Bay, ON

neck extended in flight

adult

dark dusky remiges

juvenile

yellowish bill

black neck with white stripe

adult

Common Crane
lilfordi

adult

blackish remiges

red crown

adult

black wing tips

adult

juvenile

Whooping Crane

LAPWINGS • PLOVERS Family Charadriidae

These compact birds run and stop abruptly when foraging. Shape and behavior identify plovers in general. SPECIES: 66 WORLD, 17 N.A.

Black-bellied Plover *Pluvialis squatarola* L 11½" (29 cm)

Black-and-white **breeding male** has frosty crown and nape, white vent; **female** averages less black. **Winter** and **juvenile** birds distinguished from Pacific and American Golden-Plovers by larger size, larger bill, and grayer plumage (including crown); underparts streaked rather than softly barred, but note that juvenile can be speckled with gold above. In flight, shows black axillaries and white uppertail coverts, barred white tail, and bold white wing stripe.

VOICE: Call is a drawn-out, three-note whistle, the second note lower pitched.

RANGE: Nests on Arctic tundra. Common migrant in Great Lakes region and in migration and winter at the Salton Sea, CA. Uncommon to rare elsewhere in interior.

American Golden-Plover *Pluvialis dominica* L 10¼" (26 cm)

Smaller, with a smaller bill than Black-bellied Plover; wing stripe is indistinct, and underwing is smoky gray with no black in axillaries; no contrasting white rump. Note the four evenly spaced primary tips. **Breeding male** shows broad white patches on sides of neck; underparts otherwise black. **Female** has less black but retains general pattern, including white bulging out on sides of neck. Mar. arrivals are in winter plumage; breeding plumage slowly acquired on migration north; the few seen in the West in spring are in a winterlike plumage. Most **juveniles** are rather dull, much like juvenile Black-bellied, but some are brighter, more like Pacific. Note primary projection and face pattern as well as call.

VOICE: Flight call is a plaintive *ku-wheep*.

RANGE: Fairly common migrant through Great Plains east to Mississippi River Valley, fewer east to Lake Erie. Very rare along East Coast in spring (more numerous in fall), and on western Gulf Coast in fall. In general, rare migrant, mostly fall, in West (apparently nearly all juveniles). Winters in South America.

Pacific Golden-Plover *Pluvialis fulva* L 9¾" (25 cm)

Similar to American Golden-Plover, but shorter primary tip projection with three, not four, staggered primary tips, the outer two close together; bill appears thicker, legs longer. **Breeding male** has less extensive white on sides of neck than American; white continues down sides and flanks; undertail coverts whiter; slightly larger gold markings above. **Breeding female** has less black below. **Juveniles** and **winter** birds typically appear brighter than American.

VOICE: Call is a loud, rich *chu-wheet.*

RANGE: Breeds from northern Russia to western AK. In AK, American Golden-Plover favors less vegetated slopes; Pacific favors the coast and river valleys. Winters from south Asia to Pacific islands; a few on West Coast and in central CA and very rarely at south end of Salton Sea. Casual in migration in Great Basin, AZ, and the East Coast. Some adults migrate earlier in fall than other golden-plovers.

164

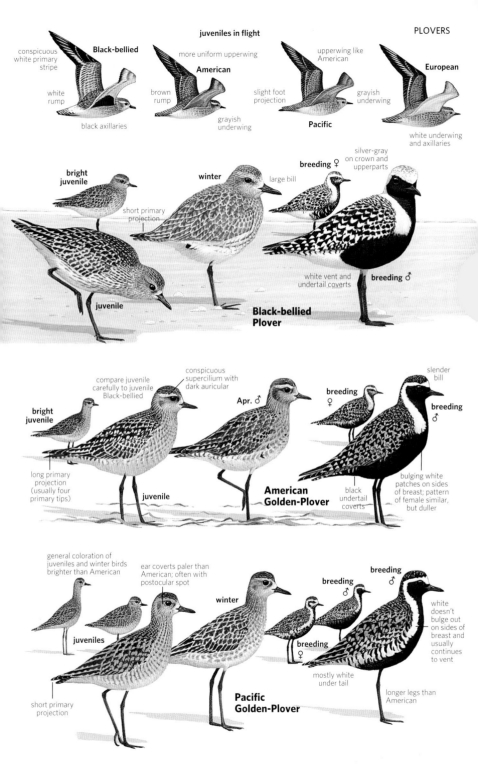

juveniles in flight

conspicuous white primary stripe

Black-bellied

more uniform upperwing

American

white rump

brown rump

black axillaries

grayish underwing

upperwing like American

slight foot projection

Pacific

grayish underwing

European

white underwing and axillaries

bright juvenile

winter

large bill

silver-gray on crown and upperparts

breeding ♀

short primary projection

juvenile

white vent and undertail coverts

breeding ♂

Black-bellied Plover

bright juvenile

compare juvenile carefully to juvenile Black-bellied

conspicuous supercilium with dark auricular

Apr. ♂

slender bill

breeding ♀

breeding ♂

long primary projection (usually four primary tips)

juvenile

American Golden-Plover

black undertail coverts

bulging white patches on sides of breast; pattern of female similar, but duller

general coloration of juveniles and winter birds brighter than American

ear coverts paler than American; often with postocular spot

winter

breeding ♂

breeding ♂

white doesn't bulge out on sides of breast and usually continues to vent

juveniles

breeding ♀

short primary projection

Pacific Golden-Plover

mostly white under tail

longer legs than American

European Golden-Plover *Pluvialis apricaria* L 11" (28 cm)

Similar to Pacific Golden-Plover; note larger size, plumper body shape, white underwing; also small bill and bolder wing bar. On **breeding males,** white nearly meets on front of breast; sides, flanks, and undertail coverts are more purely white than Pacific; note dense pattern of smaller gold spots on upperparts, unlike coarser pattern of larger spots on other golden-plovers.

VOICE: Call is a mournful, drawn-out whistle.
RANGE: Breeds from Greenland and Iceland to northwestern Russia. Winters from Europe to North Africa. Irregular spring migrant to NL; casually to eastern QC and NS. Accidental in fall to ME (9 to 11 Oct. 2008) and DE (14 to 15 Sept. 2009), both adults; in winter to southeastern AK (13 to 14 Jan. 2001, specimen).

Little Ringed Plover *Charadrius dubius* L 6" (15 cm)

A small, slim plover with conspicuous yellow eye ring; legs rather dull color. In flight, note lack of wing bar. In **breeding** adult, white line separates brown forecrown from rear of head. On winter birds and **juveniles,** brown replaces black on head and breast, and eye ring is slightly duller; juvenile often shows yellow-buff tint to pale areas on head and throat. Rather solitary.

VOICE: Call is a descending *pee-oo* that carries a long way.
RANGE: Old World species. Casual in spring to western Aleutians.

Lesser Sand-Plover *Charadrius mongolus* L 7½" (19 cm)

Bright rusty red breast; black-and-white facial pattern. **Females** are duller. **Juvenile** has broad buffy wash across breast; edged with buff above. In **winter,** white underparts except for broad grayish breast patch.

VOICE: Gives a hard and rather grating nonmusical *tirrick*.
RANGE: Asian species, rare migrant on Aleutians and on islands off western AK; casual along West Coast in fall. Casual in summer in western and northwestern AK, where it has bred. Accidental in eastern North America.

Killdeer *Charadrius vociferus* L 10½" (27 cm)

Double breast bands distinctive. Reddish orange rump is visible in flight. Downy young have only one breast band.

VOICE: Distinctive loud, piercing *kill-dee* or *dee-dee-dee,* heard mainly during the breeding season.
RANGE: Common in fields and on shores. Nests on open ground, usually on gravel. May form loose flocks and linger in north into early winter. Vagrant north of breeding range. A very early spring migrant (by late Feb.).

Wilson's Plover *Charadrius wilsonia* L 7¾" (20 cm)

Long, very heavy, black bill; broad neck band is black in **breeding male,** brown in **female** and winter male; legs grayish pink. **Juvenile** resembles adult female but note scaly-looking upperparts.

VOICE: Call is a sharp, whistled *whit.*
RANGE: Uncommon and declining on barrier islands, sandy beaches, mudflats. Recorded casually to CA, OR (once), and the Maritime Provinces and accidental far inland to Great Lakes region.

European Golden-Plover
altifrons

white underwing

brighter than juvenile American

juveniles

short primary projection

upperparts densely spotted with small gold spots

pure white on sides, flanks, and undertail coverts

breeding ♂

breeding ♀

breeding
plain dark wings

duller but complete eye ring

juvenile

pattern similar to male, but duller

breeding ♀

white throat

black forecrown bar

white forehead

rusty red breast

Lesser Sand-Plover
stegmanni

winter

white line borders forecrown

bold yellow eye ring

breeding ♂

dull legs

Little Ringed Plover

thick bill

grayish breast patch

winter

buffy breast band

juvenile

long tail and reddish orange rump

Killdeer

two breast bands

thick breast band

breeding ♂

Wilson's Plover
wilsonia

juvenile

♀

long thick bill

♀

dull fleshy legs

Common Ringed Plover
Charadrius hiaticula L 7½" (19 cm)
Almost identical to Semipalmated Plover; best distinguished by call. Breast band averages slightly broader in center than Semipalmated. White eyebrow is more distinct; eye ring is partial or lacking altogether; webbing between toes less extensive (hard to see). Bill is slightly longer, of more even thickness, and shows more orange at base; black on face meets bill where mandibles join.
VOICE: Call is a soft, fluted *pooee;* song, delivered in display flight, is a series of these notes
RANGE: Regular but rare spring migrant on western AK islands. In some years, breeds on St. Lawrence Island, AK. Casual on East Coast south to MA.

Semipalmated Plover
Charadrius semipalmatus
L 7¼" (18 cm) Dark back distinguishes this species from Piping and Snowy Plovers; bill much smaller than Wilson's Plover (page 166). Complete orangish (**breeding adults**) or buffy eye ring most obvious during breeding season. Breeding male often lacks white above eye. **Juvenile** has darker legs than adult.
VOICE: Distinctive call is a whistled, upslurred *chu-weet;* song is a series of same.
RANGE: Common on beaches, lakeshores, and tidal flats; seen throughout the continent in migration.

Piping Plover
Charadrius melodus L 7¼" (18 cm) **E**
Very pale above; orange legs; white rump conspicuous in flight. In **breeding** plumage, shows dark narrow breast band usually complete in *circumcinctus,* sometimes incomplete, especially in **females** and paler-faced East Coast birds, *melodus.* In **winter,** bill is all-dark. Distinguished from Snowy Plover by thicker bill, paler back; legs are brighter than Semipalmated Plover.
VOICE: Distinctive call is a clear *peep-lo.*
RANGE: Found on sandy beaches, lakeshores, and dunes. Endangered: generally uncommon; local and declining breeder and rare migrant in interior. In winter, *melodus* winters mainly on southern Atlantic coast and northern Bahamas; *circumcinctus* on Gulf Coast. Casual in winter to coastal southern CA (once at Salton Sea).

Snowy Plover
Charadrius nivosus L 6¼" (16 cm)
Pale above, very pale in Gulf Coast birds; thin dark bill; dark or grayish legs; partial breast band; dark ear patch. **Females** and **juveniles** resemble Piping Plover; note Snowy Plover's thinner bill, darker legs.
VOICE: Calls include a low *krut* and a soft, whistled *ku-wheet.*
RANGE: Inhabits barren sandy beaches and flats including edges of alkali lakes. Uncommon and declining on Gulf Coast. Western *nivosus,* breeding east to central Great Plains and TX, is threatened (**T**). Casual north to Great Lakes, northern Great Plains, and East Coast north to VA; accidental to YT. Recently split from Old World *C. alexandrinus,* from which it is strongly differentiated both genetically and vocally. A record of a Snowy Plover with mediocre photographs from Nome, AK, must now be regarded as equivocal to species given the split, because the areas of closest occurrence for either (Hokkaido, Japan, or WA) are nearly equidistant.

Common Ringed Plover

orbital ring absent or incomplete

bold white supercilium

bill slightly longer than Semipalmated with more extensive orange base

juvenile

breeding ♂

breeding ♀

juvenile

dark brown upperparts

winter

complete orangish orbital ring

Semipalmated Plover

always with complete breast band

juvenile

breeding ♂

breeding ♀

incomplete gray breast band

thicker bill

very pale gray upperparts

winter

paler lores

Piping Plover
melodus

breeding ♂

darker lores

breast band often broken

complete breast band

whitish uppertail coverts

orange legs

interior breeding ♂
circumcinctus

breeding ♀

Snowy Plover

very pale upperparts, like Piping

upperparts darker than Piping, paler than Semipalmated

slender bill

juvenile

♀

Gulf Coast ♂
tenuirostris

dark lateral patch

western
nivosus

♂

dark legs

Mountain Plover *Charadrius montanus* L 9" (23 cm)

In **breeding** plumage, unbanded white underparts separate this plover from all other brown-backed plovers. Buffy tinge on breast is more extensive in **winter** plumage; compare with much more patterned winter American Golden-Plover (page 164). In flight, shows white underwing; American Golden-Plover's is grayish.

VOICE: Calls heard on breeding grounds include low, drawn-out whistles and harsh notes. In migration and winter, gives a harsh *krrr* note.

RANGE: Inhabits plains; local and declining in many areas and should probably be considered a threatened species. Overall, rarely seen during migration. Gregarious in winter; usually found on short grassy or bare dirt fields, but in migration and sometimes in winter found on alkali flats in association with water, sometimes even on beaches. Formerly bred north to southern Canada. Rare migrant over much of West. Casual in Pacific Northwest. Overall, accidental in eastern North America east of TX, where recorded east to VA and FL (casual).

Eurasian Dotterel *Charadrius morinellus* L 8¼" (21 cm)

Whitish band on lower breast is somewhat obscured in **juveniles** and **winter** birds. Bold white eyebrow extends around entire head. Unlike other plovers, **females** are darker and more richly colored than males. Juvenile is darker and scalier above, much buffier below.

VOICE: Usually silent, but flight call is a soft, rolling, descending note.

RANGE: Eurasian species, very approachable; very rare, sporadic breeder in northwestern AK; casual along West Coast in fall. Has been recorded in midwinter with Mountain Plovers in the Imperial Valley, CA, and from northwestern Baja California.

Northern Lapwing *Vanellus vanellus* L 12½" (32 cm)

Most sightings of birds in **winter** plumage: dark, iridescent above, white below, with black breast; wispy but prominent crest. Wings broad and rounded, with white tips and white wing linings.

VOICE: Flight call is a whistled *pee-wit*.

RANGE: Eurasian species, casual primarily in late fall in northeast states and provinces; accidental elsewhere in East; recorded south to FL. Accidental on Shemya Island, Aleutians (12 Oct. 2006, specimen).

JACANAS Family Jacanidae

Extremely long toes and claws allow these tropical birds to walk on lily pads and other floating plants. SPECIES: 8 WORLD, 1 N.A.

Northern Jacana *Jacana spinosa* L 9½" (24 cm)

Adult's body and wing coverts are chestnut and black, frontal shield in yellow. Often raises its wings, revealing bright yellow underparts. **Immature** has white supercilium and underparts.

VOICE: Call is a series of harsh staccato notes, usually given in flight.

RANGE: Mexican and Central American species, casual visitor to ponds and marshes in south TX, where it has probably bred. Casual to southern AZ, accidental to west TX.

black lores and black forecrown bar with white forehead

Mountain Plover

uniform tan upperparts

winter

flashy white underwing

lateral brown patch

breeding

winter

winter

juvenile

bold supercilium and thin whitish breast band in all plumages

Eurasian Dotterel

gray above

juvenile

breeding ♂

winter

pattern similar to female, but duller

scaly above

juvenile

buffy belly

breeding ♀

deep chestnut underparts turning to more blackish on belly

long rounded wings with white tips

yellow flight feathers

Northern Jacana
gymnostoma

long wispy crest

yellow forehead shield

winter

long bill

blackish neck

white wing linings and tail base

broad black breast band

adult

immature

rich buff undertail coverts

winter

long toes

white supercilium and underparts

Northern Lapwing

OYSTERCATCHERS Family Haematopodidae

These chunky shorebirds have laterally flattened, heavy bills that can reach into mollusks and pry the shells open; they also probe sand for worms and crabs. SPECIES: 12 WORLD, 3 N.A.

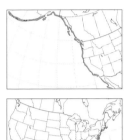

Black Oystercatcher *Haematopus bachmani* L 17½" (45 cm)

Large red-orange bill, all-dark body, pinkish legs. On immatures, outer half of bill is dusky during first year.

VOICE: Call is a loud, whistled *queep* given singly or in a fast series.

RANGE: Resident on rocky shores and islands along the Pacific coast. Some fall and winter movement away from nesting areas.

American Oystercatcher *Haematopus palliatus*

L 18½" (47 cm) Large red-orange bill. Black head and dark brown back; white wing and tail patches, white underparts. **Juvenile** appears scaly above; dark tip on bill is kept through first year. Birds feed in small, noisy flocks.

VOICE: Like Black Oystercatcher.

RANGE: Prefers coastal beaches and mudflats. Expanding northward in the East; recently established as a breeder on Cape Sable Island, NS. Very rare in southern CA *("frazari")*; most of these birds show hybrid characters with Black Oystercatcher to one degree or another. Birds from Gulf of California show less blackish flecking below hood on breast, more like "pure" American Oystercatchers. Accidental at Salton Sea, CA, and to the Great Lakes.

STILTS • AVOCETS Family Recurvirostridae

Sleek and graceful waders with long, slender bills and spindly legs. Three species inhabit North America. SPECIES: 9 WORLD, 3 N.A.

Black-necked Stilt *Himantopus mexicanus* L 14" (36 cm)

Male's glossy black back and bill contrast sharply with white underparts, long red or pink legs. Female is browner on back. **Juvenile** is brown above, with buffy edgings.

VOICE: Common call is a loud *kek kek kek,* also a loud keek call, vaguely suggestive of call of Long-billed Dowitcher.

RANGE: Breeds and winters in a wide variety of wet habitats; breeding range is spreading north. Casual north to Great Lakes and southern New England. Records increasing from these northern areas.

American Avocet *Recurvirostra americana* L 18" (46 cm)

Black-and-white above, white below; head and neck rusty in breeding plumage, gray in winter. **Juveniles** have cinnamon wash on head and neck. Avocets feed by sweeping their bills from side to side through the water. **Male** has longer, straighter bill than **female.**

VOICE: Common call is a loud *wheet,* given in an agitated series when disturbed.

RANGE: Fairly common on shallow ponds, marshes, and lakeshores. A few migrants in East are regular north to Long Island and Great Lakes; very rare farther north (to southern AK), mostly in fall.

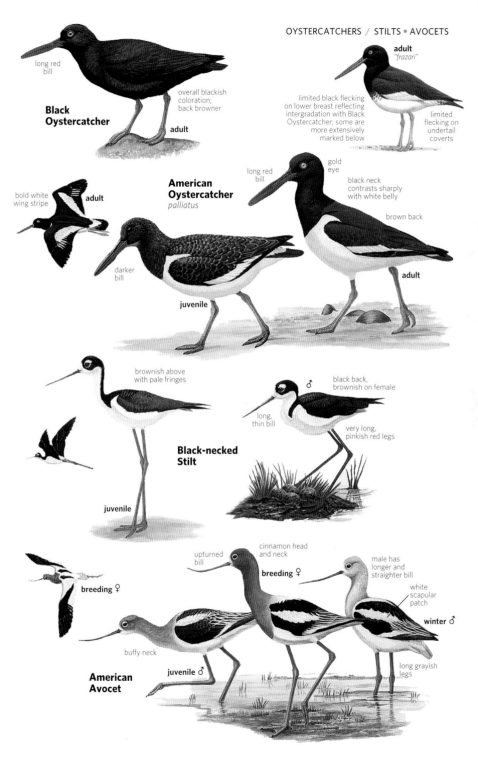

Black Oystercatcher

long red bill

overall blackish coloration; back browner

adult

adult
"frazari"

limited black flecking on lower breast reflecting intergradation with Black Oystercatcher; some are more extensively marked below

limited flecking on undertail coverts

bold white wing stripe

adult

American Oystercatcher
palliatus

long red bill

gold eye

black neck contrasts sharply with white belly

brown back

darker bill

juvenile

adult

brownish above with pale fringes

black back, brownish on female

♂

long, thin bill

very long, pinkish red legs

Black-necked Stilt

juvenile

cinnamon head and neck

upturned bill

breeding ♀

male has longer and straighter bill

white scapular patch

winter ♂

breeding ♀

buffy neck

juvenile ♂

long grayish legs

American Avocet

SANDPIPERS • PHALAROPES • ALLIES Family Scolopacidae
The majority of these shorebirds have three distinct plumages. Most begin molting to winter plumage as they near or reach their winter grounds. SPECIES: 94 WORLD, 66 N.A.

See subspecies map, page 550

Willet *Tringa semipalmata* L 15" (38 cm)
Large, plump, and grayish overall with grayish legs. In flight, note black-and-white wing pattern. Two subspecies: eastern *semipalmata* is smaller, darker, browner, and thicker billed than western *inornata* in all plumages. **Breeding** *semipalmata* is more heavily barred below with more pinkish-based bill than *inornata*. **Juvenile** *semipalmata* has more contrasting scapulars than *inornata*. Separating **winter** birds to subspecies best done by structural features and range. Winter-plumaged *semipalmata* typically not seen in North America, except for early Mar. arrivals on Gulf Coast.
VOICE: Territorial call is *pill-will-willet*, distinctly faster and higher-pitched in nominate subspecies. Other raucous calls are given year-round, all averaging higher pitched in the nominate subspecies.
RANGE: Nominate *semipalmata* nests in coastal Atlantic and Gulf salt marshes; *inornata* in interior marshes. In fall *semipalmata* is an early migrant, most departing by early Aug., and winters entirely outside North America (mainly South America). Generally an uncommon (West) to rare (East) interior migrant, except at Salton Sea, CA, where common. Some nonbreeders summer on coast.

Greater Yellowlegs *Tringa melanoleuca* L 14" (36 cm)
Legs yellow to orange. Larger than Lesser Yellowlegs; bill longer, stouter, often slightly upturned, and, in all plumages except breed-ing, two-toned. In **breeding** plumage, throat and breast are heavily streaked; sides and belly are spotted and barred with black; bill is all black. In **juvenile** birds the neck is distinctly streaked. Behavior is more active than Lesser Yellowlegs, often racing about with extended neck while pursuing prey (often very small fish).
VOICE: Call, a loud, slightly descending series of three or more *tew* notes.
RANGE: Fairly common; nests on muskeg, winters in wetland habitats. Compared to Lesser Yellowlegs, Greater is overall a later fall migrant and is much more capable of wintering at interior locations.

Lesser Yellowlegs *Tringa flavipes* L 10½" (27 cm)
Legs yellow to rarely orange. Smaller than Greater Yellowlegs; all-dark bill is shorter, thinner, and straighter. In **breeding** plumage, not nearly as heavily as marked as Greater Yellowlegs, especially on sides and flanks. **Juvenile** and **winter** birds are overall darker than Greater; juvenile Lessers are washed with grayish brown across the neck and lack the streaks that are present in Greater. A more sedate feeder than Greater, leisurely picks at prey from a more vertical position.
VOICE: Call is one or more *tew* notes, a little higher than Greater, the individual notes being more clipped and all on one pitch.
RANGE: Nests on tundra or in woodland. On nesting grounds, like Greater, often perches in trees and yelps at intruders. Common in the East (often abundant on Great Plains and Midwest); uncommon in Far West, especially in spring (more numerous in fall). Most winter in South America, a few in U.S.

short pinkish-based bill

brownish cast to upperparts

eastern *semipalmata*

contrasting scapulars

early Mar. in molt

western

eastern

heavily barred below

breeding

juvenile

Willet

striking wing pattern in flight

winter

paler above

worn breeding

western *inornata*

western Willets are larger and have a longer, darker bill

less heavily marked below

breeding

plainer and paler upperparts

winter

juvenile

bill all-dark in breeding plumage

breeding

Greater Yellowlegs

long, slightly upturned bill, grayish at base

distinctly streaked breast

heavily barred flanks

winter

yellow legs

juvenile

winter

always with slim, short, dark bill

both Lesser and Greater Yellowlegs have whitish rumps and barred tails

breeding

winter

Lesser Yellowlegs

juvenile

faintly marked sides and flanks

winter

grayish wash across chest

yellow legs

Common Greenshank *Tringa nebularia* L 13½" (34 cm)

In plumage and structure resembles Greater Yellowlegs, but less heavily streaked; legs are greenish. In flight, white wedge extends up the middle of back.

VOICE: Typical flight call is a loud *tew-tew-tew,* very much like Greater Yellowlegs, but the notes are all on one pitch.

RANGE: Common Eurasian species that annually visits western AK in spring: rare to Aleutian and Pribilof Islands; casual to St. Lawrence Island; casual in western AK in fall. Accidental to northeastern Canada and coastal northwestern CA.

Marsh Sandpiper *Tringa stagnatilis* L 8½" (21 cm)

A small and slender *Tringa* with a long needle-like bill and dispro-portionately long, greenish legs; distinct supercilium in all plumages. Plumages overall suggestive of Common Greenshank. In **breeding,** neck and sides are streaked and spotted with brown and mottled with black above. **Juvenile** is faintly streaked on sides of breast, brown-ish above with pale buff edges. **Winter adult** pale gray and uniform above and with long needle bill suggests a winter-plumaged Wilson's Phalarope (page 206); Marsh has much longer legs. In flight, white wedge extends up back; note long leg projection past tail.

VOICE: Call is a *tew* note, like call of Lesser Yellowlegs; often delivered in a series.

RANGE: Old World species. Common in Asia. Casual in central and western Aleutians (four fall records) and once on St. Paul Island, Pribilofs. One wintered recently on Oahu, HI.

Common Redshank *Tringa totanus* L 11" (28 cm)

Bright orange legs, stout bill with reddish orange base, overall brown-ish plumage with distinct eye ring. **Juvenile** and **breeding** birds are extensively streaked below; **winter** birds are diffusely mottled below. In flight, shows white dorsal wedge up back and distinctive broad white trailing edge to secondaries and inner primaries.

VOICE: Calls include a musical *tew* or more mournful *tew-hieu.* Alarm note is a series of *twek* notes.

RANGE: Eurasian species; breeds as close to North America as Iceland; numerous records for Greenland. Casual to Newfoundland in spring, once in winter.

Spotted Redshank *Tringa erythropus* L 12½" (32 cm)

Long bill with red-based lower mandible, droops at tip. **Breeding adult** is black with white spots above, underparts variably marked with white; legs dark red to blackish. **Juvenile** is brownish gray and is heavily barred and spotted below. **Winter** birds are pale gray above, spotted with white on coverts and tertials. Some fall sight-ings involve young birds in transitional plumage. Both plumages show brighter orange legs. In flight, shows white wedge on back and white wing linings.

VOICE: Distinctive call, a loud rising *chu-weet* closely resembles the call of Semipalmated Plover and completely unlike any other *Tringa.*

RANGE: Eurasian species, rare spring and fall visitor to Aleutian and Pribilof Islands; casual on Pacific and Atlantic Coasts during migration and winter; accidental elsewhere.

juvenile

long and rather stout, upturned bill; paler at base

breeding

white wedge up back

winter

not as heavily marked below as Greater Yellowlegs

Common Greenshank

greenish legs

Marsh Sandpiper

faint streaking to sides of breast; underparts otherwise white

brownish above with pale buff edges

long, slender, dark bill

streaked breast

juvenile

white wedge up back

white wedge up back

winter

winter

white secondaries and inner primaries

breeding

Common Redshank

long yellowish legs

juvenile

short red-based bill

heavily marked below, brownish cast above

juvenile

breeding

bright orange legs

extensively barred below

extensive pale spotting above

breeding

black overall

long red-based bill in all plumages with slight droop near tip

white wedge up back

very dark red legs

winter

Spotted Redshank

winter

orange legs

Wood Sandpiper *Tringa glareola* L 8" (20 cm)

Dark upperparts are heavily spotted with buff; prominent whitish eyebrow. In flight, distinguished from Green Sandpiper by paler wing linings, smaller white rump patch, and more densely barred tail. Bill is straight like Green, but shorter.

VOICE: Common call is a loud, sharp whistling of three or more notes, similar to the call of Long-billed Dowitcher.

RANGE: Eurasian species, fairly common spring and uncommon migrant and occasional breeder on the outer Aleutians; uncommon to rare on the Pribilofs; rare on St. Lawrence Island, AK. Accidental to BC, OR, CA, Baja California, and northeastern North America.

Solitary Sandpiper *Tringa solitaria* L 8½" (22 cm)

Dark brown above, heavily spotted with buffy white. White below; lower throat, breast, and sides streaked with blackish brown. Bolder white eye ring and shorter, olive legs distinguish it from Lesser Yellowlegs (page 174). In flight, shows dark central tail feathers, white outer feathers barred with black. Underwing is dark. Two subspecies: nominate *solitaria* (illustrated) from East is spotted with white in breeding plumage; spots more cinnamon-buff in western *cinnamomea*. Often keeps wings raised briefly after alighting; on the ground, often bobs its tail. Generally seen singly, occasionally in small flocks.

VOICE: Calls include a shrill *peet-weet,* higher pitched than calls of Spotted Sandpiper.

RANGE: Fairly common at shallow backwaters, pools, small estuaries, even rain puddles.

Green Sandpiper *Tringa ochropus* L 8¾" (22 cm)

Resembles Solitary Sandpiper in plumage, behavior, and calls. Structure also similar, but a little plumper, straighter billed. Note white rump and uppertail coverts, with less extensively barred tail; lacks solidly dark central tail feathers of Solitary; upperparts and wing linings are darker. Similar Wood Sandpiper has more spotting above, more barring on tail, and paler wing linings.

VOICE: Call is like Solitary, but louder.

RANGE: Eurasian species, casual in spring on outer Aleutians, Pribilofs, and St. Lawrence Island, AK.

Terek Sandpiper *Xenus cinereus* L 9" (23 cm)

Note long, upturned bill and short orange-yellow legs. In **breeding adult,** dark-centered scapulars form two dark lines on back. In flight, shows distinctive wing pattern: dark leading edge, grayer median coverts, dark greater coverts, and white-tipped secondaries.

VOICE: Flight call is a series of shrill whistled notes on one pitch, usually in threes.

RANGE: Eurasian species, rare migrant on outer Aleutians; casual on Pribilofs, St. Lawrence Island, and in Anchorage area, AK. Accidental in fall to coastal BC, CA, Baja California, and MA.

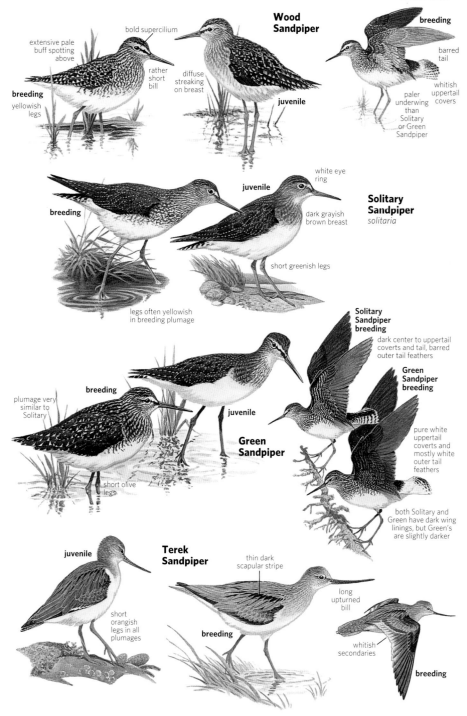

Wood Sandpiper

extensive pale buff spotting above

bold supercilium

breeding
yellowish legs

rather short bill

diffuse streaking on breast

juvenile

breeding

barred tail

whitish uppertail covers

paler underwing than Solitary or Green Sandpiper

Solitary Sandpiper
solitaria

breeding

juvenile

white eye ring

dark grayish brown breast

short greenish legs

legs often yellowish in breeding plumage

Solitary Sandpiper breeding
dark center to uppertail coverts and tail, barred outer tail feathers

Green Sandpiper breeding
pure white uppertail coverts and mostly white outer tail feathers

plumage very similar to Solitary

breeding

juvenile

Green Sandpiper

short olive legs

both Solitary and Green have dark wing linings, but Green's are slightly darker

juvenile

Terek Sandpiper

short orangish legs in all plumages

thin dark scapular stripe

breeding

long upturned bill

whitish secondaries

breeding

Wandering Tattler *Tringa incana* L 11" (28 cm)

Uniformly dark gray above; white eyebrow flecked with gray; bill dark, legs dull yellow. In **breeding** plumage, underparts are heavily barred. **Juvenile** and **winter** birds have only a dark gray wash over breast, sides, and flanks; juvenile has pale spots above. Closely resembles Gray-tailed Tattler; best distinguished by voice. Often teeters and bobs as it feeds.

VOICE: Call is a rapid series of clear, hollow whistles, all on one pitch.
RANGE: Breeds chiefly on gravelly stream banks. Winters on rocky coasts. Generally seen singly or in small groups. Casual inland during migration. Accidental to eastern North America.

Gray-tailed Tattler *Tringa brevipes* L 10" (25 cm)

Closely resembles Wandering Tattler; upperparts are slightly paler; barring on underparts finer and less extensive; whitish eyebrows are more distinct and meet on forehead. Diagnostic shorter nasal groove hard to see in field. **Juvenile** has less extensive gray on underparts; white flanks and belly; more pale spotting above.

VOICE: Best distinction is voice. Common call is a loud, ascending *too-weet,* similar to Common Ringed Plover.
RANGE: Asian species, regular spring and fall migrant on outer Aleutians, Pribilofs, and St. Lawrence Island; casual visitor to northern AK. Accidental in fall to WA and CA.

Common Sandpiper *Actitis hypoleucos* L 8" (20 cm)

Breeding adult is brown above with dark barring and streaking; white below; upper breast finely streaked. **Juvenile** and winter birds resemble Spotted Sandpiper. Note Common Sandpiper's longer tail; in juvenile, barring on edge of tertials extends along the entire feather. In flight, shows longer white wing stripe and longer white trailing edge; wingbeats not as shallow and rapid.

VOICE: Call in flight is a shrill, piping *twee-wee-wee.*
RANGE: Eurasian species, rare migrant, usually in spring, on the western Aleutians (has nested once on Attu Island), Pribilofs, and St. Lawrence Island, AK; casual to central Aleutians and Seward Peninsula.

Spotted Sandpiper *Actitis macularius* L 7½" (19 cm)

Striking in **breeding** plumage, with barred upperparts, spotted underparts, mostly pink bill. In winter, brown above, white below, sometimes with a few spots in vent region. **Juvenile** similar but with barred wing coverts; tertials plain but barred near tip. Resembles Common Sandpiper. Note Spotted Sandpiper's shorter tail; in flight, shows shorter white wing stripe, shorter white trailing edge. Both Spotted and Common Sandpipers fly with stiff, rapid, fluttering wingbeats. On the ground, both nod and teeter constantly. Generally seen singly; may form small flocks in migration.

VOICE: Calls include a shrill *peet-weet* and, in flight, a series of *weet* notes, lower pitched than Solitary Sandpiper.
RANGE: Common and widespread, found at sheltered streams, ponds, lakes, or marshes. Most winter in Central and South America. Rare in winter to southern edge of breeding range.

Wandering Tattler

breeding

juvenile

winter

gray flanks

heavily barred undertail

extensively barred below

breeding

dark gray flanks

rather short, dull yellow legs

slightly paler gray above than Wandering with more pale spotting and edging

breeding

mostly white to pure white vent and undertail

fine barring across chest and down sides

juvenile

white flanks

Gray-tailed Tattler

dark lateral breast patches

barred tertials

longer tail than Spotted

breeding

juvenile

juvenile

longer and bolder white wing stripe than Spotted

Common Sandpiper

flies with rapid, shallow, stiff wingbeats

dark lateral breast patches

1st winter

juvenile

Spotted Sandpiper

juvenile

barred wing coverts

pink bill with dark tip

heavily spotted underparts

breeding

mostly plain tertials, except near tip

Upland Sandpiper *Bartramia longicauda* L 12" (31 cm)

Small head, with large, dark, prominent eyes; long, thin neck, long tail, long wings. Legs yellow. Prefers fields, where often only its head and neck are visible above the grass. Also perches on posts on breeding grounds. In flight, wings are two-tone, blackish primaries contrast with mottled brown upperparts.

VOICE: Calls include a rolling *pulip pulip;* also a call like a wolf whistle, given in display flight on breeding grounds.

RANGE: Rare to fairly common. Nests in prairies, fallow fields, airports; migrants also found in farm fields and sod farms. Declining in some areas, especially in East and Pacific Northwest. Casual on West Coast and in the Southwest in migration.

Little Curlew *Numenius minutus* L 12" (30 cm)

Like a diminutive Whimbrel, with shorter and only slightly curved bill; note mostly pale lores, unlike the very similar, dark-lored, and probably extinct Eskimo Curlew (page 536). The slightly larger Eskimo Curlew had shorter legs, more heavily barred breast and flanks, bold Y-shaped marks on the flanks, and pale cinnamon wing linings.

VOICE: Calls include a musical *quee-dlee* and a loud *tchew-tchew-tchew.*

RANGE: Breeds in Russian Far East; winters mainly northern Australia. Casual fall vagrant to coastal central CA (four fall records involving both adult and juvenile birds) and one well-documented record in late spring for Gambell, St. Lawrence Island, AK. An accepted spring record of a flyover in western WA should perhaps be treated with some question.

Whimbrel *Numenius phaeopus* L 17½" (45 cm)

Bold, dark-striped crown; dark eye line extending through lores; long downcurved bill. In flight, North American *hudsonicus* shows dark rump and underwing; European *phaeopus,* casual vagrant to east (mainly Atlantic coast), white rump and underwing; Asian *variegatus,* rare migrant on islands in western AK, casual elsewhere in Pacific region, has whitish, variably streaked rump and underwing. Some argue that based on distinct plumage and genetic differences, the Palearctic subspecies should be split from North American Whimbrels; however, all subspecies give similar vocalizations. In North America, the western (*rufiventris,* if recognized) breeding population (migrating through Pacific states) is a little darker brown and larger than eastern birds.

VOICE: Call is a series of hollow whistles on one pitch.

RANGE: Fairly common; nests on open tundra; winters on coasts, including rocky shorelines of CA; most winter south of U.S. Generally rare in interior, except Great Lakes and in interior CA, where it can be abundant in spring in the Imperial Valley (accidental in winter, perhaps the only interior North American winter record); fairly common to common in the Antelope and Central Valleys of CA, where it is much scarcer in fall.

Upland Sandpiper

dark eye stands out in plain face

short, straight, mostly yellowish bill

adult

long neck

juvenile

long tail

juvenile

dark outer half of wing

pale lores

adult

Little Curlew

brown underwing

adult

barred flanks lack the Y-shaped marks found on Eskimo Curlew (likely extinct)

strong dark lateral crown and eye stripes

decurved bill

uniformly brown rump and tail

dark underwing

hudsonicus

juvenile

Whimbrel
hudsonicus

adult

whitish rump

whitish underwing coverts

phaeopus

intermediate *variegatus* has slightly darker rump and underwing than *phaeopus* but closer to that subspecies than to *hudsonicus*

variegatus

Bristle-thighed Curlew *Numenius tahitiensis* L 18" (46 cm)

Bright buff rump and tail (paler when worn) and extensive pattern of large buff spots on upperparts distinguish this species from slightly smaller Whimbrel. Bill is also slightly more strongly decurved near tip. Stiff feathers on the sides and flanks are very hard to see in the field.

VOICE: Main call is a loud whistled *chu-a-whit*, and all other vocalizations are completely different from Whimbrel.

RANGE: Winters South Pacific islands. Migration chiefly over water to breeding grounds in western AK. Rare spring migrant on Middleton Island and on Pribilof Islands, AK. Casual on St. Lawrence Island and on the Aleutians. Casual to West Coast in spring, after Pacific storms, as in May 1998 when more than a dozen were found from coastal WA to northern CA.

Long-billed Curlew *Numenius americanus* L 23" (58 cm)

Cinnamon-brown above, buff below, with very long, strongly down-curved bill. Lacks dark head stripes of Whimbrel. Males and **juveniles** have shorter bills. Cinnamon-buff wing linings and flight feathers, visible in flight, distinctive in all plumages. At rest closely resembles the smaller Marbled Godwit (page 186), if bill is hidden; note paler legs. Two weakly differentiated subspecies; more southerly breeding *americanus* is larger and longer billed than more northerly *parvus*, but differences are clinal. Still, bill length extremes are significant, and shorter-billed juveniles look much more Whimbrel-like in structure.

VOICE: Call is a loud musical, ascending *cur-lee*.

RANGE: Fairly common; nests in wet and dry uplands; in migration and winter found on wetlands and agricultural fields. Rare on Gulf Coast east of TX and on Atlantic coast from VA south. Casual farther north on Atlantic coast, and in Midwest.

Eurasian Curlew *Numenius arquata* L 22" (56 cm)

A large curlew that is heavily streaked below. The bill is long and strongly decurved. Distinguished from Long-billed Curlew by paler overall coloration and by white rump and wing linings, readily visible in flight; from Eurasian subspecies of Whimbrel by larger size, longer bill, and lack of dark stripes on head. Overall very similar to Far Eastern Curlew at rest, but paler, especially from lower belly to undertail; easily told in flight by white rump and wing linings.

RANGE: A widespread Eurasian species. Casual on East Coast in fall and winter. The seven records are from Newfoundland to Long Island, NY, except for one at Middle Cheyne Lake, NU.

Far Eastern Curlew *Numenius madagascariensis*

L 25" (64 cm) A large curlew with a very long, decurved bill. Closely resembles Eurasian Curlew, but overall browner, especially from lower belly to undertail coverts. In flight, underwing is heavily barred with dark; rump is the same color as remainder of upperparts.

RANGE: Despite scientific name, found nowhere near Madagascar. Breeds in Russian Far East, winters in the Sunda Isles, New Guinea, Australia, New Zealand; a few on mainland Southeast Asia. Casual in spring and early summer on Aleutians and Pribilofs, AK; accidental (Sept. 1984) from coastal BC.

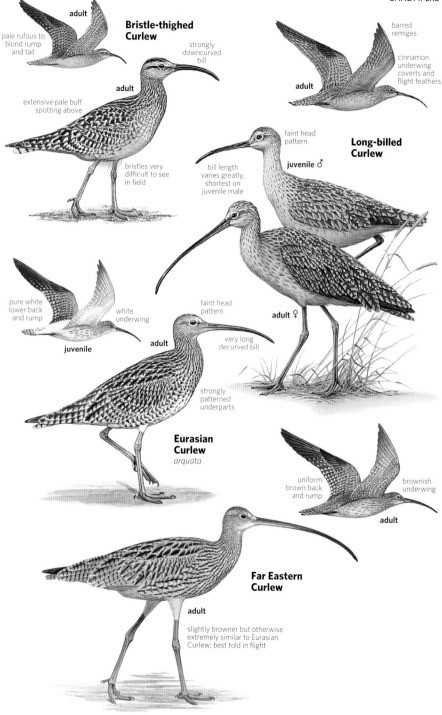

Bristle-thighed Curlew

adult

pale rufous to blond rump and tail

strongly downcurved bill

adult

extensive pale buff spotting above

bristles very difficult to see in field

barred remiges

adult

cinnamon underwing coverts and flight feathers

faint head pattern

Long-billed Curlew

juvenile ♂

bill length varies greatly, shortest on juvenile male

adult ♀

pure white lower back and rump

white underwing

juvenile

adult

faint head pattern

very long decurved bill

strongly patterned underparts

Eurasian Curlew
arquata

uniform brown back and rump

brownish underwing

adult

Far Eastern Curlew

adult

slightly browner but otherwise extremely similar to Eurasian Curlew; best told in flight

Black-tailed Godwit *Limosa limosa* L 16½" (42 cm)

Long, bicolored bill essentially straight. Tail mostly black, uppertail coverts white. **Breeding** *melanuroides* shows pale chestnut head and neck, heavily barred sides and flanks. East Coast records are of *islandica*, which in the breeding male is of a deeper and more extensive reddish color below. **Winter** birds gray above, whitish below. In all plumages, white wing linings and broad wing stripe are conspicuous in flight.

VOICE: Flight call is a rather nasal, mewing, quick *vi-vi-vi*.

RANGE: Eurasian species. Asian *melanuroides* is a rare spring migrant on western Aleutians; casual to Pribilofs and St. Lawrence Island, AK. Casual along Atlantic coast; accidental to eastern ON and LA.

Hudsonian Godwit *Limosa haemastica* L 15½" (39 cm)

Long, bicolored bill, slightly upcurved. Tail black, uppertail coverts white. **Breeding male** dark chestnut below, finely barred. **Female** larger and much duller. **Juvenile's** buff feather edges give upperparts a scaly look. Winter adult resembles Black-tailed Godwit; dark wing linings and narrower white wing stripe are distinctive in flight.

VOICE: Generally silent away from breeding grounds; call is a rather high-pitched and rising *pid-wid*.

RANGE: Breeding range not fully known. Migrates through central and eastern Great Plains in spring, much farther east in fall. Casual to Pacific states.

Bar-tailed Godwit *Limosa lapponica* L 16" (41 cm)

Long, slightly upcurved, bicolored bill. **Breeding male** reddish brown below; lacks heavy barring of Black-tailed Godwit. **Female** larger, much paler than male. In **winter** plumage, resembles Marbled Godwit but lacks cinnamon tones. Note also shorter bill and shorter legs. Black-and-white barred tail distinctive but hard to see at rest. **Juvenile** resembles winter adult but buffier overall. (Both subspecies shown in flight on page 209.) At least two subspecies of this Eurasian godwit occur in North America: *baueri*, which breeds in AK, has a heavily mottled rump and brown wing linings with white barring; European *lapponica* has whiter rump, white wing linings, brown-barred axillaries.

VOICE: Largely silent, except on breeding grounds; calls like Hudsonian, but lower in pitch.

RANGE: Alaskan *baueri* has been proven to fly nonstop in fall to southwestern Pacific wintering grounds; casual in fall on Pacific coast; also recorded in MA. European *lapponica* is a very rare migrant along the Atlantic coast.

Marbled Godwit *Limosa fedoa* L 18" (46 cm)

Long, bicolored bill, slightly upcurved. Tawny brown; mottled with black above, barred below. Barring much less extensive on **winter** birds and juveniles. Wing coverts also less patterned on juveniles. Longer legs than Bar-tailed Godwit. In flight, cinnamon wing linings and cinnamon on primaries and secondaries are distinctive. Legs black; gray in Long-billed Curlew (page 184).

VOICE: Flight call is a slightly nasal *kah-wek*; also a repeated *ga-wi-da*.

RANGE: Nests in grassy meadows, near lakes and ponds. Common on West Coast in winter, fairly common on TX Gulf Coast and in FL; rare but regular farther north in East on coast and in interior in migration.

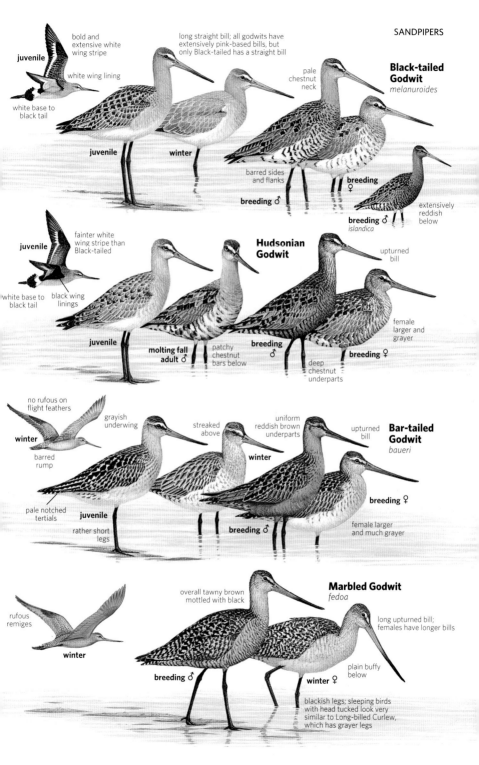

juvenile

bold and extensive white wing stripe

white wing lining

white base to black tail

long straight bill; all godwits have extensively pink-based bills, but only Black-tailed has a straight bill

pale chestnut neck

Black-tailed Godwit
melanuroides

juvenile

winter

barred sides and flanks

breeding ♀

breeding ♂

breeding ♂
islandica

extensively reddish below

juvenile

fainter white wing stripe than Black-tailed

black wing linings

white base to black tail

Hudsonian Godwit

upturned bill

female larger and grayer

juvenile

molting fall adult ♂

patchy chestnut bars below

breeding ♂

deep chestnut underparts

breeding ♀

no rufous on flight feathers

grayish underwing

winter

barred rump

streaked above

winter

uniform reddish brown underparts

upturned bill

Bar-tailed Godwit
baueri

pale notched tertials

juvenile

rather short legs

breeding ♂

breeding ♀

female larger and much grayer

Marbled Godwit
fedoa

overall tawny brown mottled with black

rufous remiges

winter

long upturned bill; females have longer bills

plain buffy below

breeding ♂

winter ♀

blackish legs; sleeping birds with head tucked look very similar to Long-billed Curlew, which has grayer legs

Ruddy Turnstone *Arenaria interpres* L 9½" *(24 cm)*

Striking black-and-white head and bib, black-and-chestnut back, orange legs mark this stout bird in **breeding** plumage. Female duller than **male.** Bib pattern, orange leg color retained in **winter** plumage; **juvenile** similar but scaly above. Uses bill to flip aside shells and pebbles. In flight, distinct pattern above identifies Ruddy and Black Turnstones.
VOICE: Distinctive call is a low-pitched, guttural rattle.
RANGE: Nests on coastal tundra; winters on mudflats, sandy beaches, rocky shores. Rare inland migrant except in Great Lakes region, where much more numerous. Rare to casual inland, migrant in West (more regular in fall); more numerous on Salton Sea, CA.

Black Turnstone *Arenaria melanocephala* L 9¼" *(24 cm)*

Black upperparts. In **breeding** plumage head is marked by white eyebrow and lore spot; white spotting visible on sides of neck and breast. Legs dark reddish brown in all plumages. Juvenile and **winter adult** are slate gray, lack lore spot and mottling. Juvenile is somewhat browner (paler), has small, round scapulars with thin and even pale edges.
VOICE: Calls include a guttural rattle, higher than Ruddy Turnstone.
RANGE: Breeds along intermediate coast of AK. Winters on rocky coasts. Very rare migrant to Salton Sea, CA; otherwise accidental migrant in the interior, east to WI and NT; casual to St. Lawrence Island, AK.

Rock Sandpiper *Calidris ptilocnemis* L 9" *(23 cm)*

Black patch on lower breast in **breeding** plumage; compare with Dunlin (page 198). Four subspecies, three in AK and *quarta* from Commander Islands and northern Kuriles may have occurred in western Aleutians, too. In *tschuktschorum*, upperparts black, edged with chestnut; in flight, shows white wing stripe, all-dark tail. Nominate subspecies breeding on Pribilofs larger, paler chestnut above, less black below, bolder white wing stripe. In all subspecies, long, slender bill slightly downcurved, base greenish yellow, legs greenish yellow. **Winter** birds variable; nominate subspecies distinctly paler gray. All but nominate separated with

See subspecies map, page 551

difficulty from Purple Sandpiper; note range, duller bill base and legs; Rock's flanks average more spotted (spots often spade-shaped), less streaked; more identification criteria needed, especially given plumage variation within basic Rock Sandpipers, even perhaps within the same subspecies. Told from Surfbird (page 190) by longer bill, smaller size, more patterned upperparts and breast.
VOICE: Gives a sharp *kwit* and a chattering call.
RANGE: Nests on tundra; most winter on rocky shores, often with Black Turnstones and Surfbirds. Migrates late in fall; Aleutian *couesi* is resident. Accidental inland in fall from BC (Atlin, 29 Oct. 1932, specimen).

Purple Sandpiper *Calidris maritima* L 9" *(23 cm)*

Long, slender bill slightly downcurved, base orange-yellow; legs orange-yellow. **Breeding adult** breast and flanks streaked with blackish brown. **Winter adult** closely resembles Rock Sandpiper.
VOICE: Call is a scratchy *keesh;* also a chatter.
RANGE: Winters on rocky shores, jetties, often with Ruddy Turnstones and Sanderlings. Migrates late fall. Rare fall migrant on Great Lakes. Casual elsewhere inland in East, in winter on Gulf Coast. Accidental in fall on north coast of AK (Point Barrow, specimen) and southwest UT.

Ruddy Turnstone

juvenile

winter

both turnstone species have upturned bills

harlequin head and breast pattern

breeding ♂

extensive rufous above; female duller

orange-red legs

striking wing pattern

breeding ♂

nearly solid slate black head and chest

white lore spot

white spots on sides breast

Black Turnstone

striking wing pattern

winter

winter

breeding

dark red legs, much darker than Ruddy

thin, slightly downcurved bill with greenish yellow base

dark gray above

winter

nominate subspecies larger and paler than other subspecies

dark postocular spot

winter

spots on flanks, often spade-shaped

yellowish legs

juvenile

blackish belly patch

Rock Sandpiper
tschuktschorum

Pribilofs breeding
ptilocnemis

breeding

juvenile

bright orange base to bill

overall dark slate gray

winter

Purple Sandpiper

heavily and extensively spotted below

orange legs

streaked sides and flanks

breeding

winter

Surfbird *Aphriza virgata* L 10" (25 cm)

Base of short, stout bill is yellow; legs yellowish green. **Breeding adult's** head and underparts are heavily streaked, spotted with dusky black; upperparts edged with white and chestnut; scapulars mostly rufous. **Winter adult** has a solid dark gray head and breast. **Juvenile's** head and breast flecked with white; back appears scaly. In flight, all plumages show a conspicuous black band at end of white tail and rump.

VOICE: Usually silent away from breeding grounds, apart from soft contact notes; occasionally gives a shrill, whistled *kee-wee-ah.*

RANGE: Nests on mountain tundra; winters along rocky beaches and reefs, usually in mixed flocks of Black Turnstones and other shorebirds. Has by far the greatest latitudinal winter range of any shorebird: Some found as far north as southeastern AK and some travel as far south as Tierra del Fuego. In spring, also found on sandy beaches. Very large number stage in fall at Bodega Bay, CA. Casual in spring on TX coast, also at Salton Sea, CA (mainly spring); several spring records elsewhere from the interior of CA; accidental FL and ON.

Great Knot *Calidris tenuirostris* L 11" (28 cm)

Larger than Red Knot, with longer bill. Compare to Surfbird and Rock Sandpiper (page 188). In **breeding** plumage, shows black breast, black flank pattern, and rufous scapulars. **Juvenile** has buffy wash and distinct spotting below; dark back feathers edged with rust. Resembles Red Knot in flight but primary coverts darker, wing bar fainter.

RANGE: Asian species, casual in spring (once in fall) to western AK. Accidental in fall to coastal OR and, remarkably, western WV.

Red Knot *Calidris canutus* L 10½" (27 cm)

Chunky and short legged. **Breeding adult** dappled brown, black, and chestnut above; buffy chestnut face and breast. In **winter,** back pale gray; underparts white. Distinguished from dowitchers (page 202) by shorter bill, paler crown, and, in flight, by whitish rump finely barred with gray. **Juveniles** similar to winter adults but have distinct spotting below, scaly-looking upperparts and when fresh, a light buff wash on breast.

VOICE: Mostly silent; sometimes gives a soft *ka-whit* in flight.

RANGE: Nests on tundra. Feeds along sandy beaches and mudflats. Eastern populations (subspecies *rufa*) crashed due to unregulated overharvesting of horseshoe crab eggs, used for fertilizer, on mid-Atlantic coast, a key staging and refueling location for Red Knots. Very rare interior migrant, though more regular from Great Lakes and Salton Sea, CA.

Sanderling *Calidris alba* L 8" (20 cm)

Palest sandpiper of **winter;** pale gray above, white below. Bill and legs black. Bold white wing stripe shows in flight. In **breeding** plumage (acquired late Apr. to late May), head, mantle, and breast are rusty. **Juveniles** blackish above, with pale edges near tips of feathers.

VOICE: Call is a *kip,* often in a series.

RANGE: Nests on tundra. Feeds on sandy beaches, running to snatch mollusks and crustaceans exposed by retreating waves, usually in flocks; sometimes found roosting on jetties with other shorebirds. Inland, a regular migrant in Great Lakes and northern Great Plains regions; also at Salton Sea, where a few winter. Rather rare elsewhere.

winter

short, sturdy bill, always with greenish yellow at base

bold white wing stripe

winter

white base to tail

extensive rufous on scapulars

Surfbird

juvenile has scaly pattern above and below

juvenile

arrow-shaped streaks on lower sides and flanks

breeding

bold arrow-shaped spots below

upperparts much more patterned than Red Knot

juvenile

Great Knot

extensive spotting below

extensive rufous on scapulars

larger and longer-billed than Red Knot

bold arrow-shaped spots below

breeding

breeding plumage suggests breeding plumage of Surfbird

whitish supercilium and dark eye line

plain gray back

winter

Red Knot

juvenile

winter

spotted chest

chunky body with short, straight bill

rufous breast and belly

breeding

pale vent and undertail coverts

subterminal dark edges with pale fringes on upperparts

pale buff wash below when fresh

whitish rump barred with gray

variably rufous upperparts, head, and breast

strongly patterned with black and white above

short, straight bill

Sanderling

juvenile

very pale above

breeding

no hind toe

winter

runs on sand

winter

dark marginal coverts

bold white wing stripe

PEEPS

These are seven species of small *Calidris* sandpipers that are difficult to identify. Collectively known as stints, by Old World English speakers, they are divided into four Old World and three New World species. Keys to identification include learning overall structure and feather topography, behavior, and the distribution patterns of each. It is essential to first learn our three common species before claiming a rare species.

Semipalmated Sandpiper *Calidris pusilla* L 6¼" (16 cm)
Black legs; tubular-looking, straight bill, of variable length. Easily confused with Western Sandpiper. In **breeding** birds, note that Semipalmated usually lacks spotting on flanks and shows only a tinge of rust on crown, ear patch, and scapulars. **Juveniles** are distinguished by stronger supercilium contrasting with darker crown and ear coverts and by more uniform upperparts. Some are brighter above than illustrated. **Winter** plumage (in North America seen most often on Gulf Coast in late Mar.) of these two species is very similar, but rounder-headed Semipalmated is plumper; note bill shape; face shows slightly more contrast; center of breast never shows the faint streaks visible on some winter Westerns. Semipalmated tends to pick at the water surface, often just at the water's edge.
VOICE: Call is a short *churk*.
RANGE: Abundant. A tundra breeder. A common migrant in eastern half of continent; generally a rare migrant in the West, south of BC; very rare in winter in south FL; no valid winter records elsewhere.

Western Sandpiper *Calidris mauri* L 6½" (17 cm)
Black legs; tapered bill, of variable length (longer in females); distal portion usually slightly drooped. Western has a blockier head, more attenuated body than Semipalmated. In **breeding** plumage, Western has arrow-shaped spots along sides, rufous at base of scapulars, and a bright rufous wash on crown and ear patch. **Juvenile** is distinguished from juvenile Semipalmated by less prominent supercilium, paler crown and face, and brighter rufous edges on back and inner scapulars. **Winter** plumage is very similar to Semipalmated. Note structure (especially bill shape). In North America, especially away from south FL, any winter-plumaged individual in fall or winter is likely to be this species.
VOICE: Call is a raspy *jeet*.
RANGE: A tundra breeder. Scarce fall-only migrant in eastern Canada and New England. Surprisingly rare in upper Midwest and casual on northern Great Plains.

Least Sandpiper *Calidris minutilla* L 6" (15 cm)
Note small size and short, thin, slightly downcurved bill. Always darker above than Western and Semipalmated Sandpipers. Legs are yellowish, but can appear dark in poor light or when smeared with mud. In **winter** plumage, has streaked brown breast band. **Juvenile** has strong buffy wash across breast. Feeds in a variety of wet habitats, but forages less in the water, often preferring to feed back from the shore's edge.
VOICE: Call is a high plaintive *kreee*.
RANGE: A tundra and taiga breeder; the likely peep inland in winter.

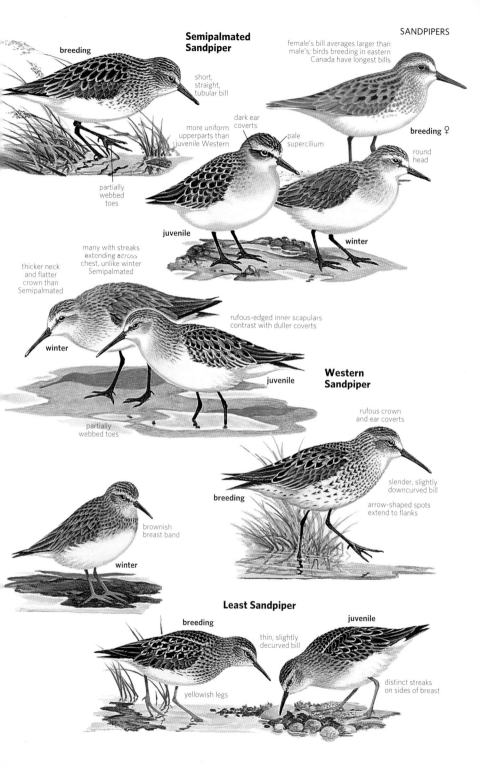

Semipalmated Sandpiper

breeding

short, straight, tubular bill

female's bill averages larger than male's; birds breeding in eastern Canada have longest bills

breeding ♀

round head

more uniform upperparts than juvenile Western

dark ear coverts

pale supercilium

partially webbed toes

juvenile

winter

many with streaks extending across chest, unlike winter Semipalmated

thicker neck and flatter crown than Semipalmated

rufous-edged inner scapulars contrast with duller coverts

winter

juvenile

Western Sandpiper

partially webbed toes

rufous crown and ear coverts

breeding

slender, slightly downcurved bill

arrow-shaped spots extend to flanks

brownish breast band

winter

Least Sandpiper

breeding

thin, slightly decurved bill

juvenile

yellowish legs

distinct streaks on sides of breast

RARE STINTS
Four Old World species have been found in North America. Red-necked Stint is the most regular, and of the remaining three, Temminck's and Long-toed Stints are almost unknown in North America away from AK. With any claim of a rare stint it is essential to get photographs. Even in migration, shorebirds hold feeding territories, and if flocks are disturbed, they soon return and sort themselves out. If the potential rarity can't be found again, chances are it was a more common species.

Red-necked Stint *Calidris ruficollis* L 6¼" (16 cm)
Rufous on throat and upper breast may be pale and indistinct; look for necklace of dark streaks on white lower breast. **Juvenile** distinguished from Little Stint by plainer wing coverts and tertials, more uniform crown pattern, and plainer breast sides.
RANGE: Asian species, regular migrant on western AK coast and islands. Breeding documented on Seward Peninsula coast of AK; breeding range on map is conjectural. Casual migrant on both coasts; accidental in interior.

Little Stint *Calidris minuta* L 6" (15 cm)
Breeding birds brightly fringed with rufous above; throat and underparts white (more suffused with color in later summer adults), with bright buff wash and bold spotting on sides of breast. Redder above than Western and Semipalmated Sandpipers (page 192); compare also with Red-necked Stint. **Juvenile** best distinguished from juvenile Red-necked by extensively black-centered wing coverts and tertials, usually edged with rufous; also note split supercilium and streaking on sides of chest.
RANGE: Eurasian species, very rare on western AK islands in spring and fall; casual on East and West Coasts, and accidental elsewhere.

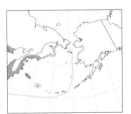

Long-toed Stint *Calidris subminuta* L 6" (15 cm)
Distinguished in all plumages from Least Sandpiper by dark forehead, pinching off prominent white supercilium before bill (dark forms J-shape on its side); also shows a split supercilium. Median coverts are white edged; note also greenish base of lower mandible; yellowish legs.
VOICE: Gives a lower-pitched call than Least Sandpiper.
RANGE: Asian species, casual in spring on St. Lawrence Island, AK, and the Pribilofs; can be fairly common on the outer Aleutians. Accidental in fall from coastal OR and CA.

Temminck's Stint *Calidris temminckii* L 6¼" (16 cm)
White outer tail feathers distinctive in all plumages. **Breeding adult** resembles the larger Baird's Sandpiper in plumage and shape (page 196), but legs are yellow or greenish yellow; note Baird's dark legs and distinct primary tip projection past tertials. In **juvenile,** upperpart feathers have a dark subterminal edge, buffy fringe.
VOICE: Call is a very distinctive, repeated, rapid dry rattle.
RANGE: Eurasian species; rare spring and fall migrant on Pribilofs, Aleutians, and St. Lawrence Island, AK. Accidental northern AK and coastal Pacific Northwest.

tertials and coverts rather dull in juvenal and breeding plumage

juvenile

breeding

breeding

some have duller rufous that is confined to lower throat

rich brick red throat

dark streaks on breast

Red-necked Stint

coverts and tertials in juvenal and breeding plumage with darker centers than Red-necked and with colored edges

Little Stint

white mantle V bolder than Red-necked

split supercilium

breeding

greenish lower mandible base on straight bill

bold white mantle V

dark spots on sides of breast with underlying buffy wash

juvenile

juvenile

yellow legs

Long-toed Stint

distinct supercilium broadens behind eye

split supercilium

white outer tail feathers

juvenile

Temminck's Stint

buffy wash with spots on sides of neck and breast

breeding

dark subterminal fringes

dark breast

juvenile

short primary projection

breeding

general coloration suggests Baird's

greenish yellow legs

white outer tail feathers

White-rumped Sandpiper *Calidris fuscicollis* L 7½" (19 cm)

Long primary tip projection beyond tertials and tail on standing bird. Similar to Baird's Sandpiper structurally, but grayer overall and usually has an entirely white rump. In **breeding** plumage, streaking extends to flanks. **Juvenile** shows rusty edges on crown and back. In winter, head and neck are dark gray, giving a hooded look.

VOICE: Call note is a very high-pitched insect-like *jeet.*

RANGE: Feeds on mudflats. Fairly common; common spring and rare fall migrant through Great Plains. Uncommon along East Coast; more numerous in fall in Northeast. Casual spring (mainly) and fall (all records are adults) in West. Spring migration late, extends to late June; juveniles don't migrate south until late Sept. Winters in southern South America; no valid North American midwinter records.

Baird's Sandpiper *Calidris bairdii* L 7½" (19 cm)

Long primary tip projection beyond tertials and tail on standing bird gives the bird a horizontal profile. Buff-brown above and across breast. Pale fringing on **juvenile's** back gives a scaly appearance. Distinguished from White-rumped by more buffy brown color and uniform plumage; in flight by dark rump. Distinguished from Least Sandpiper (page 192) by much larger size, longer and straighter bill, and primary projection.

VOICE: Call is a low raspy *kreep,* similar to call of Pectoral Sandpiper, but less rich.

RANGE: Uncommon to common; found on upper beaches and inland on lakeshores, wet fields, even beaches. Nests on tundra. Primary migration is through Great Plains. Uncommon (usually juveniles) to both coasts in fall. Winters in South America; accidental in North America in midwinter.

Spoon-billed Sandpiper *Eurynorhynchus pygmeus*

L 6¼" (16 cm) Spoon-shaped bill is diagnostic and gives bill a longer look, but spoon is sometimes hard to see with clarity at a distance; beware of other small *Calidris* with mud on bill tip. In **breeding** plumage is easily mistaken for Red-necked Stint (page 194), apart from bill. **Juvenile** has darker cheek and more contrasting supercilium than Red-necked Stint. On winter grounds in Southeast Asia, feeds farther out into the water than Red-necked Stints; probes like Western Sandpiper but often with a side-to-side motion.

RANGE: Critically endangered species. Recent estimated population on breeding grounds of fewer than 500 individuals represents a greater than 50 percent reduction in the last 15 years. Breeds on coast of Russian Far East; winters coastal Southeast Asia, especially Myanmar; casual migrant (about six records) to western and northern AK. One record of a fall migrant, a breeding-plumaged adult, from Vancouver region, BC (30 July to 3 Aug. 1978).

Broad-billed Sandpiper *Limicola falcinellus* L 7" (18 cm)

Plump body, short legs, and long, broad-based bill with drooped tip give it a distinctive profile. Note also distinctive split supercilium.

VOICE: Call is a dry and high-pitched buzzy trill; also shorter calls.

RANGE: Eurasian species, casual fall migrant on western and central Aleutians (once to Pribilofs). Accidental in fall from coastal NY. All sightings so far of juveniles.

juveniles not yet recorded in West, not seen in Northeast until late Sept.

White-rumped Sandpiper

juvenile

breeding

dull rufous edges to scapulars

white rump normally only visible in flight

juvenile

fall migrant adults often in transitional plumage; adults appear in Northeast by early Aug.

fall-molting adult

streaking extends to flanks

long wings and primary projection

like White-rumped, long wings extend beyond tail

shaped like White-rumped, but browner

whitish fringe on upperparts

juvenile

Baird's Sandpiper

breeding

juvenile

dark cheek

distinct supercilium

breeding plumage similar to breeding plumage of Red-necked Stint

very distinctive spoon-shaped bill harder to discern at a distance

juvenile

breeding

Spoon-billed Sandpiper

split supercilium in all plumages

Broad-billed Sandpiper

breeding

broad-based bill with drooped tip

short legs

bright and strongly patterned on upperparts

juvenile

juvenile

See subspecies map, page 551

Dunlin *Calidris alpina* L 8½" (22 cm)

Medium size; long bill, curved at tip; in flight, shows dark center to rump. **Breeding** plumage with reddish upperparts, black belly. Subspecies differ in size, structure, and breeding plumage: more ventrally streaked in *hudsonia*, found in eastern North America; *pacifica* in western AK and Pacific region; similar *arcticola* (breeds northern AK) and *sakhalina* (migrant western AK) winter in Asia. Three subspecies from Greenland and western Palearctic (*arctica*, *schinzii*, and *alpina*) are smaller, shorter billed, and darker above in breeding plumage; these three migrate south and then molt (adults and juveniles), unlike East Asian and North American subspecies, which molt in areas closer to the breeding grounds and then migrate south starting in mid-Sept. Both *arctica* and *alpina* are accidental on East Coast. Compare to Rock Sandpiper (page 188). In **winter** plumage, upperparts and chest brownish gray. **Juveniles** rusty above, spotted below.
VOICE: Call is a harsh, reedy *kree*.
RANGE: Nests on tundra. Otherwise found on mudflats, marshes, and pond edges. Rare on western Great Plains.

Curlew Sandpiper *Calidris ferruginea* L 8½" (22 cm)

Long, downcurved bill. In **breeding** plumage, rich chestnut underparts, mottled chestnut back, whitish area at bill base. Female slightly paler than male. Many sightings are of birds in patchy spring plumage or molting to winter plumage, showing grayer upperparts and partly white underparts. **Juvenile** appears scaly above; shows rich buff wash across breast; young birds seen in southern Canada and U.S. are in full juvenal plumage; compare with winter and shorter-legged Dunlins as well as juvenile and transitional Stilt Sandpipers. A few spring records from the Salton Sea, CA, have been in winterlike plumage (probably immatures). White rump and underwing is conspicuous in flight.
VOICE: Call is a soft, rippling *chirrup*.
RANGE: Eurasian species. Has bred on northern slope of AK. Rare migrant on East Coast. Very rare or casual on West Coast, chiefly in fall. Casual elsewhere in North America, but nearly annual in Great Lakes region.

Stilt Sandpiper *Calidris himantopus* L 8½" (22 cm)

Breeding adult has pale eyebrow, chestnut on head, slender, slightly downcurved bill, and heavily barred underparts. **Winter adult** is grayer above, whiter below; **juvenile** has more sharply patterned upperparts; the two resemble Curlew Sandpiper, but note straighter bill, yellow-green legs, and, in flight, lack of prominent wing stripe, paler tail; early juvenile has a buffy wash on breast. Feeds like dowitchers, with which it often associates, but note smaller size and disproportionately longer legs.
VOICE: Call is a low, hoarse *querp*.
RANGE: Uncommon breeder on northern slopes of AK and YT. Common migrant through Great Plains. Generally a rare migrant otherwise in West, except more numerous at Salton Sea, CA, where some winter. Fairly common in the Mississippi River Valley; fairly common farther east in fall, rare in spring.

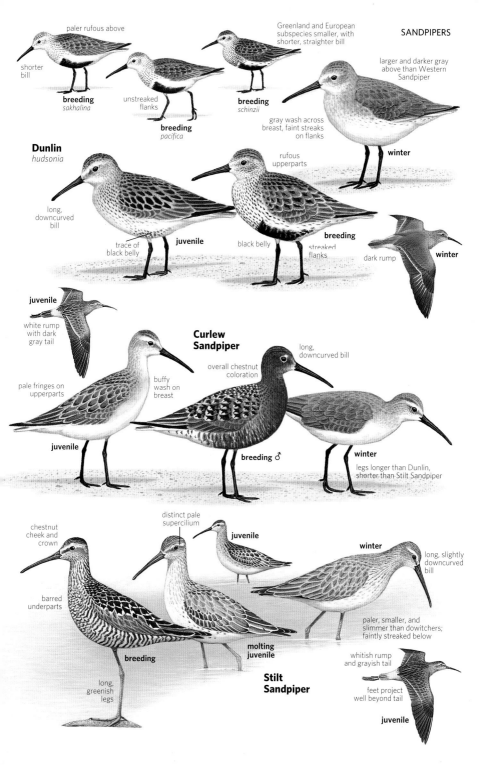

paler rufous above

Greenland and European subspecies smaller, with shorter, straighter bill

shorter bill

breeding
sakhalina

unstreaked flanks

breeding
pacifica

breeding
schinzii

gray wash across breast, faint streaks on flanks

larger and darker gray above than Western Sandpiper

winter

Dunlin
hudsonia

rufous upperparts

long, downcurved bill

trace of black belly

juvenile

black belly

breeding

streaked flanks

dark rump

winter

juvenile

white rump with dark gray tail

pale fringes on upperparts

Curlew Sandpiper

buffy wash on breast

overall chestnut coloration

long, downcurved bill

juvenile

breeding ♂

winter

legs longer than Dunlin, shorter than Stilt Sandpiper

chestnut cheek and crown

distinct pale supercilium

juvenile

winter

long, slightly downcurved bill

barred underparts

breeding

molting juvenile

Stilt Sandpiper

paler, smaller, and slimmer than dowitchers; faintly streaked below

whitish rump and grayish tail

long, greenish legs

feet project well beyond tail

juvenile

Pectoral Sandpiper *Calidris melanotos* L 8¾" (22 cm)

Prominent streaking on breast, darker in **male,** contrasts sharply with clear white belly. Male is larger than **female. Juvenile** has buffy wash on streaked breast, brighter rusty crown; compare with Baird's Sandpiper (page 196) and especially with juvenile Sharp-tailed Sandpiper.
VOICE: Call is a rich, low *churk.*
RANGE: Often feeds in wet meadows, marshes, pond edges. Common in Midwest; uncommon to fairly common on East Coast, where most numerous in fall. Scarcer in Great Plains. Scarcer farther west, where mainly juveniles are seen in fall. Breeds west on Arctic coast to Taymyr Peninsula of central Arctic Russia; winters in South America. Most return through Midwest, thus the world's farthest traveler.

Sharp-tailed Sandpiper *Calidris acuminata* L 8½" (22 cm)

Most sightings are **juveniles,** distinguished from juvenile Pectoral Sandpiper by white eyebrow that broadens behind the eye; bright buffy breast lightly streaked on upper breast and sides; streaked undertail coverts; and brighter rufous cap and edging on upperparts. **Breeding adult** is similar to juvenile, but more spotted below with dark chevrons on flanks; also distinct white eye ring.
VOICE: Call is a mellow, two-note whistle.
RANGE: Breeds in Russian Far East; casual spring and fairly common fall migrant (mostly juveniles) in western AK; rare fall migrant along entire Pacific coast. Casual across rest of continent, almost all in fall. Accidental in spring and winter in CA.

Buff-breasted Sandpiper *Tryngites subruficollis*

L 8¼" (21 cm) Dark eye stands out prominently on buffy face; underparts paler buff with dark spotting on the sides of the breast; feathering extends out lower mandible; legs orange-yellow. In flight, shows flashy white wing linings; also seen when wings raised in display on breeding grounds and even in spring migration. **Juveniles** are paler below, with scaly white fringing to feathers above.
VOICE: Mostly silent away from breeding grounds.
RANGE: Nests on tundra. Prefers plowed fields, turf farms, and wet rice fields. Migrates through the interior of the continent. In fall, very rare on the West Coast, uncommon in the East; casual in interior West. Most sightings west or east of eastern Great Plains are of juveniles. Winters in South America.

Ruff *Philomachus pugnax* ♂ L 12" (31 cm) ♀ L 10" (25 cm)

Most **breeding males** acquire dramatic ruffs in colors that range from black to rufous to white; most males are distinctly larger than females. **Female** lacks ruff and has a variable amount of black on underparts. Both sexes have a plump body, small head, and white underwing. Leg color may be greenish, yellow, orange, or red. **Juvenile** is buffy below, has prominently fringed feathers on upperparts. In flight, the U-shaped white band on rump is distinctive in all plumages.
VOICE: Mostly silent away from breeding grounds.
RANGE: Old World species. Rare migrant in western AK, along West and East Coasts, in Great Lakes region. Most birds seen on Atlantic coast are adults; most seen Apr. to May and July to Aug. Casual elsewhere. Very rare in winter in CA; once bred in AK (Point Lay, 1976).

Pectoral Sandpiper

breeding ♂

darker breast

breeding ♀

long wings and primary projection

has pale Vs on back and scapulars, is less scaly appearing above than juvenile Baird's

juvenile

pectoral band of streaks

yellowish legs

juvenile

breeding adult

ruddy crown and bold white eye ring

extensive dark chevrons on sides and flanks

streaks on undertail coverts

juvenile

bold supercilium

extensive buff on breast below streaking

broader rufous tertial edges than Pectoral

streaks on sides continue faintly across upper breast

juvenile Sharp-tailed (center) with juvenile Pectorals

Sharp-tailed Sandpiper

juvenile

juvenile

displaying adult

dark crescent

scaly upperparts

whitish underwing

juvenile

dark eye stands out in blank face

Buff-breasted Sandpiper

feathering extends out lower mandible

juvenile

spots on sides of breast

breeding adult

yellow legs

buffy underparts

Ruff

displaying breeding males

♀

♀

scaly upperparts

buffy neck and breast

white U-shaped band

juvenile ♀

summer molting ♂

back feathers often raised

juvenile ♀

small head and plump body

whitish lores and forehead

summer molting ♀

leg color variable: greenish to bright orange

winter ♂

often with colored bill base

DOWITCHERS
Medium-size, chunky, dark shorebirds, dowitchers have long, straight bills and distinct pale eyebrows. Feeding in mud or shallow water, they probe with a rapid jabbing motion. Dowitchers in flight show a white wedge from barred tail to middle of back. Calls are best distinguishing feature. By plumage, juveniles are easiest to identify and winter birds are hardest. Both species give the same song, a rapid *di di da doo,* year-round.

See subspecies map, page 551

Short-billed Dowitcher *Limnodromus griseus* L 11" (28 cm)
In flight, tail usually looks paler than Long-billed Dowitcher. **Breeding** plumage varies among the three subspecies: *griseus* (breeds in northeastern Canada), *hendersoni* (central and western Canada), and *caurinus* (AK). Unlike Long-billed, most show some white on the belly, especially *griseus,* which also has a heavily spotted breast and may have densely barred flanks; *caurinus* variable but similar to *griseus.* In *hendersoni,* which may be mostly reddish below, foreneck much less heavily spotted than Long-billed; sides have less or no barring; upperparts brighter. In all subspecies, **juvenile** brighter above, buffier and more spotted below than juvenile Long-billed; tertials and visible greater wing coverts have broad reddish buff edges and internal bars, loops, or stripes. **Winter** birds brownish gray above, white below, with gray breast; at close range note fine dark speckling on and below the breast on many.
VOICE: Call is a mellow *tu tu tu,* repeated in a rapid series in alarm call.
RANGE: Common in migration along the Atlantic coast (*griseus*); from the eastern plains to Atlantic coast from NJ south (*hendersoni*); and along the Pacific coast; in small numbers in interior CA and western NV (*caurinus*). A few *griseus* are seen on eastern Great Lakes and Gulf Coast in late spring; a few *hendersoni* north to New England in fall only. Fall migration begins earlier than Long-billed, usually in late June or early July for adults; early Aug. for juveniles. Migrant juveniles are seen through early Oct. Accidental inland in winter (Salton Sea).

Long-billed Dowitcher *Limnodromus scolopaceus*
L 11½" (29 cm) **Male's** bill is no longer than Short-billed Dowitcher, **female's** is longer. In flight, tail usually looks darker than Short-billed. **Breeding adult** is entirely reddish below; foreneck heavily spotted; sides usually barred. Bold white scapular tips in spring help separate this species from Short-billed. **Juvenile** is darker above, grayer below than Short-billed; tertials and greater wing coverts are plain, with thin gray edges and rufous tips; some birds show two pale spots near the tips. In **winter**, breast is unspotted and more extensively darker than most Short-billed. Adult Long-billeds go to favored locations in late summer to molt; Short-billeds molt on winter grounds.
VOICE: Call is a sharp, high-pitched *keek,* given singly or in a rapid series.
RANGE: Common in migration in western half of continent; less common in the East in fall, rare in spring. Fall migration begins later than Short-billed, in mid-July (West) or late July (East). Juveniles migrate later than adults; rare before Sept. Dowitchers seen inland after mid-Oct. are almost certainly Long-billed.

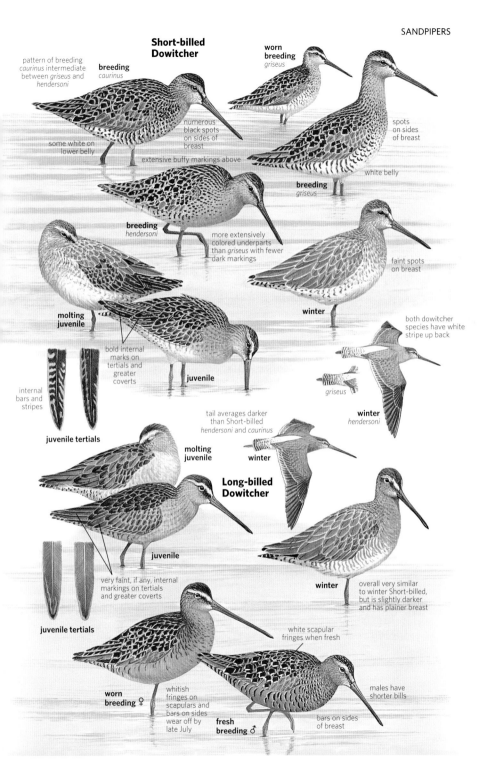

Short-billed Dowitcher

pattern of breeding *caurinus* intermediate between *griseus* and *hendersoni*

breeding *caurinus*

worn breeding *griseus*

numerous black spots on sides of breast

some white on lower belly

extensive buffy markings above

spots on sides of breast

breeding *griseus*

white belly

breeding *hendersoni*

more extensively colored underparts than *griseus* with fewer dark markings

faint spots on breast

molting juvenile

winter

both dowitcher species have white stripe up back

griseus

bold internal marks on tertials and greater coverts

internal bars and stripes

juvenile

juvenile tertials

winter *hendersoni*

tail averages darker than Short-billed *hendersoni* and *caurinus*

molting juvenile

winter

Long-billed Dowitcher

juvenile

very faint, if any, internal markings on tertials and greater coverts

winter

overall very similar to winter Short-billed, but is slightly darker and has plainer breast

juvenile tertials

white scapular fringes when fresh

worn breeding ♀

whitish fringes on scapulars and bars on sides wear off by late July

fresh breeding ♂

males have shorter bills

bars on sides of breast

Jack Snipe *Lymnocryptes minimus* L 7" *(18 cm)*

Small, chunky Eurasian species. Secretive, reluctant to flush. Flight is low, short, fluttery, on rounded wings. Bobs while feeding. Pale base to short bill, pale split eyebrow stripes with no median crown stripe, broad goldish back stripes, streaked flanks, pale underparts.

RANGE: Three late fall records for CA, two for OR, and one for Labrador. In AK, single spring record for St. Lawrence Islands and Pribilofs; once in fall for Homer.

Wilson's Snipe *Gallinago delicata* L 10¼" *(26 cm)*

Stocky, with very long bill; boldly striped head; barred flanks.

VOICE: Usually seen when flushed, as it gives a harsh two-syllable *ski-ape* call in rapid, twisting flight. Where breeding, males deliver loud *wheet* notes from perches. In swooping display flight, vibrating outer tail feathers make quavering hoots, like song of Boreal Owl.

RANGE: Nests in wet meadows, bogs, swamps; in migration and winter, also marshes, lakeshores, and muddy fields and ditches.

Common Snipe *Gallinago gallinago* L 10½" *(27 cm)*

Paler, buffier color overall than Wilson's Snipe; broader white trailing edge to secondaries, fainter flank markings; paler white-striped underwing.

VOICE: Flight-display notes of the male are lower pitched than Wilson's; harsh *ski-ape* flight call like Wilson's.

RANGE: Eurasian species, regular migrant to the western Aleutians where it has bred and been found casually in winter; rarer to the central Aleutians, Pribilofs; casual to St. Lawrence Island, AK. Accidental to Labrador.

Pin-tailed Snipe *Gallinago stenura* L 10" *(26 cm)*

Chunkier, shorter billed, and shorter tailed than Common Snipe. On ground, note barred secondary coverts and even-width pale edges on inner and outer webs of scapulars for a scalloped look. In flight, note buffy secondary covert panel, uniformly dark underwing, no pale edge to secondaries, distinct foot projection past tail. Larger Swinhoe's Snipe *(G. megala)* from Asia (unrecorded in North America) perhaps not separable in field. In hand, razor-thin outer tail feathers are diagnostic.

VOICE: Call is high, ducklike *squak.*

RANGE: Breeds in Siberia and Russian Far East; winters chiefly in Southeast Asia. Three spring specimen records from Attu Island, Aleutians.

American Woodcock *Scolopax minor* L 11" *(28 cm)*

Chunky, with long bill, barred crown, large eyes set high in head. Rounded wings. Nocturnal, secretive. Flies up abruptly; wings make a twittering sound. Males give an elaborate flight display; visible at dusk and dawn.

VOICE: Call, heard mainly in spring, is a nasal *peent.* During display also gives chirping sounds.

RANGE: Common; nests in moist woodlands. In mild winters, a few are found farther north than mapped. Casual to CO and NM; accidental to CA.

Jack Snipe

some dark oily
green above

stripes create
Vs on mantle

split pale
eyebrow

streaked
flanks

short bill
with pale
base

bobs while
feeding

displaying

**Wilson's
Snipe**

underwing

faint
trailing
edge

dark
underwing

strongly
striped head

bold pale stripes
on upperparts

strongly
striped head

heavily
barred sides
and flanks

bolder
white
trailing
edge

Common Snipe
gallinago

underwing

whitish
panel on
underwing
coverts

warmer
background
color to head,
breast, and
sides

bold pale stripes
on upperparts

flank barring
fainter than
Wilson's

**Pin-tailed
Snipe**

buffy
panel on
coverts

rather
short
bill

chunky,
with short,
rounded
wings

**American
Woodcock**

dark bands on
crown and nape

large eyes

even-width, pale
edges on scapulars,
unlike Wilson's
and Common

rich buffy
underparts

diagnostic razor-thin,
pin-shaped outer tail
feathers not visible in field

tail

darker underwing than
Common with no pale trailing
edge, much like Wilson's

underwing

PHALAROPES

These elegant shorebirds have partially lobed feet and dense, soft plumage. Feeding on the water, phalaropes often spin like tops, stirring up larvae, crustaceans, and insects. Females, larger and more brightly colored than the males, do the courting; males incubate the eggs and care for the chicks. In fall, adults and juveniles (particularly Wilson's and Red Phalaropes) rapidly molt to winter plumage; many are seen in transitional plumage farther south.

Wilson's Phalarope *Phalaropus tricolor* L 9¼" (24 cm)

Long, thin bill is distinctive. In **winter** plumage, upperparts are plain gray, lacks distinct dark ear patch; legs yellowish. Briefly held **juvenal** plumage resembles winter adult but back is browner with buffy edge to feathers, breast buffy. In flight, white uppertail coverts, whitish tail, and absence of white wing stripe distinguish juvenile and winter birds from other phalaropes.

VOICE: Calls include a hoarse *wurk* and other low, croaking notes.

RANGE: Chiefly an inland phalarope, nesting on grassy borders of shallow lakes, marshes, and reservoirs. Feeds as often on land as on water. Common to abundant in western North America; uncommon to rare in the East; rare in southern CA in winter.

Red-necked Phalarope *Phalaropus lobatus* L 7¾" (20 cm)

Chestnut on neck distinctive in **breeding female,** duller in **male.** Both have dark back with bright buff stripes along sides; bill shorter than Wilson's Phalarope, thinner than Red Phalarope. **Winter** birds are blue-gray above with whitish stripes; dark patch extends back from eye. In flight, show white wing stripe, whitish stripes on back, dark central tail coverts. Fresh **juvenile** resembles winter adult but is blacker above, with bright buff stripes.

VOICE: Call, a high, sharp *kit,* is often given in a series.

RANGE: Breeds on Arctic and subarctic tundra; winters chiefly at sea in Southern Hemisphere. Common along and off West Coast and in western portion of the interior West during migration; rare in Midwest and East; uncommon off East Coast; more numerous off ME and Maritime Provinces.

Red Phalarope *Phalaropus fulicarius* L 8½" (22 cm)

Bill shorter and much thicker than other phalaropes; yellow with black tip in breeding adult, usually all-dark in juvenile and winter adult. **Female** in breeding plumage has black crown, white face, chestnut red underparts. **Male** is duller. **Juvenile** resembles male but is much paler below; juveniles seen in southern Canada and the U.S. are **molting** to winter plumage; more closely resemble Red-necked Phalaropes. **Winter** bird is pale gray above. In flight, shows a bolder white wing stripe than Red-necked and dark central tail coverts.

VOICE: Call is a sharp *keip,* higher pitched than Red-necked.

RANGE: Breeds on Arctic shores; winters at sea. Irregularly common off West Coast during fall migration; rare to very rare inland, chiefly in fall. Very uncommon off much of East Coast, more numerous off ME and Maritime Provinces.

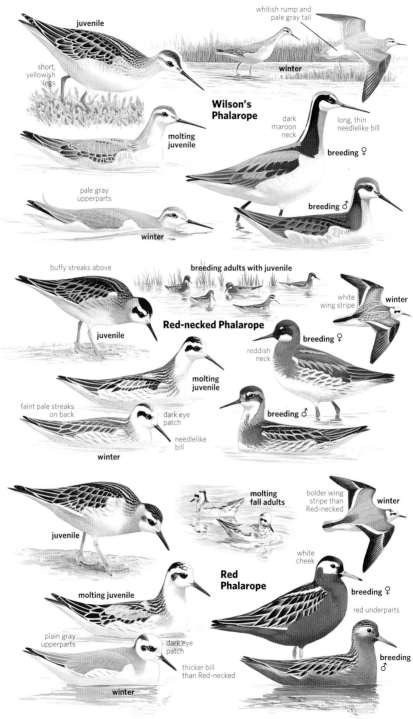

Wilson's Phalarope

juvenile

short, yellowish legs

whitish rump and pale gray tail

winter

molting juvenile

dark maroon neck

long, thin needlelike bill

breeding ♀

pale gray upperparts

winter

breeding ♂

Red-necked Phalarope

buffy streaks above

breeding adults with juvenile

juvenile

white wing stripe

winter

breeding ♀

reddish neck

molting juvenile

faint pale streaks on back

dark eye patch

needlelike bill

breeding ♂

winter

Red Phalarope

juvenile

molting fall adults

bolder wing stripe than Red-necked

winter

white cheek

molting juvenile

breeding ♀

red underparts

plain gray upperparts

dark eye patch

breeding ♂

thicker bill than Red-necked

winter

Plovers

whitish
uppertail coverts

**Snowy
Plover**

**Piping
Plover**

winter

**Little
Ringed
Plover**

breeding

**Semipalmated
Plover**

juvenile

juvenile

**Common
Ringed
Plover**

**Lesser
Sand-Plover**

winter

rufous rump and
uppertail coverts

**Wilson's
Plover**

Killdeer

**Mountain
Plover**

winter

**Eurasian
Dotterel**

juvenile

brown
rump

**Pacific
Golden-Plover**

juvenile

faint
wing stripe

brown
rump

**American
Golden-Plover**

juvenile

faint
wing stripe

**European
Golden-Plover**

juvenile

white
rump

**Black-bellied
Plover**

juvenile

bold wing
stripe

Godwits and Curlews

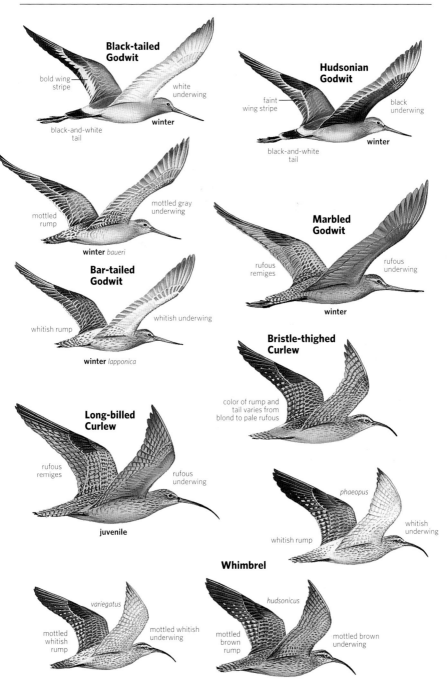

Black-tailed Godwit

bold wing stripe

white underwing

black-and-white tail

winter

Hudsonian Godwit

faint wing stripe

black underwing

black-and-white tail

winter

mottled gray underwing

mottled rump

winter *baueri*

Bar-tailed Godwit

whitish rump

whitish underwing

winter *lapponica*

Marbled Godwit

rufous remiges

rufous underwing

winter

Bristle-thighed Curlew

color of rump and tail varies from blond to pale rufous

Long-billed Curlew

rufous remiges

rufous underwing

juvenile

phaeopus

whitish underwing

whitish rump

Whimbrel

variegatus

mottled whitish rump

mottled whitish underwing

hudsonicus

mottled brown rump

mottled brown underwing

Tringa and Other Sandpipers

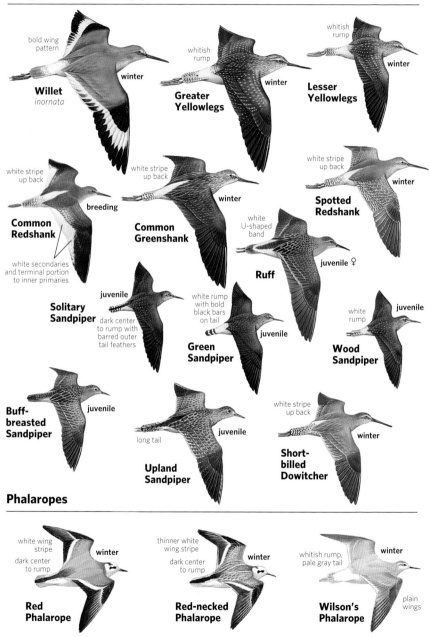

bold wing pattern

Willet
inornata

winter

whitish rump

Greater Yellowlegs

winter

whitish rump

Lesser Yellowlegs

winter

white stripe up back

Common Redshank

breeding

white secondaries and terminal portion to inner primaries

white stripe up back

Common Greenshank

winter

white U-shaped band

Ruff

juvenile ♀

white stripe up back

Spotted Redshank

winter

Solitary Sandpiper

juvenile

dark center to rump with barred outer tail feathers

white rump with bold black bars on tail

Green Sandpiper

juvenile

white rump

Wood Sandpiper

juvenile

Buff-breasted Sandpiper

juvenile

long tail

Upland Sandpiper

juvenile

white stripe up back

Short-billed Dowitcher

winter

Phalaropes

white wing stripe

dark center to rump

winter

Red Phalarope

thinner white wing stripe

dark center to rump

winter

Red-necked Phalarope

whitish rump, pale gray tail

winter

plain wings

Wilson's Phalarope

Calidris Sandpipers

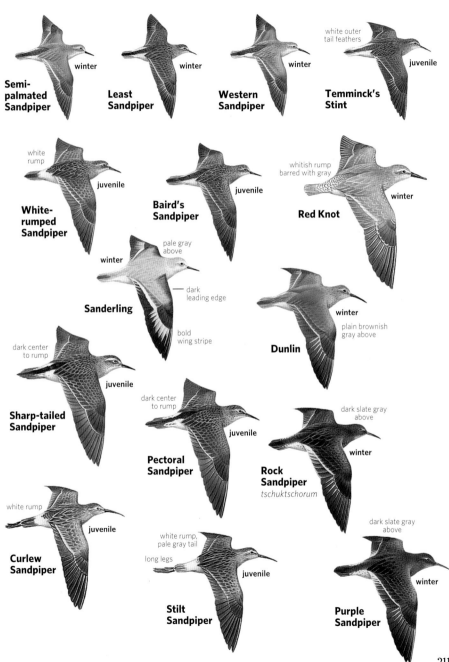

Semi-palmated Sandpiper winter

Least Sandpiper winter

Western Sandpiper winter

white outer tail feathers

Temminck's Stint juvenile

white rump

White-rumped Sandpiper juvenile

Baird's Sandpiper juvenile

whitish rump barred with gray

Red Knot winter

pale gray above

winter

Sanderling

dark leading edge

bold wing stripe

winter

plain brownish gray above

Dunlin

dark center to rump

Sharp-tailed Sandpiper juvenile

dark center to rump

Pectoral Sandpiper juvenile

dark slate gray above

winter

Rock Sandpiper *tschuktschorum*

white rump

Curlew Sandpiper juvenile

white rump, pale gray tail

long legs

Stilt Sandpiper juvenile

dark slate gray above

winter

Purple Sandpiper

GULLS • TERNS • SKIMMERS Family Laridae

A large, diverse family with strong wings and powerful flight. Some species are largely pelagic; others frequent coastal waters or inland lakes and wetlands. Gulls take from about two to four years to reach adult plumage; immatures are often variable and hard to identify. In general, male gulls are larger than females. SPECIES: 99 WORLD, 50 N.A.

Black-legged Kittiwake *Rissa tridactyla*

L 17" (43 cm) WS 36" (91 cm) Pelagic three-year gull. **Adult** has white head, nape smudged with gray in **winter;** dark eye; yellow bill; white body with gray mantle. Inner primaries are pale, wing tips inky black; legs black. **Juvenile** has dark half collar, retained into early winter; black bill; black spot behind eye; dark tail band; and, in flight, dark M across the wings. Distinguished from young Sabine's Gull by half collar, dark carpal bar. A very few young birds have pinkish legs.
VOICE: A repeated *kittiwake*; also a nasal *awk*. Calls mostly May to Aug.
RANGE: Nests in large cliff colonies; winters at sea. Seen uncommonly from shore on the West Coast, commonly in some years when a few may summer; rarely on the East Coast. Rare in fall on James Bay and Great Lakes, otherwise casual inland and south to Gulf Coast.

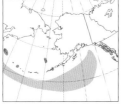

Red-legged Kittiwake *Rissa brevirostris*

L 15" (38 cm) WS 33" (84 cm) Pelagic two-year gull. **Adult** distinguished from Black-legged by coral red legs; shorter, thicker bill; darker mantle. Wings not paler on flight feathers as in Black-legged; broader white trailing edge on wings; dusky underside of primaries. In first-year, wing pattern resembles Sabine's Gull, but this is the only gull to have an all-white tail in first winter. Lacks M-pattern of Black-legged above.
VOICE: Call is higher pitched and squeakier than Black-legged.
RANGE: Breeds on cliffs, often with Black-legged Kittiwake. Rarely seen in winter. Casual on and off West Coast.

Sabine's Gull *Xema sabini* *L 13½" (34 cm) WS 33" (84 cm)*

Two-year gull with striking black-gray-and-white wing pattern in all ages. **Breeding adult** has dark gray hood with thin black ring at bottom; black bill with yellow tip; forked tail. First-summer bird is like adult but hood not complete; faint dusky nape. In **juvenal** plumage, wing pattern like adult; crown and nape soft gray-brown; bill black; tail has dark band.
VOICE: Call is a ternlike *kiew*, heard mainly on breeding grounds.
RANGE: Winters at sea mainly in the Southern Hemisphere. Common migrant off West Coast, rarely seen from shore. Rare in fall, casual in spring, in western interior. Very rare in fall east to Great Lakes region. Casual on and off East Coast and in the South.

Ivory Gull *Pagophila eburnea* *L 17" (43 cm) WS 37" (94 cm)*

Two-year arctic gull, ghostly pale and long winged. **Adults** in all plumages white with a yellow-tipped bill, black eyes, black legs. **First-winter** birds have dark face, variable amount of speckling elsewhere.
VOICE: Call is a ternlike *kew*, heard mainly on breeding grounds.
RANGE: Winters primarily in Arctic seas, closely associated with pack ice. Uncommon in northern and western AK. Casual along the Atlantic coast to NY, and inland to the northern tier of states. Accidental to southern CA, TN, and GA. Climate change imperils species.

Black-legged Kittiwake

dusky postocular bar and wash on nape

unmarked yellow bill

winter adult

short blackish legs

blackish bill

black collar

juvenile

base of primaries slightly paler than mantle

breeding adult

white underside to primaries, except for black tips

blackish M-pattern across wings

tail band

juvenile

Red-legged Kittiwake

shorter bill than Black-legged

breeding adult

very short coral red legs

pinkish legs

juvenile

base of primaries uniform with mantle, which is darker gray than Black-legged

underside of primaries medium gray, except for black tips

breeding adult

no dark M-pattern

broad white trailing edge

no tail band

1st summer

Sabine's Gull

dark bill with yellow tip

gray hood with black border

molting adult

breeding adult

large white apical spots

brownish or grayish on crown, nape, and upperparts; scaly above

dark bill

juvenile

breeding adult

forked tail

adults and juveniles have distinctive upperwing pattern

juvenile

Ivory Gull

dark eye

greenish bill with yellow-orange tip

adult

pure white

short black legs

black markings on primary tips and near tail tip

1st winter

dusky face

long, narrow, rather pointed wings

adult

Gulls

Little Gull   *Hydrocoloeus minutus*   *L 11" (28 cm) WS 24" (61 cm)*

Two- to three-year gull. **Breeding adult** has black hood, black bill, pale gray mantle, white wing tips, white underparts, red legs. **Winter adult** has dusky cap, dark spot behind eye. Wings uniformly pale gray above, dark gray to black below, with white trailing edge. Some **second-winter** birds are like adult but underwing pattern is incomplete; show some dusky slate in primaries. **First-winter** is like Bonaparte's but primaries blackish above, lack white wedge; wings show strong blackish M; crown shows more black. In all plumages, note short, rounded wings, fluttery wingbeat.

VOICE: Occasional nasal and ternlike calls. Gives rapid loud and evenly spaced notes in flight display on breeding grounds.

RANGE: Western Palearctic species, has bred irregularly around Great Lakes, and probably breeds extensively in remote Hudson Bay lowlands. Generally rare on Great Lakes—where locally more numerous, such as at Long Point and Niagara River, ON, where occasional counts into the hundreds have been made—in migration and in winter on East Coast; very rare to casual elsewhere in North America.

Bonaparte's Gull   *Chroicocephalus philadelphia*

L 13½" (34 cm) WS 33" (84 cm)   Two-year gull. **Breeding adult** has slate black hood, black bill, gray mantle with black wing tips that are pale on underside; white underparts, orange-red legs. In flight, shows white wedge on wing. **Winter** bird lacks hood. **First-winter** bird has a dark brown carpal bar on leading edge of wing, dark band on secondaries, black tail band; compare with juvenile Black-legged Kittiwake (page 212). **First-summer** bird may show partial hood; wings and tail are like first-winter. Flight is buoyant, wingbeats rapid.

VOICE: Call is a nasal or raspy ternlike *kerrr* or *gerrr*.

RANGE: Breeds at taiga ponds and marshes. Winters on inshore ocean waters, estuaries, large lakes, rivers, and sewage ponds. Uncommon inland migrant in West; common to abundant on Great Lakes, and will winter there in mild winters.

Black-headed Gull   *Chroicocephalus ridibundus*

L 16" (41 cm) WS 40" (102 cm)   Two-year gull. **Breeding adult** has dark brown hood; maroon-red bill and legs; mantle slightly paler gray than Bonaparte's Gull; black wing tips; white underparts. **Winter adult** lacks hood; bill brighter red. **First-winter** birds have orangish pink bill, pale legs, dark tail band, dark brown carpal bar. Distinguished from Bonaparte's by larger size and bill color. **First-summer** bird's hood varies from minimal, as seen on first-winter, to often nearly complete; wings and tail like first-winter. In all plumages, shows dark underside of primaries in flight, darker on adults; compare with Bonaparte's.

VOICE: Calls are ternlike, a grating or screechy *ree-ah*.

RANGE: Colonizer from Europe. Fairly common in winter in Newfoundland, where a few breed; rare off western AK; also small numbers to Maritime Provinces and coastal New England; rare south to NC and to eastern Great Lakes; casual to accidental elsewhere in North America.

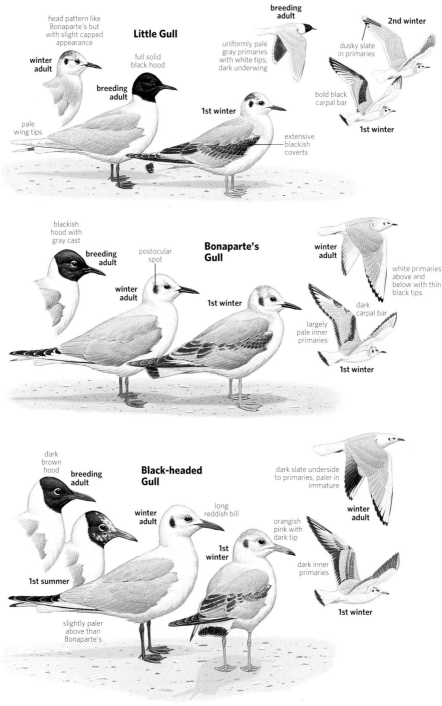

Little Gull

head pattern like Bonaparte's but with slight capped appearance

winter adult

full solid black hood

breeding adult

pale wing tips

1st winter

extensive blackish coverts

breeding adult

uniformly pale gray primaries with white tips; dark underwing

2nd winter

dusky slate in primaries

bold black carpal bar

1st winter

Bonaparte's Gull

blackish hood with gray cast

breeding adult

postocular spot

winter adult

1st winter

winter adult

white primaries above and below with thin black tips

dark carpal bar

largely pale inner primaries

1st winter

Black-headed Gull

dark brown hood

breeding adult

winter adult

long reddish bill

1st winter

1st summer

slightly paler above than Bonaparte's

dark slate underside to primaries, paler in immature

winter adult

orangish pink with dark tip

dark inner primaries

1st winter

Ross's Gull *Rhodostethia rosea* L 13½" (34 cm) WS 33" (84 cm)

Two-year gull. Variably pink below; upperwing pale gray; underwing pale to dark gray. Black collar in summer, partial or absent in winter. **First-winter** bird has black at tip of tail, dark spot behind eye; acquires black collar by first summer; in flight, shows M-pattern like Little Gull (page 214). In all plumages, note long, pointed wings; long, wedge-shaped tail; and broad, white trailing edge to wings.

VOICE: Mostly silent away from breeding Arctic, where it gives ternlike chitters and a mellow yapping.

RANGE: Arctic species of the Russian Far East, has bred in Canada (Churchill, MB) and Greenland in last two decades. Presumably winters at sea. Common fall migrant along northern coast of AK; rare and irregular in Bering Sea. Casual south to ID, NE, CO, IN, MD, and DE, Accidental to southeast CA.

Franklin's Gull *Leucophaeus pipixcan*

L 14½" (37 cm) WS 36" (91 cm) Three-year gull. **Breeding adult** has black hood, white underparts variably tinged with pink, slate gray wings with white bar and black-and-white tips on primaries. Distinguished from Laughing Gull by white bar and large white tips on primaries, pale gray central tail feathers, broader white eye crescents. All **winter** birds have a dark half hood, more extensive than any winter Laughing Gull. Second-summer has partial or no bar on primaries. **First-summer** is like winter adult but lacks white primary bar; bill and legs black. **First-winter** resembles first-winter Laughing; note white outer tail feathers, half hood, broader eye crescents, white underparts, and, in flight, pale inner primaries. Juvenile is like first-winter bird but back is brown. At all ages, distinguished from Laughing Gull by smaller size, smaller bill with less prominent hook, rounder forehead, less extensive dark on underside of primaries; shorter legs and wings give a stocky look when standing.

VOICE: Call is like Laughing Gull, but slightly higher and softer.

RANGE: Common on Great Plains. Otherwise uncommon to rare in West in migration, away from breeding areas. Also rare off West Coast and very rare to AK and northwestern Canada. In East, rare to very rare east of Mississippi River, mostly in fall. Winters on west coast of South America. Very rare in winter in coastal southern CA and Gulf Coast.

Laughing Gull *Leucophaeus atricilla*

L 16½" (42 cm) WS 40" (102 cm) Three-year gull. **Breeding adult** has black hood, white underparts, slate gray wings with black outer prima-ries. In **winter,** shows gray wash on nape. Second-summer has partial hood, some spotting on tip of tail. **Second-winter** is similar to second-summer but has gray wash on sides of breast, lacks hood. **First-winter** has extensively gray sides, complete tail band, gray wash on nape, slate gray back, dark brown wings; compare with first-winter Franklin's Gull. **Juvenile** like first-winter bird but brown on head and body.

VOICE: Calls include a crowing series of *hah* notes; also a single *kow* or *ka-ha.*

RANGE: Common along Gulf and Atlantic coasts; rare inland in East and Atlantic provinces. In West, casual on coast and inland except at Salton Sea, especially at south end and in adjacent Imperial Valley, CA, where it is fairly common, chiefly as a post-breeding visitor.

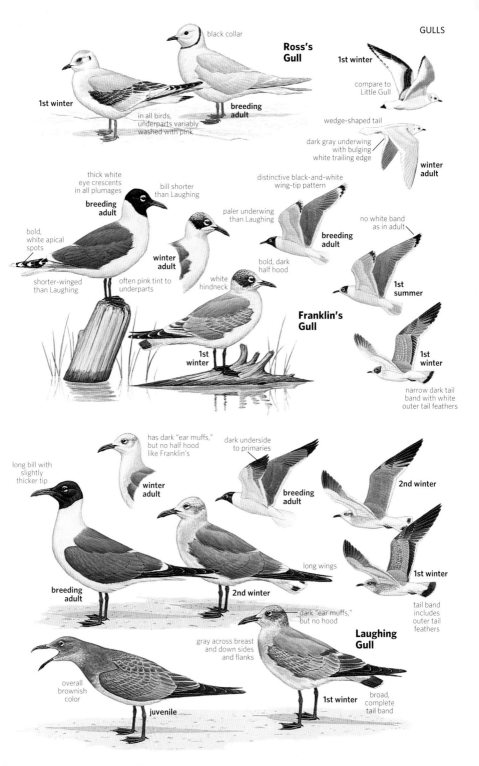

Ross's Gull

black collar

1st winter

compare to
Little Gull

1st winter

breeding
adult

in all birds,
underparts variably
washed with pink

wedge-shaped tail

dark gray underwing
with bulging
white trailing edge

winter
adult

thick white
eye crescents
in all plumages

bill shorter
than Laughing

distinctive black-and-white
wing-tip pattern

**breeding
adult**

paler underwing
than Laughing

no white band
as in adult

bold,
white apical
spots

**winter
adult**

**breeding
adult**

shorter-winged
than Laughing

often pink tint to
underparts

bold, dark
half hood

white
hindneck

**Franklin's
Gull**

1st
summer

**1st
winter**

1st
winter

narrow dark tail
band with white
outer tail feathers

has dark "ear muffs,"
but no half hood
like Franklin's

dark underside
to primaries

long bill with
slightly
thicker tip

**winter
adult**

**breeding
adult**

2nd winter

long wings

1st winter

**breeding
adult**

2nd winter

tail band
includes
outer tail
feathers

dark "ear muffs,"
but no hood

gray across breast
and down sides
and flanks

**Laughing
Gull**

overall
brownish
color

juvenile

1st winter

broad,
complete
tail band

Heermann's Gull *Larus heermanni*

L 19" (48 cm) WS 51" (130 cm) Four-year gull. **Adult** distinctive with white head, streaked gray-brown in winter; red bill; dark gray body; black tail with white terminal band; and white trailing edge on wings. Third-winter is variably intermediate between adult and second-winter. **Second-winter** bird is browner, with two-tone bill and buff tail tip. **First-winter** bird has dark brown body, lacks contrasting tail tip and trailing edge on wing. Wings are fairly long, and flight is buoyant. These dark young birds can be confused with jaegers at a distance.
VOICE: Calls include distinctive low nasal notes.
RANGE: Common post-breeding visitor from Mexican breeding grounds from early summer to early winter; smaller numbers in late winter and spring along the West Coast; small numbers have irregularly nested on the central CA coast; rare at Salton Sea, CA, where it has recently nested. Casual elsewhere inland in CA, the Southwest, NV, and to southeastern AK. Some of the interior records of adults away from the Salton Sea have been of birds in California Gull colonies but no nesting to date. Accidental to west TX, Great Lakes, southeastern AK, FL, and VA.

Black-tailed Gull *Larus crassirostris*

L 18½" (47 cm) WS 47¼" (120 cm) Three-year or four-year gull, about size of Ring-billed Gull (page 222); wings and especially bill long; legs short. Distinctive white eye crescents except on **breeding adult** and third-winter bird. Adult has black ring near red tip of bill; yellow iris, red orbital ring. Mantle dark slate gray; tail has broad subterminal band. Head of **winter adult** heavily streaked. **First-winter** bird has white on face, otherwise heavily washed with brown.
RANGE: East Asian species that is very common in Japan; casual in coastal AK, WA, and CA. In East, casual from Newfoundland to VA; accidental on TX Gulf Coast and in the interior.

Belcher's Gull *Larus belcheri* L 20" (51 cm) WS 49" (124 cm)

Medium-size, three-year gull. Plumages and bill color similar to Black-tailed Gull, but has darker mantle, dark eyes, longer legs, and thicker bill. **Winter adults** and **second-winter** birds have dark (brownish) hood, and red only on tip of bill. Adult has yellow orbital ring. **First-winter's** head and breast are smoky brown; belly white; mottled above.
RANGE: Resident on west coast of South America; accidental to CA and FL. Olrog's Gull, *L. atlanticus,* on the southern Atlantic coast of South America, was considered a subspecies of Belcher's Gull until fairly recently. Olrog's is larger, and the adult in basic plumage has a mottled, slaty head rather than a more solid-brown head like Belcher's. Vagrants to North America should be separated with care between these two species.

Heermann's Gull

red bill with small black tip

white head

breeding adult

streaked head

gray underparts

winter adult

dark wings

2nd winter

pink-based bill with dark tip

breeding adult

broad black tail band

1st winter

overall sooty brown color

Black-tailed Gull

pale eye

breeding adult

long bill with black subterminal band and red tip

overall smooth brown color with whitish in face

winter adult

breeding adult

pink-based bill with dark tip

broad, bold black tail band

long wings

rather dark slaty gray upperparts

1st winter

yellowish legs

2nd winter

thickish bill with black subterminal band and red tip

dark eye

breeding adult

Belcher's Gull

breeding adult

distinct brown hood in winter plumage

darker mantle than Black-tailed

smoky brown head and breast

broad black tail band

long yellow legs

winter adult

2nd winter

pinkish legs

1st winter

Mew Gull *Larus canus* L 16" (41 cm) WS 43" (109 cm)

Three-year gull. North American subspecies *brachyrhynchus* is the smallest of three subspecies found here; has least black on wing tips; in flight, shows much more white on primaries than Ring-billed Gull (page 222). European nominate subspecies, *canus* (**"Common Gull"**), and East Asian *kamtschatschensis* (the largest subspecies) have more extensive black on wing tips. All **adults** have white head, washed with brown in **winter**, which is more flecked and less washed in *canus;* dark gray mantle; thin yellow bill, often with a dark subterminal ring in *canus*, like Ring-billed; most have large, dark eyes. **Second-winter** bird is like adult but has two-tone bill; has less white on primaries; variably spotted tail band. **First-winter** *brachyrhynchus* is heavily washed with brown below, almost solid brown on belly; spotted with white on breast. Head and nape are washed with soft brown; mantle dark gray; primaries light brown with pale edges. Tail is almost entirely brown, with heavily barred tail coverts; wing linings evenly pale brown. **Juvenile** is like first-winter, but brown on the back and head, darker below. Second-winter and adult *kamtschatschensis* have pale eyes; first-winter birds are more like *brachyrhynchus,* but note dark tail band rather than all-dark tail, and paler tail coverts. In flight, first-winter's underwing coverts show intermediate coloration between *brachyrhynchus* (dark) and *canus* (pale). European *canus* resembles Ring-billed Gull in first winter; but note *canus*'s mostly white tail with dark subterminal band, unbarred white tail coverts, darker gray back, white wing linings mottled with brown, and more fully filled-in (with brown) covert feathers. Mew Gulls average smaller than Ring-billed Gulls, especially *brachyrhynchus*, with rounder heads, larger eyes, thinner bills.

VOICE: Calls of *brachyrhynchus* include a wheezy *kyap* and a mewing *kii-uu.*

RANGE: Nests in taiga; otherwise in coastal habitats and farm fields. North American subspecies, *brachyrhynchus*, is rare inland in Pacific states, very rare to casual elsewhere in West during migration and winter. Casual farther east to Great Lakes region. Asian subspecies, *kamtschatschensis,* rare on the Aleutians and islands in the Bering Sea. European subspecies, *canus*, breeding in North America as close as Iceland. Casual on East Coast in winter, annual in Newfoundland and almost annual in Maritime Provinces; casual south to coastal NY; accidental south to coastal NC. European *canus*, along with a closely related larger subspecies, *heinei*, from farther east, may represent a distinct species from North American *brachyrhynchus*. The situation is complicated by yet another subspecies, *kamtschatschensis* of East Asia, the largest subspecies, which shows intermediate plumage characteristics between *canus* and *brachyrhynchus*.

2nd winter

1st winter

black tail

brownish underwing

breeding adult

extensive gray on p8

winter adult

brownish wash

2nd winter

most with dark eye

short, slender, yellow bill

breeding adult

darker above than Ring-billed

long wings

Mew Gull
brachyrhynchus

overall brownish compared to juvenile Ring-billed

filled-in brown coverts with pale fringes

juvenile

1st winter

extensively dark tail

kamtschatschensis

larger size

winter adult

1st winter

pale base to tail

often with darker ring on bill

head and breast pattern on *canus* more spotted, less washed with brown as in *brachyrhynchus*

winter adult "Common Gull"
canus

covert pattern like *brachyrhynchus*

1st winter *canus*

black tail band contrasts sharply with white tail base

adult *kamtschatschensis*

adult *canus*

more black on p8 than *brachyrhynchus*

more black on p8 than *brachyrhynchus*, like *kamtschatschensis*

Ring-billed Gull *Larus delawarensis*

L 17½" (45 cm) WS 48" (122 cm) Three-year gull. **Adult** has pale gray mantle; yellow bill with black subterminal ring; pale eyes; yellowish legs; head streaked with brown in **winter. Second-winter** birds are like winter adult but bill has broader band, black on primaries is more extensive, tail usually has some blackish terminal spots. **First-winter** bird has gray back, brown wings with dark blackish brown primaries, brown-streaked head and nape; underparts white with brown spots and scalloping on breast and throat; tail has medium-wide but variable brown band and extensive mottling above band; uppertail and undertail coverts are lightly barred; secondary coverts medium gray; wing linings mostly white, with some barring. Distinguished from first-winter Mew Gull (*brachyrhynchus*; page 220) by white underparts spotted on breast and throat, tail pattern, darker primaries, heavier bill, and paler back. **Juvenal** plumage may be largely kept into early winter; resembles first-winter but back is brown, spotting below more extensive, bill has more black.

VOICE: Calls include a mewing *kee-ew,* a sharper *kyow,* and a whining *sseeaa.* Also has an extended "long call."

RANGE: Abundant and widespread; winters uncommonly outside mapped range. Rare to AK and YT and well offshore in fall. Nonbreeders oversummer south of nesting range.

California Gull *Larus californicus*

L 21" (53 cm) WS 54" (137 cm) Four-year gull. **Adult** has darker gray mantle than Herring Gull (page 224), paler than Lesser Black-backed Gull (page 232); white head, heavily streaked with brown in winter; dark eyes; yellow bill with black and red spots, black spot often smaller in breeding season; gray-green or greenish yellow legs. In flight, shows dusky trailing edge on underwing. Third-winter plumage is like adult but bill is more extensively smudged with black; wings show some brown; tail has some brown spotting. **Second-winter** has gray back, brown wings, grayish legs, two-tone bill; compare to first-winter Ring-billed Gull. **First-winter** is brown overall with veiled gray on scapulars; usually palest on throat, breast, and upper belly; legs pinkish; bill two-tone, the colors sharply defined. In flight, first-winter birds show double dark bar on inner half of wing, caused by darker secondaries and greater secondary covert bases. Distinctly smaller than Western Gull (page 228), with thinner bill. Compare first-winter birds to first- and second-winter Herring and Lesser Black-backed. **Juveniles** are variably pale below, lack pale bill base. Another subspecies, *albertaensis,* which breeds on the northern Great Plains, averages larger; in adults has a paler mantle.

VOICE: Calls include a *kyow* and a higher *kii-ow;* also a long, slightly nasal trumpeting "long call." All calls are lower pitched than Ring-billed.

RANGE: Nests at lakes; otherwise frequents a wide variety of habitats, and in winter common offshore. Uncommon to rare in much of Southwest and southwestern Great Plains and uncommon to southeastern AK (mainly late summer), rarely farther north. Nonbreeders oversummer south of breeding range. In East, casual east of Dakotas as far as East and Gulf Coasts.

spots and streaks on head

winter adult

paler underwing than Mew

1st winter

2nd winter

black subterminal band

pale eye

breeding adult

2nd winter

Ring-billed Gull

breeding adult

pale gray above

pale underside to secondaries

outer wing pattern differs from Mew

small mirror

yellow legs

dark eye

1st winter

juvenile

small dark centers to coverts

pinkish legs

tail pattern variable

dark-based greater coverts

1st winter

rather slender wings

black and red spots on bill

breeding adult

California Gull

dark secondaries

no obvious pale window

extensive black in primaries

darker gray mantle than Ring-billed or Herring

dark eye

heavy brown wash

2nd winter

breeding adult

slender pinkish bill with blackish tip

winter adult

greenish yellow legs

juvenile has blackish bill until Sept.

gray secondaries; compare to Herring and Ring-billed

1st winter

pale juvenile

dark juvenile

Herring Gull *Larus argentatus* L 25" (64 cm) WS 58" (147 cm)

Highly variable four-year gull. **Adult** has pale gray mantle; white head streaked with brown in winter; legs and feet pink; pale yellow eyes with a yellow-orange orbital ring; bill yellow with red spot. **Third-winter** plumage is like winter adult but with black smudge on bill, some brown on body and wing coverts. **Second-winter** bird has pale gray back; brown wings; pale eyes; two-tone bill. **First-winter** birds are brown overall, with dark brownish black primaries and tail band, dark eyes, dark bill, with variable pink at base; some may have bill like first-winter California Gull (page 222); but usually distinguished by darker bill, paler face and throat, and, in flight, by pale panel at base of primaries and single dark bar on secondaries. Distinguished from first-winter Western Gull (page 228) by smaller bill, paler and more mottled body plumage, and, in flight, by paler wings with pale panel, and lack of contrast between back and rump. Distinguished from first-winter Lesser Black-backed Gull (page 232) by browner, less-contrasting body plumage, usually darker belly, and, in flight, by pale primary and outer secondary coverts and less-contrasting rump pattern. Widespread North American subspecies is *smithsonianus;* Bering Sea region *vegae* has darker mantle in adult plumage, similar wing-tip pattern to Slaty-backed Gull (page 230); usually with darker eyes; head heavily streaked in winter. Western European *argenteus* and slightly darker-mantled *argentatus* from farther east are most distinct from *smithsonianus* in first-winter plumage. European birds paler and more checkered above and whiter on the rump region and tail base.

VOICE: For *smithsonianus,* a series of buglelike calls makes up "long calls." Also a full *kyow.*

RANGE: Frequents a wide variety of habitats, from far-offshore waters to coasts, farm fields, parking lots, and dumps. Common in eastern North America. Generally local in the West; common in some regions, uncommon to rare in most. European subspecies casual to Newfoundland; specimen from ON.

Yellow-legged Gull *Larus michahellis*

L 24" (61 cm) WS 57" (144 cm) Size similar to Herring Gull, but squarer head, peaked at rear of crown, and stouter, shorter bill. Adult mantle darker gray than *smithonianus* Herring. From above, wing tip darker than Herring; from below more gray, less black, on outermost primaries. Red gonys spot often extends onto upper mandible; orbital ring is redder than Herring. Yellow legs distinctive, but some *smithsonianus* Herrings may show some yellow during winter. In Yellow-legged, fainter winter head streaking restricted to face and crown, making white head stand out; by midwinter most **adults** are white headed. **First-winter** birds much paler on head and underparts than first-winter Herring; blocky head and extensive white on uppertail coverts and base of tail suggests same-age Great Black-backed (page 232). Compare also to first-winter Lesser Black-backed (page 232). Molt to first-winter occurs earlier than Herring; by fall, young Yellow-legged often appears worn. Eastern Atlantic islands subspecies, *atlantis,* is slightly darker mantled. North American records believed to be either *michahellis* or *atlantis.*

VOICE: Like *smithsonianus* Herring, but slightly deeper.

RANGE: Palearctic species. Casual winter visitor to East Coast from Newfoundland (especially) to mid-Atlantic, possibly south TX.

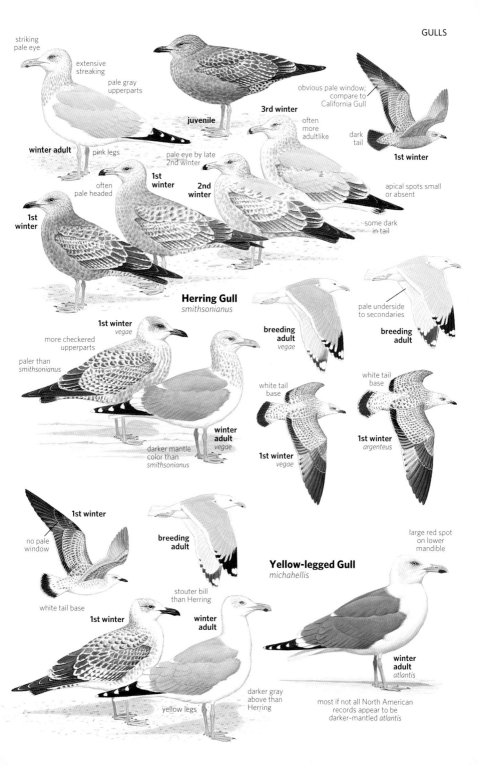

striking pale eye

extensive streaking

pale gray upperparts

juvenile

3rd winter

obvious pale window; compare to California Gull

dark tail

1st winter

winter adult

pink legs

often pale headed

pale eye by late 2nd winter

1st winter

2nd winter

1st winter

apical spots small or absent

some dark in tail

1st winter

Herring Gull
smithsonianus

pale underside to secondaries

more checkered upperparts

paler than *smithsonianus*

1st winter
vegae

breeding adult
vegae

breeding adult

winter adult
vegae

white tail base

white tail base

darker mantle color than *smithsonianus*

1st winter
vegae

1st winter
argenteus

1st winter

no pale window

white tail base

breeding adult

Yellow-legged Gull
michahellis

large red spot on lower mandible

stouter bill than Herring

1st winter

winter adult

winter adult
atlantis

darker gray above than Herring

yellow legs

most if not all North American records appear to be darker-mantled *atlantis*

Iceland Gull *Larus glaucoides* L 22" (56 cm) WS 54" (137 cm)

Highly variable four-year gull. **Adults** have white heads, suffused with brown in winter; most have yellow eyes, a few *(kumlieni)* have brown; orbital ring is purplish pink *(kumlieni)* to pink or reddish *(glaucoides)*. Late **second-winter** birds have pale eyes, gray back, two-tone bill. **First-winter** birds are buffy to mostly white; chiefly dark bill is short; eyes dark; wing tips white or irregularly washed with brown. Canadian-breeding adult *kumlieni* has wing tips variably marked with gray; a few are pure white. Greenland breeding *glaucoides* is slightly smaller and paler overall in all plumages; adults are slightly paler mantled and have pure white wing tips. First-winter birds distinguished from Thayer's by paler primaries, checkered tertials, usually paler body plumage and on some by checkered tail; from Glaucous by usually darker bill and structural features (smaller size, rounder head, and longer wings that extend beyond tail at rest).

VOICE: Mostly silent in nonbreeding season; otherwise calls close to Herring.

RANGE: Canadian *kumlieni* uncommon to rare on Great Lakes; casual to Gulf Coast, Great Plains, and the West. Most *glaucoides* migrate southeast to Iceland, a few to Europe, rare or casual to northeastern North America, perhaps a few records from the West, too.

Thayer's Gull *Larus thayeri* L 23" (58 cm) WS 55" (140 cm)

Variable four-year gull. In many **adults,** eye is dark brown but is highly variable with all degrees of shades; some are quite pale eyed, though typically not as pale as Herring Gull (page 224). Orbital ring purplish pink. Mantle slightly darker than Iceland Gull or Herring Gull; bill yellow with dark red spot; legs darker pink than Herring. Primaries pale gray below, with thin, dark trailing edge; show some black or slaty gray from above. **Second-winter** has gray mantle, contrasting gray-brown tail band, dark eye. **First-winter** variable but primaries always entirely pale below, darker than mantle above. Distinguished from Herring Gull by smaller size, paler checkered markings in plumage, and paler primaries with whitish edges; from Iceland Gull by generally darker plumage, primaries darker than mantle, and usually by unspeckled tail. Some birds are probably best left unidentified, especially when worn in late winter or spring. The degree of hybridization between Thayer's Gull and *kumlieni* Iceland Gull in eastern Arctic Canada is unknown; some treat Thayer's as subspecies of Iceland Gull. Compare to larger-size and bigger-billed Glaucous-winged Gull (page 228), in which the immatures are less mottled and the wing tips are uniform with the mantle.

VOICE: Calls similar to Herring and Iceland Gulls.

RANGE: Rare winter visitor to Great Lakes region (most numerous in fall) and through the interior to Gulf Coast. Casual on East Coast.

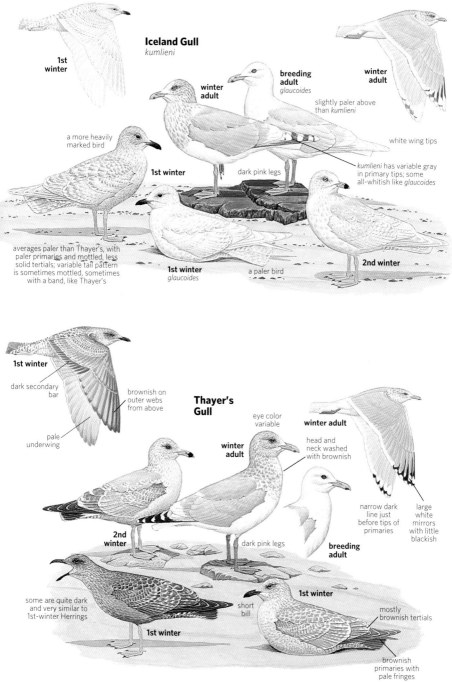

1st winter

Iceland Gull
kumlieni

winter adult

breeding adult
glaucoides

slightly paler above than *kumlieni*

winter adult

white wing tips

a more heavily marked bird

1st winter

dark pink legs

kumlieni has variable gray in primary tips; some all-whitish like *glaucoides*

averages paler than Thayer's, with paler primaries and mottled, less solid tertials; variable tail pattern is sometimes mottled, sometimes with a band, like Thayer's

1st winter
glaucoides

a paler bird

2nd winter

1st winter

dark secondary bar

brownish on outer webs from above

Thayer's Gull

pale underwing

eye color variable

winter adult

winter adult

head and neck washed with brownish

narrow dark line just before tips of primaries

large white mirrors with little blackish

2nd winter

dark pink legs

breeding adult

some are quite dark and very similar to 1st-winter Herrings

short bill

1st winter

mostly brownish tertials

1st winter

brownish primaries with pale fringes

Glaucous Gull *Larus hyperboreus*

L 27" (69 cm) WS 60" (152 cm) Heavy-bodied, four-year gull. All have translucent tips to white primaries. **Adult** has very pale gray mantle, yellow eye with orange orbital ring. Head is streaked with brown in winter. Late **second-winter** bird has pale gray back and pale eye. **First-winter** birds may be buffy or almost all-white; bill is bicolored. Distinguished from Iceland Gull by size; heavier, longer bill; flatter crown; slightly paler mantle of adults; disproportionately shorter wings, barely extending beyond tail. At all ages, distinguished from Glaucous-winged Gull by more buffy white color, contrasting pale primaries; in first-winter birds by sharply two-tone bill.
VOICE: Similar to Herring Gull; some calls actually higher pitched.
RANGE: Rare in winter south to Gulf states and southern CA. Birds on Arctic coast of northern AK are slightly smaller, and adults have slightly darker mantles, than birds from eastern Canada and Bering Sea. Occasionally hybridizes with Herring Gull (page 224).

Glaucous-winged Gull *Larus glaucescens*

L 26" (66 cm) WS 58" (147 cm) Four-year gull. **Adult** has white head, moderately streaked with brown in winter. Body is white, mantle pale gray; primaries are the same color as rest of wing above, paler below. Eyes dark, orbital ring pink; large bill yellow with red spot; legs pink. Third-winter bird is like adult but has some buff on body, bill is smudged black; some have a partial tail band. **Second-winter's** back is gray, rest of body and wings are pale buff to white with little mottling; tail evenly gray; bill mostly dark. **First-winter** bird is uniformly pale gray-brown to whitish with subtle mottling; primaries are the same color as the mantle; note young Glaucous Gull has sharply two-tone bill, pale primaries; and young Thayer's Gull (page 226) is smaller, with smaller bill, more speckled body plumage, and darker primaries. Hybridizes with Western Gull; also with Herring Gull (page 224) in south-central AK. Hybrids are extremely variable.
VOICE: Similar to Western Gull; "long call" possibly higher pitched.
RANGE: Rare well inland in Pacific states; casual east to Great Lakes region.

Western Gull *Larus occidentalis* *L 25" (64 cm) WS 58" (147 cm)*

Four-year gull. **Adults** north of Monterey, CA, have paler backs and darker eyes than southern birds. All adults have white head, dark gray back, pink legs, very large bill. In **winter,** head is lightly streaked in northern birds, white or nearly so in southern. **Third-winter** plumage resembles second-winter Yellow-footed (page 230) but tail is mostly white. **Second-winter** bird has a dark gray back, yellow eyes, two-tone bill. **First-winter** bird is one of the darkest young gulls; bill is black; in flight, distinguished from young Herring by contrast of dark back with paler rump; often sootier overall coloration, heavier bill. **Juvenile** is like first-winter but darker. Hybridizes extensively with Glaucous-winged in the Northwest; **hybrids** are seen all along the West Coast in winter; two ages are shown here. These are often confused with Thayer's Gull (page 226); note large bill, pattern of wing tips.
VOICE: A wide variety of calls, some like Herring Gull, some lower.
RANGE: Very rare to casual well inland; definite records as far east as IL and TX. Casual to southeastern AK.

Glaucous Gull

winter adult

1st winter

2nd winter

pale eye and bill tip in 2nd winter

winter adult

larger and chunkier than Iceland

very pale gray above

breeding adult

shorter wing tip projection past tail than Iceland

white wing tips

pinkish legs

1st winter

long pink bill with blackish tip

1st winter

color of 1st-winter birds varies; some by mid- to late winter are chalky white

1st winter

plumage suggests 1st-winter Thayer's; note thicker tip to bill

breeding adult

white mirrors on outer primaries

thick-tipped bill

medium gray primaries with white apical spots

breeding adult

1st winter

thick-tipped bill

Glaucous-winged Gull

1st winter

breeding adult

breeding adult

more uniform upperwing; compare to Thayer's Gull

2nd winter

1st winter

thick-tipped bill

Glaucous-winged x Western hybrids

primaries uniform in color with rest of plumage

plumage plain and unpatterned overall

1st winter

thick-tipped bill

juvenile *occidentalis*

darker mantle than *occidentalis*

northern breeding adult *occidentalis*

southern breeding adult *wymani*

1st winter

no pale window unlike 1st-winter Herring

dark sooty overall

southern winter adult *wymani*

southern 2nd winter *wymani*

Western Gull

southern 3rd winter *wymani*

1st winter *occidentalis*

Yellow-footed Gull *Larus livens*

L 27" (69 cm) WS 60" (152 cm) Four-year gull. **Adult** is like Western Gull but has yellow legs and feet; note also thicker yellow bill with red spot; dark slate gray wings; yellow eyes. **Second-winter** bird is like adult but tail looks entirely black, bill two-tone. In **first-winter** plumage, head and body are mostly white, back and wings brown, eyes dark, bill mostly dark, legs pinkish. **Juvenile** resembles first-winter Western but white belly contrasts sharply with streaked breast; upperparts are more boldly patterned; rump whiter; bill thicker.
VOICE: Calls distinctly lower pitched than Western.
RANGE: Breeds in the Gulf of CA. Fairly common post-breeding visitor in summer to Salton Sea, CA, especially the south end (common late June to Sept.); a few usually linger into winter; rarest in spring. Casual to coastal southern CA. Accidental away from Salton Sea and Imperial Valley in eastern CA, southern NV, UT, and AZ.

Slaty-backed Gull *Larus schistisagus*

L 25" (64 cm) WS 58" (147 cm) Four-year gull. **Adult** has dark slate gray back and wings, blackish outer primaries separated from mantle by a staggered row of whitish spots. Underside of primaries gray. Note broad white trailing edge to wings. Legs bright pink; striking pale eyes with red orbital ring. Head heavily mottled in winter, dark streaking often concentrated around eye. **Second-summer** bird has dark back, very pale wings. **First-year** birds show almost entirely dark tail and wing pattern like Thayer's Gull (page 226); also compare with *vegae* Herring Gull (page 224); *vegae* adult has paler upperparts, lacks broad white trailing edge; underside of primaries darker than Slaty-backed. Slaty-backed is one of our darker-mantled gulls. Only Kelp and Great Black-backed are significantly blacker mantled. Reported Slaty-backed Gulls with somewhat paler mantles might represent individual variation within the species not previously documented or, more likely, hybrids with either Glaucous-winged or *vegae* Herring Gulls.
VOICE: Similar to Glaucous-winged.
RANGE: Coastal species of northeast Asia. Uncommon to rare in coastal AK, most frequent in the Bering Sea. Very rare in winter south through Pacific states; casual elsewhere in North America but recorded from such far-flung locations as Great Lakes (multiple records), Newfoundland (multiple records), Key West, FL, and extreme south TX.

Kelp Gull *Larus dominicanus* L 23" (58 cm), WS 53" (135 cm)

Four-year gull; size, structure, and bill shape suggest Western Gull (page 228). **Adult** Kelp has black back and dull greenish legs; head streaking in winter indistinct. Eye color variable. Note restricted white in outer primaries, unlike Great Black-backed (page 232). Birds in their first year suggest the smaller, slimmer, and longer-winged Lesser Black-backed (page 232); note Kelp's much thicker bill. Change to adult plumage rapid; mantle blackish by **second summer.**
RANGE: Widespread Southern Hemisphere species; casual to Gulf Coast. A few Kelps nested on Chandeleur Islands off southeastern LA for about a decade starting in early 1990s. This resulted in pure Kelp pairings, and mixed pairings with Herring Gull that produced **hybrids.** Accidental elsewhere but recorded in TX, FL, and MD and from the Great Lakes (IN).

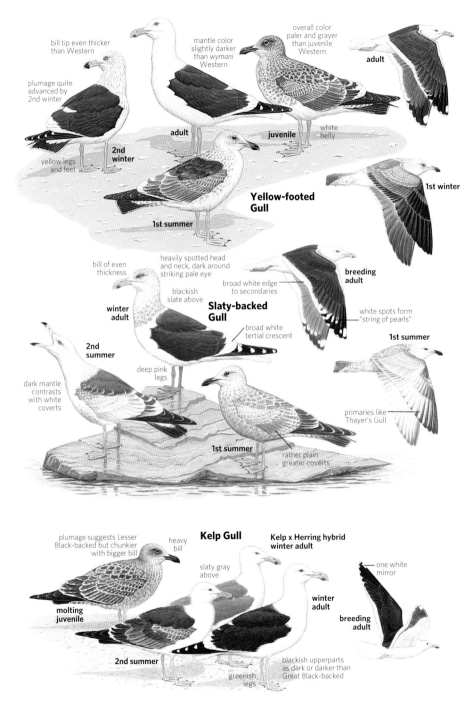

bill tip even thicker than Western

plumage quite advanced by 2nd winter

yellow legs and feet

2nd winter

mantle color slightly darker than *wymani* Western

adult

overall color paler and grayer than juvenile Western

adult

juvenile

white belly

1st summer

Yellow-footed Gull

1st winter

bill of even thickness

heavily spotted head and neck, dark around striking pale eye

blackish slate above

winter adult

Slaty-backed Gull

broad white edge to secondaries

breeding adult

white spots form "string of pearls"

broad white tertial crescent

2nd summer

deep pink legs

dark mantle contrasts with white coverts

1st summer

1st summer

rather plain greater coverts

primaries like Thayer's Gull

Kelp Gull

plumage suggests Lesser Black-backed but chunkier with bigger bill

heavy bill

slaty gray above

Kelp x Herring hybrid winter adult

winter adult

one white mirror

molting juvenile

breeding adult

2nd summer

greenish legs

blackish upperparts as dark or darker than Great Black-backed

Lesser Black-backed Gull *Larus fuscus*

L 21" (53 cm) WS 54" (137 cm) A four-year gull. **Adult** has white head, heavily streaked with brown in winter; white underparts; yellow legs. Third-winter bird has dark smudge on bill; some brown in wings. **Second-winter** bird resembles second-winter Herring Gull (page 224) but note dark gray back, much darker underwing. **First-winter** bird similar to first-winter Herring Gull but head and belly are paler, upperparts more contrastingly dark and light; bill is always entirely black. Identified in flight by darker primary and secondary coverts, more extensively dark primaries and white outer tail feathers; paler rump contrasts with back. Much smaller than Great Black-backed Gull. Smaller on average than Herring Gull, with smaller bill, but there is substantial range of overlap; also note longer wings, usually extending well beyond tail at rest. Most birds seen here are of northern European subspecies *graellsii*. A few darker mantled adults in eastern North America likely of Baltic subspecies *intermedius*.

VOICE: Like Herring Gull but slightly deeper.

RANGE: Western Palearctic species, breeding commonly as close to North America as Iceland; generally rare to locally uncommon in eastern North America with the largest numbers in mid-Atlantic and FL regions; increasing. In West, records also increasing rapidly. Rare, but a few now winter annually in eastern CO (Denver area and north) and in CA. Casual elsewhere in West and from southeastern AK. A specimen on Shemya Island in the western Aleutians on 15 Sept. 2005 has been identified as the subspecies *heuglini*, breeding on the western Russian Arctic coast, which is treated by some as a subspecies of Herring Gull or as its own species; *taimyrensis*, which breeds on the Arctic coast just east of *heuglini* (Taymyr Peninsula), is closely related but is slightly paler mantled; most winter on Pacific coast of East Asia. Both *heuglini* and *taimyrensis* molt later than other Lesser Black-backed subspecies and are larger.

Great Black-backed Gull *Larus marinus*

L 30" (76 cm) WS 65" (165 cm) Four-year gull. Huge size and large bill are distinctive. **Adult's** white head is virtually unstreaked in winter; black upperparts; white underparts; variably pale eyes; pink legs. In flight, note extensive white on outer primary that merges with white spot on second primary to form solid white area. **Third-winter** bird is like adult but shows some dark on bill, some brown in wings, sometimes dark in tail. **Second-summer** bird has pale eye, black back; wings and tail are like first-winter. Second-winter is like first-winter, but base of bill is paler, secondary coverts more evenly brown. **First-winter** bird resembles Herring Gull (page 224) but head and body are much paler, back and wings have the checkered look of young Lesser Black-backed Gull, but is paler and appears even more checkered; in flight, shows almost white rump, checkered tail band.

VOICE: Calls, such as *kyow*, are deep and hoarse. "Long call" is lower pitched and slower than Herring Gull.

RANGE: Breeding range expanding southward on the Atlantic coast. Fairly common on eastern Great Lakes, scarcer on western; rare elsewhere well inland throughout the East and on Gulf Coast; casual west to Great Plains; accidental to AK (recorded at Kodiak and Barrow) and WA.

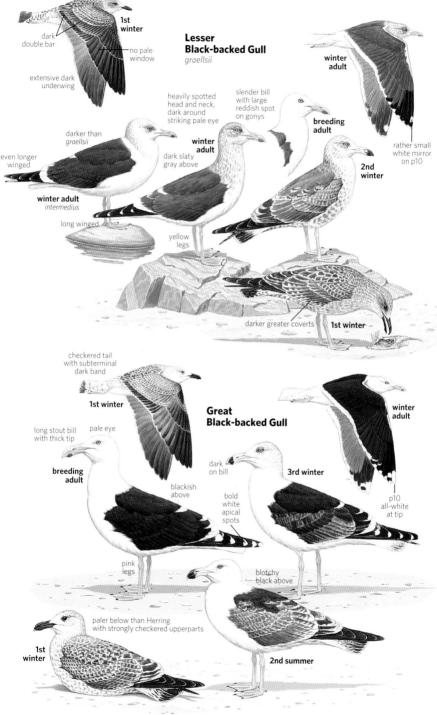

Lesser Black-backed Gull
graellsii

1st winter

dark double bar

no pale window

extensive dark underwing

winter adult

rather small white mirror on p10

heavily spotted head and neck, dark around striking pale eye

slender bill with large reddish spot on gonys

breeding adult

darker than *graellsii*

winter adult

dark slaty gray above

2nd winter

even longer winged

winter adult
intermedius

long winged

yellow legs

darker greater coverts

1st winter

checkered tail with subterminal dark band

1st winter

Great Black-backed Gull

winter adult

long stout bill with thick tip

pale eye

breeding adult

blackish above

dark on bill

3rd winter

bold white apical spots

p10 all-white at tip

pink legs

blotchy black above

paler below than Herring with strongly checkered upperparts

1st winter

2nd summer

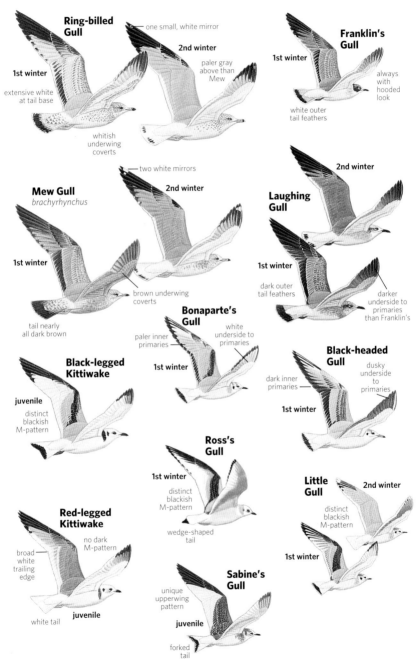

Ring-billed Gull

1st winter

extensive white at tail base

one small, white mirror

2nd winter

paler gray above than Mew

whitish underwing coverts

Franklin's Gull

1st winter

always with hooded look

white outer tail feathers

Mew Gull
brachyrhynchus

two white mirrors

2nd winter

1st winter

brown underwing coverts

tail nearly all dark brown

Laughing Gull

2nd winter

1st winter

dark outer tail feathers

darker underside to primaries than Franklin's

Bonaparte's Gull

paler inner primaries

white underside to primaries

1st winter

Black-legged Kittiwake

juvenile

distinct blackish M-pattern

Black-headed Gull

dark inner primaries

dusky underside to primaries

1st winter

Ross's Gull

1st winter

distinct blackish M-pattern

wedge-shaped tail

Little Gull

2nd winter

distinct blackish M-pattern

1st winter

Red-legged Kittiwake

no dark M-pattern

broad white trailing edge

white tail

juvenile

Sabine's Gull

unique upperwing pattern

juvenile

forked tail

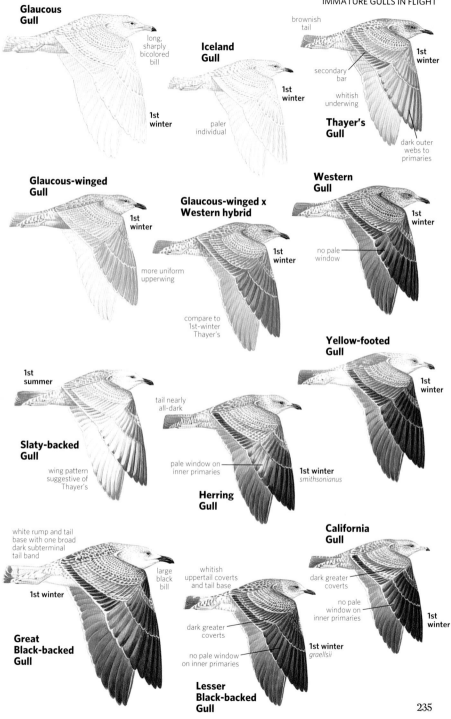

Glaucous Gull

long, sharply bicolored bill

1st winter

Iceland Gull

1st winter

paler individual

brownish tail

1st winter

secondary bar

whitish underwing

Thayer's Gull

dark outer webs to primaries

Glaucous-winged Gull

1st winter

more uniform upperwing

Glaucous-winged x Western hybrid

1st winter

compare to 1st-winter Thayer's

Western Gull

1st winter

no pale window

Yellow-footed Gull

1st winter

1st summer

Slaty-backed Gull

wing pattern suggestive of Thayer's

tail nearly all-dark

pale window on inner primaries

Herring Gull

1st winter *smithsonianus*

white rump and tail base with one broad dark subterminal tail band

large black bill

1st winter

Great Black-backed Gull

whitish uppertail coverts and tail base

dark greater coverts

no pale window on inner primaries

1st winter *graellsii*

Lesser Black-backed Gull

California Gull

dark greater coverts

no pale window on inner primaries

1st winter

235

TERNS

Distinguished from gulls by long, pointed wings and bill and by feeding technique. Most terns plunge-dive into the water after prey, primarily small fish. Most species have a forked tail.

Brown Noddy *Anous stolidus* L 15½" (39 cm) WS 32" (81 cm)

Overall dark gray-brown with whitish gray cap, blending at back; **immature** shows only a whitish line on forehead. Unlike other terns, noddies have long, wedge-shaped tail with only a small notch at tip.
VOICE: Usually silent; crowlike *karrk* call heard around breeding areas.
RANGE: Colonial nester on Dry Tortugas, FL. Casual to TX coast and off Outer Banks, NC. Accidental elsewhere on East Coast after hurricanes.

Black Noddy *Anous minutus* L 13½" (34 cm) WS 30" (76 cm)

Smaller than Brown Noddy, with shorter legs; bill is thinner and disproportionately longer; overall color is slightly blacker. In **immatures,** white area on head is very sharply defined.
RANGE: Tropical species, rare and now irregular visitor among Brown Noddies on Dry Tortugas (mostly immatures). Casual on TX coast.

Bridled Tern *Onychoprion anaethetus*

L 15" (38 cm) WS 30" (76 cm) Note white collar, brownish gray upperparts on **breeding adult.** Similar to Sooty Tern, but slimmer; wings more pointed; underwing and tail edges more extensively white; tail grayer. White forehead patch extends behind the eye, unlike Sooty. Juvenile has pale mottling above. **First-summer** appears pale headed.
VOICE: Typical call is a soft nasal *weep.*
RANGE: Nests in the West Indies; local and irregular off Florida Keys. Regular in summer well offshore in the Gulf of Mexico, and in the Gulf Stream to NC; rarely to NJ; after tropical storms, casual to New England. Many reports pertain to Sooty Terns.

Sooty Tern *Onychoprion fuscatus* L 16" (41 cm) WS 32" (81 cm)

Blackish above, white below; white forehead. Lacks white collar of Bridled Tern. Tail is deeply forked, edged with white. **Juvenile** is sooty brown overall, with whitish stippling on back; pale lower belly and undertail coverts; pale wing linings.
VOICE: Typical call is a high, nasal *wacky-wack.* Also a nasal *ipp.*
RANGE: Large breeding colony on Dry Tortugas, FL; also a few may nest on islands off TX and LA. Regular in summer well offshore in Gulf of Mexico and NC. Tropical storms can carry birds inland to Great Lakes and north to Maritime Provinces. Casual in southern coastal CA.

Aleutian Tern *Onychoprion aleuticus*

L 13½" (34 cm) WS 29" (74 cm) Dark gray above and below, with white forehead, black cap, black bill, black legs. In flight, distinguished from Common and Arctic Terns (page 240) by shorter tail, white forehead, and dark, white-edged bar on secondaries, most visible from below. **Juvenile** is buff and brown above; legs and lower mandible reddish.
VOICE: Call is a squeaky *twee-ee-ee,* unlike any other tern.
RANGE: Nests in loose colonies, sometimes with Arctic Terns. Casual or accidental in spring off BC.

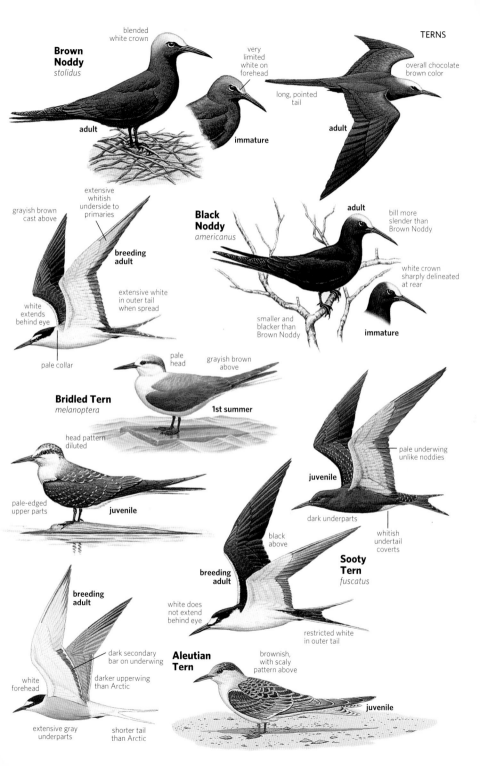

TERNS

Brown Noddy
stolidus

blended white crown

very limited white on forehead

long, pointed tail

overall chocolate brown color

adult

immature

adult

Black Noddy
americanus

extensive whitish underside to primaries

grayish brown cast above

breeding adult

extensive white in outer tail when spread

white extends behind eye

pale collar

bill more slender than Brown Noddy

adult

white crown sharply delineated at rear

smaller and blacker than Brown Noddy

immature

Bridled Tern
melanoptera

pale head

grayish brown above

1st summer

head pattern diluted

pale underwing unlike noddies

juvenile

pale-edged upper parts

juvenile

dark underparts

whitish undertail coverts

black above

Sooty Tern
fuscatus

breeding adult

breeding adult

dark secondary bar on underwing

darker upperwing than Arctic

white forehead

extensive gray underparts

shorter tail than Arctic

white does not extend behind eye

Aleutian Tern

restricted white in outer tail

brownish, with scaly pattern above

juvenile

Least Tern *Sternula antillarum* L 9" (23 cm) WS 20" (51 cm) **E**
Smallest North American tern. **Breeding adult** is gray above, with black cap and nape, white forehead, yellow bill with dark tip; underparts are white; legs yellow. By late summer, bill base is more greenish. In flight, black wedge on outer primaries is conspicuous; note also the short, deeply forked tail. **Juvenile** shows brownish, U-shaped markings; crown is dusky; wings show dark carpal bar. By first fall, upperparts are gray, crown whiter, but dark carpal bar is retained. **First-summer** birds are more like adults but have dark bill and legs, carpal bar, black line through eye, dusky primaries. Flight is rapid.
VOICE: Calls include high-pitched *kip* notes and a harsh *chir-ee-eep*.
RANGE: Nests in colonies on beaches and sandbars; also on rooftops. Fairly common but local on East and Gulf Coasts; declining, especially inland and on the CA coast. Casual to Great Lakes. Winters from Central America south.

Black Tern *Chlidonias niger* L 9¾" (25 cm) WS 24" (61 cm)
Breeding adult is mostly black, with dark gray back, wings, and tail; white undertail coverts. In flight, shows uniformly pale gray underwing and fairly short tail, slightly forked. Bill is black in all plumages. **Juvenile** and winter birds are white below, with dark gray mantle and tail; dark ear patch extends from dark crown; flying birds show dark bar on side of breast and gray flanks (distinctive for New World *surinamensis*). Some juveniles show a contrastingly paler rump. Carpal bar on upper wing is much darker than juvenile White-winged Tern. First-summer birds can be like winter adults or may have some dark feathers on head and underparts; second-summer birds are like breeding adults, but show some whitish on head; full breeding plumage is acquired by third spring. **Molting fall adults** appear patchy black-and-white as they acquire winter plumage in late summer; these birds are easily confused with the White-winged Tern.
VOICE: Calls include a metallic *kik* and a slurred *k-seek*.
RANGE: Nests on lakeshores and in marshes; declining over much of range. Migrants may be seen well offshore; rare migrant on West Coast. Winters mostly in South America.

White-winged Tern *Chlidonias leucopterus*
L 9½" (24 cm) WS 23" (58 cm) Bill and tail shorter than Black Tern; tail less deeply notched. In **breeding** plumage, bill usually black, but sometimes red; white tail, whitish upperwing coverts, and black wing linings are distinctive; upperwing shows black outer primaries. **Molting** birds are patchy black-and-white, but whitish tail and rump are distinctive; black wing linings often last until late summer. **Winter adult** has white wing linings; lacks dark breast bar of Black Tern; crown, speckled rather than solid black, not usually connected to dark ear patch. First-summer bird resembles winter adult; second-summer usually like breeding adult; adult plumage reached by third spring. **Juvenile**'s head pattern resembles Black, but browner back shows greater contrast with grayish wing coverts and whitish rump.
RANGE: Eurasian species, casual to or near East Coast; accidental to NB, QC, VT, upstate NY, ON, WI, IN, MB, AK, and CA. Fewer records in last decade.

Least Tern
antillarum

breeding adult

white forehead

yellow bill with black tip

juvenile

wingbeats very shallow and rapid

breeding adult

dark carpal bar

1st summer

dark gray above

dark ear patch extends from crown

pale underwing

molting fall adult

juvenile

dark bar on sides

gray flanks

Black Tern
surinamensis

black body

breeding adult

juvenile

white vent and undertail coverts

breeding adult

paler upperwing than Black Tern

White-winged Tern

juvenile

brownish back contrasts with white rump

black wing linings with white remiges

winter adult

tail shorter and squarer than Black Tern

bill color variable, sometimes dark red

red legs

molting adult

blackish postocular spot; head paler than Black Tern

pale rump

pure white below

Common Tern *Sterna hirundo* L 14½" (37 cm) WS 30" (76 cm)

Medium gray above; black cap and nape; paler below, though grayish in breeding plumage. Bill red, usually tipped with black. Slightly stockier than Arctic Tern, with flatter crown, longer bill and neck (head projects farther in front of wing in flight), and shorter tail. In flight, usually displays a dark wedge, variably shaped, near tip of upperwing; in late summer all outer primaries can appear dark. Early **juvenile** shows some brown above, white below, with mostly dark bill. Juvenile's forehead is white, crown and nape blackish, secondaries dark gray; compare with juvenile Forster's Tern. All immature and winter plumages have a dark carpal bar. Full **adult breeding** plumage is acquired by third spring.

VOICE: Calls are a sharp *kip* and a distinctive low, piercing, drawn-out *kee-ar-r-r-r.*

RANGE: Nests in large colonies. Common throughout breeding range. Uncommon and declining migrant on Pacific coast from WA south; away from Great Lakes largely an uncommon interior migrant, almost strictly in fall in West, west of Great Plains. An East Asian subspecies, *longipennis,* seen regularly on the islands of western AK in breeding plumage, is darker overall; bill and legs black.

Arctic Tern *Sterna paradisaea* L 15½" (39 cm) WS 31" (79 cm)

Medium gray above; black cap and nape; paler below; bill deep red. Slightly slimmer than Common Tern; rounder head; shorter neck, bill, and especially legs. In flight, upperwing appears uniformly gray, lacking dark wedge of Common; underwing shows very narrow black line on trailing edge of primaries; all flight feathers appear translucent. Note also longer tail, head does not project as far as Common. **Juvenile** largely lacks brownish wash of early juvenile Common; carpal bar less distinct; secondaries whitish and a portion of coverts whitish too, creating an effect like Sabine's Gull (page 212). Forehead white, crown and nape blackish. Full adult breeding plumage acquired by third spring.

VOICE: Call, a raspy *tr-tee-ar,* higher than Common; also a sharp *keet.*

RANGE: Migrates well offshore; casual inland during migration, especially in late spring. Has nested in MT. Winters in Antarctic and subantarctic waters.

Forster's Tern *Sterna forsteri* L 14½" (37 cm) WS 31" (79 cm)

Breeding adult is snow white below, pale gray above, with black cap and nape; mostly orange bill, orange legs and feet. Wingbeat much slower than Roseate Tern (page 242). Legs and bill longer than Common and especially Arctic Terns. Long, deeply forked gray tail has white outer edges. In flight, upperwing entirely pale. **Winter** plumage resembles Common and Arctic but is acquired by mid- to late Aug., much earlier than those species, which molt chiefly after migration out of U.S. Note also lack of dark carpal bars; most have dark eye patches not joined at nape as in Common, but many have dark streaks on nape. **Juvenile** and **first-winter** bird have shorter tails than adults and more dark color in wings. Juvenile has ginger brown cap and dark eye patch; carpal bar is faint or absent.

VOICE: Calls include a hoarse *kyarr,* lower than Common; also a higher *ket.*

RANGE: Nests in colonies in marshes. Rare in late summer and fall to New England and Atlantic Canada.

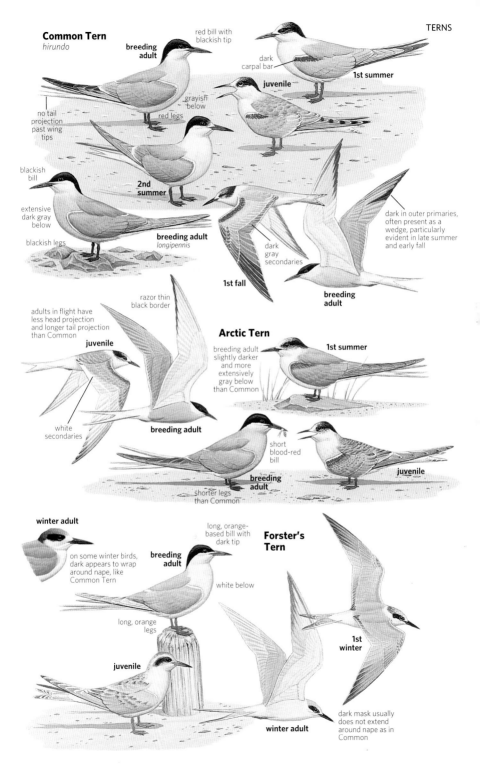

TERNS

Common Tern
hirundo

red bill with blackish tip

breeding adult

dark carpal bar

1st summer

juvenile

grayish below

red legs

no tail projection past wing tips

blackish bill

extensive dark gray below

2nd summer

blackish legs

breeding adult
longipennis

dark in outer primaries, often present as a wedge, particularly evident in late summer and early fall

dark gray secondaries

1st fall

breeding adult

razor thin black border

adults in flight have less head projection and longer tail projection than Common

juvenile

Arctic Tern

breeding adult slightly darker and more extensively gray below than Common

1st summer

white secondaries

breeding adult

short blood-red bill

breeding adult

juvenile

shorter legs than Common

winter adult

on some winter birds, dark appears to wrap around nape, like Common Tern

long, orange-based bill with dark tip

breeding adult

Forster's Tern

white below

long, orange legs

1st winter

juvenile

winter adult

dark mask usually does not extend around nape as in Common

Roseate Tern *Sterna dougallii* L 15½" (39 cm) WS 29" (74 cm) **E**
Breeding adult is white below with slight, variable pinkish cast visible in good light; pale gray above with black cap and nape. Much paler overall than Common and Arctic Terns (page 240). Lacks dark trailing edge on underside of outer wing. Bill mostly black; during summer more red appears at base. Wings shorter than Common and Arctic; flies with rapid wingbeats suggestive of Least Tern (page 238). Deeply forked all-white tail extends well beyond wings in standing bird. Legs and feet bright red-orange. **Juvenile's** brownish cap extends over forehead; mantle looks coarsely scaled, lower back barred with black; bill and legs black. **First-summer** bird has white forehead; lacks dark secondaries of immature Common. Full adult plumage is attained by second spring.
VOICE: Call is a soft *chi-weep* or *ki-vit;* alarm signal a drawn-out *zra-ap,* like ripping cloth.
RANGE: Uncommon and highly maritime, usually comes ashore only to nest. Rare on mid-Atlantic coast in late spring and summer.

Gull-billed Tern *Gelochelidon nilotica*
L 14" (36 cm) WS 34" (86 cm) **Breeding adult** white below, pale gray above, with black crown and nape, stout black bill, black legs and feet. Stockier, paler than Common Tern (page 240); wings broader; tail shorter, and only moderately forked. **Juveniles** and **winter** birds appear largely white headed apart from some fine streaking. Juvenile has pale edgings on upperparts, bill paler. Often hunts for insects over fields and marshes in direct powerful flight; does not hover or dive in water.
VOICE: Adult call is a raspy, sharp *kay-wack;* call of juvenile is a faint, high-pitched *peep peep.*
RANGE: Nests in salt marshes and on beaches. Casual on coast to New England, Atlantic Canada, and southern CA coast (north of San Diego County) to Santa Barbara County; casual or accidental from interior of North America except at Salton Sea, CA, where it nests.

Large-billed Tern *Phaetusa simplex*
L 14½" (37 cm) WS 36" (92 cm) Mantle and short tail dark gray; white below with white forehead and black cap; legs and stout bill yellow. In flight, shows striking Sabine's Gull–like pattern (page 212).
RANGE: A South American freshwater species. Accidental; recorded in late spring and summer from IL, OH, and NJ; additionally from Cuba, Bermuda, and Grenada.

Sandwich Tern *Thalasseus sandvicensis*
L 15" (38 cm) WS 34" (86 cm) Slender, black bill, tipped with yellow. **Breeding adult** is pale gray above with black crown, short black crest. In flight, shows some dark in outer primaries. White tail is deeply forked, comparatively short. Adult in **winter** plumage, seen as early as July, has a white forehead, streaked crown, grayer tail. **Juvenile's** tail is less deeply forked; bill often lacks yellow tip, in a few it is entirely yellow. By late summer, juvenile loses dark markings above.
VOICE: Calls include abrupt *gwit gwit* and *skee-rick* notes, like Elegant.
RANGE: Nests on coastal beaches and islands. Regular visitor north to NJ; casual farther north, accidental inland, particularly following tropical storms; casual to southern CA coast, where hybrids with Elegant Terns have been noted.

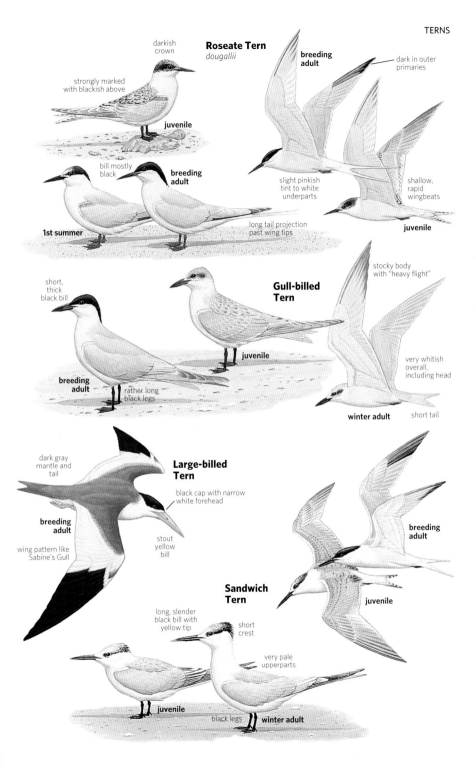

Roseate Tern
dougallii

darkish crown

strongly marked with blackish above

juvenile

bill mostly black

breeding adult

1st summer

breeding adult

dark in outer primaries

slight pinkish tint to white underparts

long tail projection past wing tips

shallow, rapid wingbeats

juvenile

Gull-billed Tern

short, thick black bill

stocky body with "heavy flight"

juvenile

breeding adult

rather long black legs

very whitish overall, including head

winter adult   short tail

dark gray mantle and tail

Large-billed Tern

black cap with narrow white forehead

breeding adult

wing pattern like Sabine's Gull

stout yellow bill

breeding adult

juvenile

Sandwich Tern

long, slender black bill with yellow tip

short crest

very pale upperparts

juvenile

black legs   **winter adult**

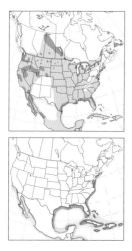

Caspian Tern *Hydroprogne caspia*

L 21" (53 cm) WS 50" (127 cm) Large, stocky; bill orange-red to coral red, much thicker than Royal Tern. In flight, shows dark underside of primaries; tail less deeply forked than Royal. **Adult** acquires black cap in **breeding** season; in **winter adult** and **juvenile,** crown streaked; never shows fully white forehead of Royal.

VOICE: Adult's calls include a harsh, raspy *kowk* and *ca-arr;* immature's a distinctive, whistled *whee-you.*

RANGE: Small colonies nest on coasts, shoals, rocky or sandy islands.

Royal Tern *Thalasseus maximus* L 20" (51 cm) WS 41" (104 cm)

Orange-red bill, thinner than Caspian. In flight, shows mostly pale underside of primaries; tail more deeply forked than Caspian. **Adult** shows white crown most of year; black cap acquired briefly early in **breeding** season. In **winter adult** and **juvenile,** black on nape does not usually extend to encompass eye, but some show more dark that includes the eye; these are best separated from smaller Elegant Terns by overall size, bill shape, and call.

VOICE: Calls include a bleating *kee-rer* and ploverlike whistled *tourreee.*

RANGE: Nests in dense colonies. Fairly common in winter on southern CA coast, where a few breed; uncommon to rare north of breeding range along Atlantic coast in late summer; casual to Bay Area in CA (formerly more regular) and in North American interior.

Elegant Tern *Thalasseus elegans* L 17" (43 cm) WS 34" (86 cm)

Bill longer, thinner than larger Royal; reddish orange in adults, yellow in some juveniles. In flight, shows mostly pale underside of primaries; compare to Caspian. **Breeding adult** is pale gray above with black crown and nape, black crest; white below, often with pinkish tinge. **Winter adult** and **juvenile** have white forehead; black over top of crown extends forward around eye; compare to Royal. Juveniles mottled above, may have orange legs; some have less black around eye, like juvenile Royal.

VOICE: Sharp *kee-rick* call very similar to Sandwich Tern.

RANGE: Breeders arrive in coastal southern CA in early Mar. Many post-breeders from western Mexico augment population, and many disperse north up coast, a few to WA, casually to BC. Lingers on southern CA coast (small numbers into Nov., casually later), but many reports of Elegant Terns in midwinter are blacker-faced Royals. Very rare spring and summer at Salton Sea, CA; casual in Southwest. Accidental from coastal areas in East (VA, FL, TX).

Black Skimmer *Rynchops niger* L 18" (46 cm) WS 44" (112 cm)

No other bird has a lower mandible longer than the upper (only exceptions are two Old World skimmer species). A long-winged coastal bird, it furrows the shallows with its red, black-tipped bill. Black above and white below; red legs and bill shape are distinctive. Female is distinctly smaller than the male. **Juvenile** is mottled dingy brown above. **Winter adults** show a white collar. Largely a crepuscular, even nocturnal, feeder and sits around much of the day, often in tightly packed flocks.

VOICE: Typical call is a nasal *ip* or *yep.*

RANGE: Nests on sandy beaches and islands. Casual north to Atlantic Canada, northwestern CA, and inland; many nest at Salton Sea, CA, and has nested in Kings County, CA.

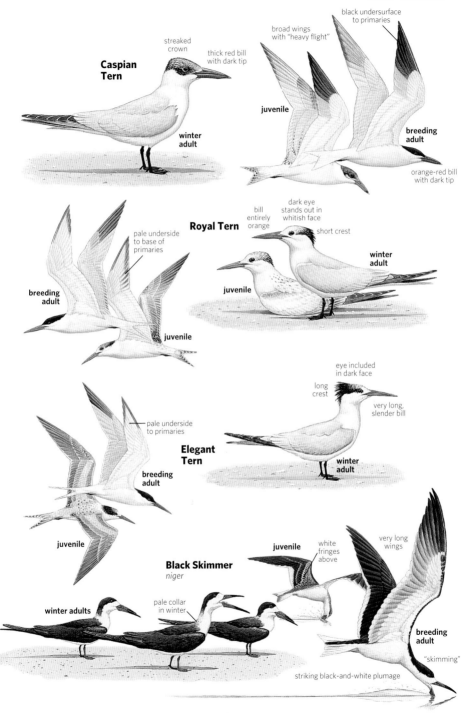

Caspian Tern

streaked crown

thick red bill with dark tip

winter adult

broad wings with "heavy flight"

black undersurface to primaries

juvenile

breeding adult

orange-red bill with dark tip

Royal Tern

pale underside to base of primaries

breeding adult

juvenile

bill entirely orange

dark eye stands out in whitish face

short crest

winter adult

juvenile

Elegant Tern

pale underside to primaries

breeding adult

long crest

eye included in dark face

very long, slender bill

winter adult

Black Skimmer
niger

juvenile

white fringes above

very long wings

winter adults

pale collar in winter

breeding adult

"skimming"

striking black-and-white plumage

SKUAS • JAEGERS Family Stercorariidae

Formerly placed with the Gulls, Terns, and Skimmers, recent molecular evidence indicates that they are most closely related to Alcidae and belong in their own family. Predatory and piratic seabirds, skuas are broader winged than jaegers. Largely silent at sea. SPECIES: 7 WORLD, 5 N.A.

Great Skua *Stercorarius skua* L 22" (56 cm) WS 54" (137 cm)

Large, heavy, and barrel-chested; wings broader and more rounded than jaegers (pages 248, 250); tail shorter and broader. Shows a distinctly hunchbacked appearance in flight and a conspicuous white bar at base of primaries; bill is heavier than jaegers. Great Skua is distinguished from South Polar Skua by overall reddish or ginger brown color and heavy streaking on back, wing coverts, and much of underparts; sometimes shows dark brown cap. **Juvenile** and immature show less streaking, especially on underparts. A small number of juvenile dark morphs have much less rufous streaking; resemble juvenile South Polar Skuas. Strong, powerful fliers, skuas pursue gulls and other seabirds and rob them of their prey.

RANGE: Uncommon; breeds in Iceland and northern U.K.; a few on Spitsbergen and northern Norway; winters in North Atlantic. Seen well offshore from Nov. to Apr.; rare in summer off Canadian east coast.

South Polar Skua *Stercorarius maccormicki*

L 21" (53 cm) WS 52" (132 cm) Like Great Skua, large, heavy, and barrel-chested; wings broader and more rounded than jaegers (pages 248, 250); tail shorter and broader. Like Great Skua, shows a distinctly hunchbacked appearance in flight, and a bold white bar at base of primaries, and a heavier bill than jaegers. In all ages, South Polar Skua shows a uniform mantle coloring and lacks the reddish tones and streaking seen on upperparts of Great Skua. In **light-morph** birds, contrastingly pale gray nape is distinctive; light morph also shows grayish head and underparts. **Dark morph** is uniformly blackish brown across mantle, with golden hackles on nape; distinguished from subadult Pomarine Jaeger (page 248) by larger size, broader and more rounded wings, more distinct white wing bar. **Juveniles** and immatures of both color morphs are darker than light-morph adults, ranging from dark brown to dark gray. In the field, birds under two years of age are generally indistinguishable from juveniles; birds over two years old are generally indistinguishable from full adults.

RANGE: Breeds in Antarctica. Winters (our summer) in the North Atlantic and North Pacific, usually from May to early Nov. Most numerous in spring and fall off the West Coast, in spring off the East Coast; casual off the southern coast of AK; accidental in ND; a record from TN after Hurricane Katrina likely this species. Very rarely seen from shore. Difficulty of identification makes range information somewhat speculative for both skua species. Several records (photographs) of birds off mid-Atlantic coast could pertain to Brown Skua (*S. antarcticus*), of Southern Hemisphere, or possibly hybrids between that species and South Polar Skua.

all Great and South Polar Skuas have broad wings and short tails with bold white flash at base of primaries

Great Skua

typical adult

buffy rufous edges on upperparts

thick bill

dull cinnamon below

short, rounded tail points

pale adult

dark adult

bold white primary flash

juvenile

bold white primary flash

uniform dark upperparts

intermediate-morph adult

short, rounded tail points

dark-morph adult

slight golden cast on nape

South Polar Skua

juvenile

most look quite dark, except for white wing patches

light-morph adults

thick bill averages slightly smaller than Great Skua

juvenile

JAEGERS

Arctic breeders. Wings are longer and slimmer than skuas. Adult plumage and long central tail feathers take three or four years to develop. Complex and variable plumages make identification extremely difficult. Most molts occur after the fall migration. All nonadult jaegers have checkered underwing coverts. Largely silent away from breeding areas.

Pomarine Jaeger *Stercorarius pomarinus*

L 21" (53 cm) WS 48" (122 cm) Body bulkier, bicolored bill longer and thicker, wingbeats slower than Parasitic Jaeger. Most birds show a distinctive second pale underwing patch at base of primaries (fainter or lacking in Parasitic). **Adult's** tail streamers, twisted at ends, form dark blobs when seen from side; length is variable, averages longer in male. Note extensive helmet on sides of head, which extends down to "jowls" area. Compare **dark-morph adults** and subadults with South Polar Skua (page 246). Some grayish brown **juveniles** are dark, some pale, but none shows the foxy red tones of most juvenile Parasitics; underwing is paler than body; pale, barred uppertail coverts forms a contrasting patch above. Central tail feathers barely project beyond outer tail feathers and have blunt tips. When well seen, its thicker, sharply two-tone bill distinguishes it from Parasitic, which has a thinner bill with a darker tip. Juvenile's primaries lack conspicuous pale tips; has strongly and evenly barred tail.

RANGE: Nests on Arctic tundra, where it feeds primarily on lemmings and young birds; otherwise frequents offshore waters. Seen less often from shore than Parasitic. Common off West Coast in spring migration and especially late in fall; often migrates in flocks. Uncommon in winter. Casual inland away from Great Lakes (rare) and Great Plains (very rare); interior migrants are seen mostly in late fall. Unlike juveniles of other two jaeger species, juvenile Pomarines are not regularly seen well south of breeding grounds until late in fall, but from mid-Nov. onward, probably the most likely jaeger species.

Parasitic Jaeger *Stercorarius parasiticus*

L 19" (48 cm) WS 42" (107 cm) Smaller size, more slender body, faster wingbeats than Pomarine Jaeger; also smaller head, thinner bill, pointed tail streamers. **Adult** lacks helmeted effect of Pomarine; is paler near bill. **Juvenile** is highly variable; shows rufous tips on primaries, sometimes very indistinct; distinctive rusty tones particularly evident on **light morphs;** tail coverts have fainter, wavier bars than Pomarine. Only juvenile Parasitic has pointed tips to central tail feathers; rounded tips in juvenile Pomarine and Long-tailed; juvenile Parasitic also has shorter central tail feathers than juvenile Long-tailed. Also compare to Long-tailed Jaeger (page 250), with which it is easily confused.

RANGE: Nests on Arctic tundra; found at sea during nonbreeding seasons. Fairly common; the jaeger species most often seen from shore in migration, often in pursuit of terns. Casual fall migrant inland; more regular on Great Lakes and Salton Sea, CA (late Aug. to early Nov.).

light-morph breeding adult

thick two-tone bill

dark below gape

thick two-tone bill

Pomarine Jaeger

light-morph juvenile

dark primaries

strong bars on vent and undertail coverts

dark-morph breeding adult

double white wing flash

light-morph breeding adult

long twisted tail feathers, often longer than shown, with thick tip

usually with breast band, which some adult males lack

light-morph juvenile

double white wing flash

light-morph 1st summer

very short and rounded central tail points

all nonadult jaegers have checkered underwing coverts

light-morph breeding adult

pale at base of forehead

slender bill

primaries with variable number of white shafts

dark-morph breeding adult

light-morph breeding adult

Parasitic Jaeger

light-morph 1st summer

dark brown above

pointed tail feathers of moderate length

rufous tips to feathers above

most have a diffuse grayish brown breast band

light-morph juvenile

light-morph juveniles

pointed tips to central tail feathers

slender bill

light-morph juvenile

most have rufous cast

fainter bars on vent and undertail coverts

pale primary tips

Long-tailed Jaeger *Stercorarius longicaudus*

L 22" (56 cm) WS 40" (102 cm) Most lightly built jaeger, with round chest, flat belly, narrow wings, and disproportionately long tail in all ages; bill rather short and thick. Flight is more graceful, ternlike. Note distinctive contrast between grayish mantle and darker flight feathers; usually has only two to three white primary shafts; no pale underwing patch except on **juvenile.** **Adult** has well-defined black cap (most restricted in extent of the three jaeger species), no breast band as in most jaegers, and usually very long, pointed central tail streamers; many fall adults have dropped them by southward migration. Juvenile's rather long central tail feathers have round, often white-edged tips; bill is half dark, half gray and appears stubby, unlike the slender, larger bill of Parasitic Jaeger (page 248). Juvenile Long-tailed's dark primaries lack the rufous-buff tips seen on most Parasitic, and body is grayer overall than Parasitic except for dark morph; fringing above whitish, never rusty. **Light-morph juveniles** show distinctive white belly and strong, even, black barring on upper- and undertail coverts; palest birds may have very pale gray heads. **Dark-morph juveniles** often lack barring on uppertail coverts.

RANGE: Nests commonly in dry, upland-tundra breeding area; migrating birds uncommon to common well off West Coast (fall); rare closer to shore and off East Coast (mainly in fall); very rare inland (mainly in early fall), and casual off Gulf Coast. Away from the Great Lakes, where Parasitic dominates, Long-tailed is as likely or more likely seen than the other two jaeger species, particularly in early fall. Feeds on small mammals, insects, even berries while on tundra; the most likely jaeger to be seen feeding on insects during migration. Winters in Southern Hemisphere oceans.

AUKS • MURRES • PUFFINS Family Alcidae

These black-and-white "penguins of the north" have set-back legs that give them an upright stance on land. In flight, wingbeats are rapid and shallow. Collectively known as "alcids." Most species are largely silent at sea, when well away from breeding grounds.
SPECIES: 24 WORLD, 22 N.A.

Dovekie *Alle alle* L 8¼" (21 cm)

Small and plump with short neck, stubby bill. **Breeding adult** is black above, white below; black upper breast contrasts sharply with white underparts, like a tiny Thick-billed Murre (page 252); white trailing edge to secondaries; dark wing linings. Usually swims tilted forward in the water. In **winter** plumage, throat, chin, and lower face are white, with white curving around behind eye.

VOICE: Largely silent, except on breeding grounds where quite vocal; most common call is a rising and falling trill, lasting one to three seconds, which can be given from the ground or in flight.

RANGE: Abundant on high North Atlantic breeding grounds. Winters irregularly in North Atlantic south to off coast of NC; rarely to FL, but occasionally in large numbers; casual inland. Uncommon and well-isolated in northern Bering Sea (Little Diomede, King, and St. Lawrence Islands). Casual in summer to northern AK, the Pribilofs, and the Aleutians. Winter grounds for this isolated population are unknown.

two to three white primary shafts in nearly all plumages

no white wing flash

Long-tailed Jaeger

breeding adult

gray upperparts contrast sharply with blackish flight feathers

1st summer

very long, pointed central tail feathers

clean blackish cap

stubby bill

breeding adult

light-morph juvenile

long slender wings

pale tips to feathers above

long tail has rounded central tail feathers with pale tips

dark-morph juvenile

stubby bill

some with pale belly and dark breast

dark primaries

typical juvenile

strong, straight bars on vent and undertail coverts

winter

blackish underwing

white tips to secondaries

breeding adult

dark neck band

black head

Dovekie

stubby bill

breeding adult

winter

whitish extends up into ear coverts

Common Murre *Uria aalge* *L 17½" (45 cm)*

Large, with a long, slender, pointed bill. Upperparts dark sooty gray, head brownish; underparts white with some dusky streaking on flanks. Some Atlantic birds have a **"bridle,"** a white eye ring and spur. In **winter** plumage, a dark stripe extends from eye across white cheek. **Juvenile** has shorter bill; distinguished from Thick-billed Murre by white facial stripe, paler upperparts, mottled flanks, and thinner bill. Two subspecies recognized in Pacific and three to five in Atlantic; Pacific subspecies generally larger, including wings and bill, than Atlantic subspecies.

VOICE: Gives guttural calls on breeding grounds that are deep and of a growling nature, *aargh.*

RANGE: Nests in dense colonies on rocky cliffs. Chick accompanies adult at sea. Off East Coast found well offshore during nonbreeding seasons; only casually seen from land. Rare south of Cape Cod to mid-Atlantic states.

Thick-billed Murre *Uria lomvia* *L 18" (46 cm)*

Stocky, with a thick, fairly short bill, arched at tip to form a blunt hook. Upperparts and throat of **adult** are darker than Common Murre; white of underparts usually rises to a sharp point on the foreneck. Most birds show a distinct white line on cutting edge of upper mandible; in Pacific birds *(arra)*, bill is slightly longer and thinner than Atlantic birds (nominate *lomvia*). In immature and **winter adult,** face and neck are more extensively dark than Common. Immature has smaller bill than adult. First-summer bird is browner above than adult; otherwise similar to winter bird.

VOICE: Gives a variety of deep growling calls on breeding grounds.

RANGE: Nests in colonies on rocky cliffs. Common on breeding grounds. On East Coast, much more regular in winter (mostly far offshore) south of Canada than Common; casual to mid-Atlantic states; recorded to FL. Casual on West Coast south of breeding range, where most records are from the Monterey area in CA. In East, casual inland, with few recent records (several from ON).

Razorbill *Alca torda* *L 17" (43 cm)*

A chunky bird, with a big head and thick neck; black above, white below. Rather long, pointed tail; heavy head; and massive, arching bill distinguish Razorbill from murres. Swimming birds often hold tail cocked up. A white band crosses the bill; in **breeding** plumage, a white line runs from bill to eye. **Immature** lacks white band; bill is smaller but still distinctively shaped.

VOICE: On nesting grounds gives a variety of growls, *knorrrr.*

RANGE: Nests on rocky cliffs and among boulders. Winters in large numbers on the Grand Banks off Newfoundland, regularly south to Long Island and irregularly to NC. Regularly seen from shore at some coastal promontories. Casual south to FL; accidental inland in East.

breeding adult

upperparts browner than Thick-billed

long, pointed bill

rounded edge to dark neck

"bridled" breeding adult

Common Murre

some dusky streaking on flanks

dark center to underwing

Common Murre breeding adult

dark postocular spur

winter

juvenile

flank streaks

bill slightly less heavy than adult

1st winter

dark face

black back

short, thick bill with curved culmen

faint gape stripe

winter adult

white flanks

breeding adult

distinct white line on cutting edge of upper mandible

white forms inverted V into blackish neck

Thick-billed Murre
arra

breeding adult

slightly smaller bill

whitish cheek with dusky postocular line

whitish cheek

immature

unique thick bill with white band

winter adult

black hood and upperparts

thick bill

long pointed tail

breeding adult

breeding adult

Razorbill

Black Guillemot *Cepphus grylle* L 13" (33 cm)

In all plumages, white axillaries and wing linings distinguish Black from Pigeon Guillemot. **Breeding adult** black overall, with large white patch on upperwing. **Winter adult** white; upperparts of *arcticus* heavily mottled with black, except on nape; wing patch less contrasty. **Juvenile** *arcticus* is sooty above; sides and wing patches mottled. First-summer birds are patchily black-and-white; wing patches mottled. Five to seven subspecies have been recognized in the Holarctic; generally five are recognized now, two found in North America. Juveniles and winter adults of the high Arctic subspecies, *mandtii* appear much paler than more southerly East Coast *arcticus*. However, in *mandtii* an all-black (in breeding plumage) color morph with various indications of melanism in winter plumage has been reported from western Greenland.

VOICE: High-pitched screaming and whistles on breeding grounds.

RANGE: Fairly common in the East; usually seen close to shore in winter on East Coast; found regularly south to MA, rarely to Long Island; casual to Carolinas; accidental inland (all inland records believed to be *mandtii*). Arctic *mandtii* is numerous south to at least Newfoundland in winter. In AK, closely associated with limits of pack ice, usually the northern Bering Sea, and also winters in the polynyas within the pack ice of the Chukchi Sea, the only wintering alcid there. Its range extends west along the Arctic coast and offshore islands of the Russian Far East. Casual from the interior of AK in late fall and winter; accidental from the Valdez Arm of Prince William Sound in summer (pair in 2005). Recent decline in AK numbers have likely resulted from climate change.

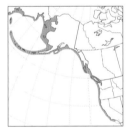

Pigeon Guillemot *Cepphus columba* L 13½" (34 cm)

Breeding and **winter adult** plumage similar to smaller Black Guillemot, but note black bar on white upperwing patch; bar often obscured in swimming bird. **Juvenile** is dusky above; crown and nape darker; wing patch marked with black edgings; breast and sides mottled gray; compare with juvenile Marbled Murrelet (page 256); some are paler and some have pale in middle of underwing, suggesting Black Guillemot. First-winter resembles winter adult but is darker. In all plumages, dusky axillaries and wing linings distinguish Pigeon from Black Guillemot. When the two species are seen together in limited area of overlap (e.g., northern Bering Sea), the size difference is obvious. Black's wings appear pointier, and it flies more rapidly with more twists and turns.

VOICE: Give a variety of very rapid, high-pitched, twittering and whistled calls on breeding grounds.

RANGE: Seen near shore when breeding. Winter range poorly known.

Cassin's Auklet *Ptychoramphus aleuticus* L 9" (23 cm)

Small, plump; wings more rounded than murrelets; bill short and stout, pale spot at base of lower mandible; pale eyes. Upperparts dark gray, shading to paler gray below; whitish belly. Prominent white crescent above eye. Juvenile paler overall; throat whitish; darker eye and black bill. When feeding at sea, often in small parties. When full, they fly weakly when flushed until out of danger and then land again.

VOICE: Call, heard only on the breeding grounds, is a harsh, rhythmic croaking, often given from safety of burrow, often in unison.

RANGE: Common; colonial breeder on islands and on isolated coastal cliffs. Highly pelagic; usually seen farther offshore than murrelets.

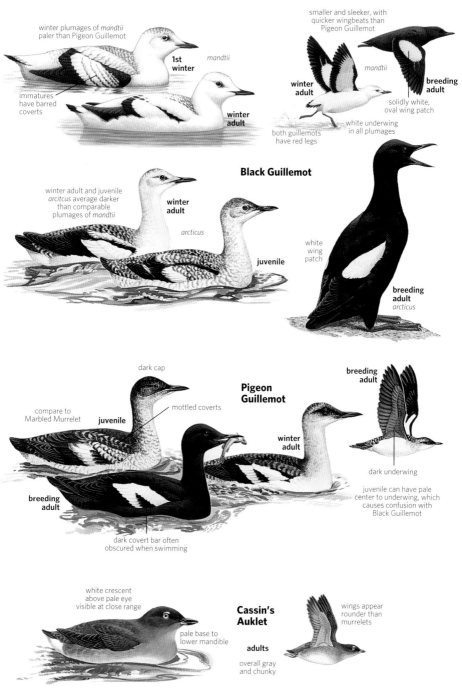

winter plumages of *mandtii* paler than Pigeon Guillemot

1st winter

mandtii

immatures have barred coverts

winter adult

smaller and sleeker, with quicker wingbeats than Pigeon Guillemot

mandtii

winter adult

breeding adult

solidly white, oval wing patch

both guillemots have red legs

white underwing in all plumages

Black Guillemot

winter adult and juvenile *arcitcus* average darker than comparable plumages of *mandtii*

winter adult

arcticus

juvenile

white wing patch

breeding adult *arcticus*

dark cap

Pigeon Guillemot

compare to Marbled Murrelet

juvenile

mottled coverts

winter adult

breeding adult

breeding adult

dark underwing

juvenile can have pale center to underwing, which causes confusion with Black Guillemot

dark covert bar often obscured when swimming

white crescent above pale eye visible at close range

Cassin's Auklet

pale base to lower mandible

adults

overall gray and chunky

wings appear rounder than murrelets

Long-billed Murrelet $Brachyramphus\ perdix$ L 11½" (29 cm)

In **winter,** lacks conspicuous white collar of similar Marbled Murrelet; shows small pale oval patches on sides of nape; in **breeding** plumage upperparts are less rufous, throat paler. While bill is larger than Marbled, it is not disproportionately so. Underwing coverts may average paler on Long-billed, but more study is needed. Formerly considered a subspecies of Marbled.

VOICE: Poorly known. A thin whistled series of *fii* notes has been described.

RANGE: Found in coastal northeast Asia. Casual throughout North America; most records from the interior are in fall of winter-plumaged birds. There are now a few summer records for the Pribilofs and one for Adak Island, Aleutians. In recent years, surveys for Marbled Murrelets in late summer on the coast of northwestern CA have detected a number of Long-billeds (nearly ten since 2003), raising the question of whether a few are nesting there, perhaps in the same areas where Marbled also nests.

Marbled Murrelet $Brachyramphus\ marmoratus$

L 10" (25 cm) **T** Bill longer than Kittlitz's Murrelet. Tail all-dark, but white on overlapping uppertail coverts. **Breeding adult** dark above, heavily mottled below. In **winter** plumage, white on scapulars distinguishes Marbled from other murrelets except Long-billed and Kittlitz's with shorter bill and breast band; white on face of Marbled is variable but less than Kittlitz's. **Juvenile** is like winter adult but mottled below; by first winter, underparts are mostly white. All murrelets have more-pointed wings and faster flight than auklets.

VOICE: Highly vocal at all times of the year; indeed, said to be the most vocal alcid at sea. Call is a series of loud, high *kree* notes.

RANGE: Nests inland, usually high on branches in trees in old-growth forest; the large-scale lumbering of this habitat has directly led to this species' being designated as threatened. In subarctic regions will nest on the ground. Fairly common but local in breeding range; rare to casual (south of northern Santa Barbara County) in southern CA.

Kittlitz's Murrelet $Brachyramphus\ brevirostris$ L 9½" (24 cm)

Bill distinctly shorter than Marbled Murrelet. Outer tail feathers white; these are best seen in flight just before landing when the tail is spread; beware of second-year (one-year-old) Marbled Murrelets, which can have extensive white on outer uppertail coverts that overhang the base of the tail. In flight, note also the very dark and contrasting (even in breeding plumage) blackish underwing coverts. **Breeding adult's** buffy grayish brown upperparts are heavily patterned; throat, breast, and flanks mottled; belly white. In **winter,** note extensive white on face, making eye conspicuous; nearly complete breast band; and white edges on secondaries. **Juvenile** distinguished from Marbled by shorter bill, paler face, white outer tail feathers.

VOICE: Low groaning and short quacking notes; far less vocal than Marbled.

RANGE: Uncommon to fairly common but local, and recent evidence of declines. Rare in northern Bering Sea. Accidental to BC and southern CA.

Long-billed Murrelet

brownish above

breeding adult

pale throat

often with pale spot on back of head

nearly complete whitish eye ring

slender bill

even line of separation between dark upperparts and white underparts

winter

white scapular patch

juvenile

Marbled Murrelet

winter

cap dips below eye

white lores

white collar

white scapular patch

breeding adult

mottled brown chin and throat

breeding adult

overall chocolate brown

winter

winter

breeding adult

white outer tail feathers

white belly

underwing contrasts darker to body, more so than breeding adult Marbled

Kittlitz's Murrelet

short, stubby bill

breeding adult

juvenile

extensive white face

color overall paler than Marbled

winter

Ancient Murrelet *Synthliboramphus antiquus* L 10" *(25 cm)*

Black crown and nape contrast with gray back. White streaks on head and nape of **breeding adult** give it an "ancient" look. Note also black chin and throat, yellowish bill. **Winter adult's** bib is smaller and flecked with white, streaks on head less distinct. **Immature** lacks head streaks; throat is mostly white; distinguished from winter Marbled Murrelet (page 256) by heavier, paler bill and by sharp contrast between head and back. In flight, Ancient Murrelet holds its head higher than other murrelets; dark stripe on body at base of wing contrasts with white underparts, white wing linings.

VOICE: Nine different calls have been identified. Call on water is a short *leep*. Around nest other calls, including a short, emphatic *chirrup*.

RANGE: Uncommon to common; breeds primarily on the Aleutians and other AK islands; winters to central CA, rarely to southern CA. Casual inland throughout North America, mainly in late fall.

scrippsi
(breeding)

hypoleucus
(breeding)

Xantus's Murrelet *Synthliboramphus hypoleucus*

L 9¾" *(25 cm)* Slate black above, white below. The more southerly breeding subspecies, nominate *hypoleucus,* has much more white in face than *scrippsi,* breeding in southern CA; intermediates have been recorded. Nominate also has a larger bill. Both subspecies distinguished from Craveri's Murrelet by lack of partial dark collar; slightly shorter, stouter bill; lack of black under bill; and white wing linings, sometimes visible when birds rise to flap before taking off. Often, though, birds rise and fly directly away and identification between *scrippsi* Xantus's and Craveri's is not possible; however, sometimes photographs reveal identity. Usually seen a few miles or more offshore.

VOICE: Call of *scrippsi,* a piping whistle or series of whistles, is heard year-round. Southern nominate subspecies gives a cricketlike rattle, similar to Craveri's.

RANGE: Nests in colonies on rocky islands, ledges, and sometimes in dense vegetation. Uncommon to fairly common. The nominate subspecies, *hypoleucus,* breeds on islets off Guadalupe Island and San Benito Islands, Baja California. The northern subspecies, *scrippsi,* breeds on islands from the San Benitos north to San Miguel Island, part of the Channel Islands, CA. Both subspecies move north after breeding as far as WA; *scrippsi* casually to southern BC; *hypoleucus* is generally rare to uncommon in North American waters; often encountered far offshore; a *hypoleucus* was found nesting once at Santa Barbara Island, CA.

Craveri's Murrelet *Synthliboramphus craveri* L 8½" *(22 cm)*

Slate black above, white below. Distinguished from Xantus's Murrelet by variably dusky gray wing linings; dark partial collar extending onto breast; slightly slimmer, longer bill; and black color of face extending under the bill. In good light, upperparts have a brownish tinge. Usually seen a few or many miles offshore.

VOICE: Call, very different from *scrippsi* Xantus's (but not unlike *hypoleucus*), is a cicada-like rattle, rising to a reedy trilling when agitated.

RANGE: Breeds on rocky islands off Baja California, at least the San Benito Islands and on islands in the Gulf of California. Regular late summer and fall post-breeding visitor to coast of southern and central CA. Few records over last 15 years; casually seen from shore.

Ancient Murrelet

breeding adult

often holds head slightly up in flight

black stripe on sides and flanks

winter adult

immature

paler throat

gray upperparts

black cap

yellowish bill

white eyebrow

breeding adult

black chin and throat

Xantus's Murrelet

slight grayish cast above

white underwing coverts

scrippsi

white crescent around eye

hypoleucus

white chin

scrippsi

Craveri's Murrelet

bill longer and thinner than Xantus's

upperparts appear darker than Xantus's

dark under bill

variable dusky underwing coverts

dark breast spur

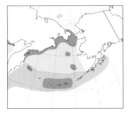

Least Auklet *Aethia pusilla* L 6¼" (16 cm)

Small and chubby, with short neck; dark above, with variably white-tipped scapulars, secondaries and greater coverts; forehead and lores streaked with white bristly feathers. Stubby, knobbed bill is dark red, with pale tip. In **breeding** plumage, acquired by Jan., a streak of white plumes extends back from behind eye; underparts are variable: heavily mottled with gray to nearly all-white. In **winter** plumage, underparts are entirely white. **Juvenile** resembles winter adult.

VOICE: On breeding grounds, gives buzzy, scratching notes.

RANGE: Abundant and gregarious, found in immense flocks. Nests on boulder-strewn beaches and islands. Often seen far from shore. Accidental in NT and coastal CA. Winters throughout Aleutians.

Parakeet Auklet *Aethia psittacula* L 10" (25 cm)

In **breeding** plumage, acquired by late Jan., broad upturned bill is orange-red; white plume extends back from behind the eye; dark slate upperparts and throat contrast sharply with white underparts; sides are mottled gray. In **winter** plumage, bill becomes duskier; underparts, including throat, are entirely white. Compare especially with larger Rhinoceros Auklet (page 262). **Juvenile** resembles winter adult, but bill smaller and darker; by first winter, bill like adult but averages darker.

VOICE: Silent except on breeding grounds, when call is a musical trill, rising in pitch.

RANGE: Fairly common on breeding grounds; nests in scattered pairs on rocky shores and sea cliffs. Found in pairs or small flocks in winter, well out to sea. Rare and irregular in winter as far south as CA, usually far offshore.

Whiskered Auklet *Aethia pygmaea* L 7¾" (20 cm)

Overall color like Crested Auklet, but note paler belly and undertail coverts. Three white plumes splay from each side of face; thin crest curls forward. In **breeding** plumage, bill is deep red with white tip. In **winter,** bill is dusky, plumes and crest less conspicuous. **Juvenile** is paler below; bill smaller; lacks crest; less striking head pattern. First-summer bird may lack crest and show reduced plumes. Often seen feeding in riptides.

VOICE: On breeding grounds, gives mewing notes.

RANGE: Fairly common but local; nests in Aleutians from Baby Islands off Unalaska Island west to Buldir, and on islands off Russian Far East.

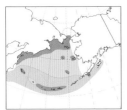

Crested Auklet *Aethia cristatella* L 9" (23 cm)

Sooty black overall; prominent quail-like crest curves forward from forehead; narrow white plume trails from behind yellow eye. **Breeding adult's** bill is enlarged by bright orange plates. In **winter,** bill is smaller and browner; crest and plume reduced. **Juvenile** has short crest, faint plume; bill smaller. Compare with smaller Whiskered Auklet. **First-summer** bird has more evident plume back from eye, but bill is still small.

VOICE: On breeding grounds, gives loud, doglike barking calls.

RANGE: Nests in crevices of sea cliffs and rocky shores. Common and gregarious. Accidental south to Baja California.

ALCIDS

stubby dark red bill

light

breeding adults

white throat

dark

Least Auklet

breeding adults

extensive whitish on underwing

winter adults

white underparts

juvenile

faint whitish scapular line

Parakeet Auklet

orange-red bill

breeding adults

unique bill shape

dark

juvenile

breeding adult

white belly

dark throat and breast

white underparts

winter adults

bill somewhat darker

white postocular line

three white head plumes

thin, curled crest

Whiskered Auklet

small dark red bill

breeding adults

breeding adult

winter adult

whitish belly

juvenile

winter adult

thick, curled crest

subdued head pattern in winter adult and especially juvenile

breeding adult

red-orange bill plates

Crested Auklet

breeding adult

winter adult

uniform dark gray underparts

white postocular line

short crest

juvenile

winter adult

much smaller bill than breeding adult; loses bill plates

1st summer

Rhinoceros Auklet *Cerorhinca monocerata* L 15" *(38 cm)*
Large, heavy-billed auklet with large head and short, thick neck. Blackish brown above; paler on sides, neck, and throat. In flight, whitish on belly blends into dark breast; compare with extensively white underparts of similar Parakeet Auklet (page 260). In **breeding** plumage, acquired by Feb., Rhinoceros Auklet has distinct white plumes and a pale yellow "horn" at base of orange bill. **Winter adult** lacks horn; plumes are less distinct, bill paler. Juvenile and **immature** lack horn and plumes; bill is dusky, eyes darker. Compare with much smaller Cassin's Auklet (page 254).
VOICE: On breeding grounds, gives a series of mooing notes.
RANGE: Common along the West Coast in fall and winter; often seen in large numbers not far offshore.

Atlantic Puffin *Fratercula arctica* L 12½" *(32 cm)*
The only East Coast puffin. **Breeding adult** identified by massive, brightly colored bill; pale face and underparts contrast with dark upperparts. **Winter adult** has smaller, duller bill, dusky face. In **juvenile** and first-winter birds, face is even duskier, bill much darker and smaller. Full adult bill takes about five years to develop. In flight, all puffins distinguished from murres and Razorbill (page 252) by red-orange legs, rounded wings, grayish wing linings, absence of white trailing edge on wing.
VOICE: On breeding grounds, gives rising and falling growling notes.
RANGE: Locally common in breeding season; winters, usually solitary, in deep water, well out at sea, a few south to off VA; casual farther south. Accidental inland to eastern Great Lakes region.

Horned Puffin *Fratercula corniculata* L 15" *(38 cm)*
A stocky North Pacific species with thick neck, large head, massive bill; underparts are white in all plumages. **Breeding adult's** face is white, bill brightly colored. Dark, fleshy "horn" extending up from eye is visible only at close range. **Winter adult's** bill is smaller, duller; face is dusky. Bill of **immature** and first-winter birds smaller and duskier than adult; full adult bill takes several years to develop. In flight, bright orange legs are conspicuous.
VOICE: Similar to Atlantic Puffin, given on breeding grounds.
RANGE: Locally common; winters well out to sea. Rare and irregular off the West Coast to southern CA, mainly in late spring and early summer.

Tufted Puffin *Fratercula cirrhata* L 16" *(41 cm)*
Stocky, with thick neck, large head, massive bill. Underparts are dark in adults. **Breeding adult's** face is white, bill brightly colored; pale yellow head tufts droop over back of neck. **Winter adult** has smaller, duller bill; face is gray, tufts shorter or absent. Juvenile has smaller, dusky bill; dark eye; white or dark underparts. First-winter bird looks like juvenile until spring molt. As in other puffins, full adult bill and plumage take several years to develop. Red-orange feet are conspicuous in flight.
VOICE: Various low grumbling notes given in breeding colonies.
RANGE: Common to abundant in northern breeding range; uncommon to rare off CA. Winters far out at sea.

white plumes
on face

"horn"

immature

**Rhinoceros
Auklet**

ALCIDS

breeding
adult

bright yellow-
orange bill

plumes
reduced and
no horn

smaller
yellowish
bill

winter
adult

breeding
adult

pale gray face

breeding
adult

uniquely shaped
and colored bill

dark collar

**Atlantic
Puffin**

darker
face

winter adult

breeding
adult

dark
underwing

smaller
bill

juvenile

red-orange
legs

"horn"

white face

breeding
adult

white
below

much smaller and
darker bill

**Horned
Puffin**

darker bill
than breeding

breeding
adult

immature

winter adult

juveniles and winter
adults darker faced

white face with
long pale yellow
head tufts

**Tufted
Puffin**

immature

breeding
adult

breeding
adult

paler belly
but variable

adults with all-black
underparts

immature

smaller yellowish-
orange bill

much darker
faced in winter,
loses tufts

thick red bill
with black
base

winter adult

PIGEONS • DOVES Family Columbidae

The larger species of these birds usually are called pigeons, the smaller ones doves. All are strong, fast fliers. Juveniles have pale-tipped feathers and lack the neck markings of adults. Pigeons and doves feed chiefly on grain, other seeds, and fruit. SPECIES: 318 WORLD, 19 N.A.

Band-tailed Pigeon *Patagioenas fasciata* L 14½" (37 cm)

Purplish head and breast; dark-tipped yellow bill, yellow legs; broad gray tail band on longish tail; narrow white band on nape, absent on juvenile. Flocks in flight resemble Rock Doves but are uniform, not varied, in plumage and lack contrasting white rump and black band at end of longer tail.

VOICE: Song is a low *whoo-whoo*.

RANGE: Locally common in low-altitude coniferous forests in the Northwest, and oak or oak-conifer woodlands in the Southwest; also increasingly common in suburban gardens, and parks. Uncommon in southeastern AK; casual in eastern North America.

Rock Pigeon *Columba livia* L 12½" (32 cm)

The highly variable city pigeon; multicolored birds were developed over centuries of near domestication. The birds most closely resembling their wild ancestors have head and neck darker than back, black bars on inner wing, white rump, and black band at end of tail. Flocks in flight show a variety of plumage patterns, unlike Band-tailed Pigeon.

VOICE: Song is a soft *coo-cuk-cuk-cuk-coooo*.

RANGE: Introduced from Europe by early settlers, now widespread and common, particularly in urban settings. Nests and roosts chiefly on high window ledges, bridges, and barns. Feeds during the day in parks and fields. Some have reverted back to nesting on rocky cliffs, the species' ancestral native habitat.

White-crowned Pigeon *Patagioenas leucocephala*

L 13½" (34 cm) A large, square-tailed pigeon of the Florida Everglades and Keys. Crown patch varies from shining white in **adult males** to grayish white in most **females** and grayish brown in juveniles. Otherwise this species looks all-black; the iridescent collar is visible only in good light.

VOICE: Songs include a loud, deep *coo-cura-cooo* or *coo-croo*.

RANGE: Flocks commute from nest colonies in coastal mangroves to feed inland on fruit. Most winter on Caribbean islands.

Red-billed Pigeon *Patagioenas flavirostris* L 14½" (37 cm)

Dark overall, with a mainly red bill.

VOICE: Distinctive song heard in early spring and summer, a long, high-pitched *cooooo* followed by three loud *up-cup-a-coo* notes.

RANGE: Perches in tall trees above a brushy understory; forages for seeds, nuts, and figs. Seldom comes to the ground except to drink. Uncommon, local, and declining in TX. Most frequently recorded along the Rio Grande from below Falcon Dam to above Zapata, more rarely farther north to southern Maverick County. Accidental on the lower TX coast (Nueces County) and from the TX Hill Country (Kerr County). Rare in winter.

rather long tail with broad pale tail band

in flight, often gives a loud wing flapping noise

ancestral natural coloration

Band-tailed Pigeon

white rump

Rock Pigeon

white band at top of nape

yellow-based bill

color variations
many other variations are seen

White-crowned Pigeon

maroon head, neck, and wing coverts; otherwise slate gray

♂

white crown

red bill with pale tip

pale eye

♀

pale eye with orange orbital ring

dark slaty gray body

red bill with pale yellow tip

Red-billed Pigeon
flavirostris

Eurasian Collared-Dove *Streptopelia decaocto* L 12½" *(32 cm)*

Slightly larger than Mourning Dove. Very pale gray-buff; black collar. Escapes of domesticated "Ringed Turtle-Dove" — formerly named *S. risoria,* derived from **African Collared-Dove** (*S. roseogrisea*), an Old World species — may form small populations, but do not do well in the wild; they are smaller, paler (but some Eurasian Collared-Doves are pale, too), with whitish undertail coverts, gray primaries; shorter tail is less black from below.

VOICE: Three-syllable song, *coo-coo-cup,* is given year-round; most similar "Ringed Turtle-Doves" give a two-syllable call.

RANGE: Eurasian species; introduced to Bahamas, spread to FL. Common; increasing and spreading throughout much of U.S. Still casual in MI and ON. A few populations are the result of local releases.

Mourning Dove *Zenaida macroura* L 12" *(31 cm)*

Trim body; long tail tapers to a point. Black spots on upperwing; pinkish wash below. In flight, shows white tips on outer tail feathers. **Juvenile** has heavy spotting; scaled effect on wings. Compare with Common Ground-Dove and Inca Dove (page 268).

VOICE: Song is a mournful *oowoo-woo-woo-woo.* Wings produce a fluttering whistle as the bird takes flight.

RANGE: Our most abundant and widespread dove, found in a wide variety of habitats. Rare fall visitor to AK, northwestern Canada.

Zenaida Dove *Zenaida aurita* L 10" *(25 cm)*

Distinguished from Mourning Dove by white on trailing edge of secondaries that shows as a squarish white spot on inner secondaries of folded wing; and by shorter, rounded, gray-tipped tail. Some, presumably males, are a very warm brown color overall; others, presumably females, are much grayer, more like Mourning Dove. Generally shy.

VOICE: Song is similar to Mourning Dove but faster.

RANGE: Primarily West Indian species; now casual, nearly accidental (three records since 1963) on the Florida Keys (where it was reported as breeding by Audubon in 1832).

White-winged Dove *Zenaida asiatica* L 11½" *(29 cm)*

Large white wing patches and shorter, rounded tail distinguish this species from Mourning Dove; also note slightly longer bill, orange-red eye. On sitting bird, wing patch shows only as a thin white line.

VOICE: Song, a drawn-out, cooing *who-cooks-for-you,* has many variations and is heard during the breeding season.

RANGE: Nests singly or in large colonies in dense mesquite, mature citrus groves, riparian woodlands, and saguaro-paloverde deserts; also found in desert towns. Expanding north on Great Plains. Has recently become well established in FL and locally along the central Gulf Coast. Very rare north to southern Canada and on East Coast north to the Maritime Provinces. Rare, mainly in fall, to West Coast. Accidental to AK (Skagway, Oct. 1981).

grayish
undertail

whitish
undertail

long
pointed tail

Eurasian
Collared-

African
Collared-

Mourning

**Eurasian
Collared-Dove**

black
collar

grayish
undertail

stockier overall than
Mourning Dove

three-tone
wing

**African
Collared-
Dove**

grayish
primaries

tapered tail
with white tips

**Mourning
Dove**

black
spotting on
upperwing

♂

♀

juvenile

long,
pointed
tail

♂

♀

Zenaida Dove
zenaida

♂

white trailing
edge to
secondaries

shorter
tail than
Mourning

squarish white
patch formed by
white tips to
secondaries

♂

red eye
with blue
orbital ring

**White-winged
Dove**

rounded tail
with white tips

crescent-shaped
white wing patch

white leading
edge to wing

Spotted Dove *Streptopelia chinensis* L 12" (31 cm)
Named for spotted collar, distinct in **adults,** obscured in **juveniles.** Wings and long, white-tipped tail are more rounded than Mourning Dove (page 266); wings unmarked; overall color more pinkish.
VOICE: Song is a rather harsh *coo-coo-croooo* and *coo-crrooo-coo,* with emphasis respectively on the last and middle notes.
RANGE: An Asian species introduced in Los Angeles in the early 1900s; formerly well established in southwestern CA but sharp declines in recent decades. Now largely restricted to southwestern Los Angeles County, Bakersfield, and Avalon, Santa Catalina Island, CA.

Oriental Turtle-Dove *Streptopelia orientalis* L 13½" (34 cm)
Large and stocky; scaly pattern above with buffy, gray, and reddish fringes on black feathers; black-and-white streaked patch on neck. North American records are nominate *orientalis,* which is dark with gray rump and tail tip that varies from whitish to pale gray. Compare also to smaller, more richly colored European Turtle-Dove (page 538).
RANGE: Asian species; casual to Aleutians, Pribilofs, and Bering Sea in spring and summer; accidental on Vancouver Island, YT, and CA.

Common Ground-Dove *Columbina passerina* L 6½" (17 cm)
Very small, with pink at base of bill; scaled effect on head and breast; short tail, often raised. Plain scapulars; bright chestnut primaries and wing linings visible in flight. **Male** has a slate gray crown, pinkish gray underparts. **Female** is grayer, more uniformly colored.
VOICE: Song is a repeated soft, ascending *wah-up.*
RANGE: Forages on open ground in the East and brushy rangeland and agricultural areas (locally in urban areas) in the West. Declining in Southeast; casual north to MI, ON, MA, NY and OR in fall and winter. Accidental WA.

Ruddy Ground-Dove *Columbina talpacoti* L 6¾" (17 cm)
Dark bill; lacks scaling of Common Ground-Dove. **Male** of the subspecies *eluta* has a gray crown, rufous color on upperparts. **Female** is mainly gray-brown overall. Both sexes show black on underwing coverts, black linear markings on scapulars. Most records in the U.S. are of west Mexican subspecies *eluta;* some TX records, however, are of east Mexican *rufipennis,* which is a richer cinnamon overall.
VOICE: Song like Common but faster and a little lower in pitch.
RANGE: Widespread in Latin America; rare to casual in Southwest, mostly in fall and winter. Has bred in southern AZ and southeastern CA. Casual to southern CA coast, southern NV, and TX.

Inca Dove *Columbina inca* L 8¼" (21 cm)
Plumage conspicuously scalloped. In flight, shows chestnut on wings like Common Ground-Dove, but note Inca Dove's longer, white-edged tail.
VOICE: Song is a double *cooo-coo,* often phrased as *no-hope.*
RANGE: Found usually near human habitations, often in parks and gardens. Slowly spreading north and, at least formerly, west. Casual wanderer north to MT, ND, ON, and MD; east to AL.

black-and-white
spotted collar

pinkish
underparts

**Spotted
Dove**
chinensis

rectangularly shaped
tail shows extensive
white from below

juvenile

adult

tail tips vary
from whitish
to pale gray

**Oriental
Turtle-Dove**
orientalis

rufous fringes give scaly
pattern above

black-and-white
streaked neck
patch

fringes
more grayish
below

gray
rump

grayish
cap

pink-based
bill

**Common
Ground-Dove**

unmarked
scapulars ♂

scaly pattern
on breast
and nape ♀

dark comma-
shaped marks

short tail

♂

rufous
outer wing

pinkish
breast

grayish cap

♀

dark
bill

**Ruddy
Ground-Dove**
eluta

males with overall
rufous coloration;
females grayish

black linear
markings on
scapulars

no scaling
on chest

♂

short tail

♂

rufous
outer
wing

scaly pattern
on head
and body

**Inca
Dove**

long tail
with white
edges

rufous
primary
patches

White-tipped Dove *Leptotila verreauxi* L 11½" *(29 cm)*

This large, plump dove has a whitish forehead and throat and dark back. In flight, white tips show plainly on fanned tail. **VOICE:** Low-pitched song is like the sound produced by blowing across the top of a bottle. Wings give a high twittering sound in flight. **RANGE:** Feeds on or near the ground, keeping close to woodlands with dense understory. Casual to south FL.

Key West Quail-Dove *Geotrygon chrysia* L 12" *(31 cm)*

Larger, longer tailed than Ruddy Quail-Dove; whitish below with a white line under the eye. Upperparts, primaries, and tail are chestnut, glossed with purple and green. **Male** highly iridescent above; female duller. **VOICE:** Song is a long, two-part cooing given from a perch in a tree. Recent North American birds have not been heard vocalizing. **RANGE:** West Indian species. Reported as breeding on Key West, FL, by Audubon in 1832. Population eliminated by about mid-19th century. Casual on the Florida Keys and south FL.

Ruddy Quail-Dove *Geotrygon montana* L 9¾" *(25 cm)*

A chunky dove. **Male's** primarily rich rufous upperparts and prominent buffy line under the eye are distinctive. **Females** brown above; plainer facial pattern. In both sexes, underparts cinnamon-buff. Quail-doves so named because they resemble quail, have a similar terrestrial lifestyle. **RANGE:** Widespread in the tropics, including the West Indies. Five records for FL and one for south TX.

LORIES • PARAKEETS • MACAWS • PARROTS Family Psittacidae

Most, if not all, in the wild in North America are descendants of escaped cage birds. Shown here are those with established populations, mostly from southern parts of CA, TX, and FL. In FL, some 15 species are breeding, but declines for many in recent years. ABA accepts introduced populations of White-winged, Monk, and Green Parakeets; Red-crowned Parrot; and Budgerigar. SPECIES: 344 WORLD, 7 N.A.

Budgerigar *Melopsittacus undulatus* L 7" *(18 cm)*

Note barred upperparts and white wing stripe visible in flight. **VOICE:** Gives a continuous pleasant warble and sharper chattering calls. **RANGE:** Australian species. Native birds green, as were populations established in southwestern FL in early 1960s; reached tens of thousands before 1980s. Only a few now present in Hernando County; count of 26 in Jan. 2011; declines perhaps because of competition for nest cavities with European Starlings and House Sparrows.

Rosy-faced Lovebird *Agapornis roseicollis* L 6¼" *(16 cm)*

A small short-tailed parrot with a bluish rump. **Adult male** with bright rose-pink face and upper breast, red band across forehead; female and immature duller. Also known as Peach-faced Lovebird. **VOICE:** Gives a short high-pitched *shreek*, singly or in a series. **RANGE:** African species. First noted in the Phoenix, AZ, region in 1987, widely established by the mid-1990s throughout the eastern half of the greater Phoenix metropolitan area; population estimated at several thousand. Continues to increase and nesting noted east of Tucson, AZ, in 2001. A colonial nester, nesting in cavities, even in saguaros.

White-tipped Dove
angelica

pale yellow eye

chunky body shape

white forehead and throat

short, broad slightly rounded tail

whitish belly

white tail tips

prominent white line under eye

Key West Quail-Dove

Ruddy Quail-Dove
montana

♀

buffy line under eye

♂

male rich rufous above, female duller

♂

chestnut upperparts glossed with purple and green

whitish below

escapes may be blue, white, or yellow

blue variant

yellow variant

Budgerigar

Rosy-faced Lovebird

red band across forehead

also known as Peach-faced Lovebird

pink face and upper breast

yellow face

barred above

green underparts

long, pointed tail

natural coloration

adult ♂

White-winged Parakeet *Brotogeris versicolurus*

L 8¾" *(22 cm)* Note white secondaries and inner primaries, yellow greater coverts, unfeathered grayish lores. Immature similar but reduced or no white on primaries and secondaries.
VOICE: Like Yellow-chevroned, but perhaps richer, less shrill.
RANGE: From northern Amazon area and south to east-central Peru. Small numbers established in Miami and Fort Lauderdale and Los Angeles and San Francisco regions by 1960s (more than 2,000 in one flock by mid-1970s at Coral Gables, FL), but have declined in recent years and now under 100 in FL; many fewer in CA.

Yellow-chevroned Parakeet *Brotogeris chiriri*

L 8¾" *(22 cm)* Lacks white on secondaries and inner primaries; body yellow-green; head and feathered lores slightly brighter yellow-green. Formerly treated with White-winged Parakeet as one species, Canary-winged Parakeet. Hybrids between the two are known from FL.
VOICE: A high and rather shrill *krere-krere* or *chiri-chiri*, given most often in flight.
RANGE: Native range from the southern Amazon to northern Argentina. Established in Miami and Fort Lauderdale areas where population recently estimated at 300 to 400, and in Los Angeles where population larger, over 1,000 and increasing.

Monk Parakeet *Myiopsitta monachus* L 11½" *(29 cm)*

Note extensively gray face and underparts.
VOICE: Call is a loud, grating *krii*.
RANGE: Native of temperate South America. The most widespread parrot in south FL, although numbers sharply decreasing in recent years. Some also found in cities of Northeast and Midwest. Nests communally in large, untidy stick nests.

Rose-ringed Parakeet *Psittacula krameri* L 15¾" *(40 cm)*

Large; slender tail, very long central feathers, bright red upper mandible. **Adult males** show black rose-edged collar. North American birds may be Indian subspecies, *manillensis.* Also known as Ring-necked Parakeet.
VOICE: Call is a loud, flickerlike *kew.*
RANGE: Old World's most widely distributed parrot; found right across tropical Africa north of the forest zone and much of south Asia. Introduced populations widely established across the globe elsewhere, including North America where small numbers are found in southwestern FL (fewer than 50) and the Los Angeles area, with many more (several thousand) in Bakersfield in the San Joaquin Valley, CA.

Nanday Parakeet *Nandayus nenday* L 13¾" *(35 cm)*

Black on head, black bill; blue wash on breast; and red thighs. Also known as Black-hooded Parakeet and in pet trade as Nanday Conure.
VOICE: Gives high and loud screeching notes.
RANGE: Native of southwestern Brazil to northern Argentina. Established in six counties on west coast of central FL from Pasco County in the north to Charlotte County in the south; overall population in excess of 1,000. Also found in the Los Angeles area and especially north on the coast from western Los Angeles to southwestern Ventura County (300 to 400 birds).

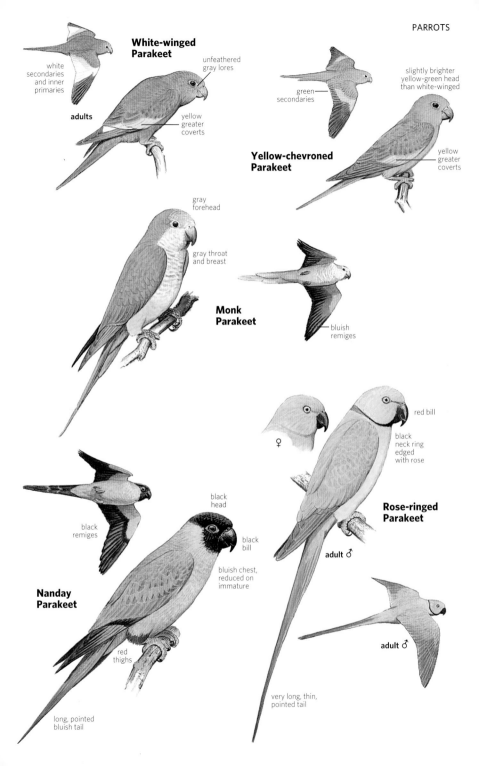

White-winged Parakeet

white secondaries and inner primaries

adults

unfeathered gray lores

yellow greater coverts

green secondaries

slightly brighter yellow-green head than white-winged

Yellow-chevroned Parakeet

yellow greater coverts

gray forehead

gray throat and breast

Monk Parakeet

bluish remiges

red bill

black neck ring edged with rose

♀

Rose-ringed Parakeet

adult ♂

black remiges

black head

black bill

bluish chest, reduced on immature

Nanday Parakeet

red thighs

long, pointed bluish tail

adult ♂

very long, thin, pointed tail

Blue-crowned Parakeet *Aratinga acuticaudata* L 14" (36 cm)
Blue face, hard to see in poor light; bicolored bill; base of long, pointed tail reddish below. Immature similar but restricted blue on head.
VOICE: Gives a loud, low, hoarse repeated *cheeeah* note.
RANGE: Native to three disjunct regions of South America. Small numbers established in Miami and Fort Lauderdale areas, the estimated population being about 200; a few known on west coast of FL and in CA (San Diego and Los Angeles areas).

Green Parakeet *Aratinga holochlora* L 13" (33 cm)
Large, nearly all-green, some with a few scattered orange feathers about the head and upper body.
VOICE: Gives sharp, squeaky, and loud harsh notes.
RANGE: Found in native range from northwestern and northeastern Mexico from southern Nuevo León and Tamaulipas south to northern Nicaragua. Three groups, sometimes treated as separate species, are recognized. Populations in south TX towns near Rio Grande belong entirely, or nearly so, to the nominate *holochlora* group, endemic to Mexico; some may represent strays as opposed to escapes. About 50 are found in the Miami and Fort Lauderdale areas.

White-eyed Parakeet *Aratinga leucophthalma*
L 12½" (32 cm) Similar to Green Parakeet with green coloration, scattered red feathers on head and at bend of wing. Best told by distinct underwing pattern: red patch at median and lesser primary coverts set off by yellow band of greater primary coverts. Immature has less red on the head and no red at bend of wing; underwing pattern indistinct.
VOICE: Varied calls, some melodious, others screeching and grating.
RANGE: In native range, the most widespread *Aratinga,* found throughout South America east of the Andes south to northern Argentina. Found in Miami and Fort Lauderdale, with estimated population about 200.

Mitred Parakeet *Aratinga mitrata* L 15" (38 cm)
Large, green, with white orbital ring; red on face, forehead, variable elsewhere on head; underwing yellow-olive. Immature with less red on head, especially face, and with a brown iris. North American individuals seem to be of the recently described subspecies, *tucumana*.
VOICE: Gives a series of strident and harsh notes.
RANGE: Native to western South America. Numerous in the Los Angeles area (over 1,000) and in Miami and Fort Lauderdale (perhaps 500 to 1,000). Also a few in New York City.

Red-masked Parakeet *Aratinga erythrogenys* L 13" (33 cm)
Similar to Mitred, but note red on leading edge of wing and underwing coverts. Adult with striking red head; immatures with much reduced red on head, pattern overlaps with Mitred; also with less red on leading edge of underwing. Often associates with Mitred Parakeets where smaller size readily apparent, although sometimes flocks are separate.
VOICE: In general, calls less harsh than Mitred.
RANGE: A native of southwestern Ecuador and northwestern Peru. Population in CA (Los Angeles, San Diego, San Francisco areas) about 500; some 200 to 300 estimated from Miami and Fort Lauderdale.

partly bluish head

adults

bicolored bill

Blue-crowned Parakeet

pale feet

long pointed tail with reddish tail base

some with a few scattered red feathers

Green Parakeet
holochlora

mainly green coloration

limited red on head and leading edge of wing

red and yellow underwing primary coverts

scattered red feathers on head and neck

adults

Mitred Parakeet

White-eyed Parakeet

red at bend of wing

smaller than Mitred

adult

scattered red head feathers

extensive red on head

red on leading edge of wing

Red-masked Parakeet

adults

long, pointed tail

extensive red on adults on leading edge of underwing, sometimes more extensive to underwing coverts

Thick-billed Parrot *Rhynchopsitta pachyrhyncha*

L 16¼" (41 cm) **Adult** green overall; red forehead, eyebrow, thighs, marginal coverts; tail long, pointed; black bill; yellow underwing bar; slow shallow wingbeats. Immature's bill paler, no red on eyebrow and wing.
VOICE: Loud, laughing calls carry far.
RANGE: Endangered. Breeds in Sierra Madre Occidental; found north to near the NM border. Northern breeders move south by winter. Former sporadic visitor primarily to Chiricahua Mountains, AZ. Last valid record in 1938; recent NM record (2003) near Rio Grande not accepted (origin). Releases into Chiricahuas in 1980s were unsuccessful.

Lilac-crowned Parrot *Amazona finschi* *L 13" (33 cm)*

Like Red-crowned, but lilac wash on crown and nape; maroon band across forehead; longer tail has green central tail feathers; cere dusky, not flesh. Has likely hybridized with Red-crowned in southern CA.
VOICE: Calls include a distinctive upslurred whistle.
RANGE: A native of west Mexico. More than 500 in Los Angeles and San Diego; fewer than 200 in south TX and Miami.

Red-lored Parrot *Amazona autumnalis* *L 13" (33 cm)*

Like Red-crowned but yellow area on face, and dark lower mandible and upper mandible tip. Immature with less yellow and red on head.
VOICE: Various loud squawks; also more pleasant notes.
RANGE: Found southern Tamaulipas to the Amazon. Over 100 now in Los Angeles and Orange Counties; a few in south TX and south FL.

Yellow-headed Parrot *Amazona oratrix* *L 14½" (37 cm)*

Large, with yellow head; immature shows less yellow; some escapes are closely related Yellow-naped (*A. auropalliata*) and Yellow-crowned (*A. ochrocephala*) Parrots, formerly treated as subspecies of Yellow-headed.
VOICE: Calls include a resonant human-like *haa-haa-haa.*
RANGE: Native of Mexico and Belize, where populations severely depleted. Small numbers established in Los Angeles area (especially in Pasadena region); a very few also in south TX and formerly FL.

Red-crowned Parrot *Amazona viridigenalis* *L 13" (33 cm)*

Pale blue on sides of head; yellowish band across tip of tail. **Adult male** has red crown; **female** averages less red; immature shows even less.
VOICE: Calls generally raucous but include a mellow, rolling *rreeoo.*
RANGE: Endemic to northeastern Mexico where population now severely depleted. Established in towns of southernmost TX where first detected in early 1970s; some may be visitors from Mexico, but most or all are probably descendants of escaped cage birds. Also well-established in CA in Los Angeles and in Orange and San Diego Counties (3,000 to 5,000) and Fort Lauderdale and West Palm Beach (300 and declining as a result of birds being caught for the pet trade).

Orange-winged Parrot *Amazona amazonica* *L 12¼" (31 cm)*

Note blue stripe and nape on yellow head; orange on outer tail feathers.
VOICE: Calls include harsh squawks and pleasant whistled notes.
RANGE: Native to South America (east of Andes) from Colombia to southeastern Brazil. Introduced and established on Barbados. Small numbers (100 to 200) established in Miami and Fort Lauderdale.

yellow underwing stripe

blackish bill

green central tail feathers

adults

red on forehead and shoulder

Thick-billed Parrot

long, pointed tail

lilac crown and nape

maroon forehead

Lilac-crowned Parrot

Red-lored Parrot

yellow on cheek

adult

Yellow-headed Parrot

yellow head

adults

pale bill

Red-crowned Parrot

yellowish terminal tail band

red crown

red forehead

♀

adult ♂

bluish on head and nape

Orange-winged Parrot

orange base to outer tail feathers

CUCKOOS • ROADRUNNERS • ANIS Family Cuculidae
Of this large family, widespread in the Old World, only a few species are seen in North America. Most are slender with long tails; two toes point forward, two back. SPECIES: 142 WORLD, 8 N.A.

Mangrove Cuckoo *Coccyzus minor* *L 12" (31 cm)*
Black mask and buffy underparts distinguish this species from other cuckoos. Upperparts grayish brown; lacks rufous primaries of Yellow-billed Cuckoo. Black tail feathers are broadly tipped with white. Species is usually considered monotypic, but Mexican birds average brighter below; two variations shown. In all juveniles, mask is paler, tail pattern muted. Like other cuckoos, perches quietly near center of tree.
VOICE: Call is a slow, guttural *gaw gaw gaw.*
RANGE: Found chiefly in mangrove swamps. An accidental vagrant along Gulf Coast from Mexico to TX and northwestern FL.

Yellow-billed Cuckoo *Coccyzus americanus* *L 12" (31 cm)*
Grayish brown above, white below; rufous primaries; lower mandible yellow. Undertail patterned in bold black and white. In **juvenal** plumage, held well into fall, tail has a much paler pattern and bill may show little or no yellow; may be confused with Black-billed Cuckoo.
VOICE: One song sounds hollow and wooden, a rapid staccato *kuk-kuk-kuk* that usually slows and descends to a *kakakowlp-kowlp* ending. Also a series of *coo* notes and a slowly repeated single *keep.*
RANGE: Common in open woods, orchards, and streamside willow and alder groves. Once numerous but now a rare breeder in CA. Formerly bred in Pacific Northwest. Rare vagrant to Atlantic Canada during fall migration; accidental AK.

Black-billed Cuckoo *Coccyzus erythropthalmus* *L 12" (31 cm)*
Grayish brown above, pale grayish below. Bill is usually all-dark. Lacks the rufous primaries of Yellow-billed Cuckoo. Note also **adult's** reddish orbital ring. Undertail patterned in gray with white tipping; compare juvenile Yellow-billed. **Juvenile** Black-billed has a buffy orbital ring; undertail is paler; underparts may have buffy tinge, especially on undertail coverts; primaries may show a little rusty brown.
VOICE: Song usually consists of monotonous *cu-cu-cu* or *cu-cu-cu-cu* phrases.
RANGE: Uncommon to fairly common; found in woodlands and along streams. Very rare breeder in northern TX, western TN, and western ID; very rare in Southeast; casual in Pacific states; accidental Southwest.

Greater Roadrunner *Geococcyx californianus* *L 23" (58 cm)*
A large, ground-dwelling cuckoo streaked with brown and white. Note the long, heavy bill, conspicuous bushy crest, and long, white-edged tail. Short, rounded wings show a white crescent on the primaries. Eats insects, lizards, snakes, rodents, and small birds.
VOICE: Song is a dovelike cooing, descending in pitch.
RANGE: Common in scrub desert and mesquite groves; less common in chaparral and open woodland.

Mangrove Cuckoo

adults

black mask

thick bill; lower mandible has yellow base

uniform brown wings

buffy underparts

large white spots on black tail

Yellow-billed Cuckoo

yellow orbital ring

thick, extensively yellow bill

adult

juvenile

spots less bold than adult

large white spots on black tail

adult

rufous primaries striking in flight

Black-billed Cuckoo

red orbital ring

slender, dark bill

slight buff tint to throat

buffy orbital ring

juvenile

adult

small white spots on grayish tail

indistinct tips

adult

uniform brown wings

crest can be raised or flattened

Greater Roadrunner

very long, graduated tail with white tips

Oriental Cuckoo *Cuculus optatus* L 12½" (32 cm)

Very difficult to distinguish from Common Cuckoo. **Adult male** and adult gray-morph female are slightly smaller and darker gray above than Common. Barring is often slightly broader, lower belly and undertail covert region buffier than Common; bill slightly thicker. In the hand, underwing primary coverts are largely white and unmarked, while they are barred in Common; in flight, white bar on underwing is much bolder than Common. **Hepatic-morph female** is rusty brown above, heavily barred on back, rump, and tail. Oriental Cuckoo (*C. saturatus*) was formerly considered polytypic, but because of vocal differences it was split into three species: Himalayan Cuckoo (*C. saturatus*), the small Sunda Cuckoo (*C. lepidus*), and Oriental Cuckoo (*C. optatus*); the latter is described here.

VOICE: Song, a variable, hollow note often delivered four times, has not been heard in North America.

RANGE: Eurasian species, casual from late spring through fall in western Aleutians, Pribilofs, St. Lawrence Island, and once on mainland AK.

Common Cuckoo *Cuculus canorus* L 13" (33 cm)

Adult male and adult gray-morph female gray above, paler below, whitish belly narrowly barred with gray. **Female** brown on sides of breast. Though variable, birds seen in AK often paler above than European birds, with fine barring below, and perhaps represent *telephonus*, though subspecies not recognized by most. **Hepatic morph** (restricted to females in Common and Oriental Cuckoos) has paler unmarked or only lightly spotted rump, unlike heavily marked hepatic-morph Oriental.

VOICE: Male's song is the familiar *cuc-coo* for which the family is named but call is rarely heard in North America.

RANGE: Old World species, rare spring and summer visitor to central and western Aleutians and Bering Sea islands. One late June Anchorage, AK, record (singing); accidental to Martha's Vineyard, MA (3 to 4 May 1981).

Groove-billed Ani *Crotophaga sulcirostris* L 13½" (34 cm)

Overall size and bill are smaller than Smooth-billed Ani. Bill does not extend above crown; lower mandible straighter. Plumage is black overall with iridescent purple-and-green overtones; long tail is often dipped and wagged. Grooves in bill are visible only at close range; these and call are distinctive. Secretive much of the time.

VOICE: Call is a liquid *tee-ho*, accented on the first syllable.

RANGE: Fairly common in summer in south TX woodlands. Rare primarily in fall and winter on Gulf Coast to FL. Casual north to MN, northern ON, west to CA, and east to VA and NJ.

Smooth-billed Ani *Crotophaga ani* L 14½" (37 cm)

Bill size variable. Black overall with iridescent bronze overtones. Long tail is often dipped and wagged. Both ani species are gregarious; several pairs usually share a nest and take turns incubating the eggs.

VOICE: Call is a whining, rising *quee-lick*.

RANGE: Found in brushy fields, scrublands; often feeds on insects stirred up by cattle. Has declined greatly in FL and is on verge of being extirpated; still numerous in West Indies including Bahamas and Cuba. Accidental north along Atlantic coast to NC; also to LA, OH.

Common

pale on underwing more barred, less contrasty than Oriental

northeast Asian birds, recognized by some as *telephonus*, are paler gray above and lightly barred below

Oriental

bill slightly thicker and more curved than Common

more contrast between whitish and slate-gray areas on underwing than Common Cuckoo

adult ♂

Oriental Cuckoo

ventral barring averages stronger

barred rump

buffy undertail coverts

delicate bill

hepatic morph ♀

adult ♀

rusty on sides of breast

adult ♂

Common Cuckoo

plain rump

nape spot

juveniles of both species are blackish brown with whitish feather fringes above

hepatic morph ♀

whitish undertail coverts

Oriental juvenile

long tail

Groove-billed Ani
sulcirostris

bill grooves visible at close range

Smooth-billed Ani

some show raised base to culmen, but many others do not

bill has no distinct grooves

slightly larger than very similar Groove-billed; most birds best distinguished by voice

BARN OWLS AND TYPICAL OWLS Families Tytonidae and Strigidae
Distinctive birds of prey, divided by structural differences into two families, Barn
Owls (Tytonidae) and Typical Owls (Strigidae). All have immobile eyes in large heads.
Fluffy plumage makes their flight nearly soundless. Many species hunt at night and
roost during the day. To find owls, search the ground for regurgitated pellets of fur and
bone below a nest or roost. Also listen for flocks of small songbirds noisily mobbing a
roosting owl. SPECIES: TYTONIDAE 15 WORLD, 1 N.A.; STRIGIDAE 191 WORLD, 23 N.A.

Barn Owl *Tyto alba* L 16" *(41 cm)*
A pale owl with dark eyes in a heart-shaped face. Rusty brown above;
underparts vary from white to cinnamon. Darkest birds are always
females, palest birds **males.** Flies with slow, shallow wingbeats.
VOICE: Typical call is a raspy, hissing screech.
RANGE: Roosts and nests in dark cavities in city and farm buildings,
cliffs, and trees. Rare to uncommon in parts of western range, uncommon to rare and declining in the East, especially in Midwest, where
mostly extirpated from northern tier of states and ON.

Long-eared Owl *Asio otus* L 15" *(38 cm)*
Slender with long, close-set ear tufts. Boldly streaked and barred on
breast and belly. Wings generally have a less prominent buffy patch
with more dark barring and a smaller black "wrist" mark than Short-
eared Owl; facial disk rusty. Hunts at night over open fields, marshes.
By day it roosts in a tree, close to the trunk.
VOICE: Generally silent except during breeding season. Common call
is one or more long *hooo* notes.
RANGE: Lives in thick woods. Uncommon. More gregarious in winter,
when often forms roosting groups. Casual to southeastern AK and
Bering Sea (likely nominate *otus* on probability).

Short-eared Owl *Asio flammeus* L 15" *(38 cm)*
Tawny; boldly streaked on breast; belly paler, more lightly streaked. Ear
tufts barely visible. In flight, long wings show buffy patch above, black
"wrist" mark below; these markings are usually more prominent than
Long-eared Owl, which also has less distinct but more individual bars on
primaries. Usually active before dark; flight wavering, wingbeats erratic.
VOICE: Typical call, heard in breeding season and sometimes in winter,
is a raspy, high barking.
RANGE: Nests on the ground. Fairly common. Somewhat irregular
and gregarious in winter; groups may gather where prey is abundant
in marshes, fields, and tundra. Those sighted on Florida Keys and
Dry Tortugas have originated from West Indian populations, likely
domingensis from Cuba.

Great Horned Owl *Bubo virginianus* L 22" *(56 cm)*
Size, bulky shape, white throat separate this owl from Long-eared Owl;
distinctive ear tufts. Chiefly nocturnal. Takes prey as large as skunks
and grouse. Widespread interior subspecies, *subarcticus,* is palest.
VOICE: Call is a series of three to eight loud, deep hoots; second and
third hoots often short and rapid. Juveniles give a raspy begging call.
RANGE: Nests in trees, caves, or on the ground. Common; habitats
vary from forest to city to open desert.

dark eyes with whitish
heart-shaped face

♂

underparts vary
from whitish
to cinnamon-buff;
males average paler

looks pale in flight with
rounded wings and no
dark carpal patches as
in Short-eared Owl

♀

long
legs

Barn Owl
pratincola

long ear tufts
that are close
together

rufous
facial disk

blackish
around eyes

**Long-eared
Owl**

overall
slender body

heavily
streaked and
barred below

slow, floppy
wingbeats

dark primary
covert patch
on underwing

buffy and
blackish wing
patches

very short
ear tufts

blackish
around eyes

heavily
spotted above

**Short-eared
Owl**
flammeus

streaked
below

prominent broad
ear tufts on
sides of head

**Great
Horned
Owl**

color of facial disk
varies geographically

white
throat

pale
overall

bulky body
shape

barred
below

subarcticus

Barred Owl *Strix varia* L 21" *(53 cm)*

A chunky owl with dark eyes, dark barring on upper breast, dark streaking below. Chiefly nocturnal; daytime roost well hidden. Easily flushed; does not generally tolerate close approach. Preys primarily on mammals and amphibians.

VOICE: Distinctive call is a rhythmic series of loud hoots: *who-cooks-for-you, who-cooks-for-you-all;* also a drawn-out *hoo-ah,* sometimes preceded by an ascending agitated barking. Much more likely than most other owls to be heard in daytime; often a pair call back and forth.

RANGE: Common in dense coniferous or mixed woods of river bottoms and swamps; also in upland woods. Northwestern portion of range is expanding rapidly; now overlaps and has hybridized with similar Spotted Owl. Probably a rare breeder in southeastern AK. Accidental to CO and NM.

Great Gray Owl *Strix nebulosa* L 27" *(69 cm)*

Our largest but not heaviest owl. Prominent rings on facial disk make the yellow eyes look small. Lacks ear tufts. Hunts for small mammals over forest clearings and nearby open country, chiefly by night but also at dawn and dusk; hunts by day during summer in northern part of range, as well as sometimes on overcast days in winter.

VOICE: Call is a series of deep, resonant *whoo* notes.

RANGE: Inhabits boreal forests and wooded bogs in the north, dense coniferous forests with meadows in the mountains farther south. Generally uncommon; rare and irregular winter visitor to limit of dashed line on map. Casual to western BC and WA. Accidental farther south and east to NE, IA, OH, PA, and Long Island, NY.

Spotted Owl *Strix occidentalis* L 18" *(46 cm)*

Large and dark-eyed, with white spotting on head, back, and underparts, rather than the barring and streaking of similar Barred Owl. Tamer than Barred. Strictly nocturnal.

VOICE: Main call is a series of four doglike barks; contact call, given mainly by females, is a hollow, upslurred whistle, *coooo-weep.*

RANGE: Inhabits thickly wooded canyons, humid forests. Uncommon; decreasing in number and range due to habitat destruction, especially in the Northwest. The threatened northern subspecies, *caurina* **(T),** is the largest and darkest subspecies, with smaller white spots. Nominate subspecies of central and southern CA is paler; Southwestern *lucida* is paler still with larger white spots.

Snowy Owl *Bubo scandiacus* L 23" *(58 cm)*

Large, white; rounded head, yellow eyes. Dark bars and spots heavier on females, heaviest on **immatures;** old males may be pure white.

VOICE: Mostly silent away from breeding grounds.

RANGE: An owl of open tundra; nests on the ground; preys chiefly on lemmings, hunting by day as well as at night. Retreats from northernmost part of range in winter; at least a few are seen annually to limit indicated by dashed line on map. In years when the lemming population plummets, many move south into northern tier of states; some may wander to central CA, eastern CO, OK, KY, and VA, exceptionally to Gulf Coast shores. These irruptives, usually heavily barred younger birds, often perch on the ground or on low stumps, buildings.

Barred Owl

dark eyes

barred breast

vertical streaks

Great Gray Owl
nebulosa

huge size

dark rings on facial disk

black and white at bottom of facial disk gives "bow tie" effect

Spotted Owl

dark eyes

color varies from darkest (Northwest) to palest (Southwest)

spotted below, including on breast

Snowy Owl

immature

round head

color varies from all-white to heavily barred depending on age and sex

Eastern Screech-Owl *Megascops asio* L 8½" (22 cm)

All three owl species shown on this page are small, with yellow eyes and pale bill tip; bill base is yellow-green on Eastern and Whiskered Screech-Owls and blackish or dark gray on Western. Ear tufts prominent if raised; when flattened, bird has a round-headed look. Underparts on all three are marked by vertical streaks crossed by dark bars: On Eastern, crossbars are spaced well apart and are nearly as wide as vertical streaks; on Western crossbars are closer together and much narrower; Whiskered is like Eastern, but markings have a bolder look. These markings are less distinct on the Eastern **rufous morph;** the latter predominates in the South, the **gray morph** on the Great Plains and in southernmost TX. Lightest and whitest, the *maxwelliae* subspecies is found in the northwestern part of the range; darker *mccallii* is found in eastern and southern portions of the Trans-Pecos. Nocturnal; best located and identified by voice.

VOICE: Two typical calls: a series of quavering whistles, descending in pitch; and a long single trill, all on one pitch. Sometimes may be heard calling inside roost hole in morning.

RANGE: Common in a wide variety of habitats: woodlots, forests, swamps, orchards, parks, and suburban gardens. Formerly classified with Western Screech-Owl as one species; range separation is not yet fully known. Both species are found at Big Bend National Park in TX: Western is uncommon, Eastern rare and may not have been recorded recently. Accidental to southeastern NM (Portales).

Western Screech-Owl *Megascops kennicottii* L 8½" (22 cm)

Generally gray overall; some birds in the humid coastal Northwest are brownish. Bill is darker and crossbars are weaker than Eastern Screech-Owl. Nocturnal; best located and identified by voice.

VOICE: Two common calls: a series of short whistles accelerating in tempo; and a short trill followed immediately by a longer trill, both on the same pitch.

RANGE: Common in open woodlands, streamside groves, deserts, suburban areas, and parks. Where range overlaps with that of Whiskered Screech-Owl, Western is generally found at lower elevations. Casual to southwestern KS.

Whiskered Screech-Owl *Megascops trichopsis* L 7¼" (18 cm)

Closely resembles gray Western Screech-Owl but slightly smaller with smaller feet and usually bolder crossbarring, similar to Eastern. Identification best done by voice and range. Nocturnal.

VOICE: Two common calls: a series of short whistles on one pitch and at a fairly even tempo; and a series of very irregular hoots, like Morse code.

RANGE: Common in dense oak and oak-conifer woodlands, at elevations from 4,000 to 6,000 feet, but mostly found around 5,000 feet. Generally at higher elevations than Western.

Eastern Screech-Owl

ear tufts

yellow eyes

pale greenish bill

rufous morph

rufous morph more numerous in Southeast, essentially unknown in West

gray morph

strong vertical and horizontal bars on underparts

gray-morph juvenile

overall pale

western Great Plains *maxwelliae*

Northwest coast *kennicottii*

dark bill

weaker crossbars than Eastern and Whiskered

Whiskered Screech-Owl *aspersus*

dark bill

Western Screech-Owl

strong crossbars like Eastern Screech-Owl

small feet

Flammulated Owl *Otus flammeolus* L 6¾" (17 cm)

Dark eyes; small ear tufts, often indistinct; variegated red-and-gray plumage. Birds in the northwestern part of the range are the most finely marked; those in the Great Basin mountains are **grayish** and have the coarsest markings; and those breeding in the southeastern part of the breeding range are **reddish.** Strictly nocturnal.

VOICE: Best located at night by its call, a series of single or paired low, hoarse, hollow hoots.

RANGE: Common in oak and pine woodlands, especially ponderosa. Nests and roosts in tree cavities, usually old woodpecker holes. Sometimes nests in loose colonies. Highly migratory. Rare in interior lowlands during migration, but likely overlooked. Accidental east to LA and FL.

Ferruginous Pygmy-Owl *Glaucidium brasilianum*

L 6¾" (17 cm) Long tail, reddish with dark or dusky bars. Upperparts gray-brown; crown faintly streaked. Eyes yellow; black nape spots look like eyes on the back of the head. White underparts streaked with reddish brown. Chiefly diurnal; active any time of day. Roosts in crevices and cavities.

VOICE: Most common whistled call is a rapid, repeated *took.*

RANGE: Found at lower elevations than Northern Pygmy-Owl. Inhabits saguaro deserts (AZ), and woodlands north to near Kingsville, TX. Rare in AZ and the slightly grayer subspecies there *(cactorum)* is considered endangered **(E).** The browner *ridgwayi* is resident in south TX.

Elf Owl *Micrathene whitneyi* L 5¾" (15 cm)

Our smallest owl. Yellow eyes; very short tail. Lacks ear tufts. Strictly nocturnal; roosts and nests in cavities in saguaros and trees.

VOICE: Call, an irregular series of high *churp* notes and chattering notes.

RANGE: Uncommon to common in desert lowlands and in foothill canyons, especially among oaks and sycamores. Eats mostly large insects. Now almost or actually eliminated in CA (pair found 2010) and NV. Casual in southwestern UT and in winter in southernmost TX. Accidental in fall to Los Angeles County.

Northern Pygmy-Owl *Glaucidium gnoma* L 6¾" (17 cm)

Long tail, dark brown with pale bars. Upperparts are either rusty brown or gray-brown; crown spotted; underparts white with dark streaks. Eyes yellow; black nape spots look like eyes on the back of the head. The grayest birds are to be found in the Rockies; those on the Pacific coast as far north as BC are browner. Chiefly diurnal; most active at dawn and dusk. An aggressive predator, this owl is a favorite target for songbirds. Birders may locate the owl by watching for mobbing songbirds.

VOICE: Call is a mellow, whistled, *hoo* or *hoo hoo,* repeated in a well-spaced series; also gives a rapid series of *hoo* or *took* notes followed by a single *took.*

RANGE: Inhabits dense woodlands in foothills and mountains. Nests in cavities. Nominate subspecies, *gnoma,* seen from southeastern AZ south through the mountains of Mexico, gives a series of double *took-took* notes with occasional single notes interspersed. Considered by some to be a separate species "Mountain Pygmy-Owl," though no marked genetic differences. Casual to Chisos Mountains, TX.

Flammulated Owl

reddish type

birds often more reddish in southeastern part of range

short ears are often indistinct

dark eyes

grayish type

Ferruginous Pygmy-Owl
cactorum

streaked crown

"false eye"

long tail with rusty bars

Elf Owl
whitneyi

rounded head with no ear tufts

yellow eyes

tiny size

very short tail

spotted crown

Northern Pygmy-Owl

"false eye"

Mountain group type

averages grayer than birds from Pacific region

Northern group type

long tail with pale bars

Northern Saw-whet Owl *Aegolius acadicus* L 8" *(20 cm)*

Reddish brown above; white below with reddish streaks; bill dark; facial disk reddish, without dark border. **Juvenile** dark brown above, tawny rust below. Strictly nocturnal; roosts during day in or near nest hole in breeding season. In winter, preferred roost is in dense evergreens. Concentrations of regurgitated pellets and "whitewash" excrement build up below favored winter roosts. Once found, this species can be closely approached. A distinctive subspecies, *brooksi*, endemic to Queen Charlotte Islands, BC, is much darker and buffier on facial disks and belly.

VOICE: Call, heard primarily in breeding season from late winter to late spring, is a monotonously repeated single-note whistle; another raspy call sounds like a saw being sharpened. Also gives a rising screech.

RANGE: Inhabits dense coniferous or mixed forests, wooded swamps, and tamarack bogs. The number of migrants that move southward in fall and winter varies from year to year.

Northern Hawk Owl *Surnia ulula* L 16" *(41 cm)*

Long tail, falconlike profile, and black-bordered facial disks identify this owl of the northern forests. Underparts are barred with brown. Flight is low and swift; sometimes hunts during daylight as well as at night. Is most often seen, however, perched high in a spruce tree. Usually can be closely approached.

VOICE: Mostly silent away from nesting sites. Gives a series of rapid whistled notes at night. Also various whistled and screeching notes.

RANGE: Mostly nonmigratory, but retreats slightly in winter from northernmost part of range. Casual south of mapped range.

Boreal Owl *Aegolius funereus* L 10" *(25 cm)*

White underparts streaked with chocolate brown. Whitish facial disk has a distinct black border; bill is pale. Darker above than Northern Saw-whet Owl. **Juvenile** is chocolate brown below. Strictly nocturnal; roosts during daylight in dense cover, usually close to tree trunk.

VOICE: Call, heard primarily in breeding season from late winter to midspring, is a short, rapid series of hollow *hoo* notes.

RANGE: Inhabits dense northern forests and muskeg. Irruptive, usually in small numbers; otherwise seldom seen south of mapped range, but this may be due to difficulty of locating them. Breeds at isolated locations high in the Rockies.

Burrowing Owl *Athene cunicularia* L 9½" *(24 cm)*

Long legs distinguish this ground dweller from all other small owls. Adult is boldly spotted and barred. **Juvenile** is buffy below. Nocturnal; flight low and undulating; often hovers like a kestrel. Perches conspicuously during daylight at entrance to burrow. FL birds *(floridana)* are darker above with more whitish spotting; less buffy below.

VOICE: Calls include a soft *coo-coooo* and a chattering series of *chack* notes.

RANGE: An owl of open country, golf courses, airports. Nests in single pairs or small colonies. Declining in much of northern Great Plains, the West, and FL; may be partly due to destruction of prairie dog towns, which it frequents, by agricultural growth. Casual vagrant (both North American subspecies) in spring and fall to East Coast, north to ME.

Northern Saw-whet Owl
acadicus

no dark border to facial disk

dark bill

rufous overall coloration

juvenile

Northern Hawk Owl
caparoch

black border to facial disk

prominent black border to facial disk

pale bill

Boreal Owl
richardsoni

dark barring on underparts

juvenile

darker body coloration than juvenile Northern Saw-whet

long tail

Burrowing Owl
western *hypugaea*

round head

juvenile

heavily spotted with buffy white above

no dark barring below

GOATSUCKERS Family Caprimulgidae

Wide mouths help these night hunters snare flying insects. Most located and identified by distinctive calls. SPECIES: 93 WORLD, 10 N.A.

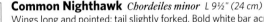

Lesser Nighthawk *Chordeiles acutipennis* L 8½" (22 cm)

Like Common Nighthawk but outer wing ("hand") shorter and inner wing ("arm") broader; wing tip usually more rounded; whitish bar across primaries closer to tip. Upperparts paler and more uniformly mottled; in Common paler wing coverts contrast more with back. Throat is white in **males,** usually buffy in **females** and **juveniles.** Underparts buffy, with faint barring. Male has white tail band. Female lacks tail band; note buffy wing bar and markings at base of primaries; wing bar indistinct in juvenile female. Molts on breeding grounds, whereas Common molts on winter grounds in South America. Seen chiefly around dusk. Flies with fluttery wingbeats.

VOICE: Call, a rapid, tremulous trill, heard only on breeding grounds.
RANGE: Fairly common; found in scrubland and desert. Very rare spring and fall migrant on Gulf Coast. Rare to casual in winter in southern CA, TX, FL. Accidental to northwestern AK, YT, ON, WV, NJ.

See subspecies map, page 552

Common Nighthawk *Chordeiles minor* L 9½" (24 cm)

Wings long and pointed; tail slightly forked. Bold white bar across primaries slightly farther from wing tip than Lesser Nighthawk. Subspecies range from dark brown in eastern birds to more grayish in the northern Great Plains subspecies, *sennetti;* color variations are subtle in adults, distinct in juveniles. Throat white in **male,** buffy in female; underparts whitish, with bold dusky bars. Female lacks white tail band. **Juvenile** shows less white on throat; more active in daylight than other goatsuckers. Roosts on the ground, on branches, posts, and roofs.

VOICE: In courtship display, male's wings make a hollow booming sound. Nasal *peent* call, given even in migration, is diagnostic.
RANGE: Seen in woodlands, suburbs, and towns. Fairly common over parts of range, but declining in East.

Antillean Nighthawk *Chordeiles gundlachii* L 8" (20 cm)

Very difficult to distinguish from variably colored Common Nighthawk, but Antillean is smaller and shorter winged; also bar across primaries averages closer to tip.

VOICE: Call, a varying *pity-pit-pit,* best identifies Antillean.
RANGE: A few in summer on Florida Keys; also on Dry Tortugas and southeastern FL mainland. Accidental to LA and Outer Banks, NC.

Common Pauraque *Nyctidromus albicollis* L 11" (28 cm)

In flight, long, rounded tail separates it from nighthawks; also shorter, rounder wings. Broad white bands on wings distinguish it from other nightjars (page 294). White tail patches conspicuous on **male,** smaller and often buffy on **female.** In close view, note chestnut ear patch. Seen just before dawn and after dusk. Flies close to the ground; often lands on roads or roadsides.

VOICE: Distinctive song is one or more low *pur* notes followed by a higher, descending *wheeer.*
RANGE: Common resident in woodland clearings and scrub.

Lesser Nighthawk
texensis

buffy bars on inner primaries

faint primary bar

juvenile

white bar closer to wing tip

rounder wing tip than Common in all plumages

"hand" shorter and "arm" broader than Common

adult ♂

molting adult ♀

molting birds in late summer are Lessers; Commons don't molt flight feathers until reaching winter grounds in South America

paler coloration than Common

rich buff bars at base of folded primaries

p10

p10 longer than p9, resulting in a pointed wing tip

white wing bar closer to primary bases and surrounded by dark

outer wing (hand) longer than Lesser

inner wing (arm) narrower than Lesser

male with white bar on tail

Common Nighthawk
hesperis

white throat

♂

juvenile

solidly dark in front of white wing bar

juvenile
sennetti

juveniles have reduced white in wings and whitish fringes on primary tips

juveniles show distinct geographic color variations; *sennetti* is silver-gray

Antillean Nighthawk

slightly shorter wing than Common and pale bar slightly closer to tip

overall similar to Common

rounded wings

wing patch; white in males, buffy in females

♂

Common Pauraque
merrilli

♂

quite buffy overall, but so are some Commons

rufous-tinged cheek

boldly fringed scapulars

long tail

♂

a few tail feathers extensively white in males, more restricted white in females

♀ ♂

Chuck-will's-widow *Caprimulgus carolinensis* L 12" *(31 cm)*
Our largest nightjar. Wings rounded. Much larger and most are redder than smaller-headed Whip-poor-will; buff-brown throat and whitish necklace contrast with dark breast. **Male'**s tail has less white than male Whip-poor-will. **Female'**s tail lacks white. Usually more shy than Eastern Whip-poor-will.
VOICE: Loud, whistling song sounds like *chuck-will's-widow.*
RANGE: Fairly common but local in oak-pine woodlands, live oak groves. Accidental west to NM, NV, coastal northern CA.

Eastern Whip-poor-will *Caprimulgus vociferus*
L 9¾" *(25 cm)* Smaller and smaller headed than Chuck-will's-widow; dark throat contrasts with white or buffy necklace and paler underparts **Male'**s tail shows much more white than male Chuck-will's-widow. **Female'**s tail has buffy tip to outer tail feathers. By day, sits on a elevated perch such as a branch; usually approachable.
VOICE: A clear, loud *whip-poor-will.* Also *cluck* notes, especially in flight.
RANGE: Found in open coniferous and mixed woodlands. Local and declining. Accidental to southeastern AK and likely from CA in late fall.

Mexican Whip-poor-will *Caprimulgus arizonae*
L 9¾" *(25 cm)* Extremely similar to Eastern Whip-poor-will; best separated by voice and perhaps behavior. Rictal bristles average longer, often with buffy base rather than entirely blackish; upper side of tail more uniform with back, rather than contrastingly paler. **Male'**s tail averages less white at tip. Will sing from tree branches, but roosts mainly on the ground; does not allow close approach like Eastern.
VOICE: Like Eastern, but *whip-poor-will* song much burrier; also gives *cluck* notes in flight.
RANGE: Breeds in conifer and mixed woodlands of the mountains of the Southwest. Casual to central and northern CA (including on coast and in winter), UT, and CO. Accidental southeastern OR.

Buff-collared Nightjar *Caprimulgus ridgwayi* L 8¾" *(22 cm)*
Note distinct buff collar across nape; otherwise like whip-poor-wills, but paler and more finely marked.
VOICE: Song is an accelerating series of *cuk* notes ending with *cukacheea.*
RANGE: Rare, irregular, and local in desert canyons of extreme southwestern NM (Guadalupe Canyon), at least formerly, and especially southeastern AZ. Accidental in southern CA (specimen, 8 June 1996, Oxnard, Ventura County). Usually roosts on the ground by day.

Common Poorwill *Phalaenoptilus nuttallii* L 7¾" *(20 cm)*
Our smallest nightjar, distinguished by short, rounded tail and wings. Outer tail feathers are tipped with white, more boldly in **male** than female. Plumage is individually and geographically variable; upperparts range from brownish gray to pale gray.
VOICE: Song is a whistled *poor-will,* with a final *ip* note audible at close range.
RANGE: Fairly common in sagebrush and on chaparral and rocky slopes; often seen on roads or roadsides after dusk or before dawn. Known to hibernate in cold weather; may winter north into normal breeding range.

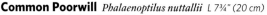

much larger than Whip-poor-will with big, flat head

most are rufous overall in coloration

gray morph

GOATSUCKERS

more restricted white than both whip-poor-will species

rufous-brown throat

some are grayer

♂

♀

rufous morph

Chuck-will's-widow

lacks white in tail

blackish throat and white collar on males

Eastern Whip-poor-will
vociferus

overall darker and less rufous than Chuck-will's-widow

female with buffy tips to outer tail feathers

♀

♂

male with much more white in outer tail than Chuck-will's Widow

longer and browner rictal bristles

♂

Mexican Whip-poor-will
arizonae

throat blackish with white collar on males

coloration very similar to Eastern Whip-poor-will

♂

male averages less white in outer tail feathers than Eastern Whip-poor-will

distinct and complete buff collar

white corners on tail hidden when perched

overall paler than Mexican Whip-poor-will

Buff-collared Nightjar

♂

male with less white than whip-poor-wills

appears large headed and small bodied

often sits on or at side of road

Common Poorwill
nuttallii

short, rounded wings

♀

short tail, other tail feathers tipped with white

♂

These fast-flying birds spend the day aloft. Long wings bend closer to the body than on similar swallows. SPECIES: 100 WORLD, 9 N.A.

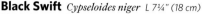

Black Swift *Cypseloides niger* L 7¼" (18 cm)
Blackish overall; long, slightly forked tail, often fanned in flight. Nests in colonies on cliffs, beneath waterfalls; also on wet sea cliffs. Wingbeats rather more leisurely than other swifts; often soars.
VOICE: *Chik* calls, sometimes in a series, delivered more slowly than Chimney.
RANGE: Uncommon and, because of nesting habitat, very local throughout breeding range. Rare over much of West in migration; most are seen in late spring during inclement or cloudy weather. Accidental in East. Thought to winter in South America.

Vaux's Swift *Chaetura vauxi* L 4¾" (12 cm)
Smaller than Chimney; usually paler below and on rump; soars less.
VOICE: Call softer, higher, and more insectlike than Chimney Swift.
RANGE: Nests in hollow trees. Fairly common in woodlands near water. In migration, regular to western NV and AZ. Rare in winter in southern CA; casual on Gulf Coast.

Chimney Swift *Chaetura pelagica* L 5¼" (13 cm)
Cigar-shaped body; short, stubby tail. Larger, usually darker below than Vaux's Swift; has longer wings, greater tendency to soar. Both species give a rocking display: Wings upraised, bird rocks from side to side.
VOICE: A louder, chattering call than Vaux's Swift.
RANGE: Nests in chimneys, barns, and hollow trees. Common in East, especially in towns; uncommon in towns on Great Plains; accidental to NT and AK. Rare and declining in southern CA. No winter records in the U.S.

White-throated Swift *Aeronautes saxatalis* L 6½" (17 cm)
Black above, black-and-white below, with long, forked tail, often held in a point. Distinguished from Violet-green Swallow (page 370) by longer, narrower wings, bicolored underparts. In poor light, may be mistaken for Black Swift but is smaller, with a more pointed rear and faster wingbeats.
VOICE: Call is a descending harsh chatter.
RANGE: Common in mountains, canyons, cliffs. Nests in crevices. Accidental to upper Midwest and to FL.

White-collared Swift *Streptoprocne zonaris* L 8½" (22 cm)
A very large, black swift; note white collar, more indistinct (sometimes incomplete) in **immatures** and possibly in some adult females, too. Slightly forked tail. Soars with wings bent down.
RANGE: Tropical species. Casual; scattered records from coastal and Great Lakes locations in the U.S. and southern Canada. Eight records, three of which are specimens: Two of those (Escambia County, FL, and Kleberg County, TX) are the Mexican subspecies, *mexicana;* the other (Broward County, FL, 15 Sept. 1994) is of the smaller West Indian subspecies, *pallidifrons,* which has more whitish about the head.

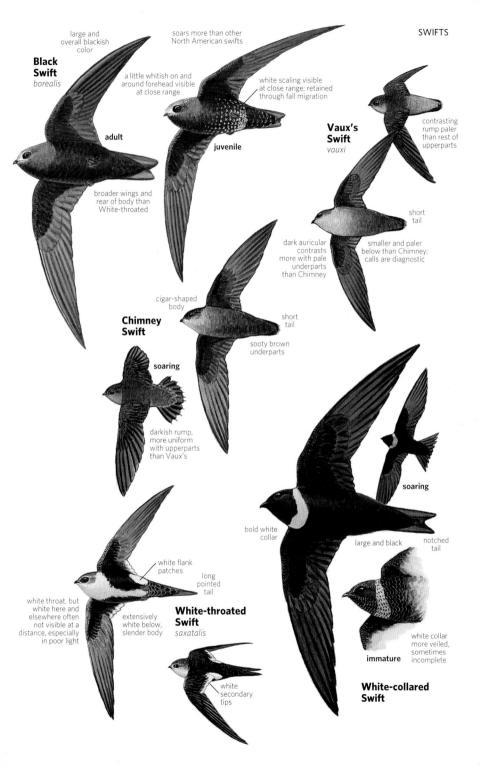

Black Swift
borealis

large and overall blackish color

soars more than other North American swifts

a little whitish on and around forehead visible at close range

white scaling visible at close range; retained through fall migration

adult

juvenile

broader wings and rear of body than White-throated

Vaux's Swift
vauxi

contrasting rump paler than rest of upperparts

short tail

smaller and paler below than Chimney; calls are diagnostic

dark auricular contrasts more with pale underparts than Chimney

Chimney Swift

cigar-shaped body

short tail

sooty brown underparts

soaring

darkish rump, more uniform with upperparts than Vaux's

white flank patches

long pointed tail

White-throated Swift
saxatalis

white throat, but white here and elsewhere often not visible at a distance, especially in poor light

extensively white below, slender body

white secondary tips

bold white collar

large and black

notched tail

soaring

white collar more veiled, sometimes incomplete

immature

White-collared Swift

White-throated Needletail *Hirundapus caudacutus*
L 8" (21 cm) Dark overall, with pale patch on back; white throat and undertail coverts; white on inner webs of tertials difficult to see in field. Tail short and stubby. A powerful flier with long, sustained periods of soaring, sometimes in wide circles, as with the three other Asian needletail species.
RANGE: Large Asian swift, casual in spring on outer Aleutians, where recorded twice each on Attu and Shemya Islands from mid- to late May.

Common Swift *Apus apus* *L 6½" (17 cm)*
Long, narrow winged, dark, with paler throat, long forked tail. Pribilofs specimen of eastern subspecies *pekinensis* is paler than nominate.
RANGE: Breeds in Palearctic; winters in Africa. Accidental in summer on Pribilofs (twice) and Miquelon Island off southern Newfoundland; records from Bermuda and probably the Northeast.

Fork-tailed Swift *Apus pacificus* *L 7¾" (20 cm)*
White rump; long tail's fork often not apparent. Overall coloration blackish with paler fringes and palish throat.
RANGE: Asian species, casual to western islands of AK and on Middleton Island, AK.

TROGONS Family Trogonidae
Colorful tropical birds with short, broad bills. SPECIES: 44 WORLD; 2 N.A.

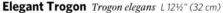

Elegant Trogon *Trogon elegans* *L 12½" (32 cm)*
Yellow bill, white breast band; undertail delicately barred. **Male** is bright green above and has bright red belly. **Female** and juvenile are browner and duller.
VOICE: Song is a series of croaking *co-ah* notes; other variations as well.
RANGE: Found in streamside woodlands, mostly at altitudes from 4,000 to 6,000 feet. Casual in west (Big Bend) and southernmost TX.

Eared Quetzal *Euptilotis neoxenus* *L 14" (36 cm)*
Wary. **Male** has bright green breast and red belly; **female** is similar but head, throat, and breast are gray and belly is paler red. Larger and with thicker body than Elegant Trogon, giving a hunchbacked look, with a disproportionately small head; bill is black or gray; lacks white breast band and barring on undertail.
VOICE: Calls include a loud upslurred squeal ending in a *chuck* note; also a loud, hard cackling. Male's song is a long, quavering series of whistled notes that increase in volume.
RANGE: Casual in mountain streamside woodlands of southeastern AZ. Records scattered throughout the year, but more records for fall. Recorded from as far north as Superstition Mountains east of Phoenix (winter), but most from Chiricahua and Huachuca Mountains; attempted to nest in latter range (upper Ramsey Canyon).

white throat and lore spot visible at close range

broad wings like a 747, often soars for extended periods

white horseshoe pattern

pale throat

short wedge-shaped tail

silvery back

white tertial tips

White-throated Needletail
caudacutus

brownish black plumage with long wings and slender body

long forked tail

Common Swift
pekinensis

soaring

palish throat

pale fringing on body apparent only at close range

Fork-tailed Swift
pacificus

long forked tail often closed, which makes tail look like a long, extended point

white spot on ear coverts

brownish on head and upperparts

Elegant Trogon
canescens

♀

yellow bill

♂

no breast band

white breast band

bright geranium red belly

bright green head and upperparts

bronze tail

blackish terminal tail band

♂   ♀

delicately barred underside of tail

Eared Quetzal

slaty gray head and breast on female

dark bill

♂

♀

both sexes lack pale marking on head; "ears" visible only at close range

all have a hunchbacked appearance and a disproportionately small head

blue-black tail when viewed from above

unlike Elegant, tail feathers extensively pure white with no fine barring

HUMMINGBIRDS Family Trochilidae

These birds hover at flowers to sip nectar with needlelike bills. Often identified to species by calls. Males' iridescent throat feathers (gorget) look black in poor light.
SPECIES: 336 WORLD, 23 N.A.

Green Violetear *Colibri thalassinus* L 4¾" (12 cm)
Green overall with dark subterminal tail band; bill slightly down-curved. **Adult male** has blue-violet patches on face and breast. Female and **immature** slightly duller, immature grayer on belly.
VOICE: Song is a repeated *tsip-tsup.*
RANGE: Tropical species. Most records in summer in the Hill Country in I X, where nearly annual; casual throughout much of eastern North America, accidental elsewhere.

Green-breasted Mango *Anthracothorax prevostii*
L 4¾" (12 cm) Has fairly thick, curved bill and purplish color in outer tail feathers. **Adult male** is deep green overall. **Female** shows broad blackish green stripe on underparts bordered in white. **Immature** is similar to female but shows cinnamon border on sides of underparts.
VOICE: Gives hard ticking chip notes, often in a series; also higher *sip* calls in flight. Often silent.
RANGE: Found from eastern Mexico to northern South America. Casual in fall and winter to south TX; accidental to NC, GA, and WI.

White-eared Hummingbird *Hylocharis leucotis*
L 3¾" (10 cm) Bill shorter than Broad-billed Hummingbird; broad white stripe extends back from eye; has black ear patch, square tail. **Adult male** has dark purple crown and chin, emerald green gorget. Female lacks purple on head; underparts spotted with green. Immature male is intermediate between adult male and female.
VOICE: Display call is a repeated, silvery *tink tink tink.* Chattering calls are loud and metallic.
RANGE: Summer visitor to mountains of Southwest. Rare in southeastern AZ; very rare in southwestern NM and west TX; accidental elsewhere in TX and in CO and MS.

Broad-billed Hummingbird *Cynanthus latirostris*
L 4" (10 cm) **Adult male** is dark green above and below, with white undertail coverts, blue gorget, and mostly red bill. Broad, forked tail is blackish blue. **Adult female** is duller above, gray below; often shows a narrow white eye stripe; has a square-tipped tail. Juveniles resemble female; by late summer, male begins to show blue and green flecks on throat, green on sides; dark, forked tail helps distinguish it from White-eared Hummingbird.
VOICE: Chattering *je-dit* call is similar to Ruby-crowned Kinglet. Male's display call is a whining *zing.*
RANGE: Common in desert canyons and low mountain woodlands. Very rare in southern CA and TX during fall and winter; casual or accidental to OR, ID, CO, and the East.

**Green
Violetear**
thalassinus

blue-violet
patches on face
and breast

adult ♂

dark subterminal
tail band

immature

long curved
bill

adult ♀

green vertical
stripe on
underparts

**Green-breasted
Mango**
prevostii

green and cinnamon
vertical stripes on
underparts

adult ♂

dark green
below

purplish tail
in all plumages

immature

**White-eared
Hummingbird**
borealis

short,
red-based
bill

bold white stripe
in all plumages

adult ♀

purple crown
and chin

adult ♂

short
tail

**Broad-billed
Hummingbird**
magicus

white stripe sometimes
bolder, never as bold as
White-eared

♀

longer billed than White-eared;
female with red at base of lower
mandible

**immature
♂**

short,
square-ended tail

male's bill
more extensively red

♂

adult ♂

longer, lobed tail
more notched

Buff-bellied Hummingbird *Amazilia yucatanensis*

L 4¼" (11 cm) Green overall with buff belly and chestnut tail; bill pinkish red with black tip. Sexes similar, but female with less intensely green throat and chest, chin mottled with buffy white, rufous usually lacking on central tail feathers, bill averages duller. Immature has more grayish on throat and breast; bill darker.

VOICE: Calls are shrill and squeaky.

RANGE: Found in woodlands and thickets; also suburban yards; readily visits feeders. Mexican species, fairly common in lower Rio Grande Valley. Rare in winter in Gulf Coast region; casual to FL; accidental AR and GA (twice on Jekyll Island).

Berylline Hummingbird *Amazilia beryllina* L 4¼" (11 cm)

Green above and below, with chestnut wings, which are visible at rest and especially in flight. Tail coppery purple overall; more purplish when fresh, more chestnut when worn. Base of lower mandible red. **Male's** lower belly is buffy gray, **female's** more grayish. Female also duller green below with whitish feathering, less extensive chestnut in wings; tail more bronzy, less purplish. Hybrids with Magnificent Hummingbird (page 304) have been noted in southeastern AZ.

VOICE: A hard, buzzy *dzzrit* given in flight or when perched. Song of male is a complex squeaky warbling with chip notes.

RANGE: Very rare summer visitor from Mexico to mountains of southeastern AZ; has bred there. Casual to southwestern NM and west TX.

Violet-crowned Hummingbird *Amazilia violiceps*

L 4½" (11 cm) Crown violet; small white postocular spot; underparts entirely and strikingly white; upperparts bronze green; tail greenish. Long bill is mostly red. Sexes similar. Immature with greatly reduced purple on crown; crown feathers with cinnamon tips; upperparts with buffy feather tipping; bill darker.

VOICE: Call is a loud, squeaky chattering; male's song is a series of sibilant *ts* notes.

RANGE: Favors streamside locations: mesquite-sycamore in lower foothills or slightly higher in oak-sycamore canyons; sometimes higher at feeding stations. Uncommon. Casual in CA and TX.

Lucifer Hummingbird *Calothorax lucifer* L 3½" (9 cm)

Bill downcurved. **Adult male** has green crown, distended but squarish; purple throat; and an especially long tail, usually held in a point, always when perched. **Female** is rich buff below; broadening pale stripe behind eye is quite broad where it meets side of neck; outer tail feathers on long tail reddish at base. **Immature male** more like female; some have a few purplish throat feathers, as do some adult females. Adult male perches high on a territorial bare branch.

VOICE: Call is a fairly hard, rather rich *chih* or *chi,* sometimes doubled; other series of notes in chases, some of which are squeaky.

RANGE: Prefers desert washes in arid foothills and up into lower mountains. Uncommon in Chisos Mountains, TX; rare in southeastern AZ, southwestern NM, and Davis Mountains, TX. Casual elsewhere in the Trans-Pecos and east to central TX.

pinkish red bill base

dull red bill base

female duller than male, with buffy belly ♀

green overall ♂

rufous wings

Berylline Hummingbird
viola

rufous

Buff-bellied Hummingbird
chalconota

green breast

rufous on wing especially visible in flight

buff belly

♂

tail extensively rufous

Violet-crowned Hummingbird
ellioti

violet crown; duller on juveniles

all with long curved bill purplish throat

adult ♂

red-based bill

striking snowy white underparts

adult ♂

Lucifer Hummingbird

pale postocular stripe gradually broadens to sides of neck

♀

very long pointed tail; fork not usually visible when bird is perched

rich buff

long tail extends past folded wings

limited rufous

immature
♂

♀

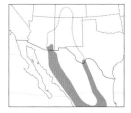

Blue-throated Hummingbird *Lampornis clemenciae*

L 5" (13 cm) A large hummingbird with a broad white eye stripe and faint white malar stripe that border dark ear patch; large black tail with broad white tips on outer feathers. **Adult male's** throat is blue; **female's** gray. Sometimes a crepuscular feeder. TX subspecies *phasmorus* is slightly greener above and on average shorter billed than *bessophilus* from AZ.
VOICE: Call is a loud, high, repeated *seep*, given perched or in flight. Song of male is a long, continuous series of *seep* notes.
RANGE: Generally uncommon; found in mountain canyons with permanent running water; very rare to NM. Casual north of mapped range; accidental to CA and in East to LA.

Magnificent Hummingbird *Eugenes fulgens* L 5¼" (13 cm)

Large with long, straight bill. **Adult male** has purple crown; metallic green throat; distinct white postocular spot; breast and upper belly black and green. Tail is dark green. **Female** is duller; squarish tail has small, grayish white tips on outer feathers; compare with female Blue-throated Hummingbird, which has much stronger eyebrow, blackish tail with broad white tips.
VOICE: Main call is a sharp chip; gives other squeaky notes.
RANGE: Uncommon to fairly common in mountains (mid to high elevation) in meadows and canyons. A few winter in southeastern AZ. Casual away from mapped range north to northern CA, NV, UT, CO (has bred), and MN; east to AR, GA, and FL.

Plain-capped Starthroat *Heliomaster constantii* L 5" (13 cm)

Large size with strikingly long, straight bill. Broad white malar stripe; white eye stripe, well outlined by dark dusky auricular patch. White patches on sides of rump are usually conspicuous; sometimes covered by wings. Throat shows variable amount of red. Sexes similar. Immature has sooty gray patch on throat with no red; buffy tipping on crown and upperparts.
VOICE: Call is a sharp chip that has been compared to Black Phoebe. Male's song, delivered from a conspicuous perch, is a series of sharp chip notes with other notes interspersed.
RANGE: Casual stray in summer and fall from Mexico to arid foothills and deserts of southeastern AZ; recorded northwest to Phoenix area.

Bahama Woodstar *Calliphlox evelynae* L 3¾" (10 cm)

Adult male has broad white collar, light purple throat; long, deeply forked tail usually held closed in a rounded point; mixed olive and rich cinnamon-buff below. In **female,** note tail projection past primary tips, cinnamon tips to outer tail feathers, slightly curved bill. Overall appearance quite suggestive of female and immature Rufous or Allen's Hummingbirds (page 308). By winter, immature males like adult males but lack purplish throat.
VOICE: Calls suggestive of Anna's Hummingbird.
RANGE: Uncommon Bahamian endemic. Prefers open woodland, beach scrub, and gardens. Four records of five birds from southeastern FL, none since 1981.

both sexes with prominent white facial stripes

Blue-throated Hummingbird
bessophilus

large size

blue throat often looks dark

♀

adult ♂

extensive white tips to outer tail feathers

usually just a white postocular spot; sometimes a short white supercilium

♀

limited white on tail

purple crown

large size

green throat

Magnificent Hummingbird
fulgens

extensive black breast

adult ♂

female highly suggestive of Rufous or Allen's

Bahama Woodstar
evelynae

♀

tail extends beyond primary tips

broad white malar stripe bordered by dark auricular

very long bill

light purplish throat with broad white collar

adult ♂

♀

white patches on sides and center of rump, sometimes covered by wings

cinnamon tips to outer tail feathers

Plain-capped Starthroat
pinicola

immature ♂

immature males like adult males, but lack purple throat

forked tail

Ruby-throated Hummingbird *Archilochus colubris*

L 3¾" (10 cm) The only regular hummingbird in East. Metallic green above. **Adult male** has a ruby red throat and black chin extending back under eye; sides and flanks dusky green; tail forked. **Female's** throat is whitish; underparts grayish white, with buffy wash on sides; tail is similar to female Black-chinned Hummingbird. Immatures resemble adult female. Some **immature males** show a red spot or two on throat by early fall. As with all hummingbirds, adult males migrate earlier than females and immatures. *Archilochus* hummingbirds seen in the Southeast in winter could be Ruby-throated or Black-chinned; females and immatures of these two species are very similar. Ruby-throated generally has a greener crown, shorter bill; note triangular dark patch on lores and more pointed shape of darker primaries, especially outermost. **VOICE:** Calls very similar to Black-chinned.
RANGE: Casual west to CA and AK. Fairly common in parts of range; found in gardens and woodland edges.

Black-chinned Hummingbird *Archilochus alexandri*

L 3¾" (10 cm) Metallic green above. In good light, **adult male** shows violet band at lower border of black throat. Underparts whitish; sides and flanks dusky green. **Female's** throat can be all-white or show faint dusky or greenish streaks. Immatures resemble adult female; **immature male** may begin to show violet on lower throat in the fall. Twitches tail more than Ruby-throated while feeding.
VOICE: Call is a soft *tchew,* sometimes mixed with squeals in chases.
RANGE: Common in lowlands and low mountains. A few winter in Southeast. Casual in late fall to Midwest and Atlantic Canada.

Costa's Hummingbird *Calypte costae* L 3½" (9 cm)

Adult male has deep violet crown and gorget extending far down sides of neck. **Female** is generally grayer above, whiter below, than female Black-chinned Hummingbird; note whitish gray feathering that comes around behind dusky auriculars; note also tail differences. Best distinguished by voice.
VOICE: Call is a high, metallic *tink,* often given in a series. Male's call is a loud *zing.*
RANGE: Fairly common in desert washes, dry chaparral. Casual north to south-coastal AK and MT and east to TX.

Anna's Hummingbird *Calypte anna* L 4" (10 cm)

Adult male's head and throat are deep rose red, the color extending a short distance onto sides of neck. **Female's** throat usually shows red flecks, often forming a patch of color. In both sexes, underparts are grayish, washed with a varying amount of green. Bill is disproportionately short. Immatures resemble female; **immature male** usually shows some red on crown. **Juveniles** lack red on throat; compare with smaller and paler female Black-chinned and Costa's Hummingbirds.
VOICE: Common call note, a sharp *chick;* chase call is a rapid dry rattling. Male's song is a jumble of high squeaks and raspy notes.
RANGE: Abundant in coastal lowlands and mountains; also in deserts, especially in winter. Casual north to south-coastal AK and in fall and winter to eastern North America (north to Maritime Provinces and QC); accidental Newfoundland (Brownsdale).

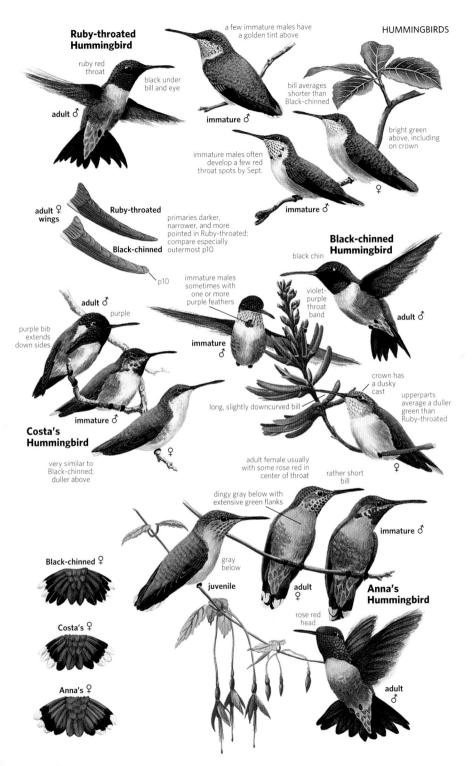

Ruby-throated Hummingbird

ruby red throat

black under bill and eye

adult ♂

a few immature males have a golden tint above

immature ♂

immature males often develop a few red throat spots by Sept.

bill averages shorter than Black-chinned

bright green above, including on crown

♀

immature ♂

adult ♀ wings

Ruby-throated

Black-chinned

primaries darker, narrower, and more pointed in Ruby-throated; compare especially outermost p10

p10

immature males sometimes with one or more purple feathers

Black-chinned Hummingbird

black chin

violet-purple throat band

adult ♂

adult ♂

purple

purple bib extends down sides

immature ♂

immature ♂

long, slightly downcurved bill

crown has a dusky cast

upperparts average a duller green than Ruby-throated

Costa's Hummingbird

very similar to Black-chinned; duller above

♀

adult female usually with some rose red in center of throat

♀

rather short bill

dingy gray below with extensive green flanks

immature ♂

Black-chinned ♀

gray below

juvenile

adult ♀

Costa's ♀

Anna's Hummingbird

rose red head

Anna's ♀

adult ♂

Calliope Hummingbird *Stellula calliope* L 3¼" (8 cm)

Very short bill and tail; primary tips extend well past end of tail. Carmine streaks on **adult male's** throat. **Female's** underparts colored like female Broad-tailed; female and immature male show a whitish lore area; very little rufous at base of tail.

VOICE: Relatively silent. Gives soft, high, slightly metallic chip notes.
RANGE: Nests in mountains. Uncommon spring migrant in desert lowlands, rare to CA coast. Fall migrant in Rockies (first adult males by mid-July). Very rare in Southeast in fall and winter; casual elsewhere in East.

Broad-tailed Hummingbird *Selasphorus platycercus*

L 4" (10 cm) All plumages show whitish around eye. **Adult male** has rose red throat, long tail with rufous-edged outer rectrices. **Female** has blended buff on underparts; similar to much smaller female Calliope, but note tail tip extends past primaries and bill much longer. Some immature males arrive back north in Apr. in an intermediate plumage with incomplete gorget and then complete their molt.

VOICE: Except during winter molt, all *Selasphorus* adult males' wingbeats produce a loud whistle; harsh and trilling in Broad-tailed. Calls include a metallic chip, distinctly different from Rufous and Allen's.
RANGE: Summers in mountains. Rare migrant through western Great Plains. Very rare in winter to Southeast. Casual to OR and coastal southern CA. Many claims pertain instead to Rufous/Allen's.

Rufous Hummingbird *Selasphorus rufus* L 3¾" (10 cm)

Adult male has rufous back, sometimes marked with green, very rarely entirely green; orange-red gorget. Immatures resemble **adult female; immature male** may show reddish brown back by winter before acquiring full gorget, which sometimes is not acquired even by spring. Female and immature closely resemble Broad-tailed, but more compact with a shorter bill; rich rufous-buff sides and flanks contrast sharply with white chest and collar.

VOICE: Buzzy wing whistle and all calls identical to Allen's. Calls include a sibilant chip, often given in a series; chase note, *zeee-chuppity-chup.*
RANGE: Common fall migrant through Rockies; a few to western Great Plains. Rare in fall in East; winters regularly in Southeast. Casual in winter in coastal southern CA. Rare in fall over much of the East.

Allen's Hummingbird *Selasphorus sasin* L 3¾" (10 cm)

Two subspecies: nominate *sasin,* breeds on central CA coast to southwestern OR; slightly larger *sedentarius,* resident on Channel Islands and on southern CA coast, where range expanding. **Adult male** usually told from Rufous by solid green back. Other plumages like Rufous, but tail feathers comparatively narrower for each sex and age class. Male has narrower rectrices than female; adult has narrower rectrices than immature. Determined in-hand when a vagrant is caught; sometimes from photographs. Width of r5 particularly important. Some immature males acquire adult malelike plumage by early fall *(sedentarius).*

VOICE: All sounds like Rufous, but male's flight display, usually a single vertical J-dive, differs from Rufous's multiple slanted J-dives.
RANGE: Allen's migrates earlier in fall and spring than Rufous. Uncommon fall migrant through southeastern AZ; casual to NV, NM, and west TX. Casual vagrant to eastern U.S. in fall and winter.

Calliope Hummingbird

adult ♂

purplish red throat streaks

overall females and immatures are very similar to Broad-tailed but are smaller, with shorter tail and bill

very short bill

whitish in supraloral region

♀

Broad-tailed Hummingbird

pale whitish eye ring in all plumages

rose red throat with narrow white line under bill

adult ♂

bluish green above

usually lacks white collar of female and immature Rufous and Allen's

blended buffy underparts

♀

long broad tail

rufous back

adult ♂

slight golden cast above

white collar

♀

Rufous Hummingbird

solid green back; a few adult male Rufous Hummingbirds also have green backs

adult ♂

Allen's Hummingbird
sasin

green-flecked ♂

immature ♂

Broad-tailed ♀

longer than Rufous or Allen's and has more restricted rufous

Rufous ♀

Rufous and Allen's have extensive rufous

Allen's ♀

narrow r5

feathers in each sex and age class narrower than Rufous

Calliope ♀

shorter than Rufous or Allen's and almost lacks rufous

KINGFISHERS Family Alcedinidae

Primarily an Old World family; only six species found in New World. Stocky and short-legged, with a large head, a large bill, and, in two North American species, a ragged crest. Look for kingfishers near woodland streams and ponds and in coastal areas. They hover over water or watch from low perches, then plunge headfirst to catch a fish. With strong bill and feet, they dig nest burrows in stream banks. SPECIES: 93 WORLD; 4 N.A.

Belted Kingfisher *Megaceryle alcyon* L 13" (33 cm)

The only kingfisher in most of North America. Both **male** and **female** have slate blue breast band; white belly and undertail coverts. Female has rust belly band and flanks; may be confused with female Ringed Kingfisher, note white belly and smaller size. Juvenile resembles adult but has rust spotting in breast band.
VOICE: Call is a loud, dry rattle.
RANGE: Common and conspicuous along rivers and brooks, ponds and lakes, and estuaries. Generally solitary. Rare in winter north into summer range. Casual to western and northern AK.

Ringed Kingfisher *Megaceryle torquata* L 16" (41 cm)

Larger than Belted Kingfisher; generally frequents larger rivers and ponds, perches on higher branches than Belted. Rufous underwing coverts of **female** distinctive in flight; white in **male**. Male is rust below with white undertail coverts. Female has slate blue breast, narrow white band, rust belly and undertail coverts. Juveniles resemble adult female, but juvenile male's breast is largely rust.
VOICE: Calls include a harsh rattle, lower and slower than Belted Kingfisher; also a single *chack* note, like Great-tailed Grackle, given chiefly in flight.
RANGE: Resident in lower Rio Grande Valley; casual elsewhere in south TX.

Green Kingfisher *Chloroceryle americana* L 8¾" (22 cm)

Smallest of our kingfishers; has a very long bill; crest inconspicuous. **Male** is green above, with white collar; white below, with rufous breast and dark green spotting. **Female** has a band of green spots across breast. Juvenile resembles adult female. Compare to larger but similarly colored Amazon Kingfisher (page 540). More retiring than other kingfishers; often hard to see and first detected by its call. Usually perches on low, sheltered branches. Flight is direct and very fast; white in outer tail feathers conspicuous in flight.
VOICE: One call, a faint but sharp *tick tick*, often ends in a short rattle; another, a squeaky *cheep*, is given in flight.
RANGE: Uncommon and often hard to see. Resident of lower Rio Grande Valley; rarer on Edwards Plateau, TX. Casual wanderer along the TX coast; also to Big Bend region. In southeastern AZ, a few pairs are resident along San Pedro River; also occasionally seen on Santa Cruz River near Nogales. One was recently discovered along the Gila River in southwestern NM.

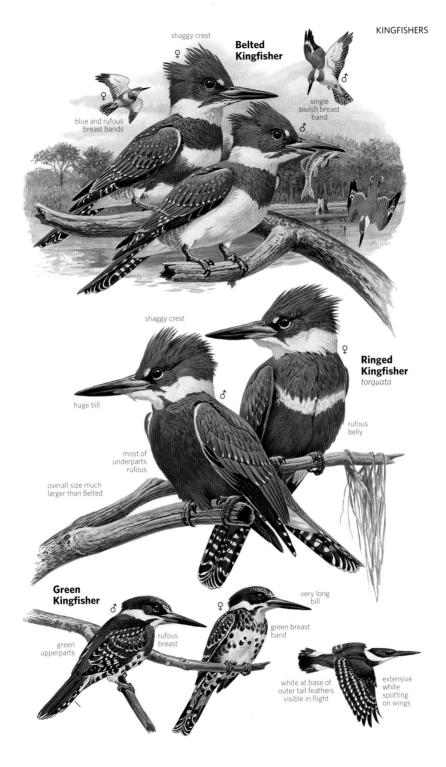

shaggy crest

Belted Kingfisher

♀

blue and rufous breast bands

single bluish breast band

♂

♂

shaggy crest

♂

huge bill

most of underparts rufous

overall size much larger than Belted

Ringed Kingfisher
torquata

♀

rufous belly

Green Kingfisher

♂

green upperparts

rufous breast

♀

green breast band

very long bill

white at base of outer tail feathers visible in flight

extensive white spotting on wings

WOODPECKERS Family Picidae

Strong claws, short legs, and stiff tail feathers enable woodpeckers to climb tree trunks. Sharp bill is used to chisel out insect food and nest holes and to drum a territorial signal.
SPECIES: 219 WORLD; 25 N.A.

Red-headed Woodpecker *Melanerpes erythrocephalus*

L 9¼" (24 cm) Entire head, neck, and throat are bright red in **adults,** contrasting with blue-black back and snowy white underparts. **Juvenile** is brownish; acquires red head during gradual winter molt; by spring closely resembles adult, but most retain secondaries with dark barring. Distinctive white inner wing patches and white rump are visible in all ages in perched and flying birds. Often seen in flight pursuing flying insects.

VOICE: In breeding season utters a loud *queark,* similar to Red-bellied Woodpecker (page 314) but harsher and sharper; uncommon to fairly common call is a guttural rattle.

RANGE: Uncommon to fairly common and declining. Inhabits a variety of open and densely wooded habitats. Now rare in the Northeast, chiefly in fall; no longer breeds there, due in part to habitat loss and competition with European Starlings for nest holes. Very rare to Maritime Provinces; casual to CA and AZ.

Acorn Woodpecker *Melanerpes formicivorus* *L 9" (23 cm)*

Inspiration for cartoon character Woody Woodpecker and his comical vocalizations based on the creator's experiences at Lake Sherwood, southern CA. Black chin, yellowish throat, white cheeks and forehead, red cap. **Female** has smaller bill than **male,** black bar on forecrown. In flight, white rump and small white patches on outer wings are conspicuous. Sociable; generally found in small, noisy colonies. Eats chiefly acorns and other nuts in winter, which they store in granaries on thick tree trunks and telephone poles, where available, by drilling holes into trunks and poles and pounding a nut into each hole. Colonies use the same "granary tree" year after year. Also takes insects, usually in flight.

VOICE: Most frequent call, *waka,* usually repeated several times.

RANGE: Common in oak woods or pine forests where oak trees are abundant. Casual to western Great Plains; accidental BC, MN.

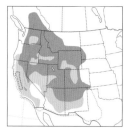

Lewis's Woodpecker *Melanerpes lewis* *L 10¾" (27 cm)*

Greenish black head and back; gray collar and breast; dark red face, pinkish belly. In flight, darkness, large size, and slow, steady wingbeats give it a crowlike appearance. **Juvenile** lacks collar and red face; belly may be only faintly pink; acquires more adultlike plumage from late fall through winter. Main food, insects, mostly caught in the air, as with Acorn Woodpecker; also eats fruit and nuts. Stores acorns, which it first shells, in tree bark crevices. Often perches on telephone poles.

VOICE: Often silent. A single squeaky sharp chip is given in interactions with other birds. Most vocal on breeding grounds.

RANGE: Uncommon to fairly common in open woodlands of interior; rare on coast. Often gregarious; fall and winter movements unpredictable. Largely eliminated in coastal Northwest. Very rare from western Great Plains to central TX. Accidental elsewhere in East, with a scattering of records. May rarely nest south of mapped range.

pure white
secondaries

adult

white primary
patches

♂

female has black
forecrown bar
♀

**Red-headed
Woodpecker**

brownish
head

juvenile

red head

barring on
secondaries

clown head
pattern

♂

adult

**Acorn
Woodpecker**

browner head
and no collar

juvenile

dark red
face

oily green upperparts,
no red on face,
and little pink on belly

gray collar

**Lewis's
Woodpecker**

pink belly

adults

slow crowlike
wingbeats

dark
wings

Golden-fronted Woodpecker *Melanerpes aurifrons*

L 9¾" *(25 cm)* Black-and-white barred back, white rump, usually an all-black tail; golden orange nape, paler in **female;** yellow feathering above bill. **Male** has a small red cap. Yellow tinge on belly not easily seen. Juvenile has streaked breast, brownish crown. In flight, all plumages show white wing patches, white rump like Red-bellied and Gila Woodpeckers; unlike them, Golden-fronted shows black, not barred, tail. Occasionally hybridizes with Red-bellied in narrow range overlap. **VOICE:** Calls, a rolling *churr-churr* and cackling *kek-kek,* are slightly louder and raspier than Red-bellied. **RANGE:** Fairly common in dry woodlands, pecan groves, and mesquite brushlands.

Red-bellied Woodpecker *Melanerpes carolinus*

L 9¼" *(24 cm)* Black-and-white barred back; white uppertail coverts; barred central tail feathers. Crown and nape red in **male. Female** has red nape only; small reddish patch or tinge on belly, usually difficult to see. **VOICE:** Call, a rolling *churr,* given in a series during breeding season, or *chiv-chiv,* is slightly softer than Golden-fronted. **RANGE:** Common in open woodlands, suburbs, and parks. Breeding range extending northward. Rare, mostly in fall and winter, to ME and Maritime Provinces; casual to NM.

Gila Woodpecker *Melanerpes uropygialis* L 9¼" *(24 cm)*

Black-and-white barred back and rump; central tail feathers barred. **Male** has a small red cap. **VOICE:** Calls, a rolling *churr* and a loud, sharp, high-pitched *yip,* often given in a series. **RANGE:** Inhabits towns, scrub desert, cactus country, and streamside woods. Accidental to eastern San Diego County, Los Angeles, and the San Francisco Bay area.

SAPSUCKERS

Drills evenly spaced rows of holes in trees and then visits these "wells" for sap and the insects they attract. Red-breasted and Red-naped Sapsuckers were formerly considered subspecies of Yellow-bellied. All species give a staggered drumming that suggests Morse code, distinct from other woodpeckers.

Williamson's Sapsucker *Sphyrapicus thyroideus*

L 9" *(23 cm)* **Male** has black back, white rump, large white wing patch; black head with narrow white stripes, bright red throat; breast black and belly yellow. **Female's** head brown; back, wings, sides barred with dark brown and white; rump white; lacks white wing patch and red chin; breast has large dark patch; belly variably yellow. Juvenile resembles adult but duller; attains adultlike plumage by Nov. Juvenile male has white throat; juvenile female lacks black breast patch. **VOICE:** Gives a harsh, shrill raptorlike call. **RANGE:** Fairly common in dry, piney forests of the western mountains; moves south or to lower elevations in winter. Rare to casual to lowland and to Great Plains, chiefly in fall and winter. Accidental in eastern North America (recorded LA, IL, NY).

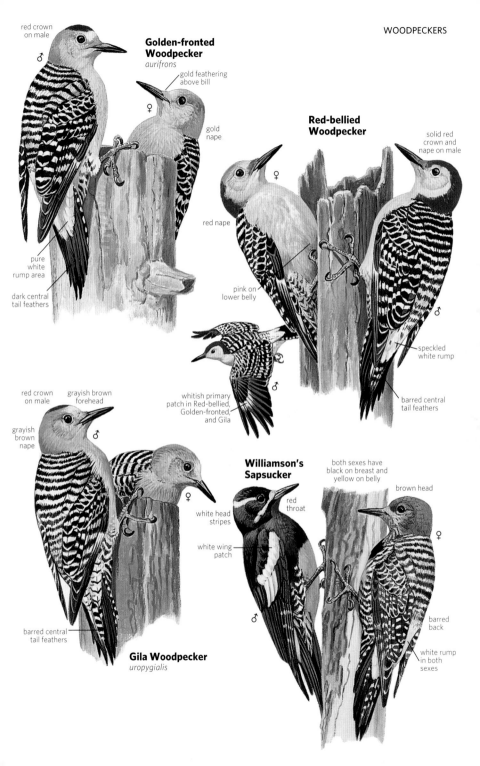

red crown on male

♂

Golden-fronted Woodpecker
aurifrons

gold feathering above bill

♀

gold nape

Red-bellied Woodpecker

solid red crown and nape on male

red nape

pink on lower belly

♂

pure white rump area

dark central tail feathers

whitish primary patch in Red-bellied, Golden-fronted, and Gila

♂

speckled white rump

barred central tail feathers

red crown on male

grayish brown forehead

grayish brown nape

♂

♀

Williamson's Sapsucker

both sexes have black on breast and yellow on belly

brown head

♀

white head stripes

red throat

white wing patch

barred central tail feathers

♂

barred back

white rump in both sexes

Gila Woodpecker
uropygialis

Red-breasted Sapsucker *Sphyrapicus ruber* L 8½" (22 cm)
This and next two species formerly considered one species. All show white wing patches. Red head, nape, and breast; large white wing patch; white rump. Back is black, lightly spotted with yellow in northern subspecies, *ruber;* more heavily marked with white in southern *daggetti.* Belly is yellow in *ruber; daggetti* has paler belly and duller head with longer white moustachial stripe. In both subspecies, briefly held (into Sept.) juvenal plumage is brownish, showing little or no red.
VOICE: Gives a querulous descending mewing call.
RANGE: Common in coniferous or mixed forests in coastal ranges, usually at lower elevations and in moister forests than Williamson's Sapsucker. Most migrate south or move to lower elevations in winter. Red-breasted frequently hybridizes with Red-naped Sapsucker; these hybrids encountered regularly to eastern NV and southeastern CA in fall and winter, and every potential easterly stray Red-breasted needs to be carefully scrutinized for signs of hybridization with Red-naped. These hybrids are more frequent in the Great Basin region and more southerly deserts east to AZ than pure Red-breasted. Apparently pure Red-breasted (mostly *daggetti; ruber* is casual) is uncommon to western Great Basin; accidental to IO (Pottawattamie County) and east TX (McLennan County).

Yellow-bellied Sapsucker *Sphyrapicus varius*
L 8½" (22 cm) Red forecrown on black-and-white head; chin and throat red in **male,** white in **female.** Back is blackish, with white rump and large white wing patch. Underparts yellowish, paler in female. Most **juveniles** retain largely brownish plumage until late in the winter; some may acquire more adultlike plumage earlier.
VOICE: Gives a querulous, descending mewing call.
RANGE: Fairly common in deciduous and mixed forests. Highly migratory; rare to very rare in the West during fall and winter. Hybridizes with Red-naped Sapsucker in narrow zone of overlap.

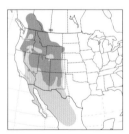

Red-naped Sapsucker *Sphyrapicus nuchalis* L 8½" (22 cm)
Very similar to Yellow-bellied, but has variable red patch on back of head; spotting on back more clearly organized into two rows. On **male,** extensive red on throat penetrates the surrounding black "frame"; on **female,** throat is partly red to almost entirely red on some birds. Juvenile is brownish overall; resembles adult by first fall except for lack of black chest. Hybridizes extensively with Red-breasted Sapsucker, less so with Yellow-bellied.
VOICE: Gives a querulous descending mewing call.
RANGE: Common in deciduous and mixed forests in the Great Basin and Rocky Mountain ranges; a few to the Sierra Nevada. Rare west of the Sierra Nevada and very rare in Pacific Northwest and west of Cascades. Casual east to western Great Plains.

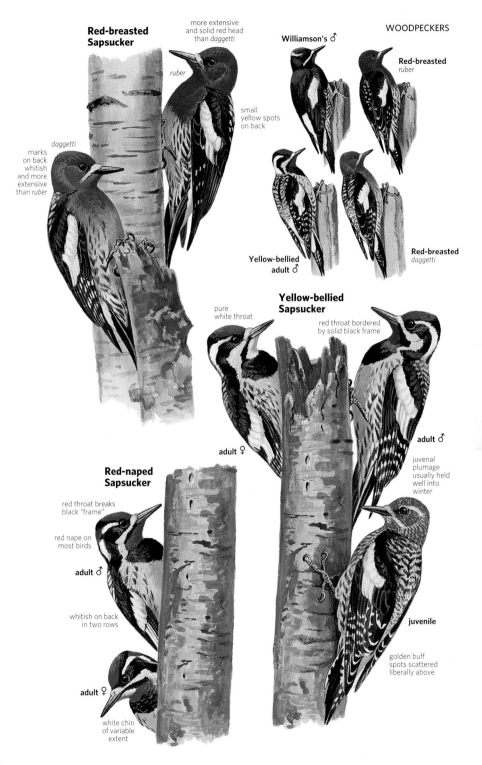

Red-breasted Sapsucker

more extensive and solid red head than *daggetti*

ruber

small yellow spots on back

daggetti

marks on back whitish and more extensive than *ruber*

Williamson's ♂

Red-breasted
ruber

Yellow-bellied adult ♂

Red-breasted
daggetti

Yellow-bellied Sapsucker

pure white throat

red throat bordered by solid black frame

adult ♀

adult ♂

juvenal plumage usually held well into winter

Red-naped Sapsucker

red throat breaks black "frame"

red nape on most birds

adult ♂

whitish on back in two rows

adult ♀

white chin of variable extent

juvenile

golden buff spots scattered liberally above

Nuttall's Woodpecker *Picoides nuttallii* L 7½" *(19 cm)*

Closely resembles Ladder-backed Woodpecker. Nuttall's shows more black on face; white bars on back are narrower, with more extensive solid black just below the nape. White outer tail feathers are sparsely spotted rather than barred; nasal tufts white. Red restricted to nape on **males,** except juveniles. May hybridize with Ladder-backed where ranges overlap in western Mojave Desert and Owens Valley, CA. Here and wherever either species is out of range, all characters, including vocalizations, should be checked for signs of hybridization. Has also hybridized with Downy Woodpecker in southern CA.

VOICE: Call is a low, rattled *prrrt,* much lower than Ladder-backed's *pik;* also gives a series of loud, spaced, descending notes.

RANGE: Prefers less arid habitat than Ladder-backed; usually seen in chaparral mixed with scrub oak and in wooded canyons and streamside trees. Casual to western NV.

Ladder-backed Woodpecker *Picoides scalaris*

L 7¼" *(18 cm)* Black-and-white barred back, spotted sides; face and underparts slightly buffy or grayish; face marked with black lines. **Male** has red crown. In CA, may be confused with Nuttall's, with which it sometimes hybridizes where ranges overlap. Ladder-backed shows less black on face; is buffy tinged rather than white below; white barring on back is more pronounced, extends to nape; white outer tail feathers evenly barred rather than spotted; nasal tufts buffy.

VOICE: Call is a crisp *pik,* very similar to Downy Woodpecker but different from the rattled *prrrt* given by Nuttall's; also a descending whinny.

RANGE: Common in dry brushlands, mesquite and cactus country; also towns and rural areas. Feeds on beetle larvae from small trees; also eats cactus fruits and forages on the ground for insects.

White-headed Woodpecker *Picoides albolarvatus*

L 9¼" *(24 cm)* Head and throat white. **Male** has a red patch on back of head. Body is black except for white wing patches. Juvenile has a variable patch of pale red on crown.

VOICE: Calls include a sharp *pee-dink* or *pee-dee-dink.*

RANGE: Nests in coniferous mountain forests, especially ponderosa and sugar pine; feeds primarily on seeds from their cones; also pries away loose bark in search of insects and larvae. Fairly common over most of range; rare and local in the North. Casual at lower altitudes in winter, including the coastal lowlands of southern CA; the westernmost Mojave Desert; the Owens Valley, CA; southern BC; and western MT.

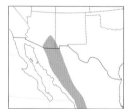

Arizona Woodpecker *Picoides arizonae* L 7½" *(19 cm)*

Solid brown back distinguishes this species from all other woodpeckers. **Female** lacks red patch on back of head. Previously known by the English name of Strickland's Woodpecker. The return to the current name, which is of even earlier origin, occurred after the smaller birds with white on the back from the high mountains around Mexico City were recognized as a full species, Strickland's Woodpecker (*P. stricklandi*).

VOICE: Call is a sharp *peek,* similar to Hairy Woodpecker but hoarser.

RANGE: Uncommon resident in foothills and mountains; generally found in oak or pine-oak forests or canyons.

on males, red restricted to rear crown; more extensive on juveniles of both sexes

black face with white borders

white nasal tufts

Nuttall's Woodpecker

narrower white bars stop on upper back

white underparts

♂

♀

outer tail feathers with fewer bars than Ladder-backed

buffy white face with narrow black frame

extensive reddish crown

buffy white nasal tufts

Ladder-backed Woodpecker

white bars extend to nape

black crown

♂

♀

buffy cast to underparts with spots on sides and flanks

tail more barred than Nuttall's

white head

White-headed Woodpecker

♂

♀

white wing patch

long narrow white patch on folded wing

brown upperparts

large spo below

♂

♀

Arizona Woodpecker

bars extend to nape

solid black patch

Ladder-backed ♂   **Nuttall's ♂**   **Arizona ♂**

Hairy Woodpecker *Picoides villosus* L 9¼" (24 cm)

White back generally identifies both Hairy and similar Downy Woodpecker. Hairy is much larger, with a larger bill; outer tail feathers are entirely white. Birds in the Pacific Northwest have pale gray-brown back and underparts. Rocky Mountain birds (such as *orius*) and other subspecies in the western lower 48 have less white spotting on wings. **Male** has red hindcrown spot, lacking in **female. Juveniles,** particularly in the Maritime Provinces *(terranovae),* have some barring on back and flanks; sides may be streaked. Juveniles on the Queen Charlotte Islands, BC, have heavily barred outer tail feathers. In young males, the forehead is spotted with white; crown streaked with red or orange.

VOICE: Calls include a loud, sharp *peek* and a slurred whinny.

RANGE: Fairly common; inhabits both open and dense forests. Uncommon to rare in the South and in FL. In the South and elsewhere, moves into forests after burns. Casual to lowlands of Southwest.

Downy Woodpecker *Picoides pubescens* L 6¾" (17 cm)

White back generally identifies both Downy and similar Hairy Woodpecker. Downy is much smaller, with a smaller bill; outer tail feathers generally have faint dark bars or spots. **Male** has red hindcrown spot, lacking in **female.** Birds in the Pacific Northwest have pale gray-brown back and underparts. Birds from the Rocky Mountains and CA (such as *leucurus,* illustrated) have less white spotting on wings.

VOICE: Call, *pik,* and whinny are softer and higher pitched than Hairy Woodpecker.

RANGE: Common; active, and somewhat unwary; often seen in suburbs, parklands, and orchards, as well as in forests. A familiar visitor to feeders. Casual in Southwest south of mapped range.

Red-cockaded Woodpecker *Picoides borealis* **E**

L 8½" (22 cm) Black-and-white barred back, black cap, and large white cheek patch identify this woodpecker; red tufts, seldom visible, on the **male**'s head. Similar Hairy and Downy Woodpeckers have solid white down backs.

VOICE: Distinctive calls, a raspy *sripp* and high-pitched *tsick,* are much more buzzy; recall a loud Brown-headed Nuthatch.

RANGE: Inhabits open, mature pine or pine-oak woodlands. Bores nest hole only in a large living pine afflicted with heartwood disease, then drills small holes around the nest opening. Pine pitch oozing down the trunk from these holes may repel predators; also makes the tree a distinctive signpost. Endangered: Populations continue to decline. Almost eliminated from VA, and has disappeared from KY and TN in the last 15 years. Accidental to northeastern IL and OH.

Great Spotted Woodpecker *Dendrocopos major*

L 9" (25 cm) Large size with large white cheek and scapular patch and red undertail coverts. Female lacks red hindcrown spot of **male.**

VOICE: Call is a sharp *kick.*

RANGE: Widespread in Eurasia. Casual to western Aleutians and Pribilofs; accidental in winter north of Anchorage, AK.

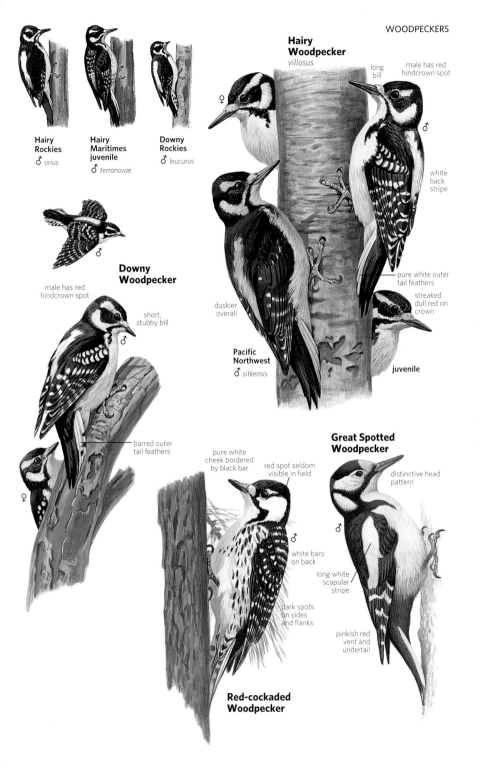

Hairy Rockies
♂ *orius*

Hairy Maritimes juvenile
♂ *terranovae*

Downy Rockies
♂ *leucurus*

Hairy Woodpecker
villosus

long bill

male has red hindcrown spot

♀

♂

white back stripe

pure white outer tail feathers

streaked dull red on crown

duskier overall

Pacific Northwest
♂ *sitkensis*

juvenile

Downy Woodpecker

male has red hindcrown spot

short, stubby bill

♂

barred outer tail feathers

♀

Great Spotted Woodpecker

pure white cheek bordered by black bar

red spot seldom visible in field

distinctive head pattern

♂

white bars on back

♂

long white scapular stripe

dark spots on sides and flanks

pinkish red vent and undertail

Red-cockaded Woodpecker

Black-backed Woodpecker *Picoides arcticus* L 9½" (24 cm)

Solid black back, heavily barred sides. **Male** has a solid yellow cap. Compare especially with *bacatus* American Three-toed Woodpecker, which has darker back than the other subspecies. Black-backed is larger; has longer, stouter bill; lacks white streak behind the eye.

VOICE: Call note is a single, sharp *pik*, lower pitched and louder than American Three-toed. Drum is low and deep and sound carries far.

RANGE: Inhabits coniferous forests; thrives in burned-over areas. Forages on dead conifers, flaking away large patches of loose bark rather than drilling into it, in search of larvae and insects. Casual south of mapped range in the East, fewer records in recent decades.

American Three-toed Woodpecker *Picoides dorsalis*

L 8¾" (22 cm) Black-and-white barring down center of back distinguishes most subspecies from similar Black-backed. Both species have heavily barred sides. **Male**'s yellow cap is blended; white barring on back is intermediate in northwestern subspecies, *fasciatus*. In Rocky Mountain subspecies, nominate *dorsalis*, back is almost entirely white. Back is much darker in eastern subspecies, *bacatus;* thinner, white submoustachial stripe helps distinguish it from Black-backed. New English and scientific names reflect a recent split: Eight Old World subspecies are now considered their own species, *P. tridactylus.*

VOICE: Call is a soft *pik* or *kimp.* Drumming becomes slightly faster and weaker toward end.

RANGE: Found in coniferous forests, especially burned-over and insect-ravaged areas. Scarcer than Black-backed in East. Accidental south to RI in winter; east to northwestern NE and southwestern KS (in summer).

Northern Flicker *Colaptes auratus* L 12½" (32 cm)

Two groups occur: **"Yellow-shafted Flicker"** in the East and far north, and **"Red-shafted Flicker"** in the West. Our flickers have brown, barred back; spotted underparts with black crescent bib. White rump conspicuous in flight; no white wing patches. Intergrades are regularly seen in the Great Plains (especially) and elsewhere. "Yellow-shafted" has yellow wing lining and undertail color, gray crown, and tan face with red crescent on nape. "Red-shafted" has brown crown and gray face, with no red crescent. "Yellow-shafted" **male** has a black moustachial stripe (red stripe in "Red-shafted" male); **females** lack these stripes.

VOICE: Call heard on the breeding ground is a long, loud series of *wick-er* notes; a single, loud *klee-yer* is given year-round.

RANGE: Common in open woodlands and suburban areas. "Yellow-shafted" is rare in the West, and "Red-shafted" casual in the East, in fall and winter. Check birds for signs of intergradation.

See subspecies map, page 552

Gilded Flicker *Colaptes chrysoides* L 11½" (29 cm)

Head pattern more like "Red-shafted" Northern, but underwing and base of tail yellow; crown more cinnamon. Note also smaller size; larger black chest patch; paler back with narrower black bars; more crescent-shaped markings below. **Female** lacks red moustachial stripe.

VOICE: Calls are like Northern.

RANGE: Inhabits low desert woodlands; favors saguaro. Hybrids with "Red-shafted" are noted in cottonwoods at middle elevations in southern AZ and the lower Colorado River.

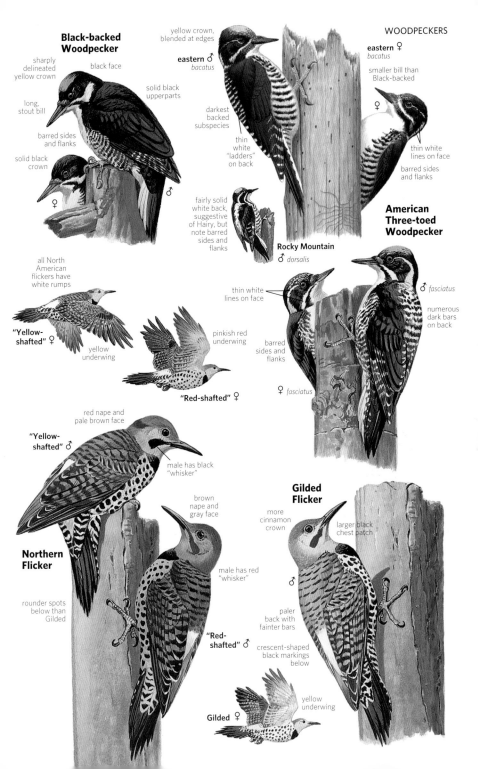

Black-backed Woodpecker

sharply delineated yellow crown

black face

long, stout bill

barred sides and flanks

solid black crown

♀

yellow crown, blended at edges

eastern ♂
bacatus

solid black upperparts

darkest backed subspecies

thin white "ladders" on back

fairly solid white back, suggestive of Hairy, but note barred sides and flanks

Rocky Mountain
♂ *dorsalis*

eastern ♀
bacatus

smaller bill than Black-backed

♀

thin white lines on face

barred sides and flanks

American Three-toed Woodpecker

thin white lines on face

barred sides and flanks

♂ *fasciatus*

numerous dark bars on back

♀ *fasciatus*

all North American flickers have white rumps

"Yellow-shafted" ♀

yellow underwing

pinkish red underwing

"Red-shafted" ♀

red nape and pale brown face

"Yellow-shafted" ♂

male has black "whisker"

brown nape and gray face

male has red "whisker"

"Red-shafted" ♂

Northern Flicker

rounder spots below than Gilded

Gilded Flicker

more cinnamon crown

larger black chest patch

♂

paler back with fainter bars

crescent-shaped black markings below

yellow underwing

Gilded ♀

Ivory-billed Woodpecker *Campephilus principalis* **E**

L 19½" (50 cm) Probably extinct. Note black chin, striking ivory bill, and the extensive white wing patches and scapular lines, visible in perched birds. **Females** have black rather than red crests. The black-and-white wing patterns of Ivory-billed and Pileated Woodpeckers in flight differ; in Ivory-billed from below, the white leading edge and the white secondaries are divided by a black bar; viewed from above, the white secondaries are diagnostic; flight of Ivory-billed is more direct and swift. Our largest woodpecker, Ivory-billed required large tracts of old-growth river forest; dead and dying trees supplied nesting sites and food: the larvae of wood-boring beetles. Destruction of habitat at the end of the 19th century and early in the 20th century and perhaps overhunting led to the disappearance of this never common species.

VOICE: Distinctive call note sounds like a toy trumpet: a high-pitched, nasal *yank,* given singly or in short series, like a loud version of the call of eastern White-breasted Nuthatch, but beware of Blue Jays giving similar calls; double-rap drum is characteristic of genus *Campephilus.*
RANGE: Formerly found north to the Ohio River, near its confluence with the Mississippi. Last definite records from U.S. were in 1944 in the Singer Tract, near Tallulah in northeastern LA. Possibly valid sightings recorded into the 1950s in FL; from 1948, perhaps into the 1980s, in Cuba (subspecies *bairdii*), but now probably extinct there. Unconfirmed sightings over the last 50 years come from eastern TX, LA, GA, and FL. In Apr. 2005 came the much publicized announcement that the species had been rediscovered more than a year earlier in the Big Woods of the White River–Cache River system of eastern AR. Documentation was provided in the form of sound recordings and brief blurred images on a videotape. However, intense searching subsequently has yet to produce more documentation, seemingly not possible in an age when most rarities discovered are photographed and those images are posted on the Internet the same day; many question the original evidence. Roger Tory Peterson, referring to his own sighting of two females in the Singer Tract in May 1942, said, "We had no trouble following the two" and "An Ivory-billed once heard is easy to find." Continued reports of sightings in FL and elsewhere may include leucistic Pileated Woodpeckers, but these sightings that lack provable evidence more likely represent wishful thinking. The finality of a species' extinction is difficult for many to accept.

Pileated Woodpecker *Dryocopus pileatus* L 16½" (42 cm)

Now likely our largest North American woodpecker. Overall black coloration with white underwing and wing patches visible in flight. Flies with deep crowlike wingbeats. **Female's** red cap is less extensive than **male.** Juvenal plumage, held briefly, resembles adult but is duller and browner overall. Generally shy.

VOICE: Call is a loud *wuck* note or series of notes, given all year, often in flight; similar call of Northern Flicker is given only in the breeding season. Loud, resonant drumming given most frequently by male.
RANGE: Prefers dense, mature forest. Found in woodlots and parklands as well as deep woods; look for the long rectangular or oval holes it excavates. Carpenter ants in fallen trees and stumps are its major food. Common in Southeast; uncommon and local elsewhere, but increasing in East. Accidental to San Joaquin Valley and Malibu, CA.

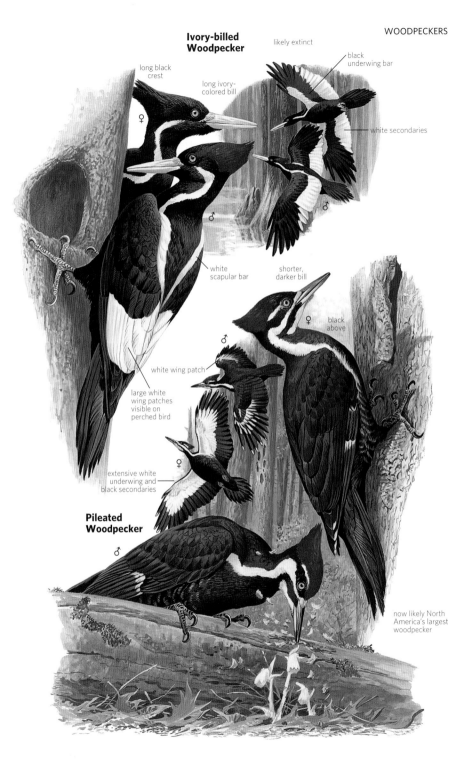

Ivory-billed Woodpecker

likely extinct

black underwing bar

long black crest

long ivory-colored bill

white secondaries

♀

♂

white scapular bar

shorter, darker bill

black above

♀

♂

white wing patch

♂

large white wing patches visible on perched bird

extensive white underwing and black secondaries

♀

Pileated Woodpecker

♂

now likely North America's largest woodpecker

TYRANT FLYCATCHERS Family Tyrannidae

A typical flycatcher darts out from a fixed perch to catch insects. Most have a large head, bristly "whiskers," and a broad-based, flat bill. SPECIES: 408 WORLD; 46 N.A.

Northern Beardless-Tyrannulet *Camptostoma imberbe*

L 4½" (11 cm) Grayish olive above and on breast; dull white or pale yellow below. Indistinct whitish eyebrow; small, slightly curved bill. Crown is darker than nape in many birds and often raised in a bushy crest. Distinguished from similar Ruby-crowned Kinglet (page 386) by buffy wing bars and lack of bold eye ring. Small, innocuous, and difficult to spot; most easily located by voice.

VOICE: Song on breeding grounds is a descending series of loud, clear *peer* notes; call is an innocuous, whistled *pee-yerp.*

RANGE: Rather uncommon in U.S. Often found near streams in sycamore, mesquite, and cottonwood groves; oak mottes in south TX.

Tufted Flycatcher *Mitrephanes phaeocercus* L 5" (13 cm)

Distinctive small, crested flycatcher with cinnamon underparts and face; brownish olive above with faint cinnamon wing bars. Behavior suggests pewees.

VOICE: Call is a whistled *tchurree-tchurree.* Sometimes given singly; also a soft *peek* like Hammond's Flycatcher.

RANGE: Widespread tropical species; partially migratory at northern end of range (northern Mexico). Casual to west TX and western and southeastern AZ. Most records are in winter.

Olive-sided Flycatcher *Contopus cooperi* L 7½" (19 cm)

Large, with rather short tail. Brownish olive above; white tufts on sides of rump distinctive but often not visible. Throat, center of breast, and belly dull white. Sides and flanks brownish olive and streaked. Bill is mostly black; center and sometimes base of lower mandible dull orange. Often perches on high, dead branches, including in migration.

VOICE: Distinctive song is a loud, clear *quick-three-beers,* the second note higher; typical call is a repeated *pip.*

RANGE: Uncommon to fairly common in coniferous forests and bogs. Casual in winter on coastal slope of southern CA.

Greater Pewee *Contopus pertinax* L 8" (20 cm)

Note longer tail, more slender crest (usually visible), and more uniformly colored underparts than Olive-sided. Worn summer birds are overall grayer than freshly molted winter ones. Unlike *Empidonax* flycatchers (pages 328–334), most *Contopus* (pewees) do not wag tails.

VOICE: Song is a whistled *ho-say ma-re-ah;* call is a repeated *pip.*

RANGE: Fairly common in mountain pine-oak woodlands. Very rare in winter in southern AZ, southern and central CA; casual in south and west TX. A few summer records for Davis Mountains of west TX and mountains of southern CA.

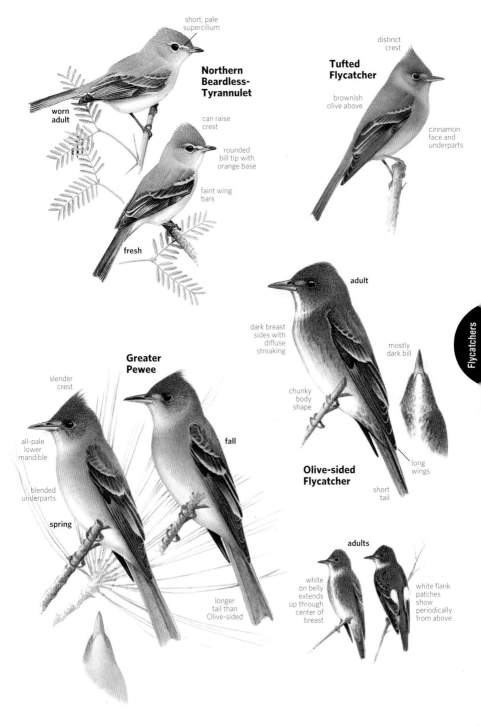

short, pale
supercilium

**Northern
Beardless-
Tyrannulet**

worn
adult

can raise
crest

rounded
bill tip with
orange base

faint wing
bars

fresh

distinct
crest

**Tufted
Flycatcher**

brownish
olive above

cinnamon
face and
underparts

adult

dark breast
sides with
diffuse
streaking

mostly
dark bill

chunky
body
shape

**Greater
Pewee**

slender
crest

all-pale
lower
mandible

blended
underparts

spring

fall

**Olive-sided
Flycatcher**

long
wings

short
tail

longer
tail than
Olive-sided

adults

white
on belly
extends
up through
center of
breast

white flank
patches
show
periodically
from above

Flycatchers

Eastern Wood-Pewee *Contopus virens* L 6¼" *(16 cm)*

Overall grayish olive above, paler on nape, whitish or pale yellow below. Bill of **adult** has black upper mandible, dull orange lower mandible. **Juvenile** and immature may have all-dark bill.

VOICE: Distinctive song is a clear, slow, plaintive *pee-a-wee,* the second note lower; this phrase often alternates with a downslurred *pee-yer.* Calls include a loud chip and clear, whistled, rising *pweee* notes; often given together, *chip pweee.*

RANGE: Common in a variety of woodland habitats. Casual in the West. No winter records in U.S.

Western Wood-Pewee *Contopus sordidulus* L 6¼" *(16 cm)*

Plumage variable; slightly darker and less greenish than Eastern Wood-Pewee; base of lower mandible usually shows some yellow-orange. Identification very difficult; best done by range and voice. Also compare to western subspecies of Willow Flycatcher (page 330). Neither wood-pewee wags tail, unlike all *Empidonax.*

VOICE: Calls include a harsh, slightly descending *peeer* and clear whistles suggestive of Eastern's *pee-yer.* Song, heard chiefly on breeding grounds, has three-note phrases mixed with the *peeer* note.

RANGE: Common in open woodlands. Uncommon migrant (possible breeder) on western Great Plains; casual elsewhere in East. No mid-winter records in U.S.

Cuban Pewee *Contopus caribaeus* L 6" *(15 cm)*

Short primary projection makes species look like *Empidonax,* but Cuban does minimal tail flicking. Note expansion of prominent white partial eye ring behind eye; dull wing bars; faint "vest."

VOICE: Call, a clear, steady *dee-dee-dee,* also a soft *dep* note.

RANGE: West Indian species; casual in south FL with four records in fall, late winter, and spring.

EMPIDONAX FLYCATCHERS

All empids are drab, with pale eye rings and wing bars. From spring to summer, plumages grow duller from wear. Some species molt before fall migration, acquiring fresh plumage in late summer. Identification depends on voice, habitat, behavior, and subtle differences in size, bill shape, primary projection, and tail length. Most flip their tails up.

Acadian Flycatcher *Empidonax virescens* L 5¾" *(15 cm)*

Olive above, with yellow eye ring, two buffy or whitish wing bars; very long primary projection. Long, broad-based bill, with mostly yellowish lower mandible. Most birds show pale grayish throat, pale olive wash across upper breast, white lower breast, and yellow belly and undertail coverts. Molts before migration; **fall** birds have buffy wing bars. Juvenile is brownish olive above, edged with buff.

VOICE: Call is a soft *peace,* extended in song to an emphatic *pee-tsup.* On breeding grounds, also gives a flickerlike *ti ti ti ti ti.*

RANGE: Found in woodlands and swamps. The only trans-Gulf migrant *Empidonax.* Rare to casual north of breeding range, including to Maritime Provinces. Accidental to NM, AZ, and BC.

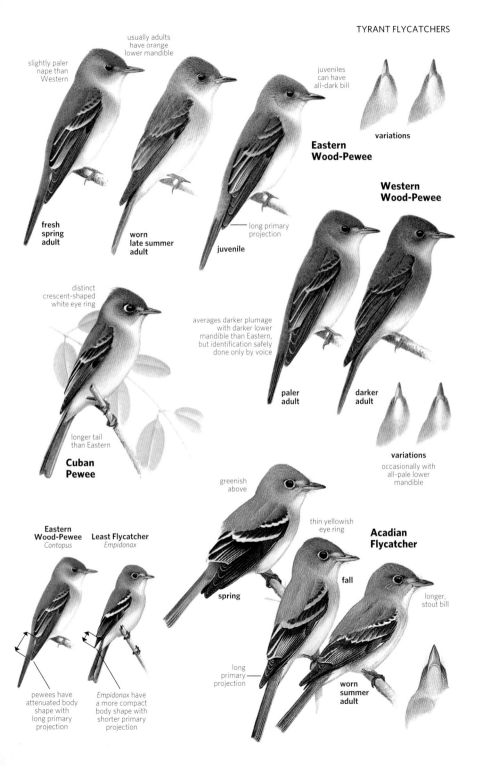

slightly paler
nape than
Western

usually adults
have orange
lower mandible

juveniles
can have
all-dark bill

variations

**Eastern
Wood-Pewee**

fresh
spring
adult

worn
late summer
adult

long primary
projection

juvenile

**Western
Wood-Pewee**

distinct
crescent-shaped
white eye ring

averages darker plumage
with darker lower
mandible than Eastern,
but identification safely
done only by voice

paler
adult

darker
adult

longer tail
than Eastern

**Cuban
Pewee**

variations
occasionally with
all-pale lower
mandible

greenish
above

thin yellowish
eye ring

**Acadian
Flycatcher**

**Eastern
Wood-Pewee**
Contopus

Least Flycatcher
Empidonax

spring

fall

longer,
stout bill

long
primary
projection

**worn
summer
adult**

pewees have
attenuated body
shape with
long primary
projection

Empidonax have
a more compact
body shape with
shorter primary
projection

Yellow-bellied Flycatcher *Empidonax flaviventris*
L 5½" (14 cm) Has rather short tail and big head. Olive above, yellow below. Broad yellow eye ring. Lower mandible entirely pale orange. Shows a more extensive olive wash across breast than Acadian Flycatcher; lacks pale area between olive and yellow belly. Also, throat is yellow, rather than whitish; bill smaller. Molts after migration; **worn fall** migrants slightly grayer above, duller below.
VOICE: Song, a liquid *je-bunk;* also a plaintive, rising *per-wee,* both heard mainly on breeding grounds. Call is a sharp, whistled *chiu,* heard in migration and on winter grounds, that sounds somewhat like Acadian.
RANGE: Nests in bogs, swamps, and damp coniferous woods. A circum-Gulf migrant, arriving in TX chiefly early to mid-May. Spring migrants still around Great Lakes through first third of June. Very rare migrant in coastal Southeast, chiefly in fall.

Alder Flycatcher *Empidonax alnorum* L 5¾" (15 cm)
Very similar to Willow, but bill slightly shorter, eye ring usually more prominent, back greener. Distinguished from eastern subspecies of Willow by darker head; from western subspecies by well-defined tertial edges, bolder wing bars, long primary projection. Also compare carefully to Least Flycatcher (page 332), which is browner above, has shorter bill with dark tip to lower mandible and very different call note.
VOICE: Best identified by voice. Call, a loud *pip,* similar to Hammond's but louder; heard year-round. Distinctive song, a falling, wheezy *weeb-ew* is heard mainly on breeding grounds, but also by some spring migrants. On breeding grounds, also gives a descending *wheer.*
RANGE: Common in brushy habitats near bogs, birch and alder thickets. Although breeds west to western AK, primarily an eastern species. A circum-Gulf migrant. Arrives in spring a bit later than Willow. Common across most parts of breeding range, but overall an uncommon and late spring migrant. Very rare migrant in coastal Southeast, mostly in fall. Likely a regular migrant in eastern MT; otherwise a casual, perhaps very rare migrant in West south of breeding range. Exact status clouded by identification problems. Winters east base of Andes south to Bolivia.

See subspecies map, page 552

Willow Flycatcher *Empidonax traillii* L 5¾" (15 cm)
Lacks prominent eye ring. Color ranges from pale gray head and greenish back of nominate eastern subspecies to darker-headed, browner *brewsteri* in Northwest. Great Basin subspecies, *adastus,* paler than *brewsteri;* endangered southwestern *extimus* **(E)** even paler. Western subspecies have duller wing bars, blended tertial edges, shorter primary projection. Told from pewees (page 328) by shorter wings, tail flicking.
VOICE: Call is a liquid *wit.* Songs, a sneezy *fitz-bew;* on breeding grounds, also a rising *brreet;* often sings in spring migration.
RANGE: Found in brushy habitats in wet areas; also pastures, mountain meadows. Quite local in southern portion of breeding range. Eastern *traillii,* a circum-Gulf migrant, arrives in south TX about 1 May, southern Great Lakes about 12 May. Most back south by late Aug. Western subspecies arrive in West after mid-May, a bit earlier for *extimus.* Fall migration into Oct. Nominate winters in southern Central America and northern South America, western subspecies to western Mexico; accidental in winter to southern CA. Casual fall migrant in coastal Southeast.

conspicuous circular yellow eye ring

short bill

1st fall

Yellow-bellied Flycatcher

worn fall adult

yellow throat and belly

olive breast

spring

Alder Flycatcher

1st fall

slight greenish cast on back

short primary projection

worn fall adult

thin eye ring

spring

strong contrast between face and throat

very dark wings with sharply contrasting tertial edges

Willow Flycatcher

worn fall adult *traillii*

1st fall *traillii*

extremely faint eye ring

1st fall *brewsteri*

long broad-based bill with pale lower mandible

darker above than *traillii*

darker face

gray face with less contrast between face and throat

spring *traillii*

extimus similar to but slightly paler than *brewsteri*

spring *extimus*

spring *brewsteri*

wing bars and tertial edges less contrasty than *traillii*

Least Flycatcher *Empidonax minimus* L 5¼" (13 cm)

Smallest eastern empid. Large-headed; rather bold white eye ring; rather short primary projection. Throat whitish; breast washed with gray; belly and undertail coverts pale yellow. Underparts usually paler than similar Hammond's Flycatcher. Bill short, triangular; lower mandible mostly pale. Molt occurs after fall migration.

VOICE: Song, a dry *che-bek* accented on the second syllable, is usually delivered in a rapid series; call, a sharp *whit,* is sometimes also given in a series. Most vocal empid in migration.

RANGE: Inhabits deciduous woods, orchards, and parks. Fairly common in East. A circum-Gulf migrant. Rare in coastal Southeast and in West, mostly in fall. Winters in Mexico and Central America, a few in FL; very rare to Gulf Coast, Southeast, and CA.

Hammond's Flycatcher *Empidonax hammondii*

L 5½" (14 cm) A small empid. Fairly large head and short tail. White eye ring, usually expanded in a "teardrop" at rear. Grayish head and throat; grayish olive back; gray or olive wash on breast and sides; belly tinged with pale yellow. Molts before migration; **fall** birds much brighter olive above and on sides of breast, yellower below. Medium-length, slightly notched tail edged with gray. Bill slightly shorter, thinner, usually somewhat darker than similar Dusky Flycatcher; primary projection longer.

VOICE: Call note, a sharp *peek.* Song, heard only on breeding grounds, like Dusky, but hoarser and lower pitched, especially on second note.

RANGE: Nests chiefly in coniferous forests. Most migrate earlier in spring and later in fall than Dusky. Casual in the East in fall and winter.

Gray Flycatcher *Empidonax wrightii* L 6" (15 cm)

Gray above, with a slight olive tinge in fresh fall plumage; whitish below, belly washed with pale yellow by late fall. White eye ring inconspicuous on pale gray face. Long bill; on most birds, lower mandible mostly pinkish orange at base, sharply divided from dark tip; on a few, entire lower mandible is pinkish orange. Short primary projection. Long tail, with thin whitish outer edge. Perched bird dips its tail down slowly, like a phoebe.

VOICE: Song is a vigorous *chi-wip* or *chi-bit,* followed by a liquid *whilp,* trailing off in a gurgle. Call is a loud *wit.*

RANGE: Fairly common in dry habitat of Great Basin, in pine or pinyon-juniper. Regular migrant on CA coast. Accidental in East in spring, fall, and winter.

Dusky Flycatcher *Empidonax oberholseri* L 5¾" (15 cm)

Grayish olive above; yellowish below, with whitish throat, pale olive wash on upper breast. White eye ring. Bill partly dark, orange at base of lower mandible blending into dark tip. Bill and tail slightly longer than Hammond's Flycatcher. Short primary projection. Molt occurs after fall migration; fresh late-fall birds are quite yellow below.

VOICE: Calls include a *wit* note, softer than Gray Flycatcher; a mournful *deehic,* heard on breeding grounds. Song has several phrases: a clear *sillit;* an upslurred *ggrrreep;* another high *sillit,* often omitted; and a clear, high *pweet.*

RANGE: Breeds in open woodlands and brush of mountainsides. Rare migrant on western Great Plains and West Coast; casual to AK; accidental in fall and winter to East.

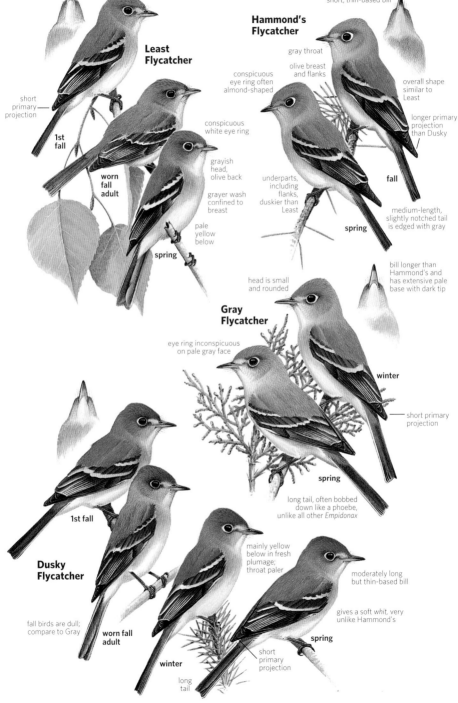

short, broad-based bill

Least Flycatcher

short primary projection

1st fall

conspicuous white eye ring

worn fall adult

grayish head, olive back

grayer wash confined to breast

pale yellow below

spring

short, thin-based bill

Hammond's Flycatcher

gray throat

olive breast and flanks

conspicuous eye ring often almond-shaped

overall shape similar to Least

longer primary projection than Dusky

underparts, including flanks, duskier than Least

fall

medium-length, slightly notched tail is edged with gray

spring

bill longer than Hammond's and has extensive pale base with dark tip

head is small and rounded

Gray Flycatcher

eye ring inconspicuous on pale gray face

winter

short primary projection

spring

long tail, often bobbed down like a phoebe, unlike all other *Empidonax*

1st fall

Dusky Flycatcher

fall birds are dull; compare to Gray

worn fall adult

mainly yellow below in fresh plumage; throat paler

moderately long but thin-based bill

gives a soft *whit*, very unlike Hammond's

spring

winter

short primary projection

long tail

Pacific-slope Flycatcher *Empidonax difficilis* L 5½" (14 cm)

Formerly considered same species as Cordilleran; known together as Western Flycatcher. Brownish green above; yellowish below with brownish tinge on breast. Broad pale eye ring, broken above, expanded behind eye; lower mandible entirely orange. Tail longer, wing tip slightly shorter than Yellow-bellied; wings and back slightly browner; less contrast in wing bars and tertial edges. Pacific-slope molts after arrival on winter grounds, so migrating **fall adults** appear more worn than spring birds. **First-fall** birds duller; wing bars buffy; variably whitish below, compare with Least Flycatcher (page 332), to which some dull birds show a close resemblance. Channel Islands subspecies *insulicola* is slightly duller.

VOICE: Call is a sharp *seet;* male gives upslurred *psee-yeet* note. Song is a complex series of notes, including call notes.

RANGE: Common in moist woodlands, coniferous forests, and shady canyons. Winters in lowlands of western Mexico. Common migrant through Southwest lowlands east to southeastern AZ. Accidental in eastern North America; other records, mostly late fall, either this species or Cordilleran.

Cordilleran Flycatcher *Empidonax occidentalis*

L 5¾" (15 cm) Formerly considered same species as Pacific-slope Flycatcher. Nearly identical but slightly larger, darker, and greener above; more olive and yellow below.

VOICE: Separable in field only by male's call, a two-note *pit peet;* some populations in western portion of breeding range give more intermediate notes. This fits with new genetic evidence that reveals a wide swath where birds have intermediate DNA. This calls into question whether the two should continue to be recognized as separate species. *Seet* note seems sharper in Cordilleran than Pacific-slope Flycatcher.

RANGE: Breeds in coniferous forests and canyons in mountains of the West. Rare in lowlands in migration, even within breeding range. Casual on the Great Plains. Winters in mountains of Mexico.

Buff-breasted Flycatcher *Empidonax fulvifrons*

L 5" (13 cm) Smallest *Empidonax* flycatcher. Brownish above; breast cinnamon-buff, paler on worn summer birds. Whitish eye ring; pale wing bars; small bill, with lower mandible entirely pale orange. Molts before migration.

VOICE: Call note is a soft *pwit.* Typical song, a quick *chicky-whew* or *chee-lick.*

RANGE: Small colonies nest in dry woodlands of canyon floors. Very local in Huachuca and Chiricahua Mountains, AZ; about a decade ago a very small, well-isolated population discovered in Davis Mountains, TX, though very few over the last two years; now casual in New Mexico, but recorded recently in Peloncillo Mountains. Formerly more widespread as a breeder. Uncommon as a migrant in the lowlands.

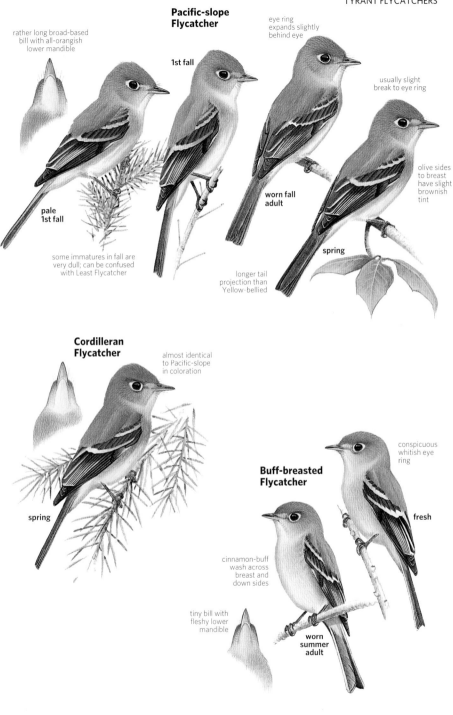

Pacific-slope Flycatcher

rather long broad-based bill with all-orangish lower mandible

1st fall

eye ring expands slightly behind eye

usually slight break to eye ring

olive sides to breast have slight brownish tint

worn fall adult

pale 1st fall

spring

some immatures in fall are very dull; can be confused with Least Flycatcher

longer tail projection than Yellow-bellied

Cordilleran Flycatcher

almost identical to Pacific-slope in coloration

spring

conspicuous whitish eye ring

Buff-breasted Flycatcher

fresh

cinnamon-buff wash across breast and down sides

tiny bill with fleshy lower mandible

worn summer adult

Eastern Phoebe *Sayornis phoebe* L 7" *(18 cm)*

Brownish gray above, darkest on head, wings, and tail. Underparts mostly white with pale olive wash on sides and breast; **fresh fall** birds are washed with yellow below. Molts before migration. All phoebes are distinguished from pewees (page 328) by their habit of pumping down and spreading their tails; Eastern Phoebe also by all-dark bill and lack of distinct wing bars. Also compare lack of eye rings and wing bars with *Empidonax* flycatchers (pages 328–334).

VOICE: Distinctive song is a harsh, emphatic *fee-be,* accented on first syllable. Typical call note is a sharp chip.

RANGE: Common in woodlands, farmlands, and suburbs; often nests under bridges, in eaves and rafters. An early spring (by early Mar.) and late fall (chiefly Oct.) migrant. Casual to rare migrant (chiefly late fall) over much of the West, and rare in winter to the Southwest and to CA.

Black Phoebe *Sayornis nigricans* L 6¾" *(17 cm)*

Black head, upperparts, and breast; white belly and undertail coverts. **Juvenal** plumage, held briefly, is browner, with two cinnamon wing bars, cinnamon rump.

VOICE: Four-syllable song is a rising *pee-wee* followed by a descending *pee-wee.* Calls include a loud *tseee* and a sharper *tsip,* slightly more plaintive than Eastern Phoebe's call.

RANGE: Common near water; casual to WA, BC, and OK. Range is expanding northward.

Say's Phoebe *Sayornis saya* L 7½" *(19 cm)*

Grayish brown above, darkest on head, wings, and tail; breast and throat pale grayish brown; belly and undertail coverts tawny.

VOICE: Song is a fast *pit-tse-ar,* often given in fluttering flight. Typical call is a plaintive, whistled *pee-ee,* slightly downslurred.

RANGE: Fairly common in dry, open areas, canyons, cliffs; perches on bushes, boulders, and fences. An early spring migrant; withdraws from wintering areas in early Mar.; returns by mid-Sept. Highly migratory; very rare in coastal Pacific Northwest. Casual in eastern North America.

Vermilion Flycatcher *Pyrocephalus rubinus* L 6" *(15 cm)*

Adult male strikingly red and brown. **Adult female** grayish brown above, with blackish tail; throat and breast white, with dusky streaking; belly and undertail coverts are peach; note also whitish eyebrow and forehead. **Juvenile** resembles adult female but is spotted rather than streaked below; belly white, often with yellowish tinge. **Immature male** begins to resemble adult by midwinter. Frequently pumps and spreads its tail down, like phoebes.

VOICE: Male in breeding season sings during fluttery display flight. Song is a soft, tinkling *pit-a-see pit-a-see;* also sings while perched. Typical call note is a sharp, thin *pseep.*

RANGE: Fairly common and approachable; found along streamsides, and near small wooded ponds. Rare winter visitor to coastal southern CA and the Gulf Coast. Casual elsewhere in eastern North America, primarily in fall.

Eastern Phoebe

worn summer adult

dark cap and face

fresh fall

pale yellow belly in fresh plumage

Black Phoebe

mostly blackish except for sharply contrasting white belly

phoebes drop their tail down

juvenile

grayish back

tawny belly

Say's Phoebe

blackish tail

dark face

immature ♀

streaked breast

pale supercilium

Vermilion Flycatcher

immature ♂

adult ♀

adult ♂

pale pinkish belly

juvenile

spotted breast

MYIARCHUS FLYCATCHERS

Myiarchus, with their longer tails and shorter wings, are less visible than kingbirds and tend to work more within the canopy. Bill size, tail pattern (of adults), and brightness of yellow belly are some of the important characteristics to scrutinize in identifying *Myiarchus*.

Brown-crested Flycatcher *Myiarchus tyrannulus*

L 8¾" (22 cm) Brownish olive above; as in all *Myiarchus* flycatchers, shows a bushy crest, rufous in primaries; bill longer, thicker, broader than Ash-throated Flycatcher. Throat and breast are pale gray; belly slightly paler yellow than Great Crested Flycatcher. Tail feathers show reddish on outer two-thirds of inner webs. TX subspecies, *cooperi,* is smaller than southwestern *magister.*

VOICE: Song is a clear musical whistle, a rolling *whit-will-do.* Call is a sharp *whit.*

RANGE: Fairly common in saguaro desert, river groves, lower mountain woodlands. Casual migrant in southern CA outside very limited breeding range; and to CA coast in fall. Very rare in LA (specimens of both *cooperi,* mostly, and *magister*) and FL, mostly in winter.

Great Crested Flycatcher *Myiarchus crinitus* L 8½" (21 cm)

Dark olive above. Gray throat and breast; bright lemon yellow belly and undertail coverts. Note broad, sharply contrasting edge to inner tertial. Outer tail feathers show entirely reddish inner webs.

VOICE: Distinctive call, a loud whistled *wheep,* sometimes given in a quick series. Song is a clear, loud *queeleep queelur qurrleep.*

RANGE: Common in a wide variety of open woods; feeds high in the canopy. Rare migrant on Great Plains away from few nesting areas. Very rare on CA coast during fall migration. Accidental to AK and northwestern Canada.

Dusky-capped Flycatcher *Myiarchus tuberculifer*

L 6¾" (17 cm) Smaller, bill larger, belly and undertail coverts usually brighter yellow than Ash-throated Flycatcher; tail shows less rufous. Secondaries have rufous edges, unlike other North American *Myiarchus.*

VOICE: Call is a mournful, descending *peeur.*

RANGE: Fairly common in wooded mountain ranges. Rare in summer in west TX. Very rare in late fall and winter in southeastern AZ and southern and central CA. Casual to OR, NV, and CO.

Ash-throated Flycatcher *Myiarchus cinerascens*

L 7¾" (19 cm) Grayish brown above; throat and breast pale gray; underparts paler than Brown-crested. Tail shows rufous on inner webs with dark tips. As in all *Myiarchus* flycatchers, briefly held **juvenal** plumage shows mostly reddish tail.

VOICE: Distinctive call, heard year-round, is a rough *prrrt.* Song, heard on breeding grounds, is a series of burry *ka-brick* notes.

RANGE: Common in a wide variety of habitats. Rare in winter in CA. Very rare, mainly fall and winter visitor to East; casual to BC.

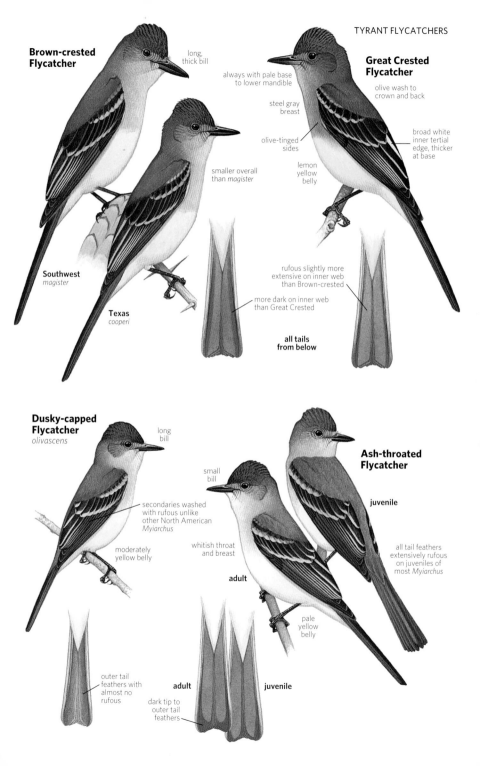

Brown-crested Flycatcher

long, thick bill

always with pale base to lower mandible

smaller overall than *magister*

Southwest *magister*

Texas *cooperi*

more dark on inner web than Great Crested

Great Crested Flycatcher

olive wash to crown and back

steel gray breast

olive-tinged sides

broad white inner tertial edge, thicker at base

lemon yellow belly

rufous slightly more extensive on inner web than Brown-crested

all tails from below

Dusky-capped Flycatcher *olivascens*

long bill

secondaries washed with rufous unlike other North American *Myiarchus*

moderately yellow belly

small bill

whitish throat and breast

adult

outer tail feathers with almost no rufous

adult

dark tip to outer tail feathers

juvenile

Ash-throated Flycatcher

juvenile

all tail feathers extensively rufous on juveniles of most *Myiarchus*

pale yellow belly

La Sagra's Flycatcher *Myiarchus sagrae* L 7½" (19 cm)
Grayish brown upperparts, mainly white underparts suggestive of Ash-throated Flycatcher, but bill longer; inner tertial edge stronger; rufous on outer tail less extensive.
VOICE: Distinctive call, a rather high-pitched *wink,* is often doubled.
RANGE: West Indian species from Bahamas, Cuba, and Caymans. Casual visitor, mainly in winter and spring, to south FL; accidental to AL.

Nutting's Flycatcher *Myiarchus nuttingi* L 7¼" (18 cm)
Similar to Ash-throated (page 338) but belly yellower; slightly more olive above; rufous primary edges blend to yellow-cinnamon secondary edges. Dark on outer webs of outer tail feathers does not extend across tip as in Ash-throated; orange, not flesh-colored, mouth lining.
VOICE: Call is a rather sharp *wheep,* different from Ash-throated.
RANGE: Tropical species; found from northwestern Mexico to Costa Rica; three certain winter records from southeastern AZ (twice) and coastal southern CA (once); once in fall from western AZ.

Piratic Flycatcher *Legatus leucophaius* L 6" (15 cm)
Dark olive-brown above; blurry olive streaking below. Distinct head pattern; dark malar streak; pale throat; stubby black bill. Black tail can show rufous edges. Often perches out in the open.
VOICE: On breeding grounds, song is a strident *whee-ee,* often followed by a rolled *ji-ji-jit;* also gives rising and falling whistles.
RANGE: Widespread tropical species. Casual in U.S. on Dry Tortugas, FL; TX records include on an oil rig in the Gulf of Mexico, Harris County, Big Bend, and Hidalgo County; and eastern NM.

Variegated Flycatcher *Empidonomus varius* L 7¼" (18 cm)
Similar to Piratic, but larger, longer bill has pale base; less distinct malar streak; more distinct streaking on upperparts, edging on wing coverts, and rufous edge on uppertail coverts and tail. Tends to perch lower than Piratic. Compare both to larger Sulphur-bellied Flycatcher.
VOICE: Call is a high, thin *pseee.*
RANGE: South American species; accidental in North America; three records in East, once in WA.

Sulphur-bellied Flycatcher *Myiodynastes luteiventris*
L 8½" (22 cm) Boldly streaked above and below. Upperparts often show an olive tinge; rump and tail rusty red; underparts pale yellow.
VOICE: Loud call is an excited chatter, like the squeaking of a rubber duck. Song is a soft *tre-le-re-re.*
RANGE: Fairly common in woodlands of mountain canyons with streams, usually at elevations between 5,000 and 6,000 feet. Inconspicuous; often perches high in the canopy. Casual to NM in summer, along the Gulf Coast (spring) and coastal CA (fall). Accidental to NJ, MA, ON, NB, and Newfoundland. All vagrants, especially in East, should be carefully separated from Streaked Flycatcher (*M. maculatus*), thus far unrecorded in North America. Streaked is similar in appearance to Sulphur-bellied but has thinner malar stripe, paler throat, and yellowish, not whitish, supercilium.

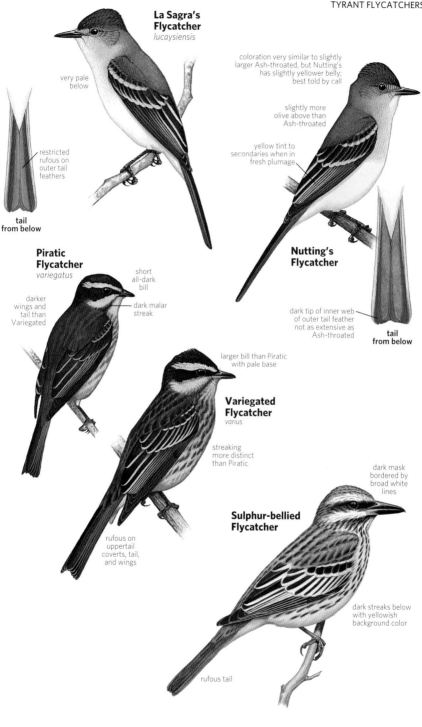

La Sagra's Flycatcher
lucaysiensis

very pale below

restricted rufous on outer tail feathers

tail from below

coloration very similar to slightly larger Ash-throated, but Nutting's has slightly yellower belly; best told by call

slightly more olive above than Ash-throated

yellow tint to secondaries when in fresh plumage

Nutting's Flycatcher

dark tip of inner web of outer tail feather not as extensive as Ash-throated

tail from below

Piratic Flycatcher
variegatus

short all-dark bill

darker wings and tail than Variegated

dark malar streak

larger bill than Piratic with pale base

Variegated Flycatcher
varius

streaking more distinct than Piratic

rufous on uppertail coverts, tail, and wings

dark mask bordered by broad white lines

Sulphur-bellied Flycatcher

dark streaks below with yellowish background color

rufous tail

Cassin's Kingbird *Tyrannus vociferans* L 9" (23 cm)

Dark brown tail; narrow buffy tips and lack of white edges on outer tail feathers help distinguish this species from Western Kingbird. Bill is much shorter than Tropical and Couch's Kingbirds. Upperparts darker gray than Western, washed with olive on back; paler wings contrast with darker back. White chin contrasts with dark gray head and breast. Belly dull yellow. **Juvenile** is duller, slightly browner above, with bold buffy edges on wing coverts; paler below.

VOICE: Call is a short, loud *chi-bew,* accented on second syllable.

RANGE: Fairly common in varied habitats; prefers denser foliage and hillier country than Western Kingbird. Scarce migrant away from breeding areas. Casual to FL. Accidental to OR, ID, and elsewhere in East.

Western Kingbird *Tyrannus verticalis* L 8¾" (22 cm)

Black tail, with white edges on outer feathers. Bill much shorter than Tropical and Couch's Kingbirds. Upperparts ashy gray, paler than Cassin's Kingbird, tinged with olive on back; dark wings contrast with paler back. Throat and breast pale gray; belly bright lemon yellow. **Juvenile** has slightly more olive on back and buffy edges on wing coverts, brownish tinge on breast, paler yellow belly.

VOICE: Common call is a sharp *whit.*

RANGE: Common in dry, open country; perches on fences, telephone lines. Some pairs breed east to Mississippi River and beyond. Scarce migrant to Pacific Northwest; casual to AK and northwestern Canada. Regular straggler in fall and early winter along the East Coast from the Maritime Provinces south; winters in small numbers in central and south FL; casual in CA.

Couch's Kingbird *Tyrannus couchii* L 9¼" (24 cm)

Almost identical to Tropical Kingbird, with thicker, broader-based bill; back slightly greener and less gray; at close range, tips of individual primaries are evenly spaced on adults. Distinguished from Western and Cassin's Kingbirds by larger bill, darker ear patch, and slightly notched brown tail. Juvenile is duller overall, with buffy edges on wing coverts.

VOICE: Distinctive calls, a shrill, rolling *breeeer;* and a more common *kip,* similar to call of Western Kingbird, given singly or in a series.

RANGE: Common in the lower Rio Grande Valley, TX, in summer; uncommon in winter. Found in groves and shrubs. Casual on Gulf Coast in fall and winter. Accidental to MI, East Coast, NM, southwestern AZ, and southern CA.

Tropical Kingbird *Tyrannus melancholicus* L 9¼" (24 cm)

Almost identical to Couch's Kingbird. Bill is thinner and longer; back slightly grayer, less green; at close range, tips of individual primaries are unevenly staggered on adults. Distinguished from Western and Cassin's Kingbirds by larger bill, darker ear patch, brighter underparts and slightly notched brown tail.

VOICE: Distinctive call is a rapid, twittering *pip-pip-pip-pip.*

RANGE: Uncommon and local in southeastern AZ and Rio Grande Valley (also a few at Big Bend), TX; found in lowlands near water; often nests in cottonwoods. Rare but regular during fall and winter along the West Coast to BC, casually to southeastern AK. Accidental to Great Lakes region and East Coast, mostly in fall.

Cassin's
adult ♂

Western
adult ♂

Tropical
adult ♂

Couch's
adult ♂

Thick-billed
1st fall

Cassin's Kingbird

juvenile

dark gray head, back, and breast

contrasting white chin

adults

brownish gray wings usually paler than back

spring adult ♂

worn fall adult

usually pale tip

Western Kingbird

pale gray head and back

juvenile

darker wings contrast with gray back

adults

black tail with white edge

spring adult ♂

worn fall adult

Couch's Kingbird

broad-based bill

spring adult ♂

yellow more extensive below than Western

adults

notched tail

even spacing of tips

adult ♂ primary tips

best identified by calls

Tropical Kingbird

longer, thinner-based bill than Couch's

spring adult ♂

notched tail

uneven spacing of tips

adult ♂ primary tips

Gray Kingbird *Tyrannus dominicensis* L 9" (23 cm)
Pale gray above, with blackish mask. Red crown patch seldom visible. Bill long and thick. Underparts mostly white, with pale yellowish wash on belly and undertail coverts. Deeply notched tail is slightly longer than Eastern and lacks white terminal band. Juvenal plumage, held well into fall, is browner above.
VOICE: Song is a buzzy *pecheer-ry*, accented on second syllable.
RANGE: Common on the FL Keys, local in mangroves on the mainland. Casual spring and fall wanderer north along the Atlantic coast to the Maritime Provinces, inland to MI and ON, and along the Gulf Coast to coastal TX; accidental BC (specimen, Cape Beale, Vancouver Island, 29 Sept. 1889).

Eastern Kingbird *Tyrannus tyrannus* L 8½" (22 cm)
Black head, slate gray back; tail has a broad white terminal band. Underparts are white, with a pale gray wash across the breast. Orange-red crown patch is seldom visible. **Juvenile** brownish gray above, darker on breast.
VOICE: Call is a harsh *dzeet* note, also given in a series.
RANGE: Common (more uncommon in West) and conspicuous in woodland clearings, farms, and orchards; often seen near water. Rare migrant to Southwest and the West Coast and to AK, where recorded north and west to Nome and Point Barrow. Winters in South America.

Fork-tailed Flycatcher *Tyrannus savana* L 14½" (37 cm)
Extremely long black tail flutters in flight. Black cap, white underparts, white wing linings distinguish it from Scissor-tailed Flycatcher. Many sightings are of **immatures,** which resemble **adult** but have a much shorter tail.
VOICE: Vagrants to U.S. mostly silent. Gives sharp *sik* notes and buzzy chattering calls.
RANGE: Widespread tropical species; casual vagrant on Atlantic coast and in TX; accidental elsewhere: recorded north to north shore of Lake Superior (ON) and west to north-coastal CA. Recorded in fall (most records) and spring in East; the few western records are in fall. There are also several winter records from TX and FL.

Scissor-tailed Flycatcher *Tyrannus forficatus* L 13" (33 cm)
Pearl gray above; whitish below with orange-buff belly. Salmon pink underwing with reddish axillaries best viewed in flight. Has very long outer tail feathers, white with black tips. **Male's** tail is longer than female; **juvenile** is paler overall, with shorter tail.
VOICE: Calls similar to Western Kingbird.
RANGE: Common; found in semi-open country. An early spring and late fall migrant. Very rare to casual wanderer in much of North America outside normal range; recorded north to AK; also very rarely nests east to VA and NC. A few winter in south TX, more in FL.

dark mask

long, stout bill

gray on head
and back

**Gray
Kingbird**

**Eastern
Kingbird**

blackish
head

small
bill

juvenile

notched
tail

black cap

gray back

adult

white
tail tip

**Fork-tailed
Flycatcher**

immature

very long,
forked black
tail

shorter
tail

**Scissor-tailed
Flycatcher**

juvenile

juvenile

pale gray
head and
back

reddish
pink
"armpit"

adult ♂

long forked tail with
extensive white;
female has shorter tail

orange-
buff belly

adult ♂

Loggerhead Kingbird *Tyrannus caudifasciatus* L 9" *(23 cm)*

Note dark crown, slight rear crest, gray upperparts, white edges on wing coverts, white underparts, white tail tip, short primary projection, long and moderately thick bill. Polytypic, seven named subspecies; at least Key West bird likely of nominate subspecies, described above. Bahama's *bahamensis* more yellow below, more brownish above. Tends to perch low and feed more within the canopy than other kingbirds.

VOICE: Nominate's calls include a loud buzzy *tireet*, often repeated. Wings produce a muffled sound in flight.

RANGE: Endemic West Indian species, resident in northern Bahamas, Greater Antilles, Cayman Islands. Two certain records: one from Key West, 8 to 27 Mar. 2007; the other at Dry Tortugas, 14 to 22 Mar. 2008 (both photographed). Other FL records unsubstantiated; one from Islamorada in early 1970s may have been the larger, rounder-headed, bigger-billed Giant Kingbird (*T. cubensis*), a now rare endemic of Cuba, but with historical extralimital records (specimens) from Great Inagua and the Caicos Islands.

Thick-billed Kingbird *Tyrannus crassirostris* L 9½" *(24 cm)*

Very thick bill. **Adult** dusky brown above, with a slightly darker head; whitish underparts, pale yellow on belly and undertail coverts. Yellow is brighter and more extensive in fresh fall adult and in **first-fall** birds, which have buffy edgings on wing coverts. Perches high in sycamores of lowland streamsides.

VOICE: Common call is a loud, high, whistled *pureet*.

RANGE: Breeds in Guadalupe Canyon and around Patagonia, AZ; rare elsewhere in southeastern AZ. Casual during fall and winter west to southern CA and in summer to west TX (Big Bend); a few scattered winter records elsewhere in TX. Accidental to BC and CO.

Great Kiskadee *Pitangus sulphuratus* L 9¾" *(25 cm)*

Yellow crown patch on black-and-white head often concealed. Brown above, with reddish brown wings and tail. Flycatches and dives for fish.

VOICE: Song is a deliberate *kis-ka-dee;* call is a loud *kreak*.

RANGE: Found chiefly in wet woodlands or near watercourses. Common. Casual vagrant north to KS and along Gulf Coast to LA (has bred). Introduced to Bermuda.

TITYRAS • ALLIES Family Tityridae

Mostly medium-size and compact with stubby bills; at least partly frugivorous. Includes becards and the tiny South American purpletufts. SPECIES: 29 WORLD, 3 N.A.

Rose-throated Becard *Pachyramphus aglaiae* L 7¼" *(18 cm)*

Rosy throat distinctive in **adult male. First-winter male** shows partially pink throat; acquires full adult plumage by its second fall. **Female** has slate gray crown, browner back. Western Mexico adult male, *albiventris*, has blackish cap, pale gray underparts. Larger *gravis* (eastern Mexico) found in TX; male darker; female more rufous than *albiventris*.

VOICE: Call, thin, mournful *seeoo*, sometimes preceded by chatter.

RANGE: Rare and local breeder and declining in southeastern AZ; foot-long nest is suspended from a tree limb. Casual (mostly in winter, but has bred) along lower Rio Grande; accidental in Trans-Pecos.

dark crown and face

long, moderately
thick bill

slightly
crested

**Loggerhead
Kingbird**
caudifasciatus

white
below

white-edged
wing coverts

short
primary
projection

dark crown and nape
with even darker mask

very thick
bill

1st fall

**Thick-billed
Kingbird**
pompalis

whitish
underparts

dark
brownish
tail

fresh fall birds
quite yellowish
on belly

**worn
summer
adult**

whitish tail tip

dark mask

bold white
supercilium

white
throat

**Rose-throated
Becard**
albiventris

1st winter
♂

thick,
stubby
bill

blackish cap

♀

paler below
than *gravis*

bright
yellow belly

darker head
than
albiventris

dark cap

adult ♂
gravis

adult
♂

rose throat

Great Kiskadee
texanus

bright rufous
wings and tail

♀
gravis

pale cinnamon
underparts

SHRIKES Family Laniidae

These masked hunters scan the countryside from lookout perches and then swoop down on insects, rodents, snakes, and small birds. Known as "butcher-birds," they mostly impale their prey on thorns. Recent research indicates that this is to mark territory and attract mates. SPECIES: 30 WORLD; 3 N.A.

Brown Shrike *Lanius cristatus* L 7½" (19 cm)

Adult male has distinct white border above black mask extending across forehead; warm brown upperparts, often brighter on rump and uppertail coverts; warm buff wash along sides and flanks. Lacks white wing patches. **Adult female** similar, but mask less solid; has some barring below on sides and flanks. **Juveniles** are barred on sides and flanks; show distinct dark subterminal edges above; dark brown mask has short whitish border above and behind eye. Much juvenal plumage is retained into fall, some even into winter. Compare all ages and plumages to much larger, longer-billed Northern Shrike. All records likely nominate *cristatus*. Old World Red-backed Shrike (*L. collurio*) and Isabelline Shrike (*L. isabellinus*) considered close relatives of Brown Shrike.

RANGE: Asian species; casual in AK where there are spring and fall records from western Aleutians, St. Lawrence Island, and Anchorage. Four fall and winter records from CA; late fall record from NS.

Loggerhead Shrike *Lanius ludovicianus* L 9" (23 cm)

Slightly smaller and darker than Northern Shrike. Head and back bluish gray; underparts white or very faintly barred. Broad black mask extends above eye and thinly across top of bill. All-dark bill, shorter than Northern Shrike, with smaller hook. Rump varies from gray to whitish. **Juvenile** is paler and barred overall, with brownish gray upperparts; acquires **adult** plumage by first fall. Seen in flight, wings and tail are darker and white wing patches smaller than Northern Mockingbird (page 404).

VOICE: Song is a medley of low warbles and harsh, squeaky notes; calls include a harsh *shack-shack.*

RANGE: Hunts in open or brushy areas; dives from low perch, then rises swiftly to next lookout. Still fairly common over parts of range; rare to very rare and declining in the eastern Midwest; has disappeared from Northeast, where now a casual visitor. The endangered subspecies *mearnsi* (**E**) is endemic to San Clemente Island off southern CA. Rare visitor to Pacific Northwest.

Northern Shrike *Lanius excubitor* L 10" (25 cm)

Larger than Loggerhead Shrike, with paler head and back, lightly barred underparts; rump whitish. Mask is narrower than Loggerhead, does not extend above eye; feathering above bill is white. Bill longer, with a more distinct hook. Often bobs its tail. **Juvenile** is brownish above and more heavily barred below than adult. **Immature** is grayer; retains barring on underparts until first spring.

VOICE: Song and calls are similar to Loggerhead.

RANGE: Uncommon; often perches high in tall trees. Southern range limit and numbers on wintering grounds vary unpredictably from year to year.

Brown Shrike
cristatus

SHRIKES

overall brown above

white supercilium above dark mask

stubby bill

juvenile

buffy-brown wash on sides and flanks

female has duller head pattern and some barring on sides and flanks

adult ♀

adult ♂

tail uniform with back

Loggerhead Shrike

darker gray back

black extends across forehead

stubby black bill

juvenile

adults

Northern Mockingbird for comparison

more extensive white patch

black wings with white primary patch

gray forehead

paler gray back

longer bill with distinct hook

faint mask with white eye ring

brownish on head and upperparts

Northern Shrike

barred underparts

immature

juvenile

VIREOS Family Vireonidae

Short, sturdy bills slightly hooked at the tip characterize these small songbirds. Vireos are closely related to shrikes. Some have "spectacles" and wing bars. Others have eyebrow stripes and no wing bars. They are generally chunkier and less active than warblers. SPECIES: 51 WORLD; 16 N.A.

Black-capped Vireo *Vireo atricapilla* L 4½" (11 cm) E

Olive above, white below, with yellow flanks and yellowish wing bars. **Male's** glossy black cap contrasts with broken white spectacles. Spectacles, smaller size, and secretive behavior distinguish **female;** immature males similar; **immature females** are more buffy. Hard to see, stays hidden in oak scrub, thickets.

VOICE: Best located by song, a persistent string of varied, twittering, two- or three-note phrases. Common call note, *tsidik,* almost identical to Ruby-crowned Kinglet.

RANGE: Endangered. Extirpated from KS by 1930s and very local in OK. Major factors are habitat destruction and brood parasitism by Brown-headed Cowbird. Accidental to NM and ON.

White-eyed Vireo *Vireo griseus* L 5" (13 cm)

Grayish olive above; white below, with pale yellow sides and flanks; two whitish wing bars; yellow spectacles. Distinctive white iris visible at close range. Juvenile is duller, with gray or brown iris. Subspecies on the FL Keys, *maynardi,* is grayer above, with less yellow below; bill larger. Southern TX subspecies, *micrus,* is colored like *maynardi,* but smaller.

VOICE: Typical song is a loud, variable five- to seven-note phrase usually beginning and ending with a sharp *chick;* also gives a chatter call suggestive of House Wren.

RANGE: Small numbers to southern ON; otherwise casual across southern Canada and the West.

Yellow-throated Vireo *Vireo flavifrons* L 5½" (14 cm)

Bright yellow spectacles, throat, and breast; white belly; two white wing bars. Upperparts olive, with contrasting gray rump. Compare with Pine Warbler (page 442), which has greenish yellow rump, streaked sides, thinner bill, and distinct but less complete spectacles.

VOICE: Song, a slow repetition of buzzy, low-pitched two- or three-note phrases separated by long pauses, often contains a rising *three-eight.* Calls include a rapid, harsh series of *cheh* notes.

RANGE: Fairly common in most of breeding range. Rare or casual in southernmost FL in winter. Most reports are misidentified Pine Warblers. Very rare vagrant in the West.

Thick-billed Vireo *Vireo crassirostris* L 5½" (14 cm)

Larger than White-eyed Vireo, with larger, slightly stouter and grayer bill. Note overall browner color, lack of gray on nape, and broken spectacles. Iris is darker than adult White-eyed.

VOICE: Song is similar but harsher; call notes slower and harsher.

RANGE: Caribbean species, casual visitor to south FL (mostly southeast) from Bahamas or Cuba. Many reports from south FL are misidentified White-eyed Vireos.

slaty cap

black cap,
white spectacles

**Black-capped
Vireo**

adult ♀

adult ♂

immature ♀

**White-eyed
Vireo**

white iris,
yellow spectacles

pale gray
nape

yellow
sides and
flanks

thin white
wing bars

Florida Keys
maynardi

broken spectacles
pattern

uniform brownish
olive upperparts

larger
bill than
White-eyed

**Thick-billed
Vireo**
crassirostris

**Yellow-throated
Vireo**

yellow
spectacles

thick
bill

yellow throat
and breast

gray
rump

white
belly

white wing bars

**Pine Warbler adult ♂
for comparison**

Blue-headed Vireo *Vireo solitarius* L 5" *(13 cm)*

Adult male's solid blue-gray hood contrasts with white spectacles and throat; hood of female and immatures partly gray. All ages have bright olive back; yellow-tinged wing bars and tertials; greenish yellow edges to dark secondaries. Distinct white on outer tail; bright yellow sides and flanks, sometimes mixed with green. Larger Appalachian *alticola* has more slaty back; only flanks are yellow.
VOICE: Song similar to Red-eyed Vireo, but slower. Call a harsh chatter.
RANGE: Fairly common in mixed woodlands; very rare to casual in West, but separation from Cassin's is difficult.

Cassin's Vireo *Vireo cassinii* L 5" *(13 cm)*

Similar to Blue-headed, but slightly smaller and duller. Less contrast between head and throat; duller, whitish wing bars and tertial edges; less white in tail. Immature female can have entirely green head; compare to Hutton's Vireo (page 354).
VOICE: Song is hoarser than Blue-headed, like Plumbeous; call similar.
RANGE: Unlike Blue-headed and Plumbeous, an early fall migrant, but scarce on western Great Plains to west TX (a few in spring); accidental farther east.

Plumbeous Vireo *Vireo plumbeus* L 5¼" *(13 cm)*

Larger, with bigger bill than Cassin's Vireo; also has sharper head and throat contrast; gray upperparts. Pattern of tail feathers similar to Blue-headed Vireo. White wing bars and flight feather edges; pale yellow, if present, only on flanks. Sides of breast gray, sometimes tinged olive. Compare worn summer birds to shorter-winged Gray Vireo.
VOICE: Song is hoarser than Blue-headed; call similar.
RANGE: Fairly common in varied woodland habitats. Rare migrant on western Great Plains and to CA coast (some winter). Casual to OR; accidental in East.

Gray Vireo *Vireo vicinior* L 5½" *(14 cm)*

White eye ring; wings brownish, with faint wing bars, the lower more prominent; short primary projection; long tail. Compare with Plumbeous and smaller West Coast subspecies *(pusillus)* of Bell's Vireo. Sticks to undergrowth; flicks tail as it forages.
VOICE: Song is a series of musical *chu-wee chu-weet* notes, faster and sweeter than Plumbeous. Calls include shrill, descending musical notes, often delivered in flight, as is song.
RANGE: Found in semiarid habitat. Largely unknown as a migrant; accidental on southern CA coast and offshore islands and to WY, western Great Plains, and, remarkably, WI (early Oct. specimen).

Bell's Vireo *Vireo bellii* L 4¾" *(12 cm)*

Endangered West Coast *pusillus* **(E)** gray above, whitish below; has indistinct white spectacles, two whitish wing bars. More easterly nominate is greenish above, yellowish below; often bobs tail. Southwestern *medius* and *arizonae* are intermediate. Active, rather secretive.
VOICE: Song is a series of scolding notes. Call notes like House Wren.
RANGE: Uncommon in moist woodlands, bottomlands, mesquite, and in eastern part of range in shrubby areas on prairies. Rarely seen in migration. Casual east and north of mapped range.

See subspecies map, page 552

352

Cassin's Vireo

more blended border than Blue-headed

duller sides

♀

Blue-headed Vireo
solitarius

white spectacles, bluish gray head

sharp contrast between auricular and throat

yellow sides with some olive

♂

Plumbeous Vireo
plumbeus

white spectacles

gray above with two white wing bars

slaty gray sides

long primary projection

shorter tail than Gray

Gray Vireo

whitish eye ring but no spectacled effect

faint wing bar

long tail, which is flipped about

short primary projection

Bell's Vireo

overall grayish color

one rather faint wing bar

rather long tail

subdued head pattern

West Coast
pusillus

bellii

Eastern *bellii* is much more olive and yellow than *pusillus;* Southwestern *medius* and *arizonae* are intermediate

Yellow-green Vireo *Vireo flavoviridis* L 6" *(15 cm)*

Similar to Red-eyed Vireo, but bill longer; head pattern more blended. Strong yellow-green wash above extends onto sides of face; extensive yellow on sides, flanks, and undertail; brightest in fall. Worn summer birds more subtle; sometimes best told by song.

VOICE: Song is a rapid but hesitant series of notes, suggesting House Sparrow.

RANGE: Very rare in fall in coastal CA (mainly late Sept. to late Oct., once singing in July in coastal San Diego County) and in summer in south TX. Casual in spring on the upper Gulf Coast and to southern AZ and NM (summer), NV, and southeastern CA (fall). Winters in South America.

Red-eyed Vireo *Vireo olivaceus* L 6" *(15 cm)*

Blue-gray crown; white eyebrow bordered above and below with black. Olive back, darker wings and tail; white underparts. Lacks wing bars. Ruby red iris visible at close range. **First-fall** bird has brown iris. Immatures and some fall adults have pale yellow on flanks and undertail coverts.

VOICE: Persistent song, sung all day, a variable series of deliberate, short phrases. Calls include a nasal, whining *quee.*

RANGE: Common in eastern woodlands; rare (possibly declining) migrant to West Coast. Winters in South America.

Black-whiskered Vireo *Vireo altiloquus* L 6¼" *(16 cm)*

Variable dark malar stripe (whisker), often hard to see. Bill larger and longer than Red-eyed Vireo. Grayish brown crown; pattern more diffuse than Red-eyed. Brownish green above; whitish below, with variable pale yellowish wash on sides and flanks.

VOICE: Song, deliberate one- to four-note phrases, less varied and more emphatic than Red-eyed.

RANGE: Common in summer in the mangrove swamps of FL Keys and along FL coasts. Casual along rest of Gulf Coast and north to Carolinas. Winters in South America.

Hutton's Vireo *Vireo huttoni* L 5" *(13 cm)*

Grayish olive above, with pale area in lores; white eye ring broken above eye. Subspecies vary from paler, grayer southwestern *stephensi* and more easterly *carolinae* to greener coastal subspecies such as *huttoni.* Separated from Ruby-crowned Kinglet by larger size, thicker bill, lack of dark area below lower wing bar.

VOICE: Song is a repeated or mixed rising *zu-wee* and descending *zoe zoo;* also a flat *chew.* Calls include a low *chit* and whining chatter; birds from interior Southwest give a harsher *tchurr-ree.*

RANGE: Fairly common in woodlands. Scarce and local but increasing in woodlands on Edwards Plateau, TX *(carolinae).* Casual to western Mojave Desert; accidental to western NV, Salton Sea, CA, and southwestern AZ. Most reports of strays to deserts represent misidentified Ruby-crowned Kinglet or immature female Cassin's Vireo. Any claim should be carefully substantiated.

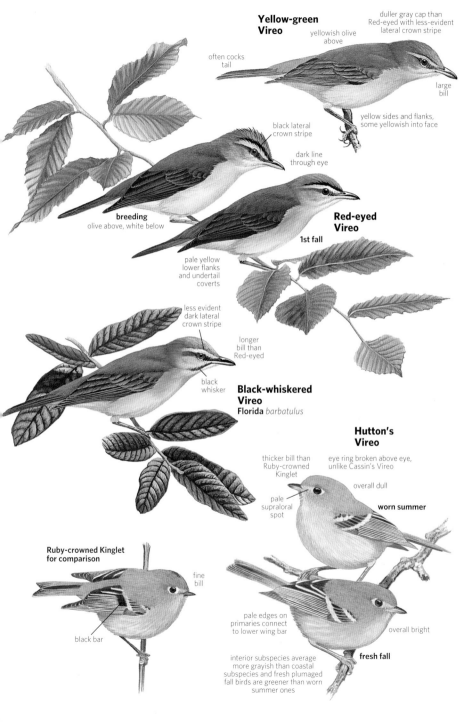

Yellow-green Vireo

duller gray cap than Red-eyed with less-evident lateral crown stripe

yellowish olive above

often cocks tail

large bill

yellow sides and flanks, some yellowish into face

black lateral crown stripe

dark line through eye

Red-eyed Vireo

1st fall

breeding
olive above, white below

pale yellow lower flanks and undertail coverts

less evident dark lateral crown stripe

longer bill than Red-eyed

black whisker

Black-whiskered Vireo
Florida *barbatulus*

Hutton's Vireo

thicker bill than Ruby-crowned Kinglet

eye ring broken above eye, unlike Cassin's Vireo

overall dull

pale supraloral spot

worn summer

Ruby-crowned Kinglet for comparison

fine bill

pale edges on primaries connect to lower wing bar

overall bright

black bar

interior subspecies average more grayish than coastal subspecies and fresh plumaged fall birds are greener than worn summer ones

fresh fall

Philadelphia Vireo *Vireo philadelphicus* L 5¼" *(13 cm)*

Adult variably yellow below, palest on belly. Greenish above, contrasting grayish cap, dull grayish olive wing bar, dull white eyebrow, dark eye line. First-fall birds and most **fall** adults often brighter yellow below. Distinguished from Warbling by extended eye line, darker cap, dark primary coverts, yellow at center of throat and breast. Similar Tennessee Warbler (page 426) has a thinner bill, white undertail coverts.

VOICE: Song very similar to Red-eyed but generally slower, thinner.

RANGE: Uncommon; found in open woodlands, streamside willows and alders. Rare migrant in coastal Southeast. Casual to mostly rare in West; chiefly occurs in fall.

Warbling Vireo *Vireo gilvus* L 5½" *(14 cm)*

Gray or olive-gray above; western birds, especially Pacific subspecies *swainsoni*, are smaller than nominate, with slighter bill. Underparts white. Dusky postocular stripe; white eyebrow, without dark upper border; brown eye. Lacks wing bars. Smaller and paler than Red-eyed Vireo; crown does not contrast strongly with back. Birds in fresh **fall** plumage tend to be greener above, pale yellow on sides of flanks.

VOICE: Song of *gilvus* delivered in long, melodious, warbling phrases; song of *swainsoni* less musical, with higher tones and break near beginning. Calls include a peevish rising *cheee* and low *chut* notes. Apparently does not interbreed where the two subspecies come together in AB.

RANGE: Common in summer; found in deciduous woods, in East especially near water. Rare migrant in coastal Southeast. Casual migrant to western AK; in winter to southern CA.

CROWS • JAYS Family Corvidae

Harsh voice and aggressive manner draw attention to these large, often gregarious birds. Powerful, all-purpose bill efficiently handles a varied diet. SPECIES: 120 WORLD; 20 N.A.

Clark's Nutcracker *Nucifraga columbiana* L 12" *(31 cm)*

Chunky gray bird with black wings and black central tail feathers. White wing patches and white outer tail feathers are conspicuous in flight. Wingbeats are deep, slow, crowlike.

VOICE: Calls include a very nasal, grating, drawn-out *kra-a-a.*

RANGE: Locally common in high coniferous forests. Every 10 to 20 years, irrupts into desert and lowland areas of the West. Accidental to East.

Gray Jay *Perisoreus canadensis* L 11½" *(29 cm)*

Fluffy with long tail, small bill. The three subspecies groups are shown here: Nominate *canadensis*, one of several subspecies common in northern boreal forests, has a white collar and forehead, with dark gray crown and nape; *capitalis*, in the southern Rockies, has a paler crown, head appears mostly white; *obscurus*, coastal resident in the Northwest from WA to northernmost CA, has a larger, darker cap extending to the crown, with underparts paler than other subspecies. **Juveniles** of all

See subspecies map, page 553

subspecies are sooty gray overall, with a faint white moustachial streak.

VOICE: Call notes include a whistled *wheeoo* and a low *chuck.*

RANGE: Largely resident. Casual in winter even slightly south of normal, resident range.

Philadelphia Vireo

dark eye line extends through lores

fall birds often quite bright yellow

pale lores

fall
gilvus

Warbling Vireo

Eastern *gilvus* distinctly larger

fall

shorter tail

dark primary coverts

spring

yellow of equal intensity extends across breast

fall
swainsoni

yellow brightest on sides

paler primary coverts

smaller than *gilvus*; comparably sized to Philadelphia

spring
gilvus

Clark's Nutcracker

long, slender bill

white outer tail feathers

white secondaries

crowlike wingbeats

adults

overall gray with whiter face and under tail; black wings

sooty overall

dark nape

whitish crown

paler on head and back than nominate subspecies

dark on nape extends to crown

small bill

juvenile

pale tail tips

pale tail tips

boreal adult
canadensis

southern Rockies adult
capitalis

pale tail tips

Northwest adult
obscurus

Gray Jay

pale gray underparts for all subspecies

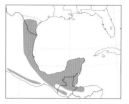

Green Jay *Cyanocorax yncas* L 10½" (27 cm)

Green-and-blue plumage blends with woodland habitat.
VOICE: Gregarious and noisy; most common call is a series of raspy *cheh-cheh-cheh* notes.
RANGE: Tropical species; range extends to south TX. Resident and locally common in brushy areas and streamside growth of the lower Rio Grande Valley; local farther north.

Brown Jay *Psilorhinus morio* L 16½" (42 cm)

Very large jay with long, broad tail. Dark, sooty brown overall with pale belly. **Adult** has black bill. **Juvenile** has yellow bill and eye ring, black by second winter; in transition, blotchy yellow-and-black bills.
VOICE: A noisy species; its harsh scream is similar to the call of Red-shouldered Hawk. Another call sounds like a hiccup.
RANGE: Tropical species. Rare and declining resident in Rio Grande woodlands below Falcon Dam, TX.

Pinyon Jay *Gymnorhinus cyanocephalus* L 10½" (27 cm)

Blue overall; blue throat streaked with white; bill long and spiky; tail short. Immature is duller. Flight is direct, with rapid wingbeats, unlike scrub-jays' undulating flight.
VOICE: Typical flight call is a high-pitched, piercing *mew,* audible over long distances. Also gives a rolling series of *queh* notes.
RANGE: Generally seen in large flocks, often numbering in the hundreds; nests in loose colonies. Common in pinyon-juniper woodlands of interior mountains and high plateaus; also yellow pine woodlands. Casual to Great Plains, west TX, and coastal CA.

Blue Jay *Cyanocitta cristata* L 11" (28 cm)

Crested jay with black barring and white patches on blue wings and tail, black necklace on whitish underparts.
VOICE: Most common of varied calls is a piercing *jay jay jay;* also gives a musical *weedle-eedle* and mimics the call of Red-shouldered Hawk.
RANGE: Common in suburbs and woodlands. Often migrates in large flocks. Casual fall and winter visitor to the West, especially the Northwest.

See subspecies map, page 553

Steller's Jay *Cyanocitta stelleri* L 11½" (29 cm)

Crested; dark blue and black overall. Some subspecies, including nominate from coast to northern Rockies, have darker backs and bluish streaks on forehead. The genetically distinct central and southern Rockies subspecies, *macrolopha,* has long crest, paler back, white streaks on forehead, white mark over eye; largest subspecies, *carlottae* (not shown), resident on the Queen Charlotte Islands off BC, is almost entirely black above. Where ranges overlap in the eastern Rockies, occasionally hybridizes with Blue Jay.
VOICE: Calls include a series of *shack* or *shooka* notes and other calls suggestive of Red-shouldered and Red-tailed Hawks.
RANGE: Common in pine-oak woodlands and coniferous forests. Bold and aggressive; often scavenges at campgrounds and picnic areas (*macrolopha* seems shyer). Rare and irregular winter visitor to lower elevations of the Great Basin, southern CA, and southwestern deserts and to western Great Plains.

Green Jay
glaucescens

blue-and-black head

body largely greenish

yellow outer tail feathers

heavy black bill

Brown Jay
palliatus

pale belly

thin yellow orbital ring

yellow bill

juvenile

very large size

adult

long, thin bill

Pinyon Jay

short tail

overall blue color; duller on immatures

Steller's Jay

black crest

bluish lines

longer crest

grayer back

stelleri

white forehead streaks and white above eye

southern Rockies
macrolopha

Blue Jay

bluish crest

blackish throat band

extensive white in wings

white tail tips

See subspecies map, page 553

Western Scrub-Jay *Aphelocoma californica* L 11" (28 cm)

Long tail; blue above; variable bluish band on chest. Coastal subspecies, including nominate *californica*, deeper blue above; contrasting brown patch; distinct white eyebrow and blue breast band; undertail coverts geographically variable; may or may not be bluish. Tame and widespread; found in urban areas. Interior subspecies (possibly a separate species) range from Great Basin's duller, slender-billed *nevadae*, to similar but slightly thicker-billed *woodhouseii*, to bluer, stouter-billed *texana*. The only reported hybridization between coastal and interior groups is from the Pine Nut Mountains on the CA-NV border south of Carson City. Interior subspecies rather shy; inhabit lower mountain woodland. All U.S. subspecies hold individual territories. Western, Island, and Florida Scrub-Jays were formerly considered one species, Scrub Jay.
VOICE: Calls include raspy *shreep,* often in a short series.
RANGE: Fairly common (interior subspecies) to common (coastal subspecies). Rare and irregular (mostly interior subspecies) to southwestern deserts and western Great Plains in fall and winter; accidental east to IN. Casual (coastal subspecies) to southwestern BC.

Island Scrub-Jay *Aphelocoma insularis* L 12" (30 cm)

Larger and with much larger bill than Western; darker blue above; always shows rich blue undertail coverts. Birds hold individual territory; takes several years for young birds to acquire territory and breed.
VOICE: Calls similar to Western, but slightly lower and deeper.
RANGE: Restricted to Santa Cruz Island, CA, where it is the only scrub-jay. Declines over last decade may have resulted from West Nile Virus, although that disease is still not documented there; perhaps other causes are involved.

Florida Scrub-Jay *Aphelocoma coerulescens* L 11" (28 cm) **T**

Distinguished from other scrub-jays by whitish forehead and eyebrow; shorter, broader bill; paler back; distinct collar; indistinct streaking below; disproportionately longer tail. Has cooperative breeding system: Fledged young remain on territory and help rear nestlings.
VOICE: Varied calls include raspy, hoarse notes.
RANGE: Restricted to FL scrub region where population declined some 90 percent in 20th century due to habitat destruction. Optimum habitat is transitional, produced by fire: consists of scrub, mainly oak, about ten feet high with small openings.

Mexican Jay *Aphelocoma wollweberi* L 11½" (29 cm)

Blue above, with slight grayish cast on back, brownish patch on center of back. Lacks crest. Distinguished from scrub-jays by absence of white throat and white eyebrow and by chunkier shape; flight is more direct. TX subspecies, *couchii,* has richer blue head. AZ **juvenile** *arizonae* retains pale bill past post-juvenal molt. Has cooperative breeding system similar to Florida Scrub-Jay.
VOICE: Calls include a loud, ringing *week,* given singly or in a series and very similar in both subspecies groups.
RANGE: Common in montane pine-oak canyons of the Southwest, where it greatly outnumbers scrub-jays. Accidental *(arizonae)* to El Paso and to near Alpine, TX *(couchii)*.

Western Scrub-Jay

thick bill

darker blue than interior subspecies

blue breast band

coastal
superciliosa

whitish flanks

heavier bill

deeper blue above than *woodhouseii* or *nevadae* with more contrasty back

thinner bill than coastal subspecies

nevadae and very similar *woodhouseii* paler blue above than coastal subspecies with duller and more blended breast band

grayish underparts

interior
nevadae

long tail

underparts paler than *woodhouseii* or *nevadae*

Texas Hill Country
texana

whitish forehead

grayish back

Island Scrub-Jay

large bill

only jay on Santa Cruz Island

larger overall

long tail

Florida Scrub-Jay

bluish undertail

darker blue above

paler blue above

Texas
couchii

adult
no breast band

Mexican Jay

Arizona
arizonae

fleshy bill

juvenile
arizonae

Black-billed Magpie *Pica hudsonia* L 19" (48 cm)

Both magpie species are black and white and have unusually long tails with iridescent green highlights. White wing patches flash in flight.
VOICE: Gregarious and noisy; typical calls include a whining *mag* and a series of loud, harsh *chuck* notes. Calls and many behavioral traits resulted in North American Black-billed Magpie being split from Old World populations of magpie, whose calls are faster and lower pitched.
RANGE: Uncommon to common inhabitant of open woodlands and thickets in rangelands and foothills, especially along watercourses. Casual south and east of normal range in winter. Some well east may be escaped cage birds.

Yellow-billed Magpie *Pica nuttalli* L 16½" (42 cm)

Separate range from similar Black-billed Magpie. Distinguished by its yellow bill and by a yellow patch of bare skin around the eye; extent of yellow variable, sometimes fully encircles eye. Both species roost and feed in flocks, usually nest in loose colonies, but Yellow-billed's behavior is more colonial than Black-billed.
VOICE: Calls are similar to Black-billed.
RANGE: Prefers oaks, especially more open oak grassland, also orchards and parks. Common resident of rangelands and foothills of central and northern Central Valley, CA, and coastal valleys south to Santa Barbara County; formerly (in 19th century) to western Los Angeles County (Conejo Valley). Not prone to wandering, but casual north almost to OR. Recent sharp declines in core range reflect losses from West Nile Virus. Yellow-billed is more closely related to Black-billed than either is to the Old World Eurasian magpie (*P. pica*).

Eurasian Jackdaw *Corvus monedula* L 13" (33 cm)

Small, black overall; gray nape and face, pale grayish eyes. Inquisitive.
VOICE: Calls include a metallic *kow* and a softer *jack* note.
RANGE: Arrived in Northeast in early 1980s, most perhaps ship assisted. Recorded from Atlantic Canada to ON and PA. Few reports by 1990s, the last one in Apr. 1999 from Newfoundland.

Tamaulipas Crow *Corvus imparatus* L 14½" (37 cm)

Smaller, glossier than American Crow; compare to larger Chihuahuan Raven (page 364), the only other crow or raven in its range.
VOICE: Call is a low, froglike *croak*.
RANGE: First appeared in U.S. in late 1960s at municipal dump near Brownsville, TX, where it has nested. Present mainly in winter in varying numbers, but sharply declining over last two decades; now almost gone.

Northwestern Crow *Corvus caurinus* L 16" (41 cm)

Nearly identical to American Crow (page 364) but slightly smaller. Considered by some to be a subspecies of American Crow.
VOICE: Call is somewhat hoarser and lower than American, but beware of juvenile American Crow.
RANGE: Inhabits northwestern coastal areas and islands, where a common scavenger along the shore. In areas of presumed range overlap with American Crow (e.g., Puget Sound), crows not identified to species. Southern and inland limits uncertain.

Black-billed Magpie

black-and-white coloration

large white wing patch

long tail

variable yellow skin around eye

yellow bill

Yellow-billed Magpie

gray nape and face with pale gray eye

Eurasian Jackdaw

Tamaulipas Crow

calling posture

low froglike calls diagnostic

small size

smaller and glossier than American Crow

range does not overlap with American or Fish Crow

Northwestern Crow

long tail extends beyond wing tips

nearly identical to American Crow

Fish Crow *Corvus ossifragus* L 15½" (39 cm)

Smaller than American Crow with smaller bill and feet and shorter legs; wings more pointed; wingbeats faster.

VOICE: Best distinguished by voice: Call, a high, nasal *uh uh,* the second note lower; also low, short *car* notes.

RANGE: Favors tidewater marshes and low valleys along eastern river systems. Interior range expanding. Often seen in winter in flocks with American Crows, when also found on farmland and in towns and dumps. Casual or very rare to MI and ON.

American Crow *Corvus brachyrhynchos* L 17½" (45 cm)

Our largest crow. Long, heavy bill is noticeably smaller than ravens. Fan-shaped tail distinguishes all crows from ravens in flight.

VOICE: Adult is readily identified by familiar *caw* call, but juvenile's higher-pitched, nasal *cah* begging call resembles the call of the similar Fish Crow.

RANGE: Generally common throughout most of its range in a wide variety of habitats. May form large foraging flocks and nighttime roosts in fall and winter.

Chihuahuan Raven *Corvus cryptoleucus* L 19½" (50 cm)

Heavier bill and wedge-shaped tail distinguish both raven species from crows. Distinguished from Common Raven by shorter wings and shorter, less wedge-shaped tail; nasal bristles extend farther out on shorter, thicker-appearing bill. Neck feathers whitish rather than grayish at base, but usually obscured; sometimes shows in windy conditions, which often dominate on Great Plains.

VOICE: Frequent call, a drawn-out croak, usually slightly higher pitched than Common Raven.

RANGE: Common in desert areas and scrubby grasslands; also farms, towns, and dumps. Formerly occurred north to western NE (19th century) until slaughter of bison herds, with which they associated; now casual in southwestern and south-central portions of NE, where it has nested, but no accepted records since 1979.

Common Raven *Corvus corax* L 24" (61 cm)

Large, with long, heavy bill and long, wedge-shaped tail. Larger than Chihuahuan Raven; note thicker, shaggier throat feathers and nasal bristles that do not extend as far out on larger bill. Unlike crows, ravens of both species are often seen soaring. Where common, often soars in flocks, sometimes in circles on a thermal. Some populations from CA, where birds are smaller, show distinct genetic differences from those elsewhere in North America and Eurasia, which are genetically close to each other. Birds with intermediate DNA have been found from WA, ID, and CA. Interestingly, these CA birds are genetically closer to Chihuahuan Ravens than to other Common Ravens.

VOICE: Most common call is a low, drawn-out croak.

RANGE: Found in a variety of habitats, including mountains, deserts, and coastal areas. Numerous in western (increasing) and northern part of range; uncommon and local, but slowly spreading, in Appalachians, New England, and northern Great Plains; casual on central Great Plains.

Fish Crow

calling posture

smaller than American Crow with slight purplish gloss

bill averages smaller and more slender than American Crow

shorter legs than American Crow

wings more pointed than American Crow and flies with more rapid flaps

American Crow

juvenile crows (all North American species) are browner than adults

begging calls can resemble Fish Crow

juvenile

larger and heavier and bigger billed than Fish Crow

wing tip more rounded than Fish Crow

white on neck shows when wind ruffles feathers

nasal bristles extend well over half the length of the culmen

Chihuahuan Raven

shorter, more rounded tail than Common Raven

longer bill than Chihuahuan and nasal bristles extend about half way out culmen

Common Raven

shaggy throat feathers

long, wedge-shaped tail

larger than Chihuahuan but beware of smaller and smaller-billed Common Ravens from California

large size, larger than Red-tailed Hawk; frequently soars

LARKS Family Alaudidae

An Old World family represented by many species, especially from Africa. Horned Lark (known as Shore Lark in the Old World) is the only established species in the New World, although Sky Lark strays to AK and is established as an introduced species on Vancouver Island, BC. Ground dwellers of open fields, larks are seed- and insect-eaters. They seldom alight on trees or bushes. On the ground, they walk rather than hop. SPECIES: 96 WORLD; 2 N.A.

Horned Lark *Eremophila alpestris* L 6¾-7¾" (17-20 cm)

Head pattern distinctive in all subspecies: black "horns"; white or yellowish face and throat with broad black stripe under eye; black bib. **Female** duller overall than **male,** horns less prominent. Conspicuous in flight is the mostly black tail with white outer feathers, brown central feathers. Briefly held **juvenal** plumage has whitish markings above, streaks below; can be confused with Sprague's Pipit (page 414). Found in both the New and Old Worlds. Over 40 subspecies have been recognized; more than two-thirds of those are from the New World. Three widespread subspecies are found in the East: **"Prairie Horned Lark,"** *praticola,* which breeds in southern Canada and the eastern U.S., is pale, with white eyebrows and throat. **"Northern Horned Lark,"** *alpestris,* is much darker, with a yellow throat. The central Arctic coast subspecies, *hoyti,* is pale like *praticola,* but larger; *giraudi* from western Gulf Coast is quite yellow below. Western subspecies vary widely in overall color; selected extremes are shown here: *enthymia* (Great Plains, not shown), the Southwest deserts' *ammophila,* and large *arcticola* (northwestern Canada, AK) are very pale; *sierrae* (northeastern CA) and *strigata* (coastal Northwest) are yellower on head and especially chest, *strigata* is also more streaked underneath; *rubea* (Central Valley, CA) is redder dorsally; *insularis* (Channel Islands, CA) is streaked below; *flava* from Palearctic (casual fall vagrant to Aleutians, Bering Sea Islands, and Middleton Island, AK) has a yellow throat and supercilium. Some subspecies are highly migratory, others are largely resident. Winter flocks of hundreds, even thousands, may comprise several subspecies.

VOICE: Calls include a high *tsee-ee* or *tsee-titi.* Song is a weak twittering, delivered from the ground or in flight.

RANGE: Common on Great Plains and much of West; less numerous and local farther east. Prefers dirt fields, gravel ridges, airports, sod farms, and shores. Flocks in winter in the eastern U.S. are mainly *alpestris.* A relatively late fall migrant and very early spring migrant. In East, by late winter pairs of *praticola* have already established breeding territories.

vagrant
to Alaska
♂ *flava*

central Arctic
coast breeder
♂ *hoyti*

widespread in
East in winter

winter
♂

"Northern
Horned Lark"
alpestris

Horned Lark

Lapland Longspur
for comparison

northwestern
Canada, Alaska
♂ *arcticola*

black tail with
white outer
tail feathers

alpestris

alpestris

coastal
Northwest
♂ *strigata*

northeastern
California
♂ *sierrae*

California
Central Valley
♂ *rubea*

praticola breeder
over much of
Midwest and East

"Prairie
Horned Lark"
praticola

Gulf Coast
giraudi

upperparts spotted
with white

juvenile
ammophila

faintly
streaked
breast

small
"horns"

winter
♂ *ammophila*

juvenile often
mistaken for
Sprague's Pipit

southwestern
deserts
♂ *ammophila*

duller than
male

♀ *ammophila*

except juveniles,
all have dark
breast band

Sky Lark *Alauda arvensis* L 7¼" (18 cm)

Old World species. Plain brown bird with slender bill; slight crest is raised when bird is agitated. Upperparts heavily streaked; buffy white underparts streaked on breast and throat. Dark eye prominent. Highly migratory Asian subspecies *pekinensis,* rare on western Aleutians and Pribilofs and casual on St. Lawrence Island, AK, is darker and more heavily streaked above; accidental in winter in WA and northern CA. All juveniles have a scaly brown mantle. In flight, shows a conspicuous white trailing edge on the inner wing and white edges on tail.

VOICE: Song is a continuous outpouring of trills and warblings, delivered in high hovering or circling song flight. Call is a liquid *chirrup* with buzzy overtones.

RANGE: Nominate *arvensis,* a widespread European subspecies introduced to Vancouver Island in the early 1900s, is resident there on open slopes and fields. The population on San Juan Islands, WA, is now extirpated. Highly migratory *pekinensis* is rare on western Aleutians and Pribilofs and casual on central Aleutians and St. Lawrence Island, AK; accidental in winter in WA and northern CA.

SWALLOWS Family Hirundinidae

Slender bodies with long, pointed wings resemble swifts, but "wrist" angle is sharper and farther from the body; flight is more fluid. Adept aerialists, swallows dart to catch flying insects. Flocks perch in long rows on branches and wires. SPECIES: 83 WORLD; 15 N.A.

Purple Martin *Progne subis* L 8" (20 cm)

Male is dark, glossy purplish blue. **Female** and juvenile are gray below. **First-spring males** have some purple below. In flight, male especially resembles European Starling (page 410); but note forked tail, longer wings, and typical swallow flight, short glides alternating with rapid flapping.

VOICE: Loud, rich gurgling and whistles; also a low *churr.*

RANGE: Locally common where suitable nest sites are available. Declining over much of North America, especially in Pacific states. Very early spring migrant in South; winters in South America. Casual to AK, including the Bering Sea region and YT.

Brown-chested Martin *Progne tapera* L 6½" (16 cm)

Smaller than Purple Martin with brownish upper parts, white below with brown sides, and brown band across breast.

RANGE: South American species. Southern subspecies, *fusca,* is an austral migrant to northern South America. Five North American records, including a specimen *(fusca)* from Monomoy Island, MA, 12 June 1983; three others on East Coast in summer and fall; once near Nogales, AZ, on 3 Feb. 2006.

Bahama Swallow *Tachycineta cyaneoviridis* L 5¾" (15 cm)

Greenish above. Deeply forked tail and white underwing coverts separate this species from similar Tree Swallow. Immatures have shorter tail fork, dusky wash on breast and wing linings.

RANGE: Endemic Bahamian species. Breeds in northern Bahamas and vicinity; casual visitor to the FL Keys, especially Big Pine Key, and nearby mainland but unrecorded since Apr. 1992.

Sky Lark
arvensis

short crest can be raised

fresh

worn

white trailing edge to secondaries

pekinensis

darker, richer, and more heavily streaked than *arvensis*

juvenile

often soars; wingbeats are slow and deep

adult ♂

eastern ♀
subis

darker overall than western subspecies

1st spring ♂

broad wings

dark purple overall

Purple Martin

eastern ♀
subis

adult ♂

paler below than nominate eastern *subis*

western ♀
arboricola

green above ♂

Bahama Swallow

adult
fusca

distinct breast band with downward point like smaller Bank Swallow

Brown-chested Martin

adults

♂

forked tail

white wing linings

Tree Swallow *Tachycineta bicolor* L 5¾" (15 cm)

Dark, glossy greenish blue above, greener in fall plumage; white below. White does not extend above eye as in Violet-green Swallow. **Juvenile** is gray-brown above; usually has more diffuse breast band than Bank Swallow (page 372). Some **first-spring females** show varying amount of adult color on crown and back; others are still brown backed. Some do not acquire adult plumage until over two years of age. Some older females look very similar to males.

VOICE: Calls and song include whistles and liquid gurgles or chirps.

RANGE: Common in wooded habitat near water, and where dead trees provide nest holes. Also nests in fence posts, barn eaves, nest boxes. Migrates in huge flocks; goes north earlier in spring and lingers farther north in fall than other swallows.

Violet-green Swallow *Tachycineta thalassina* L 5¼" (13 cm)

White on cheek extends above eye; white flank patches extend onto sides of rump; compare with larger Tree Swallow. May also be confused with White-throated Swift (page 296). Female is duller above than male. **Juvenile** is gray-brown above; white, except on rump, may be mottled or grayish.

VOICE: Gives rapid and high-pitched twittering notes.

RANGE: Common in a variety of woodland habitats. Nests in hollow trees or rock crevices, often forming loose colonies. A rather late fall migrant. Casual to western AK and in East.

Common House-Martin *Delichon urbicum* L 5" (13 cm)

Small, with forked tail. Deep, glossy blue above; mostly white below with white rump; underwing coverts pale smoky gray. Female slightly grayer below; juvenile duller. Soars for long periods. One AK specimen of eastern subspecies *lagopoda,* with more extensive white on rump.

VOICE: Call, a rough scratchy *prrit,* somewhat similar to Rough-winged Swallow.

RANGE: Old World species. Casual mainly in spring in western AK; one record for St.-Pierre, off Newfoundland (26 to 31 May 1989).

Barn Swallow *Hirundo rustica* L 6¾" (17 cm)

Long, deeply forked tail. Throat is reddish brown; upperparts blue-black; underparts usually cinnamon or buffy. Two Eurasian, white-bellied subspecies have occurred in western and northern AK: *rustica,* which has a solid dark breast band, and *gutturalis,* with incomplete breast band, which also has been found on Queen Charlotte Islands, BC. In all **juveniles,** tail is shorter but still noticeably forked; underparts pale. Flight is similar to Northern Rough-winged (page 372), with slow, floppy wingflaps and with glides. Over much of the East, indeed over much of North America, this is the most widespread swallow. Has interbred with both Cliff and Cave Swallows (page 372).

VOICE: Call is a short, sweet, single or double *vit* or *veet;* song is a series of squeaky notes.

RANGE: Common; generally nests on or inside farm buildings, under bridges, and inside culverts, in pairs or small colonies. North American *erythrogaster* is casual to central and western AK.

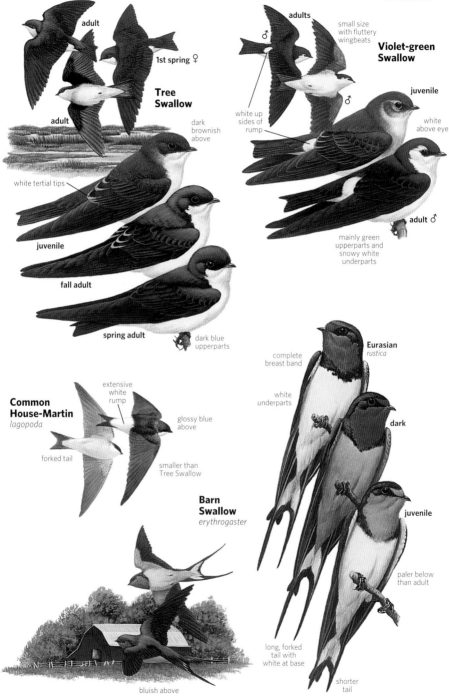

Tree Swallow

adult

1st spring ♀

adult

dark brownish above

white tertial tips

juvenile

fall adult

spring adult

dark blue upperparts

Violet-green Swallow

adults

small size with fluttery wingbeats

♂

white up sides of rump

♂

juvenile

white above eye

adult ♂

mainly green upperparts and snowy white underparts

Common House-Martin
lagopoda

extensive white rump

glossy blue above

forked tail

smaller than Tree Swallow

Barn Swallow
erythrogaster

bluish above

Eurasian
rustica

complete breast band

white underparts

dark

juvenile

paler below than adult

long, forked tail with white at base

shorter tail

Northern Rough-winged Swallow

Stelgidopteryx serripennis L 5" (13 cm) Brown above, whitish below, with gray-brown wash on chin, throat, and upper breast. Lacks Bank Swallow's distinct breast band; wings are longer, wingbeats deeper and slower. **Juvenile** has cinnamon wing bars.

VOICE: Call is a low, buzzy *zzrtt*.

RANGE: Nests in single pairs in riverbanks, cliffs, culverts, and under bridges. Migrates singly or in small flocks. A few winter to coastal CA; very rare to rare to southeastern AK and YT.

Bank Swallow *Riparia riparia* L 4¾" (12 cm)

Our smallest swallow. Distinct brownish gray breast band, often extending in a line down center of breast. Throat is white; white curves around rear border of ear patch. **Juvenile** has thin buffy wing bars; compare with juvenile Northern Rough-winged Swallow and juvenile Tree Swallow (page 370). Locally common throughout most of range. Unlike Northern Rough-winged, wingbeats are shallow and rapid; also paler rump and back contrast with darker wings.

VOICE: Call is a series of buzzy, short *dzrrt* notes, not unlike the sound made by high-tension powerlines.

RANGE: Nests in large colonies, excavating nest burrows in steep riverbank cliffs, gravel pits, and highway cuts. Often migrates in large flocks. Scarce on West Coast. Winters chiefly in South America.

Cliff Swallow *Petrochelidon pyrrhonota* L 5½" (14 cm)

Squarish tail and buffy rump distinguish this swallow from all others except Cave Swallow. Most have dark chestnut and blackish throat, pale forehead. A primarily southwestern subspecies, *melanogaster,* has cinnamon forehead like Cave Swallow, but throat is dark chestnut; has been recorded as far east as south FL. All **juveniles** are much duller and grayer than adults; throat is paler, forehead darker.

VOICE: Calls include a rough, squeaky *chri,* a nasal *trrr,* and a rattle.

RANGE: Uncommon to locally common around bridges, rural settlements, and in open country on cliffs. Range has expanded greatly in East in last two decades. Nests in colonies, building gourd-shaped mud nests. An early spring and fall migrant. Winters in South America.

Cave Swallow *Petrochelidon fulva* L 5½" (14 cm)

Squarish tail; distinguished from Cliff by buffy throat color extending through auriculars and around nape setting off dark cap; rump averages a richer color; cinnamon forehead; juvenile much paler; compare with southwestern subspecies of Cliff also with cinnamon forehead. Two subspecies in North America: Mexican *pelodoma* is larger than West Indian *fulva,* which has less buff below and a paler rump.

VOICE: Call is a rising *pweih,* much sweeter than Cliff Swallow.

RANGE: West Indian subspecies, *fulva,* is a local breeder in south FL; Mexican *pelodoma* is widespread in the Southwest. Nests in colonies in limestone caves, sinkholes, culverts, and under bridges, sometimes with Barn and Cliff Swallows. Now regular on East Coast and southern Great Lakes (apparently all or nearly all are *pelodoma*) in late fall; sometimes in flocks. Casual to southern AZ, southeastern CA.

juvenile

cinnamon
wing bars

**Northern
Rough-winged
Swallow**

long wings; flies
with slow and
floppy wingbeats

dusky wash
on throat
and breast

flies with
shallow flaps

**Bank
Swallow**
riparia

small size

juvenile

pale coloring
wraps around
back of ear
coverts

dark
breast band

has cinnamon
forehead
like Cave

dusky throat and
forehead on
juvenile, often
with some whitish
feathers

juvenile

Southwest
melanogaster

white
forehead

buffy rump

dark
throat

**Cliff
Swallow**

Cliff and Cave
Swallows both have
fairly square-ended
tails and buffy rumps

sides and flanks more
cinnamon tinged than
pelodoma

fulva

**Cave
Swallow**

buffy rump

cinnamon
forehead

rump slightly
more cinnamon
than Cliff

juvenile

darker
cinnamon on
rump than
pelodoma

pale collar
contrasts with
dark cap

West Indies
fulva

adult
pelodoma

Southwest
pelodoma

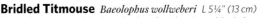

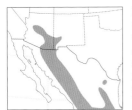

Bridled Titmouse *Baeolophus wollweberi* L 5¼" (13 cm)
Note distinct crest, black-and-white facial pattern, black throat.
VOICE: Most common call is a rapid, high-pitched variation of *chick-a-dee-dee*, similar to Juniper Titmouse. Song is a rapid and clipped series of whistled notes.
RANGE: Resident in stands of oak, juniper, and sycamore in Southwest range. Accidental to westernmost AZ (17 Feb. to 20 Mar, 1977, Bill Williams Delta, specimen).

Oak Titmouse *Baeolophus inornatus* L 5" (13 cm)
Grayish brown with a short crest. Northern subspecies, *inornatus,* is slightly smaller, paler, and smaller billed than *affabilis* (shown here) from southwestern CA and northern Baja California; birds from Little San Bernadino Mountains are paler and grayer than *affabilis.*
VOICE: Song is variable, a repeated series of syllables made up of whistled, alternating, high and low notes. Call is a hoarse *tschick-a-dee.*
RANGE: Common in warm, dry oak woodland in foothills and lower mountains. Casual to southeastern CA.

Tufted Titmouse *Baeolophus bicolor* L 6¼" (16 cm)
Note gray crest and distinct blackish forehead. **Juvenile** has brownish forehead and pale crest. In overlap zone in TX, hybrids with Black-crested Titmouse show variable brown foreheads, dark gray crests. Active and noisy.
VOICE: Typical song is a loud, whistled *peter peter peter;* less vocal than Black-crested; calls softer and less nasal.
RANGE: Deciduous woodlands, parks, and suburbs. Regularly comes to feeders. Range has expanded northward.

Juniper Titmouse *Baeolophus ridgwayi* L 5¼" (13 cm)
Like Oak Titmouse but larger, paler, and grayer; range overlaps on northern CA Modoc plateau.
VOICE: Song is a rolling series of syllables, rapid, with uniform pitch. Call is a hoarse *tschick-a-dee* similar to Bridled Titmouse. Overall, chattering call notes are more clipped and delivered much more rapidly than Oak Titmouse.
RANGE: Uncommon to fairly common in juniper or pinyon-juniper woodland. Casual to western Great Plains.

Black-crested Titmouse *Baeolophus atricristatus*
L 5¾" (15 cm) Resplit from Tufted Titmouse. **Adult** has black crest, pale forehead. **Juvenile** crown darker than upperparts; forehead dirty white.
VOICE: Calls louder, sharper than Tufted. Song is also like Tufted, but notes are slightly higher and are delivered more rapidly.
RANGE: Fairly common resident in a variety of woodland and scrub habitats.

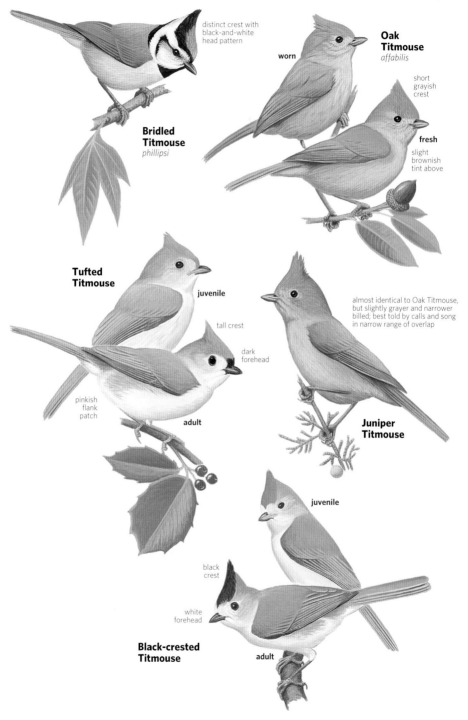

distinct crest with black-and-white head pattern

Bridled Titmouse
phillipsi

Oak Titmouse
affabilis

worn

short grayish crest

fresh

slight brownish tint above

Tufted Titmouse

juvenile

tall crest

dark forehead

pinkish flank patch

adult

almost identical to Oak Titmouse, but slightly grayer and narrower billed; best told by calls and song in narrow range of overlap

Juniper Titmouse

juvenile

black crest

white forehead

Black-crested Titmouse

adult

Black-capped Chickadee *Poecile atricapillus* L 5¼" *(13 cm)*

Black cap and bib; cheeks more extensively and purer white than similar Carolina Chickadee. Note that Black-capped Chickadee's greater wing coverts and secondaries are broadly edged in white; tertials more boldly edged, with darker centers than Carolina; flanks more olive, lower edge of black bib a bit more ragged. These differences are obscured in **worn summer** birds. Plumage is geographically variable: More northerly Black-cappeds tend to be larger and frostier, more distinct from Carolina; *occidentalis* from Pacific Northwest is darker; *nevadensis* from Great Basin is palest subspecies.

VOICE: Best distinction is voice. Call is a lower, slower *chick-a-dee-dee* than Carolina; typical song, a clear, whistled *fee-bee* or *fee-bee-ee,* the first note higher in pitch; vocalizations show considerable geographic variation.

RANGE: Common in open woodlands and suburbs. Usually forages in thickets and low branches of trees. The usual ranges of Black-capped and Carolina barely overlap, but periodic fall irruptions push Black-capped's range south of mapped range. Most frequently from eastern KY to eastern MD. Where breeding ranges overlap, the two species hybridize. In the Appalachians, Black-capped inhabits higher elevations.

Carolina Chickadee *Poecile carolinensis* L 4¾" *(12 cm)*

Very similar to Black-capped Chickadee: black cap and bib, white cheeks. Note that Carolina lacks broad white edgings on greater wing coverts; lower edge of black bib is usually neater, has less olive on flanks than Black-capped. Westernmost subspecies, *atricapilloides,* is grayer than nominate.

VOICE: Best distinction for separating species is voice. Call is a higher, faster version of *chick-a-dee-dee-dee* than Black-capped; typical song is a four-note whistle, *fee-bee fee-bay.*

RANGE: Common in open deciduous forests and suburban areas.

Mexican Chickadee *Poecile sclateri* L 5" *(13 cm)*

The only breeding chickadee in its range. Extensive black bib is distinctive, along with dark gray flanks. Lacks white eyebrow of Mountain Chickadee.

VOICE: Song is a warbled whistle; call note, a husky buzz.

RANGE: A Mexican species, fairly common resident in coniferous and pine-oak forests; found in U.S. only in Chiricahua Mountains of AZ and Animas and Peloncillo Mountains of NM.

Mountain Chickadee *Poecile gambeli* L 5¼" *(13 cm)*

White eyebrow and pale gray sides distinguish this species from other chickadees; lack of crest separates it from Bridled Titmouse (page 374). Birds of Rocky Mountain nominate subspecies *gambeli* are tinged with buff on back, sides, flanks, and have broader white eyebrow than *baileyae.*

VOICE: Call is a hoarse *chick-adee-adee-adee;* typical song, a three- or four-note descending whistle, *fee-bee-bay* or *fee-bee fee-bee.*

RANGE: Common resident in coniferous and mixed woodlands. Some irregularly descend to lower elevations in winter. Casual to Great Plains.

darkest
subspecies

**Black-capped
Chickadee**
atricapillus

worn
summer

occidentalis

palest
subspecies

nevadensis

prominent
white-edged
secondaries

uniform
white
cheeks

**fresh
fall**

warm colored
wash on flanks

atricapilloides

grayish
tinge to
rear cheek

**Carolina
Chickadee**
extimus

worn
summer

slightly shorter tail
than Black-capped

**fresh
fall**

slightly duller
flanks than
Black-capped

**Mexican
Chickadee**
eidos

extensive
black bib

broad gray
sides and
flanks

white
eyebrow

**Mountain
Chickadee**

broader white
eyebrow in
gambeli

baileyae

Rockies
gambeli

warmer-colored
flanks

Chestnut-backed Chickadee *Poecile rufescens*

L 4¾" (12 cm) Sooty brown cap, white cheeks, chestnut above and on most sides and flanks; *barlowi,* south of Golden Gate Bridge, shows almost no chestnut below.

VOICE: Call is a hoarse, rapid *tseek-a-dee-dee.*

RANGE: Found in coniferous forests, deciduous woodlands. Casual to Ventura County, CA, and east to western AL.

Boreal Chickadee *Poecile hudsonicus* L 5½" (14 cm)

Grayish brown on crown and back, with pinkish brown flanks. Note that rear portion of cheeks is heavily washed with gray.

VOICE: Call is a nasal *tseek-a-day-day.*

RANGE: Uncommon in coniferous forests. In some winters, small numbers wander south of normal eastern range, casually to IA, IL, OH, northern VA, MD, and DE, but few recent records.

Gray-headed Chickadee *Poecile cinctus* L 5½" (14 cm)

Gray-brown above, white cheek patch, and buffy sides and flanks. Distinguished from Boreal Chickadee by more extensively white cheeks, longer tail, paler flanks, and pale edges on wing coverts.

VOICE: Call is a distinctive series of *dee-deer* notes.

RANGE: Rare; found in willows and spruces edging tundra. Known as the Siberian Tit in the Old World. Casual to central AK (Fairbanks).

PENDULINE TITS • VERDINS Family Remizidae

These small, spritely birds with finely pointed bills inhabit arid scrub country, feed in brush chickadee-style, and build spherical nests. SPECIES: 13 WORLD; 1 N.A.

Verdin *Auriparus flaviceps* L 4½" (11 cm)

Adult has dull gray plumage, chestnut shoulder, yellow head; shorter tail than Bushtit. **Juvenile** lacks yellow head and chestnut shoulder.

VOICE: Song is a plaintive three-note whistle, the second note higher. Calls include rapid chip notes.

RANGE: Common in mesquite and other dense thorny shrubs. Casual on southern CA coast to Santa Barbara County.

LONG-TAILED TITS • BUSHTITS Family Aegithalidae

Tiny and long tailed. Except when breeding, feeds in large, twittering flocks. Builds an elaborate hanging nest. SPECIES: 10 WORLD; 1 N.A.

Bushtit *Psaltriparus minimus* L 4½" (11 cm)

Very small and long tailed. Gray above, paler below; female has pale eyes, male's eyes are dark. Coastal birds have brown crown; interior birds show brown ear patch and gray cap. **Juvenile male** and some adult males in the Southwest near Mexican border have a black mask (**"Black-eared Bushtit"**).

VOICE: Calls from coastal slope birds are a rapid, soft twittering, given in loud, excited chatter when a raptor appears; interior birds give sharper notes that are delivered more slowly.

RANGE: Common in a wide variety of woodlands and brushy areas; casual in fall and winter to western KS.

rich chestnut back, sides, and flanks

pure white cheeks

rufescens

Boreal Chickadee
hudsonicus

juvenile

grayish rear cheek

brownish cap

worn summer

chestnut back

Chestnut-backed Chickadee

pinkish buff flanks

plainer, more uniform greater coverts

fresh fall

gray sides

coastal central California
barlowi

pure white cheeks

black ear coverts

"Black-eared Bushtit" juvenile ♂
plumbeus

white edgings to greater coverts and secondaries

longish tail

slightly paler flanks than Boreal

Gray-headed Chickadee
lathami

gray crown

interior ♂
plumbeus

pinkish buff face

Verdin

yellow head, black eye line

sharply pointed bill

chestnut lesser coverts

juvenile

very plain, lacks yellow head

female with pale eye

long tail

interior ♀
plumbeus

Bushtit

grayish face

coastal ♂
minimus

brown crown

NUTHATCHES Family Sittidae

These short-tailed acrobats climb up, down, and around tree trunks and branches.

SPECIES: 24 WORLD; 4 N.A.

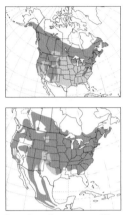

Red-breasted Nuthatch *Sitta canadensis* L 4½" (11 cm)

Black cap and eye line, white eyebrow, rust underparts; **female** and juveniles have duller head, paler underparts.

VOICE: High-pitched, nasal call sounds like a toy tin horn.

RANGE: Resident in northern and montane conifers; gleans small branches and outer twigs. Irruptive migrant; numbers and winter range vary yearly. In the East, resident range is expanding slightly southward.

White-breasted Nuthatch *Sitta carolinensis* L 5¾" (15 cm)

Larger than Red-breasted; white underparts with chestnut in vent, bluish gray above; **males** with black caps. Three distinct geographical groups, differing most in vocalizations. Eastern *carolinensis* has bluer tone to upperparts and moderate length bill; well-defined black tertial centers; males have extensive black cap. Birds of *lagunae* group from Rockies and Great Basin ranges (in North America includes *tenuissima* from Great Basin and Rockies *nelsoni*) and the Pacific slope *aculeata* group have longer, more slender bills; whiter faces; somewhat less bluish above; more blended tertial centers. These two groups very similar.

See subspecies map, page 553

VOICE: Song of eastern *carolinensis* and Pacific *aculeata* is a series of whistles on one pitch; lower and richer in *carolinensis*. Call is a low-pitched *yank* in *carolinensis,* a higher-pitched *wheer* in *aculeata* and a rising series of even higher *yida* notes in *lagunae* group. Song of latter varied, but often consists of a series of their calls.

RANGE: Fairly common; found in leafy trees in East, oaks and conifers, including pinyon-juniper, in West. Eastern *carolinensis* uncommon and local in southern portion of Southeast; found along river systems well west on Great Plains. Rockies and Great Basin *lagunae* group goes over Sierra Nevada crest to about 7,000 feet on west side; accidental coastal southern CA. Pacific *aculeata* gets up to about 4,000 feet on west side of Sierra, higher in southern CA mountains; formerly north to Tacoma, WA, region; rare (Mojave) to casual (Colorado) in fall on CA deserts.

Brown-headed Nuthatch *Sitta pusilla* L 4½" (11 cm)

Brown cap; dull buff underparts. Pale nape spot visible at close range. Narrow dark eye line borders cap.

VOICE: Call is a repeated double note like the squeak of a rubber duck. Feeding flocks also give twittering, chirping, and talky *bit bit bit* calls.

RANGE: Fairly common; found in pine woodlands. Casual to KY; accidental north to southeastern NE, WI, northern IL, OH, and NJ.

Pygmy Nuthatch *Sitta pygmaea* L 4¼" (11 cm)

Gray-brown cap; creamy buff underparts. Pale nape spot visible at close range. Dark eye line bordering cap, more indistinct in coastal *pygmaea.*

VOICE: Typical calls, a high *peep* or a rapid *peep peep;* also a piping *wee-bee;* grouped in three or more notes in nominate subspecies, *pygmaea.*

RANGE: Favors yellow-pine forest, except for birds in coastal CA pines. Roams in loose flocks. Casual fall and winter visitor to lowlands and east to Great Plains.

Red-breasted Nuthatch

grayer cap and eye line on female

♀

duller below than male

deep cinnamon underparts

prominent white supercilium

♂

well-defined black center to tertials

bluish upperparts

black cap on male

white face contrasts sharply with black cap and nape

White-breasted Nuthatch

eastern
carolinensis

grayer brown cap

♀

Brown-headed Nuthatch

longer, more slender bill

brown cap

Great Basin
♂ *tenuissima*

grayish brown cap

darker eye line

creamy buff below

Pygmy Nuthatch

CREEPERS Family Certhiidae
With curved bills, these little tree-climbers dig insects and larvae from bark. Stiff tail feathers serve as props. SPECIES: 8 WORLD; 1 N.A.

Brown Creeper *Certhia americana* L 5¼" *(13 cm)*
Camouflaged by streaked brown plumage, spirals upward from base of a tree, then flies to a lower place on another tree. Coloration of upper- and underparts is variable both geographically and individually. One subspecies, *phillipsii,* from central CA coast, is buffy gray underneath except for throat. Another subspecies, *albescens,* from mountains of southeastern AZ and southwestern NM is dark with sooty gray under-parts and contrasting white throat; blackish brown above, streaked with white. Recent genetic studies reveal this subspecies and others from on down into Middle America may be a separate species.
VOICE: Call is a soft, sibilant *see;* song, a high-pitched, variable *see see see titi see;* vocalizations of *albescens* appear distinct.
RANGE: Nests in coniferous, mixed, or swampy forests. Fairly common. Generally solitary but joins winter flocks of titmice and nuthatches. Rather rare at southern end of mapped range in East; casual to south FL.

WRENS Family Troglodytidae
Found throughout most of North America, wrens are chunky birds with slender, slightly curved bills. Tails are often uptilted. Loud song and vigorous territorial defense belie the small size of most species. SPECIES: 79 WORLD; 11 N.A.

Rock Wren *Salpinctes obsoletus* L 6" *(15 cm)*
Dull gray-brown above with contrasting cinnamon rump, buffy tail tips, broad blackish tail band. Breast finely streaked. Frequently bobs its body, especially when alarmed.
VOICE: Song is a variable mix of buzzes and trills; call is a buzzy *tick-ear* or *dzeeee.*
RANGE: Fairly common in arid and semiarid habitats, sunny talus slopes, scrublands, and dry washes. Casual in fall and winter to the East. Accidental to NT.

Canyon Wren *Catherpes mexicanus* L 5¾" *(15 cm)*
White throat and breast, chestnut belly. Long bill aids in extracting insects from deep crevices.
VOICE: Loud, silvery song, a decelerating, descending series of liquid *tee* and *tew* notes. Typical call is a sharp, buzzy *jeet.*
RANGE: Fairly common in canyons and cliffs, often near water; may also build its cup nest in stone buildings and chimneys.

Cactus Wren *Campylorhynchus brunneicapillus* L 8½" *(22 cm)*
Large; dark crown, streaked back, heavily barred wings and tail, broad white eyebrow. Breast is densely spotted with black; threatened coastal CA subspecies, *sandiegense,* is less densely spotted.
VOICE: Song, heard all year, is a low-pitched, harsh, rapid *cha cha cha cha cha.* Call is a variety of low, croaking notes, sometimes in a series.
RANGE: Uncommon to common in cactus country. Bulky nests are tucked into the protective spines of cholla cactus or thorny bushes.

blackish above with whitish spotting

thin, curved bill

brownish upperparts with pale spotting

Brown Creeper

sooty tinged underparts

white underparts

rufous rump and long, spiky tail

southwestern
albescens

eastern
americana

buffy tail tips

Rock Wren

cinnamon rump

spotted grayish crown

Canyon Wren

very long bill

white throat

rufous tail

faintly streaked breast

extensively chestnut belly

bold white supercilium

Cactus Wrens build their large ball-shaped nest usually among spiny cactus branches

densely spotted

very large size

Cactus Wren

banded tail

House Wren *Troglodytes aedon* L 4¾" (12 cm)

Separated from smaller Winter Wren by longer tail, less prominent barring on belly. Western *parkmanii* breeds east to ON; grayer above, paler below. **Juvenile** shows a bright rufous rump and darker buff below. Birds from mountains of southeastern AZ, formerly recognized as a distinct species, **"Brown-throated Wren,"** have a slightly buffier throat and breast and a bolder eyebrow. Populations farther south in Mexico much more distinct.

VOICE: Exuberant song is a cascade of bubbling whistled notes; calls include a soft *chek* and a harsh scold.

RANGE: Common in a wide variety of habitats. Winters rarely north into summer range.

Winter Wren *Troglodytes hiemalis* L 3½" (9 cm)

Very small size and short tail characterize this and Pacific Wren, formerly treated as one species. Both are rather secretive.

VOICE: Gives a *kelp* call, often doubled. Song is a beautiful series of musical trills.

RANGE: Nests in dense brush, ferns, and tree-falls, especially along stream banks, in moist coniferous woods; in winter found in woodland understory. Casual to West in late fall and winter.

Pacific Wren *Troglodytes pacificus* L 3½-4½ (9-11 cm)

Like Winter Wren but much richer colored, especially on throat and breast.

VOICE: Call is a *chimp,* often doubled, quality like Wilson's Warbler. Song is like Winter Wren, but faster, less melodic.

RANGE: Found in coniferous understory and on tundra of AK islands. Rather rare and local away from Pacific region breeding range. Rare to casual in fall and winter to Southwest.

Bewick's Wren *Thryomanes bewickii* L 5¼" (13 cm)

Long, sideways-flitting tail, edged with white spots; long white eyebrow. Subspecies differ mainly in dorsal color: Eastern *bewickii,* is reddish brown above; south TX *cryptus* (not shown) duller, but still tinged red. Widespread *eremophilus* of the western interior is the grayest; western coastal subspecies grow browner and darker as one travels north. Northwest *calaphonus* is dark, richly colored, with a rufous cast above.

VOICE: Song variable, a high, thin buzz and warble, similar to Song Sparrow. Calls include a flat, hollow *jip.*

RANGE: Found in brushland, hedgerows, stream edges, open woods, and clear-cuts in the East. Sharply declining east of the Rockies, especially east of Mississippi River, where extirpated over most of range.

Carolina Wren *Thryothorus ludovicianus* L 5½" (14 cm)

Deep rusty brown above, warm buff below; white throat and prominent white supercilium.

VOICE: Vivacious, melodious song, a loud, clear *teakettle tea-kettle teakettle* or *cheery cheery cheery.* Sings all year.

RANGE: Common in underbrush of moist woodlands and swamps, wooded suburbs. Nonmigratory, but after mild winters resident populations expand north of mapped range. After harsh winters, range limits retract. Casual to CO, NM, and AZ.

tail of moderate length

juvenile

House Wren

western
parkmanii

indistinct head pattern

eastern
aedon

"Brown-throated Wren"
southeastern Arizona

overall coloration duller and paler than Pacific Wren; vocalizations diagnostic

Winter Wren
hiemalis

larger, with much longer bill

Pacific Wren

both Pacific and Winter Wrens are tiny and have very short tails

Aleutians

darker and more richly colored above than Winter Wren

much richer buff below

pacificus

warm ruddy brown above

bold white supercilium

white bars in long tail

eastern
bewickii

grayish white underparts

Bewick's Wren

bold white supercilium

Carolina Wren

rufous-brown upperparts

white bars on long tail

grayish above

bold white supercilium

northwest
calaphonus

more richly colored above than other western subspecies

extensively rich buffy underparts

pale gray underparts

western interior
eremophilus

darker gray on belly than other subspecies

Marsh Wren *Cistothorus palustris* L 5" *(13 cm)*

Much plumage variation in eastern and western subspecies. Where ranges overlap on Great Plains, eastern birds darker, more richly colored, with black-and-white speckled neck; western birds duller.

VOICE: Songs are a mechanical-sounding mix of bubbling and trilling notes; more liquid in the East, harsher and much more variable in the West. Alarm call, a sharp *tsuk,* often doubled.

RANGE: Common in reedy marshes and cattail swamps. Football-shaped nest attached to reeds above water. Accidental to northwestern Canada; accidental to AK.

Sedge Wren *Cistothorus platensis* L 4½" *(11 cm)*

Often difficult to see. Crown and back streaked; eyebrow whitish and indistinct; underparts largely buff.

VOICE: Song begins with a few single notes followed by a weak staccato trill or chatter; call note, a rich chip, often doubled.

RANGE: Found in wet meadows or sedge marshes. Globular nest similar to that of Marsh Wren. Generally common but local on Great Plains; uncommon to rare in the East. Winters also in upper salt marshes; most numerous along western Gulf Coast. Rare and local in winter to NM. Casual to CA in late fall.

DIPPERS Family Cinclidae

Aquatic birds that wade and even swim underwater in clear, rushing mountain streams to feed. SPECIES: 5 WORLD; 1 N.A.

American Dipper *Cinclus mexicanus* L 7½" *(19 cm)*

Adult sooty gray; dark bill; tail and wings short. **Juvenile** has paler, mottled underparts and pale bill.

VOICE: Song is loud, musical, wrenlike.

RANGE: Found along mountain streams. Descends to lower elevations in winter; casual vagrant well outside mapped range.

KINGLETS Family Regulidae

Small, active birds that often hover to feed. SPECIES: 6 WORLD; 2 N.A.

Golden-crowned Kinglet *Regulus satrapa* L 4" *(10 cm)*

Orange crown patch of **male** is bordered in yellow and black; **female's** crown is yellow. Paler below than Ruby-crowned Kinglet.

VOICE: Call is a series of high, thin *tsee* notes. Song, almost inaudibly high, is a series of *tsee* notes accelerating into a trill.

RANGE: Breeds in coniferous woodlands; also found in deciduous woods in migration and winter.

Ruby-crowned Kinglet *Regulus calendula* L 4¼" *(11 cm)*

Male's red crown patch seldom visible; dusky underparts. Active; flicks wings rapidly.

VOICE: Call is a scolding *je-ditt.* Song, high, thin *tsee* notes followed by descending *tew* notes, ends with warbled three-note phrases.

RANGE: Common. Breeds in coniferous woodlands; in migration and winter also found in woodlands and thickets.

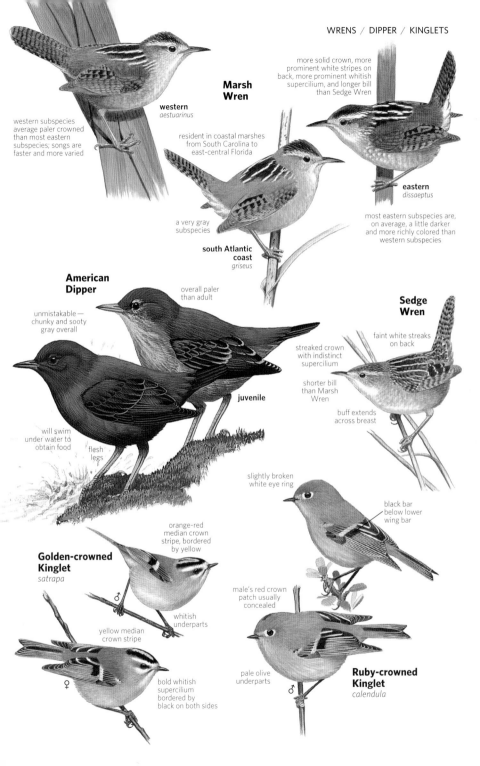

Marsh Wren

more solid crown, more prominent white stripes on back, more prominent whitish supercilium, and longer bill than Sedge Wren

western
aestuarinus

western subspecies average paler crowned than most eastern subspecies; songs are faster and more varied

resident in coastal marshes from South Carolina to east-central Florida

eastern
dissaeptus

most eastern subspecies are, on average, a little darker and more richly colored than western subspecies

a very gray subspecies

south Atlantic coast
griseus

American Dipper

unmistakable — chunky and sooty gray overall

overall paler than adult

Sedge Wren

faint white streaks on back

streaked crown with indistinct supercilium

shorter bill than Marsh Wren

buff extends across breast

juvenile

will swim under water to obtain food

flesh legs

slightly broken white eye ring

black bar below lower wing bar

orange-red median crown stripe, bordered by yellow

Golden-crowned Kinglet
satrapa

whitish underparts

yellow median crown stripe

bold whitish supercilium bordered by black on both sides

male's red crown patch usually concealed

pale olive underparts

Ruby-crowned Kinglet
calendula

SYLVIID WARBLERS Family Sylviidae

This large, almost strictly Old World family comprises 14 genera, including the one New World species, Wrentit, and one vagrant to North America. Many are neatly patterned, and many more are rather colorful. SPECIES: 76 WORLD; 2 N.A.

Wrentit *Chamaea fasciata* L 6½" *(17 cm)*
Color varies from reddish brown in **northern** populations to grayer in **southern** birds. Note distinct cream-colored eye and lightly streaked buffy breast; long, rounded tail usually cocked.
VOICE: Usually heard before seen. Male's loud song, sung year-round, begins with a series of accelerating notes and runs into a descending trill: *pit-pit-pit-tr-r-r-r.* Female's song lacks trill. Both sexes give soft, low *prr* note, often in a series.
RANGE: Common in chaparral and coniferous brushland.

GNATCATCHERS • GNATWRENS Family Polioptilidae

A New World family of small, active birds with long tails, which are usually cocked. Gnatcatchers are mostly shades of blue, white, and gray. SPECIES: 15 WORLD; 4 N.A.

Blue-gray Gnatcatcher *Polioptila caerulea* L 4¼" *(11 cm)*
Long tail with white outer tail feathers is not graduated. **Male** is bluish above, in **breeding** plumage has black line on sides of crown. **Female** is grayer. Active, often joins mixed-species feeding flocks.
VOICE: Call is a querulous *pwee.* Varied song of high, thin notes, chips and buzzy notes; harsher in slightly duller western *amnoenissima.*
RANGE: Favors woodlands, thickets, and chaparral.

Black-capped Gnatcatcher *Polioptila nigriceps*
L 4¼" *(11cm)* Separated from Blue-gray and Black-tailed Gnatcatchers by more graduated white outer tail feathers and longer bill; **breeding male's** black cap extends below eye. **Female** and winter male best identified by tail shape and pattern and by voice.
VOICE: Calls like California Gnatcatcher or Bewick's Wren.
RANGE: Usually found a little higher than Black-tailed in foothill canyons. West Mexican species. Very rare in U.S.; numbers fluctuate.

Black-tailed Gnatcatcher *Polioptila melanura* L 4" *(10 cm)*
White terminal spots on graduated tail feathers; short bill. **Breeding male** has glossy black cap, contrasting with eye ring. **Female** washed with brown.
VOICE: Calls include rasping *cheeh* and hissing *ssheh;* song is a rapid series of *jee* notes; all vocalizations harsher than Blue-gray.
RANGE: Desert resident; partial to washes.

California Gnatcatcher *Polioptila californica* L 4¼" *(11 cm)* **T**
Similar to Black-tailed, but is darker with less white in outer tail feathers, less distinct eye ring.
VOICE: Call is a rising and falling, kitten-like *zeeer;* song is a series of *jzer* or *zew* notes.
RANGE: Local resident in sage scrub of southwestern CA. Dark northern nominate subspecies now threatened, due to habitat destruction.

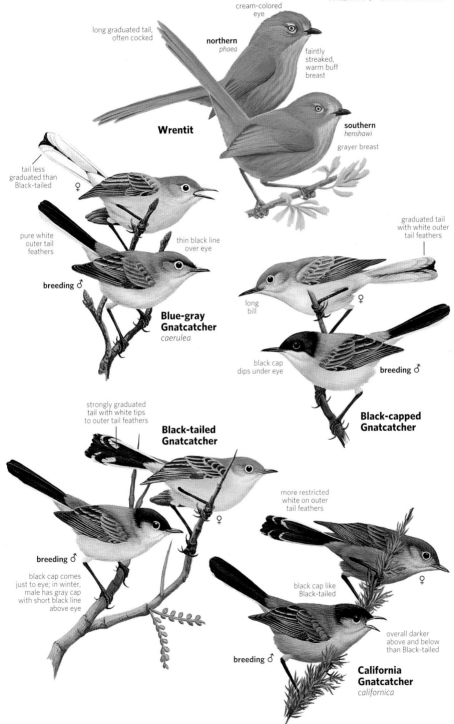

cream-colored
eye

long graduated tail,
often cocked

northern
phaea

faintly
streaked,
warm buff
breast

Wrentit

southern
henshawi

grayer breast

tail less
graduated than
Black-tailed

♀

pure white
outer tail
feathers

breeding ♂

thin black line
over eye

**Blue-gray
Gnatcatcher**
caerulea

graduated tail
with white outer
tail feathers

long
bill

♀

black cap
dips under eye

breeding ♂

**Black-capped
Gnatcatcher**

strongly graduated
tail with white tips
to outer tail feathers

**Black-tailed
Gnatcatcher**

♀

more restricted
white on outer
tail feathers

breeding ♂

black cap comes
just to eye; in winter,
male has gray cap
with short black line
above eye

black cap like
Black-tailed

♀

overall darker
above and below
than Black-tailed

breeding ♂

**California
Gnatcatcher**
californica

LEAF WARBLERS Family Phylloscopidae
This large Old World family of small, mostly greenish birds includes the *Phylloscopus* and the more colorful *Seicerus* genera. Many are difficult to identify. SPECIES: 69 WORLD; 6 N.A.

Yellow-browed Warbler *Phylloscopus inornatus* L 4½" (11 cm)
Small billed with bold supercilium, two pale wing bars, sharply defined edges on tertials.
VOICE: Call, upslurred *swee-eet,* suggests male Pacific-slope Flycatcher.
RANGE: Breeds northeastern Palearctic. Casual in fall to Attu, St. Paul, and St. Lawrence Islands, AK.

Willow Warbler *Phylloscopus trochilus* L 4½" (11 cm)
Like Arctic Warbler, but smaller, with smaller bill and a plainer wing.
VOICE: Whistled *hoo-eet* call note is very unlike Arctic's buzzy note.
RANGE: A highly migratory Old World species. Several fall records for St. Lawrence Island, AK.

Dusky Warbler *Phylloscopus fuscatus* L 5½" (14 cm)
Dusky brown color and lack of wing bar distinguish this species from Arctic Warbler; distinct eyebrow; dull white in front of eye.
VOICE: Call, a hard, sharp *tschick,* suggests Lincoln's Sparrow.
RANGE: Asian species, casual on islands off western AK, and in fall on Middleton Island, AK, and in CA.

Arctic Warbler *Phylloscopus borealis* L 5" (13 cm)
Distinct supercilium; olive above, pale wing bar, stouter bill than Yellow-browed. Breeding subspecies in AK is smaller, small-billed *kennicotti*; Aleutian specimens of migrants are larger *xanthrodyas* with broader bill.
VOICE: Song is a long, loud series of toneless buzzy notes. Calls include a buzzy *dzik.*
RANGE: Breeds in willow thickets. Four fall records for coastal CA.

GRASSBIRDS Family Megaluridae
Old World family of medium- to large-size, mostly skulking birds. SPECIES: 52 WORLD; 2 N.A.

Middendorff's Grasshopper-Warbler
Locustella ochotensis L 6" (15 cm) Chunky with whitish-tipped, wedge-shaped tail. Indistinct dark markings above; yellowish buff below, faintly streaked breast, rustier above. Very secretive.
RANGE: East Asian species, casual migrant to westernmost Aleutians primarily in fall; also recorded in Bering Sea region from Nunivak and St. Lawrence Islands and the Pribilofs.

Lanceolated Warbler *Locustella lanceolata* L 4½" (11 cm)
Smaller than Middendorff's; streaks extend to feather tips; clear brown fringe on tertials. Extensively streaked below. Highly secretive. Walks and runs; flicks its wings.
VOICE: Distinctive call, a metallic *rink-tink-tink,* delivered infrequently; also an excited *chack.* Song, a thin, reeling sound, like a fishing line.
RANGE: Mainly Asian species. Many occurred in spring and summer of 1984 on Aleutian island of Attu; a few since. Accidental in fall in CA.

small bill

bold wing bars

distinct whitish tertial edges

Yellow-browed Warbler
1st fall *inornatus*

Willow Warbler
adult *yakutensis*

overall plain with no wing bars

small bill

St. Lawrence Island birds have been more olive above, yellowish below

long primary projection

small fine bill

long pale supercilium

Dusky Warbler

brownish upperparts

long, pale supercilium

faint pale wing bar

olive upperparts

breeding Alaska subspecies

spring
kennicotti

larger bill

Arctic Warbler

more distinct whitish supercilium than Lanceolated

dull rufous rump

faint back streaks

fall
xanthodryas

yellower underparts

Middendorff's Grasshopper-Warbler

spring

Middendorff's and Lanceolated, like all *Locustella* and most members of this family, are very skulking

wedge-shaped tail with pale tips

overall more richly colored than in spring; can be yellow-buff tinged below

fall

Lanceolated Warbler

strong back streaks

faint supercilium

pale tertial fringes

wedge-shaped tail

adult

distinctly streaked below, but variable; streaking duller on immatures

OLD WORLD FLYCATCHERS Family Muscicapidae

Short-legged birds that perch upright and obtain insects primarily through flycatching. May flick wings or tail. Species of genus *Ficedula* nest in cavities; genus *Muscicapa* build exposed nests. Not related to New World tyrant flycatchers. SPECIES: 271 WORLD; 14 N.A.

Narcissus Flycatcher *Ficedula narcissina* L 5¼" (13 cm)

Adult male overall black and yellow-orange; most orange on eyebrow and throat. Has yellow rump; white patch on inner secondary coverts. **First-spring male** similar, but duller. **Female** drab; brownish olive above, with green on rump; contrasting reddish-tinged uppertail coverts and tail; whitish throat; brownish mottling on breast. First-fall male similar to female.

RANGE: East Asian species; two spring records of males on Attu Island.

Taiga Flycatcher *Ficedula albicilla* L 5¼" (13 cm)

Distinct white oval patches at base of outer tail feathers visible in flight, barely visible on folded tail from below; prominent eye ring. **Breeding male** with reddish throat. **Females** and winter males have whitish throats; grayish wash on breast. All show extensive patch of black on uppertail coverts. Formerly considered a subspecies of Red-breasted Flycatcher (*F. parva*). Perches low; often drops to ground to catch prey, then returns to perch. Frequently flicks tail up.

VOICE: Gives rattled *trrt* call; also a metallic *tic* and harsh *ze-it*.

RANGE: Asian species; casual in late spring to western Aleutians; one spring and one fall record on St. Lawrence Island, AK; one fall record from CA (25 Oct 2006, Putah Creek).

Dark-sided Flycatcher *Muscicapa sibirica* L 5¼" (13 cm)

Dark grayish brown upperparts and wash on sides and flanks; center of breast diffusely streaked. Whitish half collar; brownish supraloral spot; short bill; long primary projection; dark centers on undertail coverts may be concealed. Northern nominate subspecies, *sibirica*, is darker and more diffusely streaked than southern subspecies.

RANGE: Asian species; casual to western Aleutians; four spring records for Pribilofs; one fall record for Bermuda.

Asian Brown Flycatcher *Muscicapa dauurica* L 5¼" (13 cm)

Grayish brown above; largely whitish below; grayish wash across chest or, rarely, some diffuse streaks. Bill larger than Dark-sided or Gray-streaked, extensively flesh-colored at base of lower mandible; primary projection shorter; supraloral area paler.

RANGE: Asian species; two spring records: one from Attu Island, Aleutians; the other from St. Lawrence Island, AK.

Gray-streaked Flycatcher *Muscicapa griseisticta*

L 6" (15 cm) Larger and with smaller head than Dark-sided; primary projection longer. Note distinctive, but variable, streaking below; paler supraloral spot; more distinct submoustachial stripe usually shows some markings; undertail coverts white.

RANGE: East Asian species; casual to western Aleutians in late spring and fall; also several Pribilof records in spring and fall.

unmistakable black and yellow-orange with white wing patch

olive above

Narcissus Flycatcher

Narcissus ♀

adult ♂

spring ♀

reddish-fringed tail

1st spring ♂

Taiga Flycatcher

spring ♀

grayish wash on breast

brownish sides and flanks

Dark-sided Flycatcher
sibirica

reddish throat

spring

breeding adult ♂

long primary projection

1st fall

1st fall

♀

black uppertail coverts and tail with white oval patches, best viewed in flight, in all plumages

Asian Brown Flycatcher
dauurica

distinct whitish eye ring

spring

unstreaked underparts

pinkish-based bill

1st fall

shorter primary projection

spots form distinct streaks on white underparts

spring

1st fall

very long primary projection

Gray-streaked Flycatcher

Siberian Rubythroat *Luscinia calliope* L 6" (15 cm)

Male has a ruby red throat and broad, white submoustachial stripe. **Female** has white throat, often with some pink on adult and buffy on immature; compare with smaller Bluethroat, which has rufous tail patches, dark breast band, paler underparts.

VOICE: Calls include a deep, low *chuck* and a loud, whistled *quee-ah;* song, long and complex, includes warbles and lisping notes.

RANGE: Asian species; rare spring and fall migrant on western Aleutians, very rare on Pribilofs, casual on St. Lawrence Island, AK. Accidental in ON in late fall.

Bluethroat *Luscinia svecica* L 5½" (14 cm)

Colorful throat pattern distinguishes **breeding male** from all other birds. In all plumages, rufous patches at base of tail conspicuous in flight. In **female** and immature, note dark breast band. Runs on ground, usually with tail cocked. Generally furtive.

VOICE: In courtship, males sing from high perches and in elaborate display flight. Varied, melodious song often begins with a crisp, metallic *ting ting ting;* call, *tchak,* often given in a series; mimics other species.

RANGE: Uncommon; nests in tundra thickets. Regular migrant on St. Lawrence Island; casual on Pribilofs and western Aleutians. Accidental in CA in fall.

Red-flanked Bluetail *Tarsiger cyanurus* L 5½" (14 cm)

Note bluish tail, often flicked down; orangish flanks. **Adult male** has bright blue upperparts, but much individual variation; brightest birds may be several years old. Immature males closely resemble **females** until second fall. Rather secretive.

VOICE: Calls include a *hueet* and dry *keck-keck.*

RANGE: Primarily Asian species; casual to western Aleutians and Pribilofs, mainly in spring; one fall record for Farallones, off CA.

Northern Wheatear *Oenanthe oenanthe* L 5¾" (15 cm)

Tail pattern distinctive: white rump, tail with dark central and terminal band. Greenland and eastern Canadian Arctic subspecies, *leucorhoa,* averages a little larger, richer buff below; western birds whitish, with buff tinge. **Males** in fall and winter resemble females. Active; bobs tail.

VOICE: Calls include *chak* and whistled *wheet.* Song, a scratchy warbling mixed with call notes, often given in flight with tail spread.

RANGE: Prefers open, stony habitats. Uncommon; very rare along Atlantic coast in fall, casual in late spring. Casual to accidental elsewhere in North America. Single winter records TX and LA.

Stonechat *Saxicola torquatus* L 5¼" (13 cm)

Compact body; pale spot on inner coverts; paler rump. Black on head of **adult male** obscured by fresh pale feather tips in **fall;** orange-buff wash on breast; extensive white on sides of neck, belly, rump. All records from eastern *maurus* group of subspecies, known as "Siberian Stonechat." **Female** and **first-fall male** have pale throat; pale buffy rump.

RANGE: Favors open country. Eurasian species; casual from scattered locations in AK; accidental in fall from NB and CA.

Siberian Rubythroat

ruby red throat bordered by black and white stripes

uniform-colored tail

adult ♂

1st fall ♂

pink-throated adult ♀

shorter, fainter supercilium than Bluethroat

♀

dark outline to white throat

long supercilium

♀

blue throat with rufous spot

1st fall ♀

some color on throat

winter adult ♂

Bluethroat
svecica

breeding ♂

narrow whitish supraloral stripe

white throat

♀

rufous at base of outer tail feathers in all plumages conspicuous in flight

Red-flanked Bluetail
cyanurus

blue tail often flicked down

blackish mask

orange sides and flanks

♀

adult ♂

pale supercilium

spring ♀
oenanthe

pale gray upperparts

fall adult ♂
oenanthe

Northern Wheatear

pale supercilium

blackish wings

♀

long wings

breeding adult
♂ *oenanthe*

black inverted T-shape on short, white tail

1st fall
leucorhoa

long wings

breeding adult
♂ *leucorhoa*

black face and throat, white nape

orangish chest

spring ♂

faint supercilium

small whitish wing patch

1st fall

Stonechat
maurus group

spring ♀

fall adult ♂

whitish rump

THRUSHES Family Turdidae

Eloquent songsters of many habitats that feed mainly on insects and fruit. Some species, especially in the genus *Catharus,* are difficult to identify. SPECIES: 181 WORLD; 26 N.A.

Eastern Bluebird *Sialia sialis* L 7" (18 cm)

Chestnut throat, sides of neck, breast, sides and flanks; contrasting white belly, white undertail coverts. **Male** is uniformly deep blue above; **female** grayer. The subspecies resident in the mountains of southeastern AZ, *fulva,* is paler overall. All subspecies distinguished from Western Bluebird by chestnut on throat and sides of neck and by white, not grayish, belly and under tail.

VOICE: Call note is a musical, rising *chur-lee,* extended in song to *chur chur-lee chur-lee.*

RANGE: Found in open woodlands and orchards. Nests in holes in trees and posts; also in nest boxes. Accidental BC (Fort Nelson).

Western Bluebird *Sialia mexicana* L 7" (18 cm)

Male's upperparts and throat are deep purple-blue; breast, sides, and flanks chestnut; belly and undertail coverts grayish. Most birds show some chestnut on shoulders and upper back. **Female** duller, brownish gray above; breast and flanks tinged with chestnut, throat pale gray.

VOICE: Call note a mellow *few,* given in a brief series for song.

RANGE: Nests in holes in trees and posts; also in nest boxes. Common in woodlands, farmlands, orchards; in desert areas during winter, found in mesquite-mistletoe groves. Casual on western Great Plains.

Mountain Bluebird *Sialia currucoides* L 7¼" (18 cm)

Male is sky blue above, paler below, with whitish belly and undertail coverts. **Female** is brownish gray overall, with white belly and undertail coverts. In fresh fall plumage, female's throat and breast tinged with red-orange; brownish rear flank contrasting with white undertail coverts distinguishes it from female Eastern Bluebird, which has reddish flank. Note also longer, thinner bill and longer primary tip projection of Mountain Bluebird. Often hovers above prey.

VOICE: Call is a thin *few;* song, a low, warbled *tru-lee.*

RANGE: Nests in tree cavities and buildings and nest boxes. Inhabits open rangelands, meadows, generally at elevations above 5,000 feet; in winter, found primarily in open lowlands and desert. Highly migratory. Winter movements unpredictable; in some years many to southwestern deserts, in other years a few or almost absent; casual in the East during migration and winter.

Townsend's Solitaire *Myadestes townsendi* L 8½" (22 cm)

Large and slender; gray overall, with bold white eye ring. Buff wing patches and white outer tail feathers are most conspicuous in flight. Often seen on a high perch.

VOICE: Call note is a high-pitched *eek;* song, heard all year, is a loud, complex, melodious warbling.

RANGE: Nests on the ground. Fairly common in coniferous forests on high mountain slopes; in winter, also in wooded valleys, canyons, wherever juniper berries are available. Highly migratory; very rare along Pacific coast; casual in fall and winter in Midwest and Northeast.

juveniles of all
bluebirds are
spotted

juvenile

rufous in both
sexes wraps around
sides of neck and
includes throat

**Eastern
Bluebird**
sialis

♀

overall paler
than nominate
sialis

white belly and
undertail coverts

southwestern
♂ *fulva*

gray sides
to neck

♀

all-blue
head

some dark rufous
scapulars

**Western
Bluebird**

gray
undertail
coverts

♂

**Mountain
Bluebird**

thin bill

♀

gray
flanks

long primary
projection

overall sky blue
coloration

♂

boldly spotted
plumage

juvenile

overall gray color
with prominent
white eye ring

**Townsend's
Solitaire**

prominent buffy
wing stripe

buffy wing
patches

mostly white
outer tail
feathers

Gray-cheeked Thrush *Catharus minimus* L 7¼" (18 cm)

Overall, colder coloration than Swainson's with faint, incomplete eye ring. Breeding *minimus* on Newfoundland can be slightly warmer colored above, more like Bicknell's.

VOICE: Thin, nasal song is somewhat like Veery, but first and last phrases drop, middle one rises; call, a sharp *pheu,* similar to Veery, but higher pitched and not descending.

RANGE: Favors coniferous or mixed woodlands in taiga for breeding. Uncommon in East on migration. Casual in West south of breeding grounds and west of Great Plains.

Bicknell's Thrush *Catharus bicknelli* L 6¼" (16 cm)

Identifying silent migrants from Gray-cheeked very difficult. Smaller, warmer brown above, lower mandible averages more yellow.

VOICE: Song usually comes in three parts, the first and last rising.

RANGE: Nests in stunted conifer (krummholz) vegetation above 3,000 feet and very locally on coast. Declining. Most winter on Hispaniola in montane forest. Usually only safely identified on nesting grounds.

Veery *Catharus fuscescens* L 7" (18 cm)

Reddish brown above, white below, with gray flanks, grayish face, incomplete and indistinct gray eye ring. Upperparts duller, breast more spotted in more westerly *salicicola* than eastern *fuscescens.*

VOICE: Song is a descending series of *veer* notes; call, a sharp, descending, whistled *veer.*

RANGE: Fairly common; found in dense, moist woodlands and streamside thickets. Winters in South America. Casual in Southwest.

Swainson's Thrush *Catharus ustulatus* L 7" (18 cm)

Brownish above, buffy lores and eye ring; buffy breast with dark spots; brownish gray sides, flanks. Pacific coast subspecies such as *ustulatus* reddish brown above, less distinctly spotted below; distinguished from Veery by face pattern, buffy brown sides and flanks, and voice. No evidence of interbreeding with *swainsoni* group just to east; this group may represent a separate, cryptic species.

VOICE: Song, an ascending spiral of whistles; call, a liquid *whit* in Pacific coast subspecies, a sharper *quirk* in others; at night a peeping *queep.*

See subspecies map, page 554

RANGE: Fairly common; found in moist woods and swamps. Pacific subspecies winter from western Mexico south to Central America; migrate in spring through southwestern CA deserts, in fall coastally and offshore. Others are trans-Gulf migrants, winter in South America.

Hermit Thrush *Catharus guttatus* L 6¾" (17 cm)

Complete, often whitish eye ring; reddish tail. Upperparts vary from rich brown to gray-brown. Eastern subspecies such as widespread *faxoni* have buff-brown flanks. Larger, paler western mountain subspecies, such as *auduboni,* and smaller, darker North Pacific subspecies, such as *guttatus,* have grayish flanks. Often flicks wings and slowly raises tail.

VOICE: Song is a serene series of clear, flutelike notes, the similar phrases repeated at different pitches. Calls include a deeper *chuck,* often doubled, and a whiny, upslurred *wee.*

RANGE: Fairly common; breeds in coniferous or mixed woodlands; in migration in winter, also found in thickets, chapparal, and even gardens.

Gray-cheeked Thrush

faint partial eye ring around rear of eye

grayish lores

1st fall
aliciae

aliciae

almost identical to Gray-cheeked, but slightly smaller and averages warmer above, especially on tail

minimus

grayish brown flanks

grayish brown flanks

Bicknell's Thrush

face pattern like Gray-cheeked with faint partial eye ring around rear of eye

Veery

western
salicicola

eastern
fuscescens

faintly spotted chest

gray flanks

buffy eye ring and supraloral line

upperparts with a russet cast, like many Veeries

Pacific coast
ustulatus

buffy eye ring and supraloral line

Alaska
incanus

olive-gray upperparts

buffy flanks

Swainson's Thrush

back more olive, less grayish, than *incanus*

swainsoni

spots on breast blacker than *ustulatus*

brownish flanks

Hermit Thrush

all juvenile *Catharus* have spotted plumage

juvenile
faxoni

faxoni

thin eye ring

brownish buff flanks

warm brown upperparts with contrasting rufous tail

grayish upperparts

auduboni

grayish flanks

pale rufous tail

smaller, darker, and with deeper rufous tail than *auduboni*

guttatus

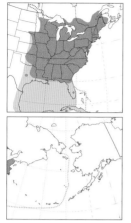

Wood Thrush *Hylocichla mustelina* L 7¾" (20 cm)

Reddish brown above, brightest on crown and nape; rump and tail brownish olive. White eye ring conspicuous on streaked face. Large dark spots on whitish throat, breast, and sides.

VOICE: Loud, liquid song of three- to five-note phrases, each phrase usually ending with a complex trill. Calls include a rapid *pit pit pit.*

RANGE: Fairly common in moist deciduous or mixed woods. Declining in some regions. Casual in the West.

Eyebrowed Thrush *Turdus obscurus* L 8½" (22 cm)

Brownish olive above, with distinct white eyebrow. Belly is white, sides pale buffy orange. **Male** has dark gray throat and breast; **female's** throat is white and streaked; browner head shows little contrast with rest of upperparts. Wing linings pale gray.

VOICE: Flight call is a high, piercing, drawn-out *dzee.*

RANGE: Asian species. Regular spring migrant on the Aleutians, casual in fall; casual on St. Lawrence Island and in northern AK; accidental CA (late spring).

Dusky Thrush *Turdus naumanni* L 9½" (24 cm)

In *eunomus,* white eyebrow conspicuous on blackish head. Wings mostly rufous, underwing almost entirely rufous. Below, white edgings give a scaly look to dark breast and sides. Note also distinctive white crescent across breast. Female and immatures average duller overall. Several sight records of redder nominate subspecies for western AK islands; however, intergrades between these two subspecies are not rare in Asia.

VOICE: Call is a series of *shack* notes; also a shrill, wheezy *shrree* similar to European Starling.

RANGE: Asian species. Casual spring migrant on westernmost Aleutians; accidental on St. Lawrence Island, Point Barrow, AK, and in winter south to coastal WA.

Fieldfare *Turdus pilaris* L 10" (25 cm)

Large thrush with gray head and rump contrast with purplish brown upper back, blackish tail. Below, dark arrowhead-shaped spots pattern the buffy breast and extend along sides. White wing linings flash in flight.

VOICE: Song is a noisy twittering; call note is a series of *shack* notes, like Dusky Thrush; also gives a thin *seeh.*

RANGE: Eurasian species. Breeds from Greenland to Russian Far East. Casual vagrant, mainly in winter, to northeastern North America; most records from Atlantic Canada. Casual to AK. Accidental to ON and MN.

Redwing *Turdus iliacus* L 8¼" (21 cm)

Distinctive whitish to buffy eyebrow; boldly streaked below, with rusty red flanks; rusty red wing linings visible in flight.

VOICE: Call is a thin, penetrating *seeeh,* usually heard in flight; also a hard *kuk* note.

RANGE: Eurasian species; casual visitor to Newfoundland, mainly in winter; accidental south to Long Island, NY, and PA; once coastal WA.

Wood Thrush

nearly complete bold, white eye ring

rich rufous above, especially on crown and nape

speckled cheeks

boldly spotted with black below

Eyebrowed Thrush

white eyebrow

olive upperparts

♀

browner lower throat

dark gray throat

extensive white belly

♂

broad whitish supercilium

black upper breast band bordered by white crescent

rusty rufous wings

♀ *eunomus*

extensive black markings below

♂ *eunomus*

rich buffy supercilium

Dusky Thrush

reddish underparts

♂ *naumanni*

Fieldfare

flashy white wing linings

gray head

purplish brown back

buffy breast with bold arrow-shaped spots

gray rump

immature

Redwing

distinct whitish buff supercilium

rusty red wing linings

streaked underparts

rusty red flanks

immature

American Robin *Turdus migratorius* L 10" (25 cm)
Gray-brown above, with darker head; bill yellow; underparts brick red; vent white. Most western birds paler and duller overall than eastern nominate *migratorius*, which breeds to western AK; in most, tail has white corners, visible in flight. Northwestern subspecies, *caurinus,* is equally dark but most lack white tail spots; breeds north to southeastern AK. **Juvenile's** underparts are tinged cinnamon, heavily spotted.
VOICE: Loud, liquid song, is a variable *cheerily cheer-up cheerio.* Calls include a rapid *tut tut tut;* a high, thin *ssip* in flight.
RANGE: Common, widespread. Often seen on lawns, searching for earthworms; also eats insects and berries. Nests in shrubs, trees, and eaves. In winter, found in woodlands, suburbs, and parks. Numbers vary greatly from winter to winter.

Clay-colored Thrush *Turdus grayi* L 9" (23 cm)
Brownish olive above; tawny buff below; pale buffy throat is lightly streaked with olive. Lacks white around eye conspicuous in American Robin. Rather secretive but will come to feeders.
VOICE: Calls include a slurred *reeeur-ee,* a clucking note, and, in flight, a high, thin *ssi;* song resembles American Robin but is slower, clearer.
RANGE: Prefers dense thickets, streamside brush, and woodlands. Widespread in tropics; rare and increasing in southernmost TX; accidental to Big Bend.

Rufous-backed Robin *Turdus rufopalliatus* L 9¼" (24 cm)
Distinguished from American Robin by rufous back and wing coverts, uniformly gray head with no white around eye, and more extensively streaked throat. Somewhat secretive.
VOICE: Calls include a plaintive, drawn-out, whistled *teeeuu,* a clucking series of *chuk* notes, and in flight, a high, thin *ssi.*
RANGE: Prefers dense shrubbery. West Mexican species, very rare winter visitor to southern AZ, casual from southern and southwestern TX to southern CA.

White-throated Thrush *Turdus assimilis* L 9½" (24 cm)
Distinct white collar; throat with dark brown streaking and white bib; head and upperparts brownish; often shows yellow orbital ring; underparts gray. Compare to Clay-colored Robin with less marked throat, lack of white bib, overall tawnier color, more extensively yellow bill.
VOICE: Call is a nasal *rreeuh,* often doubled.
RANGE: Tropical species; casual to southernmost TX in winter.

Aztec Thrush *Ridgwayia pinicola* L 9¼" (24 cm)
Male is blackish brown above, with white patches on wings, white uppertail coverts; tail broadly tipped with white; contrasty underparts. **Female** is browner. **Juvenile** is streaked above with creamy white; underparts whitish and heavily scaled with brown.
VOICE: Calls are a quavering *wheeerr,* a metallic *wheer,* and a clear *sweee-uh.*
RANGE: Mexican species, rare and irregular to southeastern AZ, mainly late summer; casual to south TX.

juvenile

heavily
spotted
underparts

bold white
broken eye ring

white spot

yellow bill

♀

♂

**American
Robin**
migratorius

brick red
underparts

gray head
with no white
around eye

brownish head
and upperparts

rufous back and
wing coverts

long
streaks
on throat

unlike American
Robin, no pale
markings around eye

greenish
bill

**Clay-colored
Thrush**
tamaulipensis

white extends
up to lower
breast in a point

tawny buff
underparts

**Rufous-
backed
Robin**
rufopalliatus

dark head with yellow
orbital ring

**Aztec
Thrush**

spotted plumage;
note white wing
pattern

heavy black
streaks on
throat

white
bib

blackish head,
breast, and back

dark brown
upperparts

♂

grayish
underparts

**White-throated
Thrush**
suttoni

♀

browner breast
than male

extensive
white on
wing

white
belly

juvenile

white rump
and tail tip

Varied Thrush *Ixoreus naevius* L 9½" *(24 cm)*

Male has grayish blue nape and back, orange eyebrow; underparts orange with black breast band; buffy orange bar on underwing prominent in flight. **Female** distinguished from American Robin (page 402) by orange eyebrow and wing bar, dusky breast band. **Juvenile** resembles female but has white belly, scalier-looking throat and breast. In a very rare variant morph, all orange color is replaced by white.

VOICE: Call is a soft, low *tschook;* song is a slow series of variously pitched notes, rapidly trilled.

RANGE: Found in dense, moist woodlands, especially coniferous forests. Generally feeds in trees. Very rare in winter as far east as New England and south to VA. Numbers vary from year to year in southern part of mapped winter range.

MOCKINGBIRDS • THRASHERS Family Mimidae

Notable singers, unequaled in North America for the rich variety and volume of their song. Some mimic the songs of other species. SPECIES: 34 WORLD; 12 N.A.

Gray Catbird *Dumetella carolinensis* L 8½" *(22 cm)*

Dark gray with black cap and chestnut under tail. Juvenile with paler cap and under tail.

VOICE: Song is a mixture of melodious, nasal, and squeaky notes interspersed with catlike *mew* notes; some are good mimics. Most readily identified by harsh, downslurred *mew* call; also gives a low *quirt* and a clucking noise.

RANGE: Generally common but rather secretive in thickets.

Blue Mockingbird *Melanotis caerulescens* L 10" *(25 cm)*

Adult deep slaty blue with black mask and red eye. Immature slightly duller, brownish tinge to wings, darker eye. Secretive.

VOICE: Song and call notes vary widely.

RANGE: Mexican species which moves altitudinally. Casual in winter to southeastern AZ. Records from Long Beach, CA (adult), and NM were judged questionable on origin; records from Rio Grande Valley, TX, were perhaps also questionable but were accepted.

Northern Mockingbird *Mimus polyglottos* L 10" *(25 cm)*

White outer tail feathers and white wing patches flash in flight.

VOICE: Song is a mixture of original and imitative phrases, each repeated several times. Often sings at night. Imitates other species' songs and calls. Both sexes sing in fall, claiming feeding territories. Call is a loud, sharp *check.*

RANGE: Found in a variety of habitats, including towns. Casual well north of mapped range, as far as AK.

Bahama Mockingbird *Mimus gundlachii* L 11" *(28 cm)*

Browner than Northern Mockingbird, with streaking on flanks; white only on tail tip. Lacks white wing patches; flight more direct.

VOICE: Song is varied but not known to include imitations; call slightly harsher, more downslurred than Northern.

RANGE: Prefers dense cover. Caribbean species; casual south FL.

Varied Thrush
meruloides

juvenile

orange wing bars and markings on wing

orange supercilium

♀

♂

black breast band

blackish cap

dark chestnut undertail coverts

short, slender dark bill

Gray Catbird

overall steel gray

Blue Mockingbird

overall blue, looks dark in poor light

adults have red iris

paler blue on crown and throat

black mask

adult

juvenile

white outer tail feathers

Northern Mockingbird
polyglottos

extensive white wing patch

Bahama Mockingbird
gundlachii

distinct streaks on lower sides, flanks, and undertail coverts

lacks white wing patch of Northern Mockingbird

darker tail than Northern Mockingbird

whitish tail tips

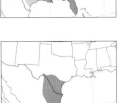

Brown Thrasher *Toxostoma rufum* L 11½" (29 cm)

Reddish brown above, heavily streaked below. Immature's eyes darker. Compare to Wood Thrush (page 400).

VOICE: Sings a series of varied melodious phrases, each phrase usually given only two or three times. Seldom imitates other birds. Calls include a sharp *spuck* and a low *churr.*

RANGE: Common in hedgerows and woodland edges. Rare to very rare to south TX, the West, and Maritime Provinces; casual to Newfoundland. Accidental to northern AK in fall (Point Barrow) and NT.

Long-billed Thrasher *Toxostoma longirostre* L 11½" (29 cm)

Closely resembles Brown Thrasher but much grayer above, with longer, more strongly curved bill; also has darker malar stripe, blacker streaking below, shorter primary projection.

VOICE: Song similar to Brown Thrasher. Gives *tsuck* call like Brown; other calls, a mellow *kleak,* and a loud, whistled *cheeooep.*

RANGE: Inhabits dense bottomland thickets. Very rare in west TX; casual in NM and CO.

Sage Thrasher *Oreoscoptes montanus* L 8½" (22 cm)

Yellow eye, white wing bars, white-cornered tail. Grayish above, boldly streaked below. **Worn** late summer birds show much less streaking, can resemble Bendire's Thrasher. Juvenile has streaked head and back.

VOICE: Song is a long series of warbled phrases. Calls include a *chuck* and a high *churr.*

RANGE: Found in sagebrush plains. A very early spring migrant (a few by late Feb. or early Mar.). Rarely winters north into breeding range. Very rare to Pacific coast and offshore islands. Casual vagrant to eastern North America.

California Thrasher *Toxostoma redivivum* L 12" (31 cm)

Dark above, with pale eyebrow, dark eye, dark cheeks. Pale throat is outlined by dark malar and contrasts with brownish breast; belly and undertail coverts tawny buff. Darker overall than Crissal Thrasher, with which it is often confused at locations where their ranges approach each other. Can be separated by vocalizations, which are distinct. Crissal may actually be more closely related to LeConte's Thrasher; Crissal responds readily to LeConte's playback.

VOICE: Calls are a low, flat *chuck* and *chur-erp.* Song is loud and sustained, with mostly guttural phrases, often repeated once or twice. Imitates other species and sounds.

RANGE: Common in chaparral-covered foothills and other dense brushy habitats. Locally found at oases and canyons in westernmost CA deserts. Although California and Crissal ranges closely approach one another, they have not overlapped. For instance, at Yaqui Well, in Anza-Borrego State Park, there are resident California Thrashers while just a few miles away in the Borrego Valley, a few Crissal Thrashers can still be found. Casual to southern OR.

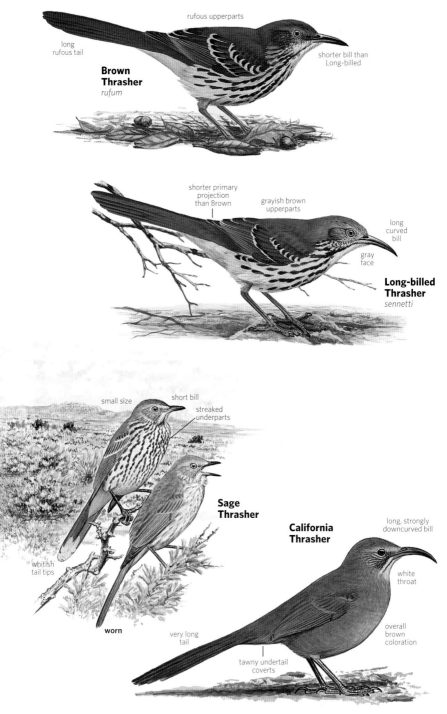

long rufous tail

rufous upperparts

Brown Thrasher
rufum

shorter bill than Long-billed

shorter primary projection than Brown

grayish brown upperparts

long curved bill

gray face

Long-billed Thrasher
sennetti

small size

short bill

streaked underparts

Sage Thrasher

whitish tail tips

worn

very long tail

tawny undertail coverts

California Thrasher

long, strongly downcurved bill

white throat

overall brown coloration

Curve-billed Thrasher *Toxostoma curvirostre* L 11" (28 cm)

Breast mottled; bill all-dark, longer, heavier, and usually more strongly curved than Bendire's Thrasher. Breast spots indistinct in the western-most subspecies, *palmeri*. Subspecies from extreme southeastern AZ to south TX, *oberholseri*, shows clearer spotting below; has pale wing bars; conspicuous white tips on tail. **Juvenile** has shorter bill. Genetic sampling indicates that the *curvirostre* group, of which *oberholseri* is a subspecies, and the *palmeri* group may represent separate species, but studies did not include samples from the region of overlap immediately to the west or south of Chiricahua Mountains, AZ.

VOICE: Distinctive call, a sharp upslurred *whit-wheet*, sometimes three-noted *(palmeri)* or even-pitched *whit-whit (oberholseri)*. Song is elaborate and melodic, and includes low trills and warbles.

RANGE: Common in canyons, semiarid brushlands. Some seasonal movement in Great Plains population. Casual *(palmeri)* to southeastern CA; accidental to west to southern CA coast, north to MT and northern Great Plains, east to the upper Midwest, and the FL Panhandle.

Bendire's Thrasher *Toxostoma bendirei* L 9¾" (25 cm)

Breast mottled; bill shorter and usually less curved than Curve-billed; base of lower mandible pale; color a little buffier. White tail tips are similar to *oberholseri* subspecies of Curve-billed. Distinctive arrowhead-shaped spots on breast are not present in **worn** summer plumage.

VOICE: Song is a sustained, melodic warbling, each phrase repeated one to three times. Low *chuck* call is seldom heard.

RANGE: Uncommon and local; found in open farmlands, grasslands, Joshua trees, and brushy desert. Casual to southern CA coast in late summer, fall, and winter but fewer records in last two decades.

Le Conte's Thrasher *Toxostoma lecontei* L 11" (28 cm)

Palest thrasher, with pale grayish brown upperparts, darker tail; tawny undertail coverts. Bill and eye are dark. Not as elusive as Crissal, but a fast runner. Often detected by listening for whistled call.

VOICE: Song, heard chiefly at dawn and dusk, is loud and melodious. Calls include a diagnostic ascending, whistled *tweeep*, often delivered from a bush.

RANGE: Prefers arid, sparsely vegetated habitats. Appears partial to several species of saltbush (especially in San Joaquin Valley, CA), including *Atriplex* species and cholla cactus. Uncommon over most of range. Development for housing and especially agriculture as well as overgrazing for livestock have extirpated some populations (e.g., in parts of San Joaquin Valley). A few records in atypical habitat and slightly out of range (Morongo and Moreno Valleys, CA).

Crissal Thrasher *Toxostoma crissale* L 11½" (29 cm)

Large and slender, with a distinctive chestnut undertail patch and a dark malar streak. Paler and grayer than geographically separated California Thrasher (page 406), which has buffy, not chestnut, undertail coverts.

VOICE: Song is varied and musical. Calls include a repeated *chideery* and a whistled, even-pitched *toit-toit-toit.*

RANGE: Very secretive, hiding in underbrush; indeed, one of our most secretive passerines. Found mainly in dense mesquite and willows along streams and washes; sometimes on lower mountain slopes.

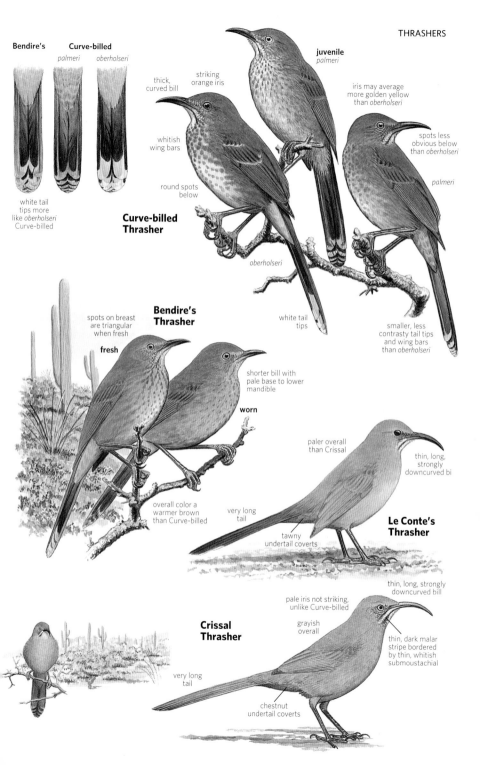

Bendire's

Curve-billed

palmeri

oberholseri

white tail
tips more
like *oberholseri*
Curve-billed

thick,
curved bill

striking
orange iris

juvenile
palmeri

iris may average
more golden yellow
than *oberholseri*

whitish
wing bars

spots less
obvious below
than *oberholseri*

round spots
below

palmeri

**Curve-billed
Thrasher**

oberholseri

white tail
tips

smaller, less
contrasty tail tips
and wing bars
than *oberholseri*

spots on breast
are triangular
when fresh

**Bendire's
Thrasher**

fresh

shorter bill with
pale base to lower
mandible

worn

overall color a
warmer brown
than Curve-billed

very long
tail

paler overall
than Crissal

thin, long,
strongly
downcurved bi

tawny
undertail coverts

**Le Conte's
Thrasher**

thin, long, strongly
downcurved bill

pale iris not striking,
unlike Curve-billed

grayish
overall

**Crissal
Thrasher**

thin, dark malar
stripe bordered
by thin, whitish
submoustachial

very long
tail

chestnut
undertail coverts

BULBULS Family Pycnonotidae
Noisy, active Old World family of the tropics and subtropics. SPECIES: 123 WORLD; 1 N.A.

Red-whiskered Bulbul *Pycnonotus jocosus* L 7" (18 cm)
Red ear spot and undertail coverts distinctive. **Juvenile** lacks ear patch; undertail coverts are paler.
VOICE: Utters a chattering series of notes.
RANGE: Asian species. Escaped cage birds first noted in early 1960s in Miami, FL; now established as a small population in suburbs and parklands south of Miami. Some also in Los Angeles area.

STARLINGS Family Sturnidae
Widespread Old World family. Chunky and glossy birds; most species are gregarious and bold. American Birding Association recognizes only two; Hill Myna not yet accepted by ABA. SPECIES: 117 WORLD; 3 N.A.

Common Myna *Acridotheres tristis* L 10" (25 cm)
Dark brown, with black head and white undertail coverts; yellow bill and skin around eye; white tail tip, patch at base of primaries, and white wing linings distinctive in flight. Juveniles have more brownish heads. Found in urban areas; also open country in native range.
VOICE: Calls include gurglings, whistles, and screeches.
RANGE: South Asian species; introduced elsewhere, including HI, where it is common. Established in south FL, where it is spreading.

Hill Myna *Gracula religiosa* L 10½" (27 cm)
Glossy black; orange-red bill; yellow wattles and legs; white wing patch.
VOICE: Call is a loud, piercing two-note whistle, *ti-ong*. An excellent mimic, and captive birds can be fine talkers.
RANGE: Asian species, popular as a cage bird. A small number of escaped birds, first noted in 1960s, persists but is very local in Miami.

European Starling *Sturnus vulgaris* L 8½" (22 cm)
Adult in **breeding** plumage is iridescent black, with a yellow bill with blue base in male, pink in female. In fresh **fall** plumage, feathers are tipped with white and buff, giving a speckled appearance; bill brownish. In flight, note short, square tail, stocky body, and short, broad-based, pointed wings that appear pale gray from below. **Juvenile** is gray-brown, with brown bill.
VOICE: Song includes squeaks, warbles, chirps, and twittering; also imitates songs of other species. Otherwise, often silent, though gives various harsh calls in interactions with others and has a soft, breezy flight call.
RANGE: A Eurasian species introduced in NY in 1890-91, it soon spread across the continent. Abundant, bold, aggressive, it often competes successfully with native species for nest holes. Outside nesting season, usually seen in large flocks, sometimes mixed with blackbirds. Casual to central AK and YT; one collected on Shemya Island, western Aleutians, likely originated from North America.

Red-whiskered Bulbul

long black crest

small red auricular spot

black breast mark

juvenile

red undertail coverts

white tail tips

Common Myna
tristis

yellow skin around eye

blackish head

yellow bill

brown above and below

white wing patch

white undertail coverts

white tail tip

Brown-headed Cowbird for comparison ♂

short tail

triangular wings

Hill Myna
intermedia

yellow wattle

thick orange bill

white wing patch

fall

plumage heavily spotted with whitish

European Starling
vulgaris

winter

slender dark bill

short tail

breeding ♂

yellow bill

glossy plumage

juvenile

overall grayish brown

dull, blurred streaks on belly

ACCENTORS Family Prunellidae

Small Eurasian family, most species found in mountainous country. One species strays to North America. SPECIES: 13 WORLD; 1 N.A.

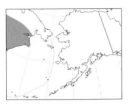

Siberian Accentor *Prunella montanella* L 5½" (14 cm)

Bright tawny buff below; dark crown and cheek patch; buffy eyebrow broadens behind head; gray patch on side of neck.
VOICE: Call is a high, thin series of *see* notes.
RANGE: Rare fall migrant to St. Lawrence Island, AK. Casual elsewhere in fall (once spring) in AK and Pacific Northwest (recorded BC, WA, ID, and MT).

WAGTAILS • PIPITS Family Motacillidae

Slender-billed birds. Most species pump their tails as they walk. Wagtail flight is strongly undulating. SPECIES: 66 WORLD; 10 N.A.

Eastern Yellow Wagtail *Motacilla tschutschensis*

L 6½" (17 cm) Olive above, yellow below; tail shorter than other wagtails. In **breeding** plumage, AK nesting *tschutschensis* has a speckled breast band. Asian *simillima,* recognized by some, seen on Aleutians and Pribilofs, averages greener above, yellower below. **Females** duller, **immatures** whitish below.
VOICE: Call, a loud, buzzy *tsweep,* similar to Eastern Kingbird.
RANGE: Generally common on AK breeding grounds; casual fall (Sept.) migrant on CA coast.

Gray Wagtail *Motacilla cinerea* L 7¾" (20 cm)

Gray above, with greenish yellow rump, yellow below; whitish tertial edges, long tail. **Breeding male** has black throat. **Female** and winter birds have whitish throat, paler below.
VOICE: Call, a metallic *chink-chink,* is similar to White Wagtail.
RANGE: Eurasian species. Very rare spring migrant on western Aleutians; casual on Bering Sea islands; accidental south to BC and CA.

White Wagtail *Motacilla alba* L 7¼" (18 cm)

Breeding adult (*ocularis*) has black nape, gray back; eye line, throat, bib, and usually chin are black. In breeding adult male *lugens* upperparts are black, wings mostly white; chin usually white; female duller. **Winter adults** retain distinct wing pattern. In nominate *alba,* face is white in all plumages. Juveniles of all subspecies are brownish above with two faint wing bars. **Immature** closer to adult but retains most of juvenile wing; immature *ocularis* has darker bases to median coverts than *lugens,* but separation problematic.
VOICE: Calls include a two-note *chizzik* given in flight and a whistled *chee-wee* given from perch.
RANGE: Northeast Asian *ocularis* breeds sparingly in western AK; *lugens* breeds in coastal East Asia (south of *ocularis*), has nested and hybridized with *ocularis* in western AK, was formerly treated as a separate species, Black-backed Wagtail. Both subspecies casual on West Coast; accidental elsewhere. Nominate *alba,* breeding as close as Iceland and Greenland, is accidental on Atlantic coast.

412

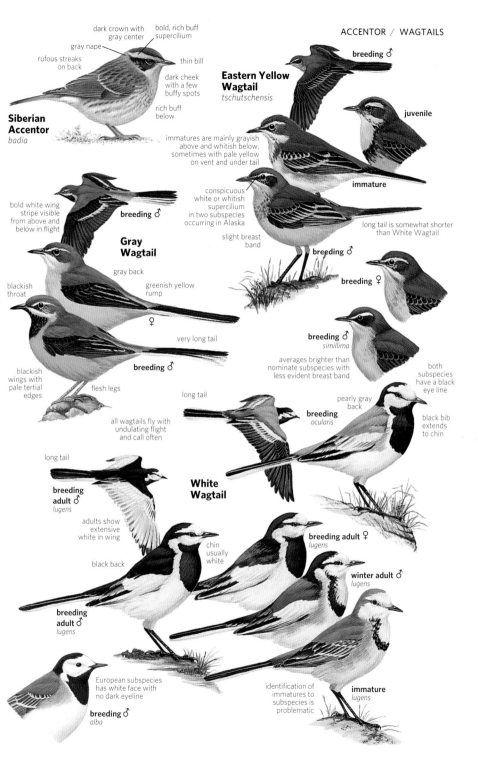

dark crown with gray center

bold, rich buff supercilium

gray nape

rufous streaks on back

thin bill

dark cheek with a few buffy spots

rich buff below

Siberian Accentor
badia

Eastern Yellow Wagtail
tschutschensis

breeding ♂

juvenile

immatures are mainly grayish above and whitish below, sometimes with pale yellow on vent and under tail

immature

bold white wing stripe visible from above and below in flight

breeding ♂

conspicuous white or whitish supercilium in two subspecies occurring in Alaska

long tail is somewhat shorter than White Wagtail

Gray Wagtail

gray back

greenish yellow rump

slight breast band

breeding ♂

blackish throat

♀

breeding ♀

very long tail

blackish wings with pale tertial edges

breeding ♂

breeding ♂
similima

both subspecies have a black eye line

flesh legs

long tail

averages brighter than nominate subspecies with less evident breast band

all wagtails fly with undulating flight and call often

breeding
ocularis

pearly gray back

long tail

White Wagtail

black bib extends to chin

long tail

breeding adult ♂
lugens

adults show extensive white in wing

chin usually white

breeding adult ♀
lugens

black back

winter adult ♂
lugens

breeding adult ♂
lugens

European subspecies has white face with no dark eyeline

identification of immatures to subspecies is problematic

immature
lugens

breeding ♂
alba

American Pipit *Anthus rubescens* L 6½" (17 cm)

Breeding birds grayish above, faintly streaked below, except for the *alticola* subspecies, from the Rockies and high mountains of CA, which has richly colored underparts with fewer or no streaks. In **winter** becomes browner above and more streaked below. Bill mostly dark; legs dark or tinged with pink. Tail has white outer feathers. An Asian subspecies, *japonicus,* is more boldly streaked below, with pink legs, white wing bars.
VOICE: Call, given in flight, is a sharp *pip-pit;* song, a rapid series of *chee* or *cheedle* notes, often given in flight on breeding grounds.
RANGE: Common; nests on tundra in the far north, mountaintops farther south. Winter flocks are found in fields. Asian *japonicus* subspecies is rare in western AK, casual in fall to coastal CA.

Sprague's Pipit *Anthus spragueii* L 6½" (17 cm)

Dark eye prominent in pale buff face. Pale edges on rounded back feathers give a scaly look; rump is streaked. Faint necklace on breast. Legs pinkish. Outer tail feathers are more extensively white than American Pipit. Uncommon, secretive, and somewhat solitary. Does not pump tail. Compare to juvenile Horned Lark (page 366).
VOICE: Call is a loud, squeaky *squeet,* usually repeated. Song, given continuously in high flight, is a descending series of musical *tzee-a* notes.
RANGE: Nests in prairies. Winter in grassy fields. Uncommon. Very rare in fall and winter to CA; accidental to Pacific Northwest and eastern North America.

Olive-backed Pipit *Anthus hodgsoni* L 6" (15 cm)

Grayish olive back, faintly streaked. Eyebrow orange-buff in front of eye, white behind. Broken white stripe borders dark ear spot. Throat and breast rich buff, with rather large black spots on breast. Belly pure white; legs pink.
VOICE: Call is a buzzy *tsee.*
RANGE: Asian species. Rare migrant on western Aleutians; casual to Pribilofs and St. Lawrence Island, AK; accidental to CA and NV.

Pechora Pipit *Anthus gustavi* L 5½" (14 cm)

Shows distinct primary projection. Resembles immature Red-throated Pipit, but has richly patterned back plumage, extending onto the nape; black centers with dull rufous edges contrast with white lines, or "braces," on the sides. Also a yellowish wash across the breast contrasts with the whitish belly. Quite secretive.
VOICE: Call is a hard *pwit* or *pit;* often silent when flushed.
RANGE: Asian species. Casual in spring on the western Aleutians and St. Lawrence Island, AK (mostly fall).

Red-throated Pipit *Anthus cervinus* L 6" (15 cm)

Note unpatterned nape. Pinkish red head and breast are distinctive in **breeding male,** less extensive in **breeding female** and fall adults. Fall **immatures** and some breeding females show no red.
VOICE: Call, given in flight, is a high, piercing *tseee,* dropping in pitch. Loud, varied song is delivered from the ground or in song flight.
RANGE: Regular migrant on islands in Bering Sea; rare fall migrant along CA coast; casual inland and in Northwest. A few spring records in West (coastal and interior).

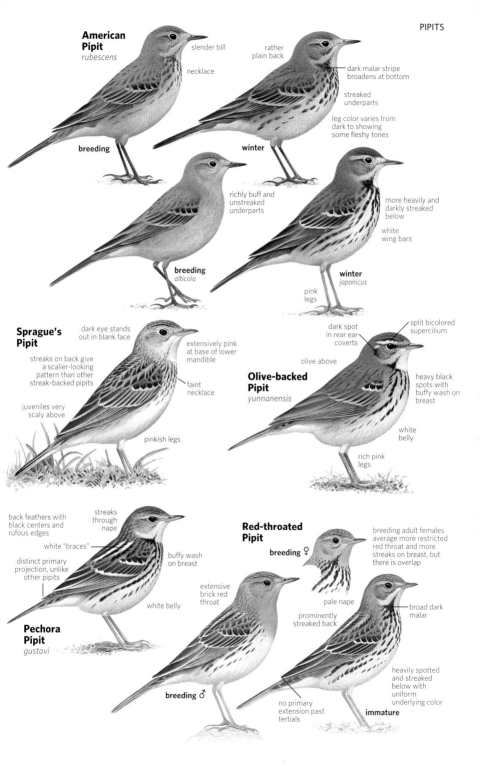

American Pipit
rubescens

slender bill

necklace

breeding

rather plain back

dark malar stripe broadens at bottom

streaked underparts

leg color varies from dark to showing some fleshy tones

winter

richly buff and unstreaked underparts

breeding
alticola

more heavily and darkly streaked below

white wing bars

winter
japonicus

pink legs

Sprague's Pipit

dark eye stands out in blank face

streaks on back give a scalier-looking pattern than other streak-backed pipits

extensively pink at base of lower mandible

faint necklace

juveniles very scaly above

pinkish legs

dark spot in rear ear coverts

split bicolored supercilium

olive above

Olive-backed Pipit
yunnanensis

heavy black spots with buffy wash on breast

white belly

rich pink legs

back feathers with black centers and rufous edges

streaks through nape

white "braces"

distinct primary projection, unlike other pipits

Pechora Pipit
gustavi

buffy wash on breast

white belly

Red-throated Pipit

breeding ♀

breeding adult females average more restricted red throat and more streaks on breast, but there is overlap

extensive brick red throat

pale nape

prominently streaked back

broad dark malar

heavily spotted and streaked below with uniform underlying color

no primary extension past tertials

breeding ♂

immature

WAXWINGS Family Bombycillidae

Red, waxy tips on secondary wing feathers are often indistinct, and sometimes they are absent altogether. All waxwings have sleek crests, silky plumage, and yellow-tipped tails. Where berries are ripening, waxwings come to feast in amiable, noisy flocks.
SPECIES: 3 WORLD; 2 N.A

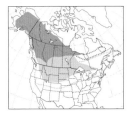

Bohemian Waxwing *Bombycilla garrulus* L 8¼" (21 cm)

Larger and grayer than Cedar Waxwing; underparts gray; undertail coverts cinnamon. White and yellow spots on wings. In flight, white wing patch at base of primaries is conspicuous. **Juvenile** browner above, streaked below, with pale throat.

VOICE: Distinctive call, a buzzy twittering, lower and harsher than Cedar Waxwing.

RANGE: Nests in open coniferous or mixed woodlands; often seen perched on top of a black spruce. Winter range varies widely and unpredictably; large flocks visit scattered locations, feeding on berries and small fruits. Also eats insects, flower petals, and sap. Irregular winter wanderer to the Northeast, usually in small numbers; annual in ME, Maritime Provinces, and Newfoundland. Casual to southern CA and northern portions of AZ, NM, and TX. Individuals are sometimes seen in flocks of Cedar Waxwings.

Cedar Waxwing *Bombycilla cedrorum* L 7¼" (18 cm)

Smaller and browner than Bohemian Waxwing; belly pale yellow; undertail coverts white. Lacks yellow spots on wings. **Juvenile** is streaked; lacks white wing patches of juvenile Bohemian. Because this species usually nests late in summer, juvenal plumage is seen well into fall. Highly gregarious in migration and winter.

VOICE: Call is a soft, high-pitched, trilled whistle.

RANGE: Found in open habitats where berries are available; also eats insects, flower petals, and sap. Very rare to central AK.

SILKY-FLYCATCHERS Family Ptilogonatidae

This New World tropical family of slender, crested birds is closely related to the waxwings. The family's common name describes their soft, sleek plumage and agility in catching insects on the wing. SPECIES: 4 WORLD; 2 N.A.

Phainopepla *Phainopepla nitens* L 7¾" (20 cm)

Male is shiny black; white wing patch conspicuous in flight. In both sexes, note distinct crest, long tail, red eyes. Juvenile resembles **female;** but has browner eyes; both have gray wing patches. Young males acquire patchy black in fall. Flight is fluttery but direct, and often very high.

VOICE: Distinctive call note is a low-pitched, whistled, querulous *wurp?* Song is a brief warble, seldom heard.

RANGE: Nests in early spring in mesquite brushlands, feeding chiefly on insects and mistletoe berries. In late spring they move into cooler, wetter habitat and raise a second brood. In fall some wander to CA coast and offshore islands. Casual to Pacific Northwest and CO. Accidental in eastern North America.

Bohemian Waxwing
pallidiceps

juvenile

crest

streaked
underparts

gray
belly

cinnamon
undertail
coverts

white tips of primary
coverts conspicuous
in flight

juvenile

streaked
underparts

crest

**Cedar
Waxwing**

yellowish
belly

yellow
tail tip

white
undertail
coverts

conspicuous crest

adults with red iris;
brownish iris of juvenile
retained well into fall

glossy black
color

females and
juveniles overall
grayish

♂

♀

♂

Phainopepla

prominent white
wing patches
visible in flight;
pale gray in female

LONGSPURS • SNOW BUNTINGS   Family Calcariidae

Recent molecular work using mitochondrial and nuclear DNA has shown that long-spurs and snow buntings are well differentiated at the molecular level from Emberiz-idae (mostly our sparrows) and belong in their own family. They are gregarious in the nonbreeding season and prefer open country. Some are secretive; others less so. They often flush in groups to avoid predators. SPECIES: 6 WORLD; 6 N.A.

Smith's Longspur *Calcarius pictus* L 6¼" (16 cm)

Outer two feathers on each side of tail are almost entirely white. Bill is thinner than other longspurs. Note long primary projection, a bit shorter than Lapland, but much longer than Chestnut-collared or McCown's (page 420); shows rusty edges to greater coverts and tertials. **Breeding male** has black-and-white head, rich buff nape and underparts; white patch on shoulder, often obscured. **Breeding female** and all **winter** plumages are duller, crown streaked, chin paler. Dusky ear patch bordered by pale buff eyebrow; pale area on side of neck often breaks through dark rear edge of ear patch. Underparts are pale buff with thin reddish brown streaks on breast and sides. Females have much less white on lesser coverts than males.

VOICE: Typical call is a dry, ticking rattle, harder and sharper than Lapland and McCown's Longspurs. Song, heard in spring migration and on the breeding grounds, is delivered only from the ground or a perch. It consists of rapid, melodious warbles, ending with a vigorous *wee-chew*.

RANGE: Generally uncommon and secretive, especially in migration and winter. Nests on open tundra and damp, tussocky meadows. Winters in open, grassy areas; sometimes seen with Lapland Longspurs. Regular spring migrant in the Midwest, east to western IN; irregularly western OH. Casual to East Coast region from MA to GA; also casual to AL and in West south of breeding range (recorded south to CA, NV, and Big Bend, west TX).

Lapland Longspur *Calcarius lapponicus* L 6¼" (16 cm)

Outer two feathers on each side of tail are partly white, partly dark. Note also, especially in winter plumages, the reddish edges on the greater coverts and on the tertials. The reddish edges of the tertials form an indented, or notched, shape. **Breeding male's** head and breast are black and well outlined: a broad white or buffy stripe extends back from eye and down to sides of breast; nape is reddish brown. **Breeding female** and all **winter** plumages are duller; note bold dark triangle outlining plain buffy ear patch; dark streaks (female) or patch (male) on upper breast; dark streaks on side. On all winter birds, note broad buffy eyebrow and buffier underparts; belly and under tail are white, unlike Smith's Longspur; also compare head and wing patterns. **Juvenile** is yellowish and heavily streaked above and on breast and sides. Often found amid flocks of Horned Larks and Snow Buntings; look for Lapland's darker overall coloring and smaller size.

VOICE: Song, heard mostly on the breeding grounds, is a rapid warbling, frequently given in short flights. Calls include a musical *tee-lee-oo* or *tee-dle* and, in flight, a dry rattle distinctively mixed with whistled *tew* notes.

RANGE: Rare to common (e.g., Great Plains). Breeds on Arctic tundra; winters in grassy fields, grain stubble, and on shores.

Smith's Longspur

distinctive black-and-white head pattern

breeding ♂

fine streaks on buffy underparts

breeding ♀

long primary projection, but shorter than Lapland

more white in outer tail feathers than Lapland

white lesser coverts often best noted in flight

winter ♂

Lapland Longspur

chestnut nape

extensive black head bordered by white

breeding ♂

breeding ♀

well-defined dark border to auriculars

dark chest band

very long primary projection

winter ♂

restricted white in outer tail feathers

dark lateral crown stripes surround paler center to crown

juvenile

rufous-edged greater coverts and tertials

whitish belly

buffy fall ♀

winter ♀

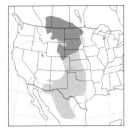

Chestnut-collared Longspur *Calcarius ornatus*

L 6" (15 cm) White tail marked with blackish triangle. Very short primary projection; primary tips barely extend to base of tail. **Breeding adult male's** black-and-white head, buffy face, and black underparts are distinctive; a few have chestnut on underparts. Lower belly and undertail coverts whitish. Upperparts black, buff, and brown, with chestnut collar, whitish wing bars. **Winter males** are paler; feathers edged in buff and brown, obscuring black underparts. Male has small white patch on shoulder, often hidden; compare with Smith's Longspur (page 418). Breeding female resembles **winter female** but is darker, usually shows some chestnut on nape. Juvenile's pale feather fringes give upperparts a scaled look; tail pattern and bill shape distinguish juvenile from juvenile McCown's Longspur. Fall and winter birds have grayer bills than McCown's.

VOICE: Song, heard only on breeding grounds, is a pleasant, rapid warble, given in song flight or from a low perch. Distinctive call, a two-syllable *kittle,* repeated one or more times. Also gives a soft, high-pitched rattle and a short *buzz* call.

RANGE: Fairly common; nests in moist upland prairies. Somewhat shy; generally found in dense grass; gregarious in fall and winter. Casual during migration to eastern North America and Pacific Northwest; more regularly to CA.

McCown's Longspur *Rhynchophanes mccownii* L 6" (15 cm)

White tail marked by dark inverted-T shape. Note also stouter, thicker-based bill than bills of other longspurs. (Given this and genetic evidence, it has been restored to its own genus.) Primary projection slightly longer than Chestnut-collared Longspur; in perched bird, wings extend almost to tip of short tail. **Breeding male** has black crown, black malar stripe, black crescent on breast; gray sides. Upperparts streaked with buff and brown, with gray nape, gray rump; chestnut median coverts form contrasting crescent. **Breeding female** has streaked crown; may lack black on breast and show less chestnut on wing. By fall, bill is pinkish with dark tip; feathers are edged with buff and brown. **Winter female** is paler than female Chestnut-collared, with fewer streaks on underparts and a broader buffy eyebrow; overall suggestive of female House Sparrow (page 528). Some **winter males** have gray on rump; variable blackish on breast; retain chestnut median coverts. **Juvenile** is streaked below; pale fringes on feathers give upperparts a scaled look; paler overall than juvenile Chestnut-collared. Often found amid large flocks of Horned Larks. Look for McCown's chunkier, shorter-tailed shape, slightly darker plumage, mostly white tail, thicker bill, and undulating flight.

VOICE: Song, heard only on breeding grounds, is a series of exuberant warbles and twitters, generally given in song flight. Calls include a dry rattle, a little softer and more abrupt than Lapland Longspur; also gives single finchlike notes.

RANGE: Locally fairly common but range has shrunk significantly since the 19th century. Nests in dry shortgrass plains; in winter, also found in plowed fields and dry lake beds. Very rare visitor to interior CA and NV. Casual in coastal CA and southern OR and BC; accidental to the East Coast.

Chestnut-collared Longspur

chestnut collar

short primary projection

breeding males

black breast and belly

winter ♂

veiled black breast and belly

small darkish bill

faint streaks on breast

winter ♀

dark triangle on white tail

McCown's Longspur

black crown

breeding ♂

black chest patch

breeding ♀

black inverted-T on white tail

plainer face than Chestnut-collared with buffy supercilium

unstreaked buffy breast

thick pinkish bill

short tail

winter ♀

veiled blackish chest patch

chestnut median coverts

juvenile

slightly longer primary projection than Chestnut-collared

winter ♂

Snow Bunting *Plectrophenax nivalis* L 6¾" (17 cm)

Black-and-white breeding plumage acquired by end of spring by wear. Bill is black in summer, orange-yellow in winter. In all seasons, note long black-and-white wings. **Males** usually show more white overall than **females,** especially in the wings. **Juvenile** is grayish and streaked, with buffy eye ring; very similar to juvenile McKay's Bunting. **First-winter** plumage, acquired before migration, is darker overall than adult.

VOICE: Calls include a sharp, whistled *tew;* a short buzz; and a musical rattle or twitter. Song, heard only on the breeding grounds, is a loud, high-pitched musical warbling.

RANGE: Fairly common; breeds on tundra, rocky shores, and talus slopes. During migration and winter, found on shores, especially sand dunes and beaches, in weedy fields and grain stubble, and along roadsides, often in large flocks that may include Lapland Longspurs and Horned Larks. A few are found in winter on Atlantic coast to northernmost FL. Accidental to southern CA, AZ, and west TX.

McKay's Bunting *Plectrophenax hyperboreus* L 6¾" (17 cm)

Breeding male is mostly white, having a pure white back and less black on wings and tail than Snow Bunting; **female** darker with fine dark markings on crown and back; separated from Snow Bunting by wing tip and tail patterns in flight (McKay's has more white) and by solid white panel on greater wing coverts. **Winter** plumage is edged with rust or tawny brown, but male is whiter overall than Snow Bunting; female more similar to male Snow Bunting. Again, look carefully at wing tip and tail patterns and at the greater coverts, but beware that male Snow Bunting is similar. Certain identification of winter female McKay's problematic, especially when introducing the issue of hybrids. McKay's briefly held juvenal plumage (not shown) is significantly paler than Snow Bunting both dorsally and especially ventrally; rather than being dark gray across the chest, juveniles are whitish and lightly streaked.

VOICE: Calls and song similar to Snow Bunting.

RANGE: Known to breed only on St. Matthew Island and the much smaller nearby Hall Island in the Bering Sea, not far from where they winter. They arrive on breeding territories in spring earlier (at least a few by the latter half of Mar.) than Snow Buntings arrive at their own breeding sites. A few sometimes present in late spring on St. Lawrence Island, AK; summer rarely on Pribilofs. Rare to uncommon in winter along western coast of AK; casual in winter south on coast to OR, in interior of AK and on Aleutians. Some breeding males seen on Bering Sea islands have white backs but show extensive black on scapulars, sometimes fine back streaking. These may well be hybrids, but perhaps one-year-old McKay's are not as pristine white above.

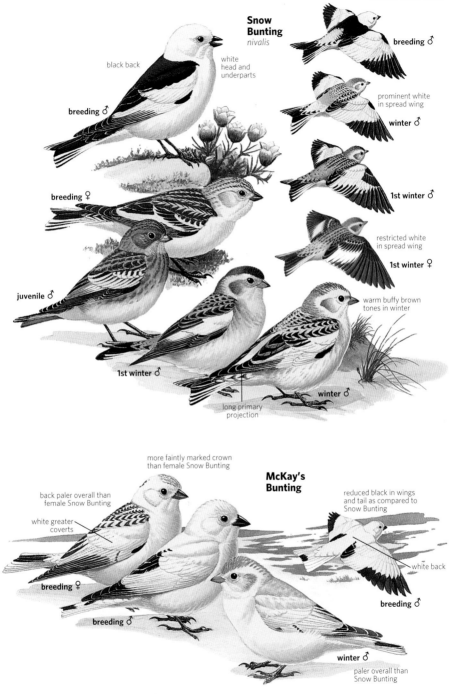

Snow Bunting
nivalis

black back

white head and underparts

breeding ♂

breeding ♀

juvenile ♂

1st winter ♂

long primary projection

breeding ♂

prominent white in spread wing

winter ♂

1st winter ♂

restricted white in spread wing

1st winter ♀

warm buffy brown tones in winter

winter ♂

McKay's Bunting

more faintly marked crown than female Snow Bunting

back paler overall than female Snow Bunting

white greater coverts

breeding ♀

breeding ♂

reduced black in wings and tail as compared to Snow Bunting

white back

breeding ♂

winter ♂

paler overall than Snow Bunting

WOOD-WARBLERS Family Parulidae

A colorful New World family. About half of its numerous species occur in North America and most of those are highly migratory, with a few reaching central South America. SPECIES: 114 WORLD; 57 N.A.

Blue-winged Warbler *Vermivora cyanoptera* L 4¾" (12 cm)

Male has bright yellow crown and underparts, white or yellowish white undertail coverts, black eye line, blue-gray wings with two white wing bars. **Female** duller overall. In both sexes, bill is long and slender; extensive white on tail is visible from below.

VOICE: Main song is a wheezy *beee-bzzz*, the second note lower; alternate song is longer and more complex. Call is a dry, sharp chip; in flight, call is a thin *zit*.

RANGE: Locally common; inhabits brushy meadows, second-growth woodlands, and power-line cuts; nests on the ground. Rare to Atlantic Canada in fall; very rare vagrant to western U.S., but increasing. Prefers a greater diversity of habitat than Golden-winged Warbler. Range is expanding at northern edge; gradually replacing Golden-winged. In general, a 50-year rule has been postulated: The time from when Blue-wingeds first arrive in Golden-wingeds' range and when the last hybrids are seen (sometimes "Lawrence's Warbler," see below) and only Blue-wingeds remain is about 50 years.

Golden-winged Warbler *Vermivora chrysoptera*

L 4¾" (12 cm) **Male** has black throat; black ear patch bordered in white; yellow crown and wing patch. **Female** similar but duller. In both sexes, extensive white on tail is conspicuous from below; underparts are grayish white; bill long and slender. In areas where Golden-winged and Blue-winged overlap, hybrids are frequent. These hybrids may vary considerably from parent species in amount of black on head and throat, amount of yellow below, and size and color of wing bars. Some variations are shown here of the two main types, the more frequent **"Brewster's Warbler"** and the rare **"Lawrence's Warbler"** backcross; both of these were originally described as distinct species. "Lawrence's" is most often produced by crossing a first-generation hybrid with one of the parent species, which can result in the recessive traits showing.

VOICE: Main song is a soft *bee-bz-bz-bz;* also gives an alternative song similar to Blue-winged Warbler. Calls also similar to Blue-winged. Hybrids' songs usually sound like one of the parent species.

RANGE: Prefers overgrown pastures, briery woodland borders. Overall prefers earlier successional habitats than Blue-winged, which is more catholic in its choices; Blue-winged males arrive at breeding territory a little earlier, thus perhaps explaining that species' dominance. Nests on the ground. Uncommon to rare and declining. Very rare in fall to Maritime Provinces; very rare and declining vagrant to western U.S. Found farther north and at higher elevations in the Appalachians than Blue-winged. Winters in southern Mexico and Central America.

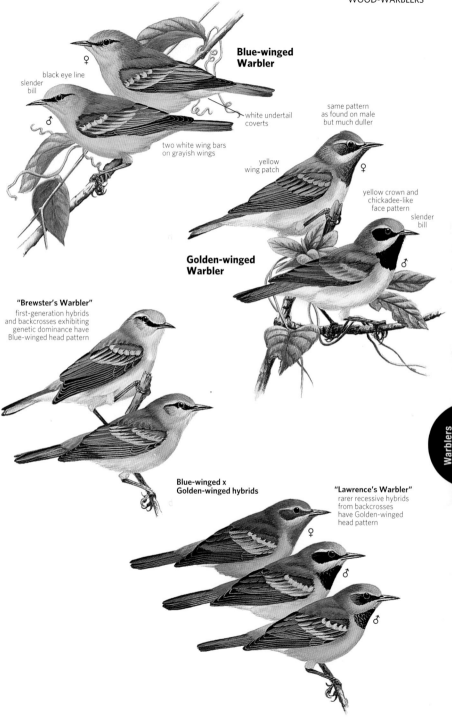

Blue-winged Warbler

black eye line
slender bill

♀

♂

white undertail coverts

two white wing bars on grayish wings

same pattern as found on male but much duller

yellow wing patch

♀

yellow crown and chickadee-like face pattern

slender bill

♂

Golden-winged Warbler

"Brewster's Warbler"
first-generation hybrids and backcrosses exhibiting genetic dominance have Blue-winged head pattern

Blue-winged x
Golden-winged hybrids

"Lawrence's Warbler"
rarer recessive hybrids from backcrosses have Golden-winged head pattern

♀

♂

♂

Warblers

Tennessee Warbler *Oreothlypis peregrina* L 4¾" (12 cm)

Plump, with short tail and long, straight bill. **Male** in spring is green above with gray crown, bold white eyebrow; white below. **Female** is tinged with yellow or olive overall, especially in fresh fall plumage. Adult male in fall resembles spring adult female but shows more yellow below. Immature also yellowish below; resembles young Orange-crowned, but is greener above and has a shorter tail and usually white undertail coverts. Spring male may be confused with Red-eyed (page 354) and Warbling (page 356) Vireos; note especially Tennessee Warbler's slimmer bill, greener back.

VOICE: Distinctive two- or three-part song; in three-part version, several rapid two-syllable notes are followed by a few higher single notes, usually ending with a staccato trill. Call is a sharp chip; flight call is a thin *seet*.

RANGE: Fairly common in interior; uncommon along East Coast. Found in coniferous and mixed woodlands in summer, mixed open woodlands and brushy areas during fall migration. Somewhat scarcer in recent decades. Nests on the ground; generally feeds high in trees. Rare migrant in West; very rare in winter in coastal CA.

See subspecies map, page 554

Orange-crowned Warbler *Oreothlypis celata* L 5" (13 cm)

Olive above, paler below. Yellow undertail coverts and faint, blurred streaks on sides of breast separate Orange-crowned from similar Tennessee Warbler. Note also Orange-crowned's thinner, slightly downcurved bill and longer tail. Plumage varies from the smaller, brighter, yellower birds of western U.S., such as Pacific *lutescens,* to the duller *orestera* (not shown; affinities appear closer to *celata,* including juvenal and immature plumages) of the Great Basin and Rockies; to the dullest, *celata,* which breeds across AK and Canada and winters primarily in southeastern U.S.; *celata* is the latest fall migrant of the warblers. Another subspecies, *sordida* (not shown), of the Channel Islands and adjacent mainland in southern CA, is like *lutescens* but darker and more streaked ventrally. Tawny orange crown, absent in some **females** and **immatures,** is seldom discernible in the field. Immature *celata* can be particularly drab; young *celata* is similar to immature Tennessee but shows yellow undertail coverts and grayer upperparts.

VOICE: Song is a high-pitched staccato trill, faster in *lutescens;* quite variable in *sordida.* Call note, a somewhat metallic chip; also a thin *seet*.

RANGE: Inhabits open, brushy woodlands, forest edges, and thickets. Nests on the ground; generally feeds in low branches, often in dead leaf clumps. Common in the West; rarer in the East, especially scarce on East Coast north of Southeast region.

Nashville Warbler *Oreothlypis ruficapilla* L 4¾" (12 cm)

Bold white eye ring, gray head, olive upperparts, and white area below legs. **Female** is duller than **male.** Rump brighter on longer-tailed western subspecies *ridgwayi,* which more often wags its tail.

VOICE: Song of eastern *ruficapilla* is a series of high *see-weet* notes and a lower short trill; call, a dull *chink.* In *ridgwayi,* song is sweeter, call is sharper.

RANGE: Common; found in second-growth woodlands, brushy areas, and spruce bogs. Rare migrant on Great Plains, most likely *ruficapilla*. Accidental in fall on St. Lawrence Island, AK.

slender, straight spike-like bill

fall

breeding ♀

bright green back, faint wing bars

bright green upperparts

whitish undertail coverts on most, sometimes pale yellow

fall

gray cap and dark eye line

white belly

Tennessee Warbler

short tail

breeding ♂

grayish head

immature ♀

slender bill slightly downcurved

grayish underparts

split eye ring

longer tail than Tennessee

even drab birds have yellowish undertail coverts

Orange-crowned Warbler
celata

♂

blurry streaks on breast

brighter overall and more of a lemon yellow below

♂ *lutescens*

grayish head with complete white eye ring

olive upperparts

adult ♂

Nashville Warbler
ridgwayi

adult ♂
ruficapilla

yellow-tinged rump characteristic of *ridgwayi*

extensively yellow underparts

pale vent

olive-edged remiges

yellowish chin and throat

shorter tail than *ridgwayi*

immature ♀

rump duller and more olive than *ridgwayi*

Colima Warbler *Oreothlypis crissalis* L 5¾" (15 cm)
Larger and browner than Virginia's Warbler; rufous crown patch usually visible.
VOICE: Song is a trill similar to Orange-crowned Warbler. Call is a loud note similar to Virginia's but not quite as sharp.
RANGE: Mexican species; breeding range extends to oak woodlands of Chisos Mountains, Big Bend National Park, TX. Casual to Davis Mountains farther north, where a small population of presumed hybrids with Virginia's exists at high elevations, just below the summit of Mount Livermore. The English name refers to a state in Mexico, part of its winter range and where the species was discovered.

Virginia's Warbler *Oreothlypis virginiae* L 4¾" (12 cm)
Bold white eye ring on gray head; upperparts gray. Yellow patch on breast, yellow undertail coverts. Female is duller overall. Fall **immature** is browner; little or no yellow on breast. Often wags its long tail.
VOICE: Song is a rapid series of thin notes, often ending with lower notes; call is a sharp *chink.*
RANGE: Common in mountain brushlands and stunted oaks. Rare to coastal CA in fall, casual in winter. Casual to OR and may be a rare breeder in mountains of southeastern OR. Very rare migrant along western Great Plains. Casual, mostly accidental in East but recorded as distantly (from normal range) as NB and Goose Bay, Labrador.

Lucy's Warbler *Oreothlypis luciae* L 4¼" (11 cm)
Pale gray above, whitish below, with a short tail, which is periodically bobbed. **Male's** reddish crown, patch, and rump distinctive. Female and **immatures** duller; can be confused with juvenile Verdin, which has different bill shape and face pattern and longer tail.
VOICE: Lively song is a short trill followed by lower, whistled notes. Call is a sharp *chink.*
RANGE: Fairly common in mesquite and cottonwoods along watercourses mostly in lowlands, but locally up in foothill canyons too; nests in tree cavities. Very rare to coastal CA, mainly in fall, casually in winter; also casually in winter along Rio Grande in west TX. Accidental to OR, south TX, LA, and MA. Winters in western and southwestern Mexico.

Crescent-chested Warbler *Oreothlypis superciliosa*
L 4¼" (11 cm) Bluish gray head with broad white eyebrow; green back; no wing bars or white in tail. Chestnut crescent distinct on **adult male;** reduced on female and immature male; absent or an orange wash on **immature female.**
VOICE: Call is a high, sharp *sik,* similar to Orange-crowned Warbler, but softer. Song is a very fast, buzzy trill on one pitch.
RANGE: Resident from northern Mexico to northern Nicaragua; casual to southeastern AZ, where recorded from scattered locations at all times of the year from Huachuca Mountains northwest to Santa Rita Mountains. Most records are in pine-oak at mid-level elevations in mountains, but recorded from Patagonia, AZ, in winter; one sight record for Chisos Mountains, TX.

Colima Warbler

brownish wash to back and sides

apricot-colored rump and undertail

long tail

overall brownish gray with yellow undertail coverts

Virginia's Warbler

whitish eye ring

immature

yellow breast patch

tail longer than Nashville, frequently bobbed

gray-edged remiges

yellow undertail coverts

♂

Lucy's Warbler

immature ♀

dark eye stands out in blank face

reddish chestnut cap

rusty rump

very short tail

reddish chestnut rump

♂

breast band very faint or lacking

bluish head with bold, broadening white supercilium

immature ♀

green back

adult ♂

no wing bars

obvious chestnut breast band reduced on adult female and immature male

Crescent-chested Warbler

very plain, lacks yellow head of adult

juvenile Verdin for comparison

longish tail

Northern Parula *Setophaga americana* L 4½" (11 cm)

Short-tailed warbler, gray-blue above with yellowish green upper back, two bold white wing bars. Throat and breast bright yellow, belly white. In **adult male,** reddish and black bands cross breast. In **female** and immature male, bands are fainter or absent.

VOICE: One song is a rising buzzy trill, ending with an abrupt *zip* in eastern birds; no clear final note in primary song and trill rate is slightly slower in more westerly birds. Calls include a clear chip.

RANGE: Common. Nests in coniferous or mixed woods, especially near water and where Spanish moss (South) and old-man's beard lichen (in North) are present; these are used in building nests. Rare (mainly spring) throughout the West in migration; very rare in summer throughout West, and a number of nesting records for coastal CA. Casual in winter in CA and the Southwest.

Tropical Parula *Setophaga pitiayumi* L 4½" (11 cm)

Dark mask and lack of distinct white eye ring distinguish Tropical from Northern Parula. Also yellow of throat extends farther onto sides of face; **male** has more blended orange breast band.

VOICE: Song like western types of Northern Parula. Chip notes also like Northern.

RANGE: Rare in south TX. Very rare to west TX; accidental to northeast CO. The few records in southeast AZ could in part pertain to west Mexican *pulchra,* which has thicker white wing bars.

Yellow Warbler *Setophaga petechia* L 5" (13 cm)

Plump, yellow overall; short tail, prominent dark eye; reddish streaks below distinct in **male,** faint or absent in **female; immatures** duller. Much geographic variation: Red streaks on western subspecies fainter than widespread eastern *aestiva,* but *aestiva* males are individually variable. Northern subspecies greener above; green extends forward through crown; immatures of *amnicola* breeding in northeastern Canada olive overall; birds from northwestern North America also dull, can be more brownish. Southwest *sonorana* pale with faint red streaks below; resident *gundlachi* of southernmost FL from West Indian **"Golden"** group; note green crown (some adult males have dull chestnut, as can some *aestiva* males; subspecies farther south typically with brighter chestnut), short primary projection. Resident subspecies in mangroves from Mexico south known as **"Mangrove Warbler"**; in nearly all "Mangrove" subspecies, adult males have chestnut heads and width of streaking varies; immatures of this and "Golden" are very dull.

VOICE: Song, rapid, variable, is sometimes written *sweet sweet sweet I'm so sweet.* Call is a sweet, rich chip, often repeated in a rapid series when agitated. The flight call is a breezy *zeet.*

RANGE: Favors wet habitats, especially willows and alders; open woodlands, orchards. In East, *aestiva* an early fall migrant, large numbers can be seen migrating west on Gulf Coast in Aug.; *amnicola* migrates later. Small numbers winter from coastal CA to southern AZ. Small numbers of "Mangroves" (*oraria*) resident in mangroves in coastal south TX. "Golden" also found in mangroves. "Mangrove" accidental to southern CA (San Diego and south end of Salton Sea) and southern AZ (Roosevelt Lake). These likely involve *castaneiceps* from southern Baja California Sur or *rhizophorae* from coast of west Mexico, or both.

Northern Parula

bluish above with bronze-green back

no eye ring

yellow extends into partly bluish face

immature ♀

broken white eye ring

banded chest

adult ♂

blended orange breast

thick white wing bars

♀

adult ♂

extensively yellow on underparts

Tropical Parula
nigrilora

adult
♂ *aestiva*

dark eye stands out in blank face

Yellow Warbler
aestiva

immature
♀ *aestiva*

red streaks on underparts

pale tertial edges

immature
♀ *gundlachi*

♀ *aestiva*
faint streaks

whitish underparts

short tail with yellow tail spots

immature
♀ *amnicola*

overall duller and darker than *aestiva*

"Golden" adult
♂ *gundlachi*

gundlachi found in Cuba and south Florida

immature
rubiginosa

immatures of northern subspecies are duller and darker and are washed with olive or brownish

adult
♂ *sonorana*

sonorana found in Southwest

"Mangrove" adult
♂ *oraria*

oraria found in eastern Mexico and coastal south Texas

Chestnut-sided Warbler *Setophaga pensylvanica*

L 5" (13 cm) **Breeding male** has yellow crown, black eye line, black whisker stripe; extensive chestnut on sides; **female** has greenish crown, less chestnut. Fall adults and **immatures** are lime green above, with white eye ring, whitish underparts, yellowish wing bars. Often cocks its tail.

VOICE: Song is a whistled *please please pleased to meetcha;* call is a chip note like Yellow Warbler but is not repeated rapidly.

RANGE: Fairly common in second-growth deciduous woodlands. Rare migrant in West; has nested along Front Range in CO. Casual in winter in CA and southern AZ. Casual to AK.

Magnolia Warbler *Setophaga magnolia* L 5" (13 cm)

Male is blackish above, with white eyebrow, white wing patch, yellow rump; broad white tail patches. Underparts yellow, streaked on breast and sides; undertail coverts white; under tail white except for black band at tip. Female has two wing bars; some **first-spring females** have dull white eye ring; often confused with rare Kirtland's Warbler (page 440). **Fall adults** and **immatures** are drabber, with grayish olive upperparts; white eye ring; faint gray band across breast. Compare immature Prairie Warbler (page 440). Does not bob tail.

VOICE: Song is a short, whistled *weety-weety-weeteo.* Call, given rather infrequently, is a unique, weak, nasal *tchif* or *wenk.*

RANGE: Fairly common to common breeder in moist coniferous forests. Casual in winter in south FL. Rare throughout the West, including southeastern AK in migration; casual in winter in CA and southern AZ.

Cape May Warbler *Setophaga tigrina* L 5" (13 cm)

Most plumages have yellow on face, the color usually extending to sides of neck. Note also short tail and yellow or greenish rump; the thin bill, slightly downcurved, is quite atypical for a *Setophaga.* **Breeding male's** chestnut ear patch and striped underparts distinctive; wing patch white. **Female** drabber, grayer, with two narrow white wing bars. **Immature male's** ear patch is less distinct. **Immature female** can be extremely drab, with gray face and only a tinge of yellow below and on rump. Note dark ventral streaking; always has greenish edges on flight feathers. One of our most aggressive warblers, often vigorously defending a food source, often nectar.

VOICE: One song is a high, thin *seet seet seet seet;* call, a very high, thin *sip.*

RANGE: Breeds in black spruce forests, where often uncommon except during spruce budworm outbreaks. Rare west to TX in migration; very rare (formerly) to casual throughout the West. Winters chiefly in the West Indies; a few birds winter in southernmost FL. Populations have declined in recent decades, likely reflecting control efforts to prevent spruce budworm outbreaks.

usually cocks tail

greenish crown

Chestnut-sided Warbler

breeding adult ♀

yellow crown

chestnut sides

breeding adult ♂

gray face with white eye ring

lime green upperparts

yellowish wing bars

grayish white underparts

immature

Magnolia Warbler

gray head, white eye ring

immature

grayish breast band

silver gray crown with bold supercilium

yellow rump in all plumages

yellow throat with extensive black streaking below

1st spring ♀

extensive white vent and undertail coverts

black back, white wing patch

white tail patch

white wing bars

fall adult ♂

breeding adult ♂

all have greenish edges to remiges

very slender, slightly downcurved bill

breeding adult ♀

yellow surroun chestnut chee

greenish rump

immature ♀

breeding adult ♂

white wing patch

immature ♂

all plumages with streaked underparts

Cape May Warbler

Black-throated Blue Warbler *Setophaga caerulescens*

L 5¼" (13 cm) **Male's** black throat, cheeks, and sides separate blue upperparts from white underparts. Bold white patch at base of primaries. Appalachian males south of Susquehanna drainage *(cairnsi)* average darker above, but are only weakly differentiated. **Female's** pale eyebrow is distinct on dark face; upperparts brownish olive; underparts buffy; wing patch smaller, occasionally absent on immature females.

VOICE: Typical song is a slow series of four or five wheezy notes, the last note higher: *zwee zwee zwee zweeee* or a slower *zur zurr zreee.* Call is a single sharp *dit,* like Dark-eyed Junco.

RANGE: Inhabits deciduous forests; prefers lower or mid-level branches. Very rare migrant west of the Mississippi River to TX and Great Plains. Very rare in fall and casual in spring and winter in West. Accidental to southeastern AK. A few winter in south FL; most migrate to the West Indies in winter; *cairnsi* winters on more westerly Greater Antilles.

Yellow-rumped Warbler *Setophaga coronata* L 5½" (14 cm)

Yellow rump, yellow patch on side, yellow crown patch, white tail patches. In northern and eastern birds, **"Myrtle Warblers,"** note white eyebrow, white throat and sides of neck, contrasting cheek patch. Western birds, **"Audubon's Warblers,"** have yellow throat, except for a few immature females. Some males in the mountains of the Southwest show more black. All **females** and fall males are duller than **breeding males** but show same basic pattern.

VOICE: Song, a slow warble, usually rising or falling at the end in "Audubon's," a musical trill in one song of "Myrtle." Call note of "Myrtle" is lower, flatter.

See subspecies map, page 554

RANGE: Common to abundant in coniferous or mixed woodlands. The most northerly of our wintering warblers ("Myrtle" in particular); Yellow-rumped's digestive system enables it to digest berries from bayberry, wax myrtle, and poison oak, unlike other warblers. Also unlike other warblers, Yellow-rumped is a facultative migrant (moves around during the winter) likely as a result of shifting food sources. "Myrtle" is fairly common in winter on West Coast, scarce in the Southwest. "Myrtle" and "Audubon's" hybridize frequently, chiefly through the Canadian Rockies of BC and western AB. "Audubon's" is uncommon to Great Plains, central TX; casual in the East.

Black-throated Gray Warbler *Setophaga nigrescens*

L 5" (13 cm) **Adult** plumage is basically the same year-round: black-and-white head; gray back streaked with black; white underparts, sides streaked with black; small yellow spot between eye and bill. Lacks central crown stripe of the Black-and-white Warbler (page 444); undertail coverts are white. Immature male resembles adult male; immature female is brownish gray above, throat white.

VOICE: Varied songs include a buzzy *weezy weezy weezy weezy-weet.* Call is a flat *tchip,* slightly duller than Townsend's. Flight call is like other related species in this group (all species on page 436).

RANGE: Inhabits woodlands, brushlands, chaparral. Very rare migrant on western Great Plains. Rare in winter in lower Rio Grande Valley, TX. Very rare during migration and in winter along the Gulf Coast. Casual otherwise in eastern North America.

whitish supercilium
with dark cheek

whitish
eye arc

buffy
underparts
♀

most females
show whitish patch
at base of primaries

dark blue
crown

**Black-throated
Blue Warbler**

black throat

black stippling
on back

♂ *caerulescens*

white patch larger
in adult males

Appalachians
♂ *cairnsi*

"Audubon's Warbler"

more extensively
black overall

Southwest breeding ♂

most have pale
yellow in rounded
throat patch

**Yellow-rumped
Warbler**

yellow crown patch

"Myrtle Warbler"
coronata

fall ♀

yellow
throat

angled
whitish
throat

breeding ♀

yellow
patches
on sides
of breast

breeding ♂
auduboni

breeding ♂

browner above than
fall "Audubon's"

distinct
whitish
supercilium

yellow rump

whitish patch
sharply angled
on sides of
throat

fall ♀

gray upperparts

♀

bold white
supercilium

yellow
supraloral spot
in all plumages

broad white
submoustachial

black
throat

adult ♂

**Black-throated
Gray Warbler**

Black-throated Green Warbler *Setophaga virens*
L 5" (13 cm) Bright olive green upperparts; yellow face with greenish ear patch. Underparts are white, tinged with yellow on sides of vent and often on breast. **Male** has black throat and upper breast and black-streaked sides. **Female** and immatures show much less black below; **immature female** generally has dark streaking only on sides.
VOICE: One song is a hoarse *zeee zeee zee-zo-zee;* the other is often written as *trees, trees, whispering trees.* Call is a sharp, flat *tip* or *tsik,* much like the other species on this page. The flight note is a thin, non-buzzy *see.*
RANGE: Fairly common in coniferous or mixed forests in summer. A few winter in south TX. Fall migration in East very extended (July to early Nov.). Very rare migrant in West, mostly in late fall. Casual in winter in coastal CA; accidental to southeastern AK.

Golden-cheeked Warbler *Setophaga chrysoparia*
L 5½" (14 cm) **E** Dark eye line, unmarked yellow ear patches, and lack of any yellow on underparts distinguish this species from similar Black-throated Green Warbler. **Male** black above, with black crown, black bib, black-streaked sides. **Female** and immature male duller, upperparts olive with dark streaks; chin yellowish or white; sides of throat streaked. **Immature female** shows less black on underparts.
VOICE: Song, *bzzzz layzee dayzee,* ends on a high note. Call notes are like Black-throated Green.
RANGE: Endangered; local in mixed cedar-oak woodland of the Edwards Plateau in central TX. In TX, males appear on breeding territory in early Mar., depart in July. Almost unknown as a migrant in U.S. (many TX reports are probably questionable), although recorded regularly in the Sierra Madre Oriental of northeastern Mexico. Accidental to FL, NM, and CA (Farallones), all well-documented records.

Hermit Warbler *Setophaga occidentalis* L 5½" (14 cm)
Yellow head, with dark markings extending from nape onto crown. **Male** has black throat; in **female** and immatures, chin yellowish, throat shows less blackish. **Immature female** shows more olive above.
VOICE: Song is a high *seezle seezle seezle seezle zeet-zeet.* Call notes like Black-throated Green or Townsend's.
RANGE: Fairly common in mountain forests; nests in tall conifers. During migration, also seen in lowlands, especially in spring; also regular migrant through mountains of Southwest. Uncommon to rare in CA winter range. Casual in spring to southwestern BC. Casual in the East.

Townsend's Warbler *Setophaga townsendi* L 5" (13 cm)
Dark crown, dark ear patch bordered in yellow. Yellow breast with streaked sides. **Adult male's** throat is black; **female** and immature male have streaked lower throat. **Immature female** is duller, lacks streaking on back. Frequently hybridizes with Hermit; with **hybrids,** genetic dominance usually shows birds with yellowish, streaked underparts of Townsend's and yellow head of Hermit.
VOICE: Variable song, a series of hoarse *zee* notes. Call note, *tchip,* much like the other species on this page, but harder than Black-throated Gray. Flight call for all five of these related species is a thin, non-buzzy *see.*
RANGE: Found in coniferous forests. Rare fall migrant on western Great Plains. Casual in the East.

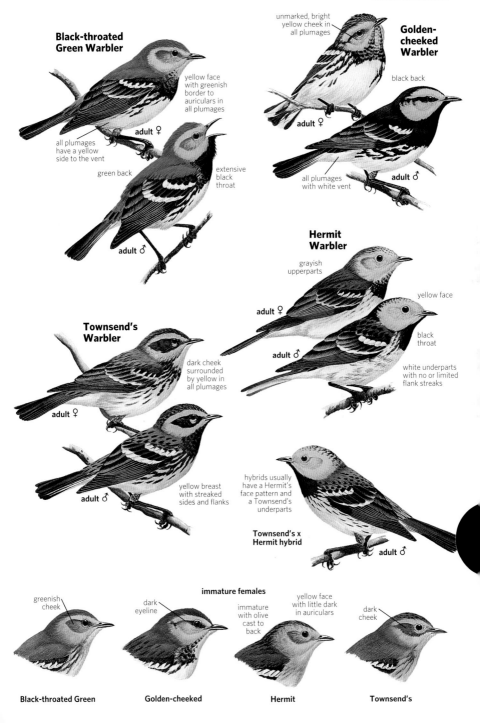

Black-throated Green Warbler

unmarked, bright yellow cheek in all plumages

Golden-cheeked Warbler

black back

adult ♀

yellow face with greenish border to auriculars in all plumages

all plumages have a yellow side to the vent

green back

adult ♀

extensive black throat

adult ♂

all plumages with white vent

adult ♂

Hermit Warbler

grayish upperparts

adult ♀

yellow face

black throat

adult ♂

white underparts with no or limited flank streaks

Townsend's Warbler

dark cheek surrounded by yellow in all plumages

adult ♀

adult ♂

yellow breast with streaked sides and flanks

hybrids usually have a Hermit's face pattern and a Townsend's underparts

Townsend's x Hermit hybrid

adult ♂

immature females

greenish cheek

dark eyeline

immature with olive cast to back

yellow face with little dark in auriculars

dark cheek

Black-throated Green

Golden-cheeked

Hermit

Townsend's

Blackburnian Warbler *Setophaga fusca* L 5" (13 cm)

Fiery orange throat, broad white wing patch, triangular ear patch conspicuous in **adult male. Female** and immature male have paler throat, **immature female** paler still; note also two white wing bars, streaked back, and bold yellow or buffy eyebrow, broader behind the eye, that curls around onto side of neck. Some almost whitish below with whitish supercilium can be easily mistaken for immature female Cerulean, which is a rare migrant over most of East, especially in fall. Orange or yellow forehead stripe and white in outer tail feathers distinct in all males.

VOICE: One song is a short series of high notes followed by a squeaky, ascending trill, ending on a very high note; another is a series of high, two-part phrases. Calls include a sharp *tckik;* flight call is a buzzy *zzee.*

RANGE: Fairly common breeder in coniferous or mixed forests; also pine-oak woodlands in Appalachians. Rare in fall to coastal CA; casual elsewhere west of dashed line on map in spring and fall.

Cerulean Warbler *Setophaga cerulea* L 4¾" (12 cm)

Small, with short tail, two wide white wing bars. **Adult male** bluish above with dark streaks; dark breast band, thinner, even broken, on first-spring male, which also shows indication of whitish supercilium. **Female** has greenish mantle, blue-green or bluish crown; pale eyebrow broadens behind the eye but does not connect to sides of neck like Blackburnian; breast and throat pale yellowish. Immature male like female, but shows some bluish and dark streaks above.

VOICE: Song is a short, fast, accelerating series of buzzy notes on one pitch, ending with a long, single buzz note. Rather infrequently heard chip is slurred; flight call is a buzzy *zzee,* like Blackburnian.

RANGE: Declining in the heart of its range. Found in tall trees in swamps, bottomlands, mixed woodlands near water. Fall migration begins from the second week of July. Range is expanding slightly in Northeast. Casual migrant north to Atlantic Provinces, west to Great Plains and CA; accidental elsewhere in West.

See subspecies map, page 555

Palm Warbler *Setophaga palmarum* L 5½" (14 cm)

Upperparts olive. **Breeding adult** of eastern subspecies, *hypochrysea,* known as "Eastern Palm" or "Yellow Palm," has chestnut cap, yellow eyebrow, and entirely yellow underparts, with chestnut streaking on sides of breast. Fall adults and immatures lack chestnut cap and streaking; yellow is duller. Western nominate subspecies, *palmarum,* known as "Western Palm," has whitish belly and darker streaks on sides of breast; less chestnut. **Fall adults** and immatures are drab; some are washed with pale yellow below. Habitually wags its tail as it forages.

VOICE: Song is a rapid, buzzy trill. Call is a sharp *tsik.* Flight call is a high *seet* or low *see-seet.*

RANGE: Fairly common; nests in brush at edge of spruce bogs. During migration and winter, found in woodland borders, open brushy areas, and marshes. Eastern *hypochrysea* winters in Southeast (north of central FL) and migrates north and south through Atlantic states (casual west of Appalachians); *hypochrysea* is significantly earlier in spring and later in fall than western *palmarum,* which has a much broader winter range, including the entire West Indies away from U.S. range. Regular on West Coast in fall and winter *(palmarum); hypochrysea* casual to CA.

Blackburnian Warbler

breeding ♀

bold white wing patch

all Blackburnians have pale mantle lines

fiery orange throat

adult males have buffy belly

fall adult ♂

breeding adult ♂

dark triangular auricular patch

bold, broad supercilium connects to pale sides of neck

immature ♀

bold, broad supercilium does not connect to sides of neck

unstreaked greenish back

Cerulean Warbler

cerulean blue head and upperparts

bold, whitish wing bars

pale blue above, brightest on crown

immature ♀

short tail

adult ♂

blackish breast band

adult ♀

rufous cap

western breeding
palmarum

duller midsection contrasts with yellow throat and under tail

more olive above and more uniformly yellow below in all plumages than *palmarum*

both subspecies constantly bob tail

white tail spots

Palm Warbler

all yellow underparts with chestnut streaks on sides of breast

distinctive pale supercilium and dark eye line

eastern breeding
hypochrysea

underparts and supercilium yellowish throughout

earlier spring and later fall migrant than *palmarum*

western fall
palmarum

western fall
palmarum

some are tinged yellow throughout underparts

eastern fall
hypochrysea

yellow undertail coverts

Grace's Warbler *Setophaga graciae* L 5" *(13 cm)*

Black-streaked gray back; throat and upper breast bright yellow; rest of underparts white, with black streaks on sides; short bill; yellow eyebrow becomes white behind eye. **Female** slightly duller; immature browner above.

VOICE: Song is a rapid, accelerating trill. Call is a sweet chip.

RANGE: Inhabits coniferous forests of southwestern mountains, especially yellow pines. Usually forages high in the trees. Very rare to southern CA casual to northern CA; accidental IL.

flavescens

Yellow-throated Warbler *Setophaga dominica*

L 5½" *(14 cm)* Plain gray back; large, white patch on each side of head. **Male** has black crown and face; in female, black is less extensive on crown. Throat and upper breast bright yellow; black streaks on sides; bold white eyebrow. Eastern subspecies *dominica* and *stoddardi* (of eastern Gulf Coast, not shown) have yellow supraloral area, unlike most of more westerly *albilora; stoddardi* and birds from Delmarva Peninsula have very long bills. Usually forages high, creeping methodically along the branches.

VOICE: Song is a series of clear, downslurred whistles ending with a rising note. Call is a rich chip.

RANGE: Fairly common in live oak and pine woodlands, cypress, and sycamores. Rare north to southern ON in spring and Newfoundland in fall. Very rare to casual in West during migration; casual in winter.

Kirtland's Warbler *Setophaga kirtlandii* L 5¾" *(15 cm)* **E**

Blue-gray above, strongly black-streaked on back; yellow below, streaked on sides; white eye ring, broken at front and rear; two indistinct wing bars. Often confused with first-spring female Magnolia Warbler (page 432). **Adult female** is slightly duller; **immature female** brownish above. Constantly wags its tail.

VOICE: Song is a loud series of low notes followed by slurred whistles; call is a low, forceful chip, similar to Prairie.

RANGE: Endangered: The annual breeding census counted 1,773 singing males in 2010, up from the historic low of 167 singing males in 1987. Nests in northern MI, where controlled plantings and fires produce the required habitat: young jack pines. Very rare in summer outside MI, with recent breeding records from southern ON and especially WI. Very rarely seen in migration. Winters in the Bahamas (most recent records on Eleuthera). Accidental MO, IL, and ME.

Prairie Warbler *Setophaga discolor* L 4¾" *(12 cm)*

Olive above, with faint chestnut streaks on back; yellow patch below eye; bright yellow below, streaked with black on sides; indistinct wing bars. **Female** and immature male are slightly duller. **Immature female** is duller still, grayish olive above; compare to fall Magnolia Warbler (page 432). Usually forages in lower branches and brush, twitching its tail.

VOICE: Distinctive song, a rising series of buzzy *zee* notes. Call is a flat *tsuk*.

RANGE: Generally common in open woodlands, scrublands, overgrown fields, and mangrove swamps. Casual in the West, except in coastal CA, where it is rare in fall. Also rare in fall to Atlantic Canada and Newfoundland. Declining in upper Midwest.

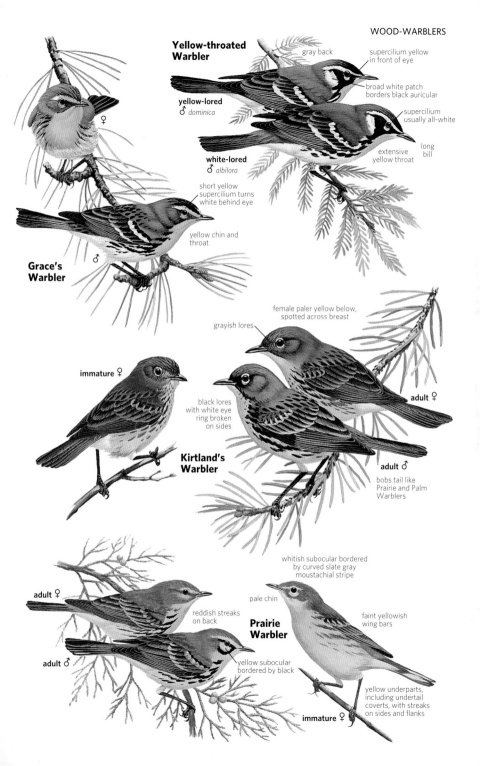

Yellow-throated Warbler

gray back

supercilium yellow in front of eye

yellow-lored
♂ *dominica*

broad white patch borders black auricular

supercilium usually all-white

white-lored
♂ *albilora*

extensive yellow throat

long bill

♀

short yellow supercilium turns white behind eye

yellow chin and throat

Grace's Warbler

♂

female paler yellow below, spotted across breast

grayish lores

immature ♀

black lores with white eye ring broken on sides

Kirtland's Warbler

adult ♀

adult ♂

bobs tail like Prairie and Palm Warblers

adult ♀

reddish streaks on back

Prairie Warbler

whitish subocular bordered by curved slate gray moustachial stripe

pale chin

faint yellowish wing bars

adult ♂

yellow subocular bordered by black

yellow underparts, including undertail coverts, with streaks on sides and flanks

immature ♀

Bay-breasted Warbler *Setophaga castanea* L 5½" (14 cm)

Breeding male has chestnut crown, throat, and sides; black face; creamy patch at each side of neck; two white wing bars. **Female** is duller. Some first-spring females lack chestnut below and are dull above; note pale on sides of nape; not as streaked overall as spring female Blackpoll. **Fall adults** and **immatures** resemble Blackpoll Warbler and Pine Warbler. Bay-breasted is brighter green above, wing bars are thicker; underparts show little or no streaking and little yellow; flanks usually show some buff or bay color; legs usually entirely dark; undertail coverts are buffy or whitish. Both Bay-breasted and Blackpoll have short tail projection past undertail coverts and show white tips to primaries. **VOICE:** Song consists of high-pitched double notes. Calls include a sharp chip note and a buzzy *zeet*. Flight call similar to Yellow Warbler. **RANGE:** Fairly common; nests in coniferous forests. Migrates slightly earlier in fall than Blackpoll. Very rare migrant in the West; a few winter records from southern CA.

Blackpoll Warbler *Setophaga striata* L 5½" (14 cm)

Solid black cap, white cheeks, and white underparts identify **breeding male;** back and sides boldly streaked with black. Compare Black-and-white Warbler (page 444). **Female** is duller overall, variably greenish above and pale yellow below; some are gray; note streaking. In **fall,** all birds resemble Bay-breasted and Pine Warblers. Mostly pale greenish yellow below, with dusky streaking on sides; legs pale on front and back, dark on sides; undertail coverts long and usually white. Crown and nape are more of an olive green, not yellow-green; wing bars are a little thinner, bill is a bit thinner, and eye line is slightly more developed. **VOICE:** Song is a series of high *tseet* notes. Calls include a sharp chip and a buzzy *zeet*. Flight call like Bay-breasted. **RANGE:** Nests in varied habitats. Uncommon to common in northern breeding range. Migrates later in fall than Bay-breasted Warbler. Rare migrant over much of West. Rare in fall in much of southern Midwest and very rare in most of Southeast because much migration is well off East Coast. Uncommon in spring on western Great Plains north of CO, otherwise rare in West, chiefly coastal CA; fewer in recent decades.

Pine Warbler *Setophaga pinus* L 5½" (14 cm)

Relatively large bill; long tail projection past undertail coverts; throat color extends onto sides of neck, setting off a well-defined dark cheek patch; in Bay-breasted and Blackpoll, cheek blends into the throat. **Male** is greenish olive above, without streaking; throat and breast yellow, with dark streaks on sides of breast; belly and undertail coverts white. **Female** is duller. **Immatures** are brownish or brownish olive above, with whitish wing bars and brownish tertial edges; male is dull yellow below, female largely white; both have brown wash on flanks. **VOICE:** Song is a twittering musical trill, varying greatly in speed. Call is a flat, sweet chip. **RANGE:** Common in pines in summer; also in mixed woodlands in winter. An early spring migrant. Very rare in fall (Oct.) and winter in CA, chiefly southern CA; casual elsewhere in West; rare to Atlantic Canada in fall and early winter. Winter birds in Southeast often forage on ground in flocks with other species, including Yellow-rumped Warblers ("Myrtle"), Eastern Bluebirds, and Chipping Sparrows.

Bay-breasted Warbler

more of yellow cast above than Blackpoll, especially on nape

slightly broader wing bars than Blackpoll

short tail like Blackpoll

large buffy patch on side of neck

breeding ♀

largely unstreaked below

immature ♀

usually with pale buffy under tail

dark legs and feet

breeding adult ♂

Blackpoll Warbler

short tail with long undertail coverts and white tail spots

breeding ♀

faint streaks on sides of breast

fall

breeding ♂

black cap and malar stripe borders white cheek

yellow soles to feet and usually to back of legs

yellow legs

unstreaked back in all plumages, unlike Blackpoll and Bay-breasted

brownish above

adult ♀

pale wraps around auriculars

dark auriculars contrast sharply with throat

immature ♀

whitish below

immature ♂

adult ♂

Pine Warbler
pinus

white lower belly and undertail coverts

long tail projection past undertail coverts

Black-and-white Warbler *Mniotilta varia* L 5¼" (13 cm)

The only warbler that regularly creeps along branches and up and down tree trunks like a nuthatch. Boldly striped on head, most of body, and undertail coverts. **Male's** throat and cheeks are black in breeding plumage; in winter, chin is white. **Female** and **immatures** have pale cheeks; female diffusely streaked on buffy flanks; buffy wash particularly bright on immature female; fall immature male is clean white below with strong black streaks on sides and flanks.

VOICE: Song is a long series of high, thin *wee-see* notes. Calls include a sharp chip and high *seep-seep*.

RANGE: Common in mixed woodlands. Rare in West south of breeding range; occurs mostly as a migrant, but a few winter too, especially in CA. Casual to AK.

American Redstart *Setophaga ruticilla* L 5¼" (13 cm)

Male glossy black, with bright orange patches on sides, wings, and tail; belly and undertail coverts white. **Female** is gray-olive above, white below with yellow patches. Immature male resembles female; by **first spring,** lores are usually black, breast has some black spotting; adult male plumage is acquired by second fall. Like redstarts of the genus *Myioborus,* often fans its tail and spreads its wings when perched.

VOICE: Variable song, a series of high, thin notes usually followed by a wheezy, downslurred note. Call is a rich, sweet chip.

RANGE: Common in second-growth woodlands. Rare migrant in CA and the Southwest; formerly more numerous. Casual in western AK.

Worm-eating Warbler *Helmitheros vermivorum*

L 5¼" (13 cm) Bold, dark stripes on buffy head; upperparts brownish olive; underparts buffy; long, spikelike bill.

VOICE: Song is a series of sharp, dry chip notes, like Chipping Sparrow but faster. Common call is a buzzy *zeep-zeep.*

RANGE: Found chiefly in dense undergrowth on wooded slopes. Often feeds in clusters of dead leaves. Rare in spring to southern ON and southern MI (has nested). Casual vagrant to CA, the Southwest, the Great Plains, and Atlantic Canada; accidental elsewhere in West.

Swainson's Warbler *Limnothlypis swainsonii* L 5½" (14 cm)

Pale eyebrow, conspicuous between brown crown and dark eye line. Brown-olive above, grayish below. Bill very long and spiky. Secretive. Walks or shuffles on the ground and swivels while picking up dead leaves.

VOICE: Song is a series of thin, slurred whistles like beginning of song of Louisiana Waterthrush; often ends with a rising *tee-oh.* Various renditions include *whee whee whip-poor-will* or *ooh ooh stepped in poo.* Calls include a loud, dry chip.

RANGE: Uncommon. Found in undergrowth in swamps and cane-brakes; rare and local in mountain laurel and rhododendron. Walks or shuffles on the ground and shivers its entire body while picking up dead leaves. Casual north to Ontario and Nova Scotia and west to eastern CO, NM, and west TX. Accidental to east-central AZ.

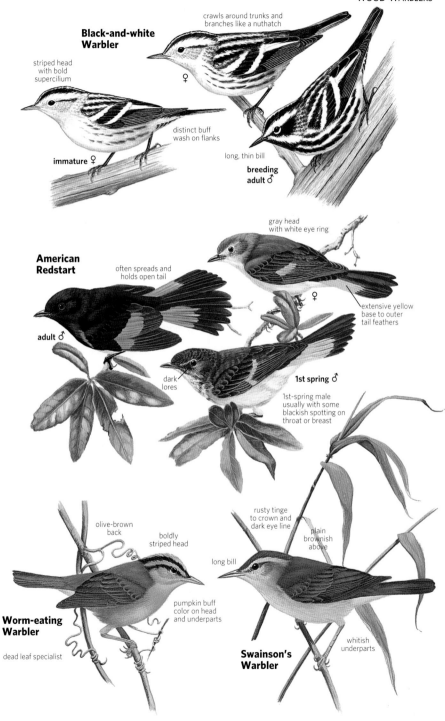

Black-and-white Warbler

crawls around trunks and branches like a nuthatch

striped head with bold supercilium

♀

distinct buff wash on flanks

immature ♀

long, thin bill

breeding adult ♂

American Redstart

gray head with white eye ring

often spreads and holds open tail

♀

extensive yellow base to outer tail feathers

adult ♂

dark lores

1st spring ♂

1st-spring male usually with some blackish spotting on throat or breast

olive-brown back

boldly striped head

rusty tinge to crown and dark eye line

plain brownish above

long bill

pumpkin buff color on head and underparts

Worm-eating Warbler

dead leaf specialist

Swainson's Warbler

whitish underparts

Prothonotary Warbler *Protonotaria citrea* L 5½" (14 cm)

Plump, short tailed, and long billed. Eyes stand out on **male's** golden-yellow head; white undertail coverts and tail patches; plain blue-gray wings. **Female** duller, head less golden.
VOICE: Song is a series of loud, ringing *zweet* notes; gives a dry chip note and buzzy flight call.
RANGE: Fairly common. The only eastern warbler that nests in tree or other cavities and crannies; usually selects a low site along streams or surrounded by sluggish or stagnant water. Casual to rare north of mapped range in East and throughout the West (most frequently recorded from CA), where more frequently recorded in fall.

Ovenbird *Seiurus aurocapilla* L 6" (15 cm)

Russet crown bordered by dark stripes; bold white eye ring. Olive above; white below, with bold streaks of dark spots; pinkish legs. Generally seen on the ground; walks, with tail cocked, rather than hops.
VOICE: Typical song is a loud *teacher teacher teacher,* rising in volume. Calls include a sharp, dry chip.
RANGE: Common in mature forests. Rare in the West. Rare in winter along Gulf and Atlantic coasts to NC. Overall a rare migrant in the West.

Louisiana Waterthrush *Parkesia motacilla* L 6" (15 cm)

Distinguished from Northern Waterthrush by contrast between white underparts and salmon buff flanks, which are pale on some and bright on others, especially fresh-plumaged late summer birds; bicolored eyebrow, pale buff in front of eye, white and much broader behind eye; larger bill; bubblegum pink legs. Most have pure white throats; a few have spots on lower throat. A ground dweller; walks, rather than hops, bobbing its tail constantly but usually slowly.
VOICE: Call note, a sharp *chick,* is slightly flatter than Northern Waterthrush. Song begins with three or four shrill, slurred notes followed by a brief, rapid jumble.
RANGE: Uncommon; found along mountain streams in dense woodlands, also near ponds and in swamps. Returns in early spring (by early Apr., even at northern end of breeding range); many have already departed from south by midsummer, nearly all by late Aug. Rare migrant and winter visitor in southeastern AZ. Casual elsewhere in West (most records from CA) and to NS.

Northern Waterthrush *Parkesia noveboracensis*

L 5¾" (15 cm) Distinguished from Louisiana Waterthrush by lack of contrast in color between flanks and rest of underparts; buffy eyebrow of even width throughout or slightly narrowing behind eye; smaller bill; drabber leg color. Some birds are washed with pale yellow below, whereas others are whiter below, with whiter eyebrow. Most have spotted throats, but some are unspotted. A ground dweller; walks, rather than hops, bobbing its tail constantly and usually rapidly.
VOICE: Call note, a metallic *chink,* is slightly sharper than Louisiana Waterthrush. Song begins with loud, emphatic notes and ends in lower notes, delivered more rapidly.
RANGE: Found chiefly in woodland bogs, swamps, and thickets; wintering birds also found in mangroves. Rare migrant in CA and the Southwest, where a few may winter.

WOOD-WARBLERS

long, stout
spiky bill

adult ♂

brilliant
yellow head
and
underparts

short tail with extensive
white tail spots

white
undertail coverts

contrasting
gray wings

♀

**Prothonotary
Warbler**

**Louisiana
Waterthrush**

long supercilium,
buffy in front of eye,
pure white behind

upperparts average
grayer than Northern

large bill

bobs tail
slowly from
side to side

usually
unstreaked
throat

contrasting
salmon-buff flanks

Ovenbird

bright pink
legs on most

rufous crown
with blackish lateral
crown stripes

olive above

bold white
eye ring

extensively
spotted below

some have a more
whitish supercilium and
underparts color

**Northern
Waterthrush**

smaller
bill than
Louisiana

supercilium and background
color to underparts always
same color — usually buffy
or buffy-yellow

some have
unmarked throat

usually
streaked
throat

Northern Waterthrush
rapidly bobs its tail up
and down

paler

duller flesh legs
than Louisiana

Mourning Warbler *Geothlypis philadelphia* L 5¼" *(13 cm)*

Lack of bold white eye ring distinguishes **adult male** from Connecticut Warbler. **Adult female** and especially **immatures** may show a thin, nearly complete eye ring, but compare with Connecticut. Immatures generally have more yellow on throat than MacGillivray's; compare also with female Common Yellowthroat (page 452). Immature males often show a little black on breast. Mourning Warblers hop rather than walk.

VOICE: Call is a flat, hollow chip, not unlike one call of Bewick's Wren. Song is a series of slurred two-note phrases followed by two or more lower phrases.

RANGE: Uncommon in dense undergrowth, thickets, moist woods, and second-growth woodland; nests on the ground. Most spring migration, which is late, is west of the Appalachians. Very rare or casual in the West. Accidental in winter in coastal southern CA. Winters in Central America and northern South America.

MacGillivray's Warbler *Geothlypis tolmiei* L 5¼" *(13 cm)*

Bold white crescents above and below eye distinguish all plumages from male Mourning and all Connecticut Warblers. Crescents may be very hard to distinguish from the thin, nearly complete eye ring found on female and immature Mourning Warblers. **Immature** MacGillivray's generally have grayer throat than immature Mournings and a fairly distinct breast band above yellow belly. Field identification is often very difficult; often best determined by call notes. Like Mourning, MacGillivray's hops rather than walks.

VOICE: Call is a sharp, harsh *tsik,* distinctly different from Mourning. Two-part song is a buzzy trill ending in a downslur.

RANGE: Fairly common; found in dense undergrowth. Rare migrant through western Great Plains. Casual to western AK, in winter in coastal CA and in the East.

Connecticut Warbler *Oporornis agilis* L 5¾" *(15 cm)*

Large eye with bold white eye ring conspicuous on **male's** gray hood and **female's** brown or gray-brown hood. Eye ring is sometimes slightly broken on back side. **Immature** has a brownish hood and brownish breast band. A large, stocky warbler, noticeably larger than Mourning and MacGillivray's Warblers. Like Mourning, long undertail coverts give Connecticut a short-tailed, plump appearance. Walks rather than hops.

VOICE: Loud, accelerating song repeats a brief series of explosive *beech-er* or *whip-ity* notes. Most frequently given call is a buzzy *zeet,* like Yellow or Blackpoll Warbler. Rarely heard call note is a nasal *chimp* or *poitch.*

RANGE: Uncommon; found in spruce bogs, moist woodlands; nests on the ground; generally feeds on the ground or on low limbs. Spring migration is very late (after mid-May in Midwest) and almost entirely west of the Appalachians. Fall migrants uncommon in the East; very rare (CA) or casual in the West, where there are only a few very late spring records. Fall migration is a little later than Mourning. Like Blackpoll, most migrate well off Atlantic coast. Winter range more poorly known than any North American passerine; believed now to be in the vicinity of the Pantanal of central South America.

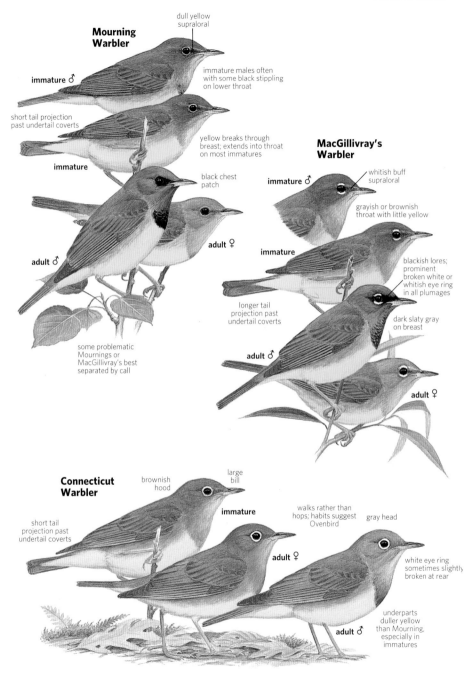

Mourning Warbler

dull yellow supraloral

immature ♂

immature males often with some black stippling on lower throat

short tail projection past undertail coverts

yellow breaks through breast; extends into throat on most immatures

immature

black chest patch

adult ♂

adult ♀

some problematic Mournings or MacGillivray's best separated by call

MacGillivray's Warbler

immature ♂

whitish buff supraloral

grayish or brownish throat with little yellow

immature

blackish lores; prominent broken white or whitish eye ring in all plumages

dark slaty gray on breast

longer tail projection past undertail coverts

adult ♂

adult ♀

Connecticut Warbler

brownish hood

large bill

immature

walks rather than hops; habits suggest Ovenbird

gray head

short tail projection past undertail coverts

adult ♀

white eye ring sometimes slightly broken at rear

adult ♂

underparts duller yellow than Mourning, especially in immatures

Kentucky Warbler *Geothlypis formosa* L 5¼" (13 cm)

A short-tailed, long-legged warbler. Bold yellow spectacles separate black crown from black on face and sides of neck; underparts are entirely yellow, upperparts bright olive. Black areas are duller on **female,** olive on immature female. Hops on ground much of the time.

VOICE: Song is a series of rolling musical notes, *churry churry churry,* much like Carolina Wren. Call is a low, sharp *chuck.*

RANGE: Common in rich, moist woodlands; nests and feeds on the ground in dense undergrowth. Rare to Maritime Provinces and southern ON; casual to Newfoundland. Very rare vagrant to CA and the Southwest.

Canada Warbler *Cardellina canadensis* L 5¼" (13 cm)

Black necklace on bright yellow breast identifies **male;** note also bold yellow spectacles. In **female,** necklace is dusky and indistinct. Male is blue-gray above, female duller. All birds have white undertail coverts.

VOICE: Song begins with one or more short, sharp chip notes and continues as a rich and highly variable warble. Call is a sharp *tick.*

RANGE: Uncommon in dense woodlands and brush. Usually forages in undergrowth or low branches, but also seen flycatching. Winters in South America. Rather rare in eastern Gulf region and FL, mostly in fall. Casual vagrant in the West (annually in CA in fall).

Wilson's Warbler *Cardellina pusilla* L 4¾" (12 cm)

Olive above, yellow below; yellow lores. Long plain tail, often cocked. **Male** has black cap; in **female,** cap blackish or absent, forehead yellowish. Yellow lores and lack of white in tail help distinguish female from female Hooded Warbler. Coloration varies geographically from bright *chryseola* of Pacific states to duller, olive-faced nominate *pusilla* of East (breeds west to NT); *pileolata* from AK and Rockies intermediate. Western females average more black on crown.

VOICE: Song is a rapid, variable series of *chee* notes, falling into a weak trill in *pusilla;* Common call is a sharp *chimp;* also a *tsip,* the frequent flight call.

RANGE: Fairly common, much more numerous in the West than in the East; nests in dense, moist woodlands, bogs, willow thickets, and streamside tangles. Small numbers winter in coastal CA and southern AZ.

Hooded Warbler *Setophaga citrina* L 5¼" (13 cm)

All ages have dark lores, unlike Wilson's Warbler; also bigger bill and larger eye. Extensive black hood identifies **male. Adult female** shows blackish or olive crown and sides of neck; sometimes has black throat or black spots on breast; **immature female** lacks black. Note that in both sexes tail is white below; seen from above, white outer tail feathers are conspicuous as the bird flicks its tail open; often secretive.

VOICE: Song is loud, musical, whistled variations of *ta-wit ta-wit ta-wit tee-yo.* Call is a flat, metallic chink.

RANGE: Fairly common in swamps, moist woodlands; prefers hiding in dense undergrowth and low branches. Rare migrant in the Southwest and CA, where it has nested; also has nested in CO; casual in other western states. Rare in fall to the Maritime Provinces; casual to Newfoundland.

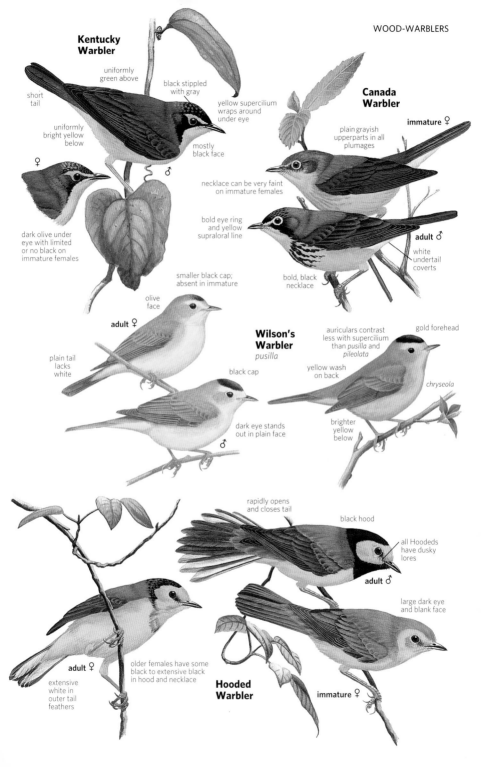

Kentucky Warbler

uniformly green above

black stippled with gray

yellow supercilium wraps around under eye

short tail

uniformly bright yellow below

mostly black face

♀

♂

dark olive under eye with limited or no black on immature females

Canada Warbler

immature ♀

plain grayish upperparts in all plumages

necklace can be very faint on immature females

bold eye ring and yellow supraloral line

adult ♂

white undertail coverts

bold, black necklace

smaller black cap; absent in immature

olive face

adult ♀

Wilson's Warbler
pusilla

plain tail lacks white

black cap

dark eye stands out in plain face

♂

auriculars contrast less with supercilium than *pusilla* and *pileolata*

yellow wash on back

gold forehead

chryseola

brighter yellow below

rapidly opens and closes tail

black hood

all Hoodeds have dusky lores

adult ♂

large dark eye and blank face

adult ♀

older females have some black to extensive black in hood and necklace

extensive white in outer tail feathers

Hooded Warbler

immature ♀

Common Yellowthroat *Geothlypis trichas* L 5" *(13 cm)*

Adult male's broad black mask is bordered above by gray or white, below by bright yellow throat and breast; undertail coverts yellow. **Female** lacks black mask; has whitish eye ring. Subspecies vary geographically in color of mask border and extent of yellow below. Southwestern subspecies, *chryseola,* is brightest below and shows the most yellow. **Immatures** are duller and browner overall; immature males have some black in face. Often cocks tail.

VOICE: Variable song; one version is a loud, rolling *wichity wichity wichity wich.* Calls include a raspy *tidge.*

RANGE: Common; stays low in grassy fields, shrubs, and marshes.

Fan-tailed Warbler *Basileuterus lachrymosus* L 5¾" *(15 cm)*

Large, with long, graduated, white-tipped tail held partly open and pumped sideways or up and down. Head pattern distinct with broken white eye ring; white lore spot; yellow crown patch. Note tawny wash on breast. Often walks or shuffles on ground; secretive.

VOICE: Song of rich, loud slurred notes; call is a penetrating *schree.*

RANGE: Tropical species. Found low in canyons or ravines. Casual, mainly late spring, to southeastern AZ. Accidental to Big Bend, TX (fall), and east-central NM (spring). Also recorded from Baja California Norte (Dec.).

Gray-crowned Yellowthroat *Geothlypis poliocephala*

L 5½" *(13 cm)* Large, with a long, graduated tail; thick, bicolored bill with curved culmen; split white eye ring; lores blackish in **males,** slaty gray in **females.** Rather shy.

VOICE: Song is a rich, varied warble; call is a rising *chee dee.*

RANGE: Tropical species. Favors grassland with scattered bushes. Former resident of the Brownsville area in south TX; population eliminated in early 20th century. Recently, several certain records in the lower Rio Grande Valley; other reports uncertain.

Yellow-breasted Chat *Icteria virens* L 7½" *(19 cm)*

Our largest warbler, with long tail, thick bill, and white spectacles. Lores black in **males,** gray in **females.** Eastern *virens* is shorter tailed and more olive above and has more restricted white in the malar region than *auricollis* of West. Rather shy.

VOICE: Unmusical song, a jumble of harsh, chattering clucks, rattles, clear whistles, and squawks, sometimes given in hovering display flight. Call is a sharp, drawn-out *chow;* also a nasal *air* and a sharp *cuk* or *cuk-cuk-cuk.*

RANGE: Inhabits dense thickets and brush. In East, partial to second-growth forest, whereas in West, distribution tied to riparian habitat. Rather shy. Regular straggler in fall to Maritime Provinces and Newfoundland; multiple fall records for Greenland. Rare in winter on the East Coast; casual on West Coast.

Common Yellowthroat
occidentalis

narrow whitish eye ring

yellowish under tail

yellowish tinge on neck

color below varies

prominent black mask; bordered by whitish in western subspecies

adult ♂ southwestern
chryseola

underparts nearly solid yellow

adult ♂

bright yellow; some interior birds have much more white on belly

black mask on adult males bordered by grayish in most eastern birds

long tail

immature male has some black in face

immature ♂

adult ♂
trichas

Fan-tailed Warbler
tephra

yellow crown patch bordered by black

uniformly gray above

spreads open graduated tail with white tips; bobs tail up and down or side to side

whitish supraloral spot and broken eye ring

gray head with broken white eye ring and dark lores

tawny wash on breast

thick bill with pinkish base

♀

long tail

♂

Gray-crowned Yellowthroat
ralphi

Yellow-breasted Chat

eastern ♂
virens

white supraloral

less black in face than male

grayer above

extensive bright yellow underparts in all plumages

thick bill

longer white submoustachial

eastern ♀
virens

western ♂
auricollis

♀
auricollis

Painted Redstart *Myioborus pictus* L 5¾" (15 cm)

Bright red lower breast and belly; black head and upperparts; bold white wing patch. White outer tail feathers conspicuous as the bird fans its tail. **Juvenile** has sooty, not red, breast; acquires full adult plumage by end of summer.

VOICE: Song is a series of rich, liquid warbles; call is a clear, whistled *chee*.

RANGE: Found in pine-oak canyons. Very rare visitor to southern CA; a scattering of records elsewhere north to BC and scattered throughout East.

Red-faced Warbler *Cardellina rubrifrons* L 5½" (14 cm)

Adult's red-and-black face pattern distinctive; back and long tail gray, rump and underparts white. Immature is duller, face pinkish. Flips tail much like Wilson's and Canada Warblers (both *Wilsonia;* page 450).

VOICE: Song is a series of varied, ringing *zweet* notes. Call is a sharp *tchip,* suggestive of Black-throated Gray Warbler.

RANGE: A warbler of high mountains, generally found above 6,000 feet. Uncommon to fairly common, especially in fir and spruce mixed with oaks. Nests on the ground. Rare to west TX; casual to southern CA and NV; accidental to northern CA, WY, CO, south TX, LA, and GA.

Slate-throated Redstart *Myioborus miniatus* L 6" (15 cm)

Head, throat, and back are slate black, breast dark red. Chestnut crown patch visible only at close range. Lacks white wing patch of similar Painted Redstart; white on outer tail feathers less extensive; tail strongly graduated. Found in pine-oak canyons, forests.

VOICE: Song is a variable series of sweet *s-wee* notes, often accelerating toward the end. Call, a chip note, is very different from Painted Redstart.

RANGE: Middle and South American species, casual in southeastern AZ, southeastern NM, and west and south TX. All confirmed records are in spring.

Golden-crowned Warbler *Basileuterus culicivorus*

L 5" (13 cm) Resembles Orange-crowned Warbler (page 426) but crown shows a distinct yellow or buffy orange central stripe, bordered in black; note also yellowish green eyebrow.

VOICE: Gives five to seven clear whistled notes, ending with a distinct upslur. Call is a rapidly repeated *tuck.*

RANGE: Tropical species, casual in south TX, chiefly in winter; accidental in east-central NM (spring).

Rufous-capped Warbler *Basileuterus rufifrons* L 5¼" (13 cm)

Rufous crown, bold white eyebrow, throat extensively bright yellow. Long tail, often cocked. Generally stays low in the undergrowth.

VOICE: Song begins with musical chip notes and accelerates into a series of dry, whistled warbles; call, a *tik,* is often doubled or in a rapid series.

RANGE: Mexican species. Inhabits brush and woodlands of foothills or low mountains. Casual to southern Edwards Plateau and Big Bend, TX; also southeastern AZ (has nested); recently documented from southwestern NM in winter (Guadalupe Canyon).

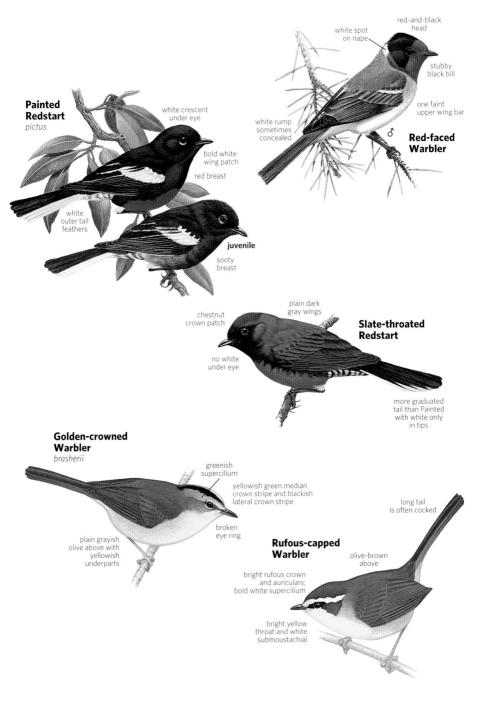

Painted Redstart
pictus

white crescent under eye

bold white wing patch

red breast

white outer tail feathers

juvenile

sooty breast

red-and-black head

white spot on nape

stubby black bill

one faint upper wing bar

white rump sometimes concealed

Red-faced Warbler

♂

plain dark gray wings

chestnut crown patch

Slate-throated Redstart

no white under eye

more graduated tail than Painted with white only in tips

Golden-crowned Warbler
brasherii

greenish supercilium

yellowish green median crown stripe and blackish lateral crown stripe

broken eye ring

plain grayish olive above with yellowish underparts

long tail is often cocked

Rufous-capped Warbler

olive-brown above

bright rufous crown and auriculars; bold white supercilium

bright yellow throat and white submoustachial

OLIVE WARBLER Family Peucedramidae
Recently placed in its own family because relationships are uncertain.

Olive Warbler *Peucedramus taeniatus* L 5¼" *(13 cm)*
Dark face patch broadens behind eye. Long, thin bill; two broad white wing bars; outer tail feathers extensively white. **Adult male's** head tawny brown, **female's** more yellowish. Juveniles and **first-fall** birds resemble female but are paler or whitish below; crown is gray. Adult plumage acquired by second fall.
VOICE: Typical song is a loud *peeta peeta peeta,* similar to Tufted Titmouse; distinctive call is a soft, whistled *phew.*
RANGE: Favors open coniferous forests at elevations above 7,000 feet. Nests and forages high in trees. A few remain on breeding grounds in winter, mostly at slightly lower elevations. Casual to west TX.

BANANAQUIT Incertae sedis (of uncertain taxonomic placement)
Family affiliation of this species is uncertain. Prefers nectar.

Bananaquit *Coereba flaveola* L 4½" *(11 cm)*
Note thin, downcurved bill. **Adult** has conspicuous white eyebrow, yellow rump; underparts white, with yellow breast; small white wing patch. **Juvenile** is much duller, but shows same basic pattern.
VOICE: Call is a high-pitched *sint* or *tsip;* song is geographically variable, even within West Indies; Bahama birds give several ticks followed by rapid clicking.
RANGE: Tropical species; casual visitor from the Bahamas to southeastern FL coast and upper FL Keys; one record for Fort De Soto Park in west FL. Of the approximately 35 records, only about a half dozen are in the last two decades. The records are mostly in winter, but extend into May.

TANAGERS Family Thraupidae
A large colorful Neotropical family, mostly frugivorous. 197 WORLD; 1 N.A.

Western Spindalis *Spindalis zena* L 6¾" *(17 cm)*
Formerly known as Stripe-headed Tanager. This essentially West Indian species now comprises five subspecies from Bahamas (two), Cuba, Cayman Islands, and Cozumel Island. **Males** are strikingly patterned: Most are black-backed (*zena*). A few records appear to be of *townsendi,* restricted to Grand Bahama and Abaco, which shows greenish orange back; occasionally black with dull orange edgings. (Abaco birds average more black on back than Grand Bahama birds.) Recent winter record in Key West of male *pretrei,* with a yellow-green back, from Cuba. **Females** of all subspecies grayish olive, with pale eyebrow, pale greater covert patch, and distinct white spot at base of primaries.
VOICE: Call is a thin, high *tsee,* given singly or in a series. Variable song is a very high-pitched, thin whistle, with some ventriloquial qualities.
RANGE: West Indian species; very rare visitor from Bahamas to southeastern FL and Florida Keys, scarcest in midsummer. One recent nesting record from Pine Key, Everglades National Park (Aug. 2009). Accidental on west coast of FL.

adult ♂

tawny head

Olive Warbler
arizonae

adult ♀

1st spring ♂

long
thin bill

pale patch at base
of primaries

pale center to
dark mask

1st fall

bold white
supercilium

Bananaquit
bahamensis

yellowish rump

endemic to Cuba;
salvini from Grand
Cayman looks
very similar

curved bill with
reddish gape

bright yellow-
green back

juvenile

black does not
cross throat as in
zena and *townsendi*

yellow
breast

white patch at base
of primaries

adult

white tail
spots

♂ *pretrei*

Western
Spindalis

striking head
pattern

Abaco birds average
more black on back
than those from
Grand Bahama

nearly all
males seen in
Florida have
blackish backs

♂ *zena*

orange
breast

♂ *townsendi*

coloration typical
of Abaco birds

prominent
buffy white
supercilium
borders dark auricular

greenish
orange
back

♀ *townsendi*

white spot
at base of
primaries

♂ *townsendi*

coloration typical of
Grand Bahama birds

EMBERIZIDS Family Emberizidae

All have conical bills. This large family includes the towhees, sparrows, and *Emberiza* buntings. SPECIES: 329 WORLD; 55 N.A.

White-collared Seedeater *Sporophila torqueola*
L 4½" *(11 cm)* Tiny, with thick, short, strongly curved bill and rounded tail. **Adult male** has black cap; white crescent below eye; incomplete buffy collar; white wing bars; white patch at base of primaries. **Females** are paler, lack cap and collar; wing bars duller. A strongly polytypic species; *sharpei,* a comparatively dull subspecies, is the one found in northeastern Mexico and south TX.

VOICE: Song is pitched high, then low, a variable *sweet sweet sweet sweet cheer cheer cheer.* Calls include a distinct, high *wink.*

RANGE: In U.S. favors cane brakes and other riverside vegetation. Formerly more widespread in Rio Grande Valley; its range extended farther east and numbers were greater. Now found mainly from Zapata to northern Webb County, TX.

Black-faced Grassquit *Tiaris bicolor* L 4½" *(11 cm)*
Adult male mostly black below, dark olive above; head is black. **Female** and immatures pale gray below, gray-olive above.

VOICE: Song is a buzzing *tik-zeee;* call is a high, lisping *tst.*

RANGE: Found nearly throughout West Indies except Cayman Islands; mostly absent from Cuba; also found in northern South America; casual stray to south FL. Fewer than ten records, with no strong seasonal pattern.

Yellow-faced Grassquit *Tiaris olivaceus* L 4¼" *(11 cm)*
Adult male of mainland subspecies, *pusillus,* shows extensive black on head, breast, and upper belly; golden yellow supercilium, throat, and crescent below eye; olive above. Adult **female** and immature male have traces of same head pattern; olive above. Adult male of West Indian subspecies, *olivaceus,* shows less black; black on breast more of a bib. Female lacks any black below.

VOICE: Song is thin, insectlike trills; call is a high-pitched *sik* or *tsi.*

RANGE: Tropical species. West Indies *olivaceus* found on Cuba, Jamaica, and the Cayman Islands. Mexican *pusillus* is found north to southern Tamaulipas. About ten records, nearly equally divided between each subspecies and scattered throughout the year, in south FL and southernmost TX.

Olive Sparrow *Arremonops rufivirgatus* L 6¼" *(16 cm)*
Dull olive above, with brown stripe on each side of crown. Lacks reddish cap of similar Green-tailed Towhee. **Juveniles** are buffier, with pale wing bars; faintly streaked on neck and breast.

VOICE: Calls include a dry chip and a buzzy *speeee.* Song is an accelerating series of chip notes.

RANGE: Tropical species, fairly common in southernmost TX in dense undergrowth, brushy areas, live oak; less numerous north to southern portion of Edwards Plateau.

short, stubby bill with strongly curved culmen

tiny size

thin, pale wing bars

buffy below

blackish head

White-collared Seedeater
sharpei

pale buffy white collar

distinctly patterned wings

1st winter ♂

adult ♂

grayish head and underparts

Black-faced Grassquit
bicolor

olive above

♀

Yellow-faced Grassquit

West Indies *olivaceus*

black bib

adult ♂

black head and underparts

adult ♂

♀

mainland *pusillus*

bright yellow supercilium and throat

extensive black breast

adult ♂

Sparrows

prominently striped head

olive upperparts

juvenile

blurred breast streaks

pale underparts

Olive Sparrow
rufivirgatus

Eastern Towhee *Pipilo erythrophthalmus* L 7½" (19 cm)

Male's black upperparts and hood contrast with rufous sides and white underparts. Distinct white patch at base of primaries and distinct white tertial edges. White in outer tail feathers is conspicuous in flight, or seen from below when bird is perched. Most have red eyes. **Females** are similarly patterned, but black areas are replaced by brown. **Juveniles** are brownish and show distinct streaks below. The nominate subspecies is largest and shows most extensive white in tail. Wing length, and extent of white in wings and tail, declines from the northern part of the range to the Gulf Coast, while the size of bill, legs, and feet increases. FL peninsula subspecies, *alleni,* is smaller in all measurements, paler, and duller; has less white in wings and tail; has straw-colored eyes. The *rileyi* subspecies (not shown), from northernmost FL to east-central NC, shows intermediate characteristics, eyes either red or straw colored; eye color particularly variable in birds from southern GA and coastal SC. Like other towhees, Eastern scratches with its feet together.

VOICE: Full song has three parts, often rendered as *drink your tea,* or shortened to two parts: *drink tea.* Northeastern birds' call is a slightly upslurred *chwee;* in *alleni,* a clearer, even-pitch or upslurred *swee.*

RANGE: Partial to second growth with dense shrubs and extensive leaf litter; southern subspecies, especially *alleni,* favors coastal scrub or sand dune ridges and pinelands. Nominate subspecies is partly migratory; casual west to CO and AZ. Has declined from the northeastern part of its range by as much as 90 percent in recent decades. Other subspecies are largely resident. Eastern and Spotted Towhee have each been restored to full species status; formerly considered one species, Rufous-sided Towhee. The two interbreed along rivers in the Great Plains, particularly the Platte and its tributaries.

See subspecies map, page 555

Spotted Towhee *Pipilo maculatus* L 7½" (19 cm)

Distinguished from similar Eastern Towhee by white spotting on back and scapulars; also on tips of median and greater coverts, which forms white wing bars. In general, **females** differ less from **males** than Eastern, with *arcticus* from Great Plains showing the greatest difference. In both sexes the amount of white spotting above and white in tail shows marked geographical variation, with *arcticus* displaying most white. Subspecies, principally *montanus,* from the Great Basin and Rockies shows less white. The Northwest coast's *oregonus* is darkest and shows least white of all the subspecies. White increases southward to *megalonyx* of southern CA and *falcinellus* (not shown) of the Central Valley region of CA.

VOICE: Song and calls also show great geographical variation. Interior subspecies give introductory notes, then a trill. Pacific coast birds sing a simple trill of variable speed. Call of *montanus* is a descending and raspy mewing. Great Plains *arcticus* and all coastal subspecies give an upslurred, questioning *queee.*

RANGE: Some populations are largely resident, while others are migratory; *arcticus* is the most migratory and is casual in eastern North America.

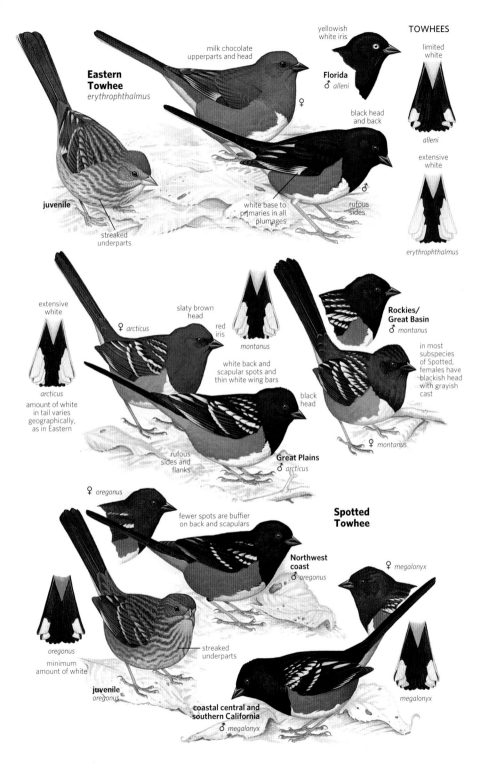

TOWHEES

Eastern Towhee
erythrophthalmus

milk chocolate upperparts and head

yellowish white iris

Florida
♂ *alleni*

limited white

alleni

black head and back

black head and back

extensive white

erythrophthalmus

juvenile

streaked underparts

white base to primaries in all plumages

rufous sides

♀

♂

extensive white

arcticus

amount of white in tail varies geographically, as in Eastern

♀ *arcticus*

slaty brown head

red iris

montanus

white back and scapular spots and thin white wing bars

Rockies/ Great Basin
♂ *montanus*

in most subspecies of Spotted, females have blackish head with grayish cast

black head

rufous sides and flanks

Great Plains
♂ *arcticus*

♀ *montanus*

♀ *oregonus*

fewer spots are buffer on back and scapulars

Spotted Towhee

♀ *megalonyx*

Northwest coast
♂ *oregonus*

oregonus

minimum amount of white

streaked underparts

juvenile
oregonus

coastal central and southern California
♂ *megalonyx*

megalonyx

Green-tailed Towhee *Pipilo chlorurus* L 7¼" (18 cm)

Olive above with reddish crown, distinct white throat bordered by dark stripe and white stripe. **Juvenile** has two faint olive wing bars; plumage is streaked overall; upperparts tinged with olive; lacks reddish crown.

VOICE: Clear, whistled song begins with *weet-chur,* ends in raspy, buzzy trill. Calls include a catlike *mew.*

RANGE: Fairly common in dense brush, chaparral, on mountainsides and high plateaus. Rare on CA coast and Great Plains. Casual in migration and winter throughout the East.

California Towhee *Melozone crissalis* L 9" (23 cm)

Brownish overall; crown slightly warmer brown than rest of upperparts. Buff throat is bordered by a distinct broken ring of dark brown spots; no dark spot on breast as in Canyon Towhee. Lores are same color as throat and contrast with cheek; undertail coverts warm cinnamon. **Juvenile** shows faint wing bars.

VOICE: Call is a sharp, metallic *chink* note; also gives some thin, lispy notes and an excited, squealing series of notes, often delivered as a duet by a pair. Song, accelerating *chink* notes with stutters in the middle, is heard mostly in late afternoon.

RANGE: Resident in chaparral, parks, and gardens. The subspecies *eremophilus* (**T**) of Inyo County, CA, is threatened. With Canyon Towhee, California was formerly considered one species, Brown Towhee.

Canyon Towhee *Melozone fusca* L 8" (20 cm)

Similar to California Towhee. Canyon is paler, grayish rather than brown, with shorter tail; more contrast in reddish crown gives a capped appearance; crown is sometimes raised as short crest. Larger whitish belly patch with diffuse dark spot at junction with breast; paler throat bordered by finer streaks; lores the same color as cheek; distinct buffy eye ring. Juveniles are streaked below. Closest genetic relative is White-throated Towhee *(M. albicollis)* of Mexico.

VOICE: Call is a shrill *chee-yep* or *chedep.* Song, more musical, less metallic than California; opens with a call note, followed by sweet slurred notes. Also gives a duet of lisping and squealing notes, like California.

RANGE: Favors arid, hilly country; desert canyons. Largely resident within range; casual southwestern KS (Morton County); no range overlap with California Towhee.

Abert's Towhee *Melozone aberti* L 9½" (24 cm)

Black face; upperparts cinnamon-brown, underparts paler, with cinnamon undertail coverts. Closest genetic relative is California Towhee.

VOICE: Call is a sharp *peek;* song is a series of *peek* notes.

RANGE: Common within its range, but somewhat secretive. Inhabits desert woodlands and streamside thickets, generally at lower altitudes than similar Canyon Towhee, but the two species are found together at many locations. Also found in suburban yards and orchards. No range overlap with California Towhee.

Green-tailed Towhee

olive wings and tail

dull rufous crown

white supraloral and submoustachial

white throat

gray chest

spring

fall

juvenile

extensively streaked below

California Towhee

juvenile

crown fairly uniform with rest of head

blurry streaks

no breast spot

Canyon Towhee

pale rufous crown

buffy throat

whitish belly with fairly conspicuous breast spot

Abert's Towhee

overall warm brown coloration

pale bill

blackish around bill

Rufous-winged Sparrow *Peucaea carpalis* L 5¾" *(15 cm)*

Two blackish streaks on sides of face; pale rufous crown streaked with black; pale median stripe. Distinctive reddish lesser wing coverts often covered. Long, rounded tail. **Juvenile's** facial stripes less distinct; bill dark; breast and sides lightly streaked; plumage seen as late as Nov.

VOICE: Distinctive call note, a sharp, high *seep.* Variable song, several chip notes followed by an accelerating trill of chip or *sweet* notes.

RANGE: Fairly common but local; found in flat areas of tall desert grass mixed with brush and cactus. A few were recently discovered in Guadalupe Canyon, NM.

Cassin's Sparrow *Peucaea cassinii* L 6" *(15 cm)*

A large, drab sparrow, with large bill, fairly flat forehead. Long, rounded tail has white tips on outer feathers, most conspicuous in flight. Streaked upperparts show distinct anchor-shaped marks. A few dark streaks are usually on lower flanks. **Juvenile** is streaked below; paler overall than juvenile Botteri's. In fresh fall plumage, black-centered, white-fringed tertials.

VOICE: Best located and identified by song, often given in brief, fluttery song flight: typically a soft double whistle, a loud, sweet trill, a low whistle, and a final, slightly higher note; or a series of chip notes ending in a trill or warbles. Also gives a trill of *pit* notes.

RANGE: Secretive; inhabits arid grasslands with scattered shrubs, cactus, and mesquite. Casual vagrant to the East and Far West.

Bachman's Sparrow *Peucaea aestivalis* L 6" *(15 cm)*

Shaped like Cassin's and Botteri's. Adult gray above, heavily streaked with chestnut or dark brown; sides of head buffy gray; a thin dark line extends back from eye. Breast and sides buff or gray. Subspecies range in overall brightness from reddish *illinoensis* of western part of range to grayer and darker *aestivalis* of FL. Birds from central part of range, *bachmani* (not shown), are intermediate. **Juvenile** has distinct eye ring; streaked throat, breast, sides. Like Cassin's and Botteri's, quite secretive outside breeding season.

VOICE: Best located and identified by song: one clear, whistled introductory note, followed by a variable trill or warble on a different pitch. Male sings from open perch; often heard in late summer.

RANGE: Inhabits dry, open woods, especially pines; scrub palmetto. Northern range has markedly declined over last several decades.

Botteri's Sparrow *Peucaea botterii* L 6" *(15 cm)*

Size and shape like Cassin's and Bachman's. Tail lacks white tips and central barring of Cassin's. Upperparts streaked with dull black, rust or brown, and gray; underparts unstreaked; breast and sides buff. Southeastern AZ *arizonae* redder above; *texana* of far south TX slightly grayer. **Juvenile** buffy below, finely streaked. Generally secretive.

VOICE: Best located and identified by song: several high sharp *tsip* or *che-lik* notes, often given alone or followed by a short, accelerating, rattly trill.

RANGE: Inhabits grasslands (particularly sacaton grass for *arizonae*) dotted with mesquite, cactus, and brush. TX subspecies *texana* declining because of habitat loss; now uncommon and local. A pair present June 1997 in Presidio County, west TX, not identified to subspecies.

light rufous crown with fine black streaks

broad pale gray eyebrow with black eye line

distinct blackish malar and moustachial stripes

bicolored bill

Rufous-winged Sparrow

all *Peucaea* sparrows have large bills and long, rounded tails

pattern suggestive of adult but duller and streaked below

juvenile

rufous lesser coverts usually hidden

white underparts

all *Peucaea* sparrows but Rufous-winged tend to be secretive, except for singing territorial males

Cassin's

whitish tail corners

whitish eye ring

overall appears like a large Brewer's

brighter rufous above than nominate subspecies

dark "anchors" on upperpart feathers

whitish tertial fringes

Bachman's Sparrow
aestivalis

illinoensis

faint flank streaks

Cassin's Sparrow

streaked below

buffy breast, sides, and flanks

prominent head and back streaks

juvenile

arizonae

redder above than *texana*

streaked below

juvenile

head pattern more poorly defined than Bachman's

Botteri's Sparrow

streaked upperparts

slightly larger bill than Cassin's

buffy, unstreaked rear flanks

texana

streaked below

juvenile
texana

Rufous-crowned Sparrow *Aimophila ruficeps* L 6" (15 cm)

Gray head, dark reddish crown, distinct whitish eye ring, rufous line extending back from eye, single black malar stripe. Gray-brown above, with reddish streaks; gray below; long, rounded tail. Subspecies range in overall color from paler, grayer *eremoeca*, found over most of eastern interior range, to widespread paler, reddish southwestern subspecies, *scottii*; smaller, darker Pacific coastal subspecies slightly show variable amounts of reddish above. **Juvenile** buffier above; breast and crown streaked; may show two pale wing bars. Not gregarious.

VOICE: Calls include a distinctive sharp *dear,* usually given in a series; song is a rapid, bubbling series of rising and falling chip notes.

RANGE: Largely resident. Locally common on rocky hillsides and steep brushy or grassy slopes.

GENUS *SPIZELLA*

Spizella **sparrows are mostly small and slim, and they have long, notched tails. Highly gregarious.**

American Tree Sparrow *Spizella arborea* L 6¼" (16 cm)

Our largest *Spizella*. Note distinctively bicolored bill. Gray head and nape crowned with rufous; rufous stripe behind eye. Gray throat and breast, with dark central spot, rufous patches at sides of breast. Back and scapulars streaked with black and rufous. Outer tail feathers thinly edged in white on outer webs. Underparts grayish white with buffy sides and flanks. **Winter** birds are buffier; rufous color on crown sometimes forms a central stripe. **Juvenile** is streaked on head and underparts. In all plumages, western *ochracea* paler overall.

VOICE: Calls include a musical *teedle-eet;* also a thin *seet.* Song usually begins with several clear notes followed by a variable, rapid warble.

RANGE: Fairly common. Uncommon to rare west of Rockies. Casual to southern CA, most of Southwest away from mapped range, northern portion of Gulf States, and Bering Sea Islands. Accidental to Gulf Coast and offshore oil rigs. Breeds along edge of tundra, in open areas with scattered trees and brush. Winters in open areas in weedy fields, corn stubble, marshes, and groves of small trees. A late fall migrant, usually not seen until late Oct., even in upper Midwest and northern Great Plains; some remain until late Apr., a very few into last third of May.

Field Sparrow *Spizella pusilla* L 5¾" (15 cm)

Gray face with reddish lateral crown and grayish median crown stripe, distinct whitish eye ring, bright pink bill. Back is streaked except on gray-brown rump. Breast and sides in *pusilla* are buffy, bright buffy when fresh; legs pink. **Juvenile** streaked below; wing bars buffy. Birds in westernmost part of range, *arenacea,* average paler and grayer; extremes are shown here. Some dull grayish birds, possibly *arenacea,* seen in Midwest in fall and winter.

VOICE: Song is a series of clear, plaintive whistles accelerating into a trill; hard chip note similar to Orange-crowned Warbler.

RANGE: Fairly common in open, brushy woodlands, overgrown fields. A rather early spring and late fall migrant (mid-Oct. to early Nov., peak late Oct.). Uncommon to rare in Maritime Provinces; casual west of mapped range but recorded west to CA.

Rufous-crowned Sparrow

dull rufous crown; prominent white eye ring

interior
eremoeca

dark malar stripe

more reddish above, buffer below than *eremoeca*

grayish underparts

coastal

long rounded tail

streaked crown and white eye ring

juvenile
-eremoeca

American Tree Sparrow

largest *Spizella*

breeding
arborea

bicolored bill

rufous patch at breast sides and buffy flanks

winter
ochracea

breast spot

juvenile
ochracea

Field Sparrow

grayish median crown stripe and white eye ring

pink bill

western
arenacea

grayer overall than nominate subspecies

eastern juvenile
pusilla

buffy breast

eastern
pusilla

Chipping Sparrow *Spizella passerina* L 5½" (14 cm)

Breeding adult has chestnut crown, distinct white eyebrow; black eye line extends through lores to base of bill in all plumages; no moustachial stripe; gray unstreaked rump. **Winter adult** has browner cheek; streaked crown shows some rufous. **First-winter** bird averages less rufous on crown; buffier below. In **juvenal** plumage, often held into Oct., especially in West, underparts are prominently streaked.

VOICE: Song is rapid trill of dry notes on one pitch. Call is *tsik;* flight call, also given perched, is a high, hard *seep.*

RANGE: Widespread and common. Found on lawns and in fields, woodland edges, and pine-oak forests. Very rare in winter north of mapped range. Casual in fall to St. Lawrence Island, AK.

Clay-colored Sparrow *Spizella pallida* L 5½" (14 cm)

Blackish streaked crown with distinct pale median stripe. Broad, whitish eyebrow; pale lores; brown cheek outlined by dark postocular and moustachial stripes; conspicuous pale submoustachial stripe. Nape gray; unstreaked rump is same color as back, unlike Chipping. **Adult** in fall and winter buffier overall; gray nape and pale stripe on sides of throat stand out more; in **juvenile,** breast and sides streaked.

VOICE: Song is a brief series of insectlike buzzes; like Brewer's, song heard in late winter and spring migration. Flight call is a thin *sip.*

RANGE: Fairly common to common in western breeding range. Less common and more local farther east; breeds north to James Bay, ON; breeding range has recently spread very locally to Northeast. Prefers brushy fields, groves, and streamside thickets. Fairly common migrant on western Great Plains. Winters primarily from Mexico south, uncommon in south TX. Rare in fall, very rare in winter and spring on both coasts and in AZ. Rare in winter in south FL. Casual to YT and AK, where recorded west to Gambell, St. Lawrence Island.

Brewer's Sparrow *Spizella breweri* L 5½" (14 cm)

Brown crown with fine black streaks; lacks clearly defined, pale median crown stripe and strongly contrasting head pattern of Clay-colored. Distinct whitish eye ring; grayish white eyebrow; ear patch pale brown with darker borders; pale lores; dark malar stripe; rump buffy brown, may be lightly streaked. **Juvenile** lightly streaked on breast and sides. In fall, somewhat buff below. Canadian *taverneri* larger, bigger billed, colder in coloration; has stronger back streaking, sometimes also flank streaking, and more strongly outlined face, more like Clay-colored. Juvenile *taverneri* heavily streaked, blackish below; some believe this **"Timberline Sparrow"** may be a separate species.

VOICE: Song is a series of varied bubbling notes and buzzy trills at different pitches; often sings in spring on winter grounds and on migration. Song of *taverneri* differs: overall higher pitched, less buzzy, lacks descending *sweet* notes. Call is a thin *sip,* like Clay-colored Sparrow.

RANGE: Common; *breweri* breeds in mountain meadows, sagebrush flats; scarce migrant along Pacific coast (mainly fall) and through western Great Plains; accidental farther east. Larger *taverneri* breeds in alpine zone of Canadian Rockies to east-central AK. This subspecies largely unknown otherwise, but specimens of migrants exist from AZ, NM, TX, WA; only one winter specimen known for U.S. (San Diego County, CA), suggesting *taverneri* winters in Mexico.

rufous crown with well-defined white supercilium

dark line runs through lores to base of bill in all plumages, diagnostic for Chipping

Chipping Sparrow

lacks defined dark moustachial stripe, unlike Clay-colored and Brewer's

breeding adult

trace of rust on crown

juvenile

grayish nape

winter adult

streaky juvenal plumage held into Oct.

often buffy below

1st winter

gray rump diagnostic for Chipping, but often hidden by wings

bold head pattern, including strong median crown stripe and buffy cast to plumage

buffy auricular

breeding

Clay-colored Sparrow

grayish nape contrasts with rest of buffy plumage

all have pale lores, like Brewer's, but unlike Chipping

dark moustachial and broad, pale submoustachial stripes

streaking on underparts

juvenile

quite buffy below

fall

rather distinct white eye ring and pale lores

juvenile
breweri

head pattern like Clay-colored but more muted and without contrastingly pale median crown stripe

breweri overall a sandy color, less rich than Clay-colored

pale brown rump

Brewer's Sparrow

breeding
breweri

larger than *breweri* with markings stronger and darker, more like Clay-colored

streaking on underparts, often rather faint

grayish auricular

"Timberline"
taverneri

juvenile

streaks below heavier and darker than juvenile *breweri*

breeding

Black-chinned Sparrow *Spizella atrogularis* L 5¾" (15 cm)

Gray overall; streaked with rusty above; bill bright pink; long tail. **Breeding male** has black chin. **Female** has less or no black. Winter birds lack any black on face. **Juveniles** like female but light streaks below.

VOICE: Song, an accelerating series of sweet notes; call, a high, thin *seep*.

RANGE: Inhabits brushy arid slopes in foothills and mountains. Rather local. Moves into burn areas, at least in CA. Only casually seen in migration. Casual north to OR.

Black-throated Sparrow *Amphispiza bilineata* L 5½" (14 cm)

Triangular black throat contrasts with white eyebrow, white submoustachial stripe. **Juvenal** plumage, often held well into fall, lacks black throat, but note bold white eyebrow; breast and back finely streaked. In all, white extends up along edge of outer tail feather.

VOICE: Song is rapid and high pitched: two clear notes followed by a trill. Calls are faint, tinkling notes.

RANGE: Fairly common in desert, especially on rocky slopes. Irregular interior Pacific Northwest, casual to BC and AB. Also casual to East, primarily in fall and winter.

Five-striped Sparrow *Amphispiza quinquestriata*

L 6" (15 cm) Dark brown above; breast and sides gray; white throat bordered by black and white stripes. Dark breast spot. Juvenile unstreaked.

VOICE: When singing, males often on rather conspicuous perches. Song is a series of phrases, each preceded by a few introductory notes. Call is a loud, hollow *tchep*.

RANGE: West Mexican species barely reaching southeastern AZ in tall, dense shrubs on rocky, steep hillsides. Very local; usually seen when singing; few winter records; but likely overlooked; rather secretive.

Sage Sparrow *Amphispiza belli* L 6¼" (16 cm)

White eye ring, white supraloral, broad white submoustachial stripe bordered by malar stripe ranging from indistinct to distinct; dark central breast spot; dusky streaking on sides; white fringe to tail feathers. **Juvenile** duller overall, more streaked. On **interior** (largest and palest) subspecies *nevadensis*, back buffy brown with dusky streaks. Smaller CA **coastal** *belli* much darker, lacks streaks on back, has stronger malar stripe, as does intermediate *canescens*. All subspecies run on ground with tail cocked.

VOICE: From a low perch, male sings a geographically variable jumbled series of rising and falling phrases. Twittering call consists of thin, juncolike notes. Song of *nevadensis* a strongly defined, pulsating pattern of rising and falling phrases, seemingly in a minor key. Song of *belli* richer, consisting of more jumbled notes, like a cross of Blue Grosbeak and Rufous-crowned Sparrow. Song of *canescens* intermediate, but closer to *belli*.

RANGE: Interior *nevadensis* and slightly darker *canescens* favor alkaline flats in sagebrush, saltbush. Coastal *belli* found in mountain chaparral. Ongoing studies may reveal that this complex represents two or three species; *canescens* comes nearly into contact with *nevadensis* in vicinity of Bishop, Inyo County, CA. There is thus far little evidence of hybridization between these two, which seem better differentiated both genetically and vocally than *canescens* is from *belli*.

See subspecies map, page 555

Black-chinned Sparrow

breeding ♀

breeding ♂

pink bill

dark chin

streaked rufous back

medium gray head and underparts

juvenile

Black-throated Sparrow
deserticola

bold white supercilium and submoustachial stripe

large black throat patch

bold white supercilium

white throat

white edge to outer tail feathers and small white tail tip

juvenile

Five-striped Sparrow
septentrionalis

short white supercilium

distinctive black and white throat stripes

brownish upperparts

dark gray breast

dark breast spot

white belly

plain back

central breast spot

Sage Sparrow

streaks on flanks

coastal
belli

short whitish supercilium

faint back streaks

runs on the ground with tail up

interior
juvenile
nevadensis

interior
nevadensis

canescens

intermediate between *nevadensis* and *belli*

Lark Bunting *Calamospiza melanocorys* L 7" (18 cm)

Stocky; short tail; whitish wing patches; thick bluish gray bill. **Breeding male mostly black. Female** streaked below; buffy sides, brown primaries. **Winter male** similar but has black primaries; immature male darker, has some black around bill and chin. Gregarious in migration and winter. In flight, looks short, round winged; shallow wingbeats.
VOICE: Call, a whistled *hoo-ee.* Song consists of rich whistles and trills.
RANGE: Common; nests in dry plains and prairies, especially in sagebrush. Rare in fall and winter to West Coast; casual to Pacific Northwest and in the East in fall, winter, and especially spring.

Vesper Sparrow *Pooecetes gramineus* L 6¼" (16 cm)

White eye ring; dark ear patch bordered in white along lower and rear edges; white outer tail feathers. Lacks bold eyebrow of Savannah Sparrow. Distinctive chestnut lesser coverts usually hidden by scapulars. Eastern nominate slightly darker overall than widespread *confinis.* Northwest coast breeding *affinis* (not shown) buffy-tinged below.
VOICE: Song is rich and melodious: two long, slurred notes followed by two higher notes, then a series of short, descending trills. One call is a thin, *Spizella*-like *seep.*
RANGE: Uncommon to fairly common in dry grasslands, farmlands, forest clearings, and sagebrush; declining in the East.

Lark Sparrow *Chondestes grammacus* L 6½" (17 cm)

Head pattern distinctive in adults; note dark central breast spot. **Juvenile's** colors are duller; breast, sides, and crown streaked. In all ages, white-cornered tail is conspicuous in flight.
VOICE: Song begins with two loud, clear notes, followed by a series of rich, melodious notes and trills and unmusical buzzes. Call is a sharp *tsip,* sometimes in a rapid series.
RANGE: Gregarious, found in various types of open country, often along roads. Formerly bred as far east as NY and MD; now rare in the East, seen mainly in fall. Accidental to YT and AK.

Savannah Sparrow *Passerculus sandwichensis* L 5½" (14 cm)

Highly variable. Eyebrow yellow or whitish; pale median crown stripe; strong postocular stripe. The numerous subspecies vary geographically; extremes shown here. West Coast subspecies show increasingly darker color from north to south, with AK and interior subspecies paler, widespread *nevadensis* of West (not shown) and *savanna* of East are palest; *beldingi* of southern CA coastal marshes darkest. The *rostratus* subspecies, **"Large-billed Sparrow,"** which winters on the edge of Salton Sea, rarely in coastal CA, is very dull with a large bill. In the East, the degree of darkness is somewhat reversed: Arctic subspecies are darker than more southerly Canadian and U.S. subspecies. Large, pale *princeps,* **"Ipswich Sparrow,"** breeds on Sable Island, NS; winters on East Coast beaches. All are gregarious and can skulk, but will readily flush up into bushes, small trees, or fence lines for good viewing.
VOICE: Song begins with two or three chip notes, followed by two buzzy trills. Distinctive flight call is a thin *seep.* Song of *rostratus* has short, high introductory notes, followed by about three rich, buzzy *dzeeee* notes; call is a soft, metallic *zink.*
RANGE: Common in a variety of open habitats, marshes, grasslands.

Lark Bunting

early spring ♂

all with thick, bluish bill

short tail with white terminal spots

on winter males, black surrounds bill

streaking below darker than on females

winter ♂

all with pale wing patches, buffy white in females

breeding ♂

indistinct face pattern ♀

pale wraps around dark-bordered auricular

bold white eye ring; no strong eye line as in Savannah

Vesper Sparrow

streaking on breast less extensive than smaller Savannah

eastern
gramineus

Lark Sparrow
grammacus

distinctive head pattern

confiding behavior

chestnut coverts usually concealed

dark breast spot

whitish at base of primaries

western
confinis

paler, grayer than nominate subspecies

white outer tail feathers

juvenile

dark eye line

variably yellow supercilium

extensive streaks

extensive white in tail

subdued head pattern

streaked below

savanna

short tail

strongly yellow supercilium

darkest eastern subspecies

subdued head pattern and faint streaks on back

"Large-billed"
rostratus

whitish supercilium

large bill

darkest subspecies with rich yellow supercilium

Savannah Sparrow

whitish supercilium

labradorius

grayish-brown streaks below

large and pale overall

pale secondary panel

"Ipswich"
princeps

fainter streaks below

beldingi

local resident in coastal southern California salt marshes

GENUS *AMMODRAMUS*
Sparrows of the genus *Ammodramus* tend to be large headed and large billed; they are also usually secretive.

Grasshopper Sparrow *Ammodramus savannarum*
L 5" (13 cm) Small and chunky, with short tail and flat head. Buffy breast and sides, usually without obvious streaking. Dark crown has a pale central stripe; note also white eye ring and, on most birds, a yellow-orange spot in front of eye. Lacks broad buffy orange eyebrow and pale blue-gray ear patch of Le Conte's Sparrow (page 476). Compare also with female Orange Bishop and Savannah Sparrow (page 472). **Juvenile's** breast and sides are streaked with brown. **Fall** birds are buffier below but never as bright as Le Conte's. Subspecies vary in overall color from dark FL subspecies, *floridanus* (**E**), to *ammolegus* of southeastern AZ, which is more reddish above and has fine rufous streaks on breast; however, individuals of both *pratensis* and *perpallidus* can be reddish too. Eastern *pratensis* is slightly more richly colored than western *perpallidus,* which spreads east through the Great Plains.
VOICE: Typical song is one or two high chip notes followed by a brief, grasshopperlike *buzz;* also sings a series of varied squeaky and buzzy notes.
RANGE: Found in pastures, grasslands, palmetto scrub, and old fields. Somewhat secretive; feeds and nests on the ground. Declining in East.

Baird's Sparrow *Ammodramus bairdii* L 5½" (14 cm)
Orange tinge to head (duller on worn summer birds), usually with less distinct median crown stripe than Savannah Sparrow (page 472); note especially the two isolated dark spots behind ear patch and lack of postocular line. Widely spaced, short dark streaks on breast form a distinct necklace; also shows chestnut on scapulars. **Juvenile's** white fringes give a scaly appearance to upperparts. Very secretive, especially away from breeding grounds.
VOICE: Song consists of two or three high, thin notes, followed by a single warbled note and a low trill.
RANGE: Uncommon, local, and declining. Found in grasslands and weedy fields. Casual east to WI and from CA (most records from the Farralones). Accidental from East Coast and coastal LA and NV.

Henslow's Sparrow *Ammodramus henslowii* L 5" (13 cm)
Large flat head; large gray bill. Resembles Baird's Sparrow but head, nape, and most of central crown stripe are greenish; wings extensively dark chestnut. **Juvenile** is paler, yellower, with less streaking below; compare with adult Grasshopper Sparrow. Secretive, but after being flushed several times may perch in the open for a few minutes before dropping back into cover.
VOICE: Distinctive song, a short *se-lick,* accented on second syllable.
RANGE: Uncommon, local; now occurs only rarely in the Northeast; accidental to NM (mid-Oct.). Two sight records for northeastern CO. Found in wet shrubby fields, weedy meadows, and reclaimed strip mines. In winter, found also in the understory of pine woods.

supraloral spot paler

extensive fine streaking below with underlying buffy wash

juvenile
perpallidus

short, spiky tail as in other *Ammodramus*

richly colored supraloral spot; rest of supercilium gray

full eye ring and strong median crown stripe

large bill

buffy breast

Grasshopper Sparrow

much more blackish above, down through uppertail coverts

fresh fall
perpallidus

floridanus

streaked back

thick fleshy bill

Orange Bishop ♀ for comparison

short, blunt tail

buffiest subspecies overall and the most rufous one above, although a few *perpallidus* and *pratensis* are rufous above too

a few reddish streaks on sides of breast

ammolegus

no postocular stripe as in Savannah Sparrow; median crown stripe inconspicuous

two dark spots near auricular stand out on orangish buff head

short necklace

Baird's Sparrow

scaly pattern above

juvenile

pea-soup green head with blackish lateral crown and postocular stripes

large bill

rich dark chestnut on upperparts

necklace

Henslow's Sparrow

juvenile

largely unstreaked breast

Saltmarsh Sparrow *Ammodramus caudacutus* L 5" *(13 cm)*

Similar to Nelson's, but bill longer and head flatter; orange-buff face triangle contrasts strongly with paler, crisply streaked underparts. Also dark markings around eye and head are more sharply defined; eyebrow streaked with black behind eye. **Juvenile's** crown is blacker than juvenile Nelson's; cheek darker; streaks below more widespread and distinct. Hybridizes with Nelson's in southern coastal ME.

VOICE: Song softer, more complex than Nelson's.

RANGE: Found in grassy tidal marshes; accidental inland.

Le Conte's Sparrow *Ammodramus leconteii* L 5" *(13 cm)*

White central crown stripe, becoming orange on forehead, chestnut streaks on nape, and straw-colored back streaks distinguish Le Conte's from Saltmarsh and Nelson's Sparrows. Bright, broad, buffy orange eyebrow, grayish ear patch, thinner bill, and orange-buff breast and sides separate it from Grasshopper Sparrow (page 474). Sides of breast and flanks have dark streaks. **Juvenal** plumage, seen on breeding grounds and in fall migration, is buffy; crown stripe tawny; breast heavily streaked.

VOICE: Song is a short, high, insectlike buzz.

RANGE: A bird of wet grassy fields, marsh edges. Fairly common but secretive; scurries through matted grasses like a mouse. Casual migrant in the Northeast and in the West, where it has wintered.

See subspecies map, page 555

Nelson's Sparrow *Ammodramus nelsoni* L 4¾" *(12 cm)*

Distinguished from Le Conte's by gray median crown stripe; whitish or gray streaks on scapulars; gray, streakless nape. **Juvenile** has fainter median crown stripe; duller nape; variably thicker eye line; less contrast above; lacks streaking across breast. Plumage variable: *nelsoni,* of interior, has orange-buff triangle on face; streaked buffy breast contrasts with white belly; back strongly marked with black and white stripes; *subvirgatus,* of the Maritime Provinces and coastal ME, is duller overall; has diffuse streaking below; grayer upperparts. In *alterus* (not shown) from James and Hudson Bays, brightness is intermediate, streaks blurred.

VOICE: Song, a wheezy *p-tssssshh-uk,* ends on a lower note.

RANGE: Not often detected as a migrant in the interior, and casual in western interior. Very rare in winter on CA coast. Spring migration is late (late May).

See subspecies map, page 557

Seaside Sparrow *Ammodramus maritimus* L 6" *(15 cm)*

Long, spikelike bill with thick base, thin tip. Tail is short, pointed. Yellow supraloral patch. **Juveniles** are duller, browner, than adults. Seaside Sparrows vary widely in overall color. Most subspecies, like the widespread *maritimus,* are grayish olive above. The greener *mirabilis* (**E**), formerly called "Cape Sable Sparrow," inhabits a small area in southwestern FL. Gulf Coast subspecies such as *fisheri* have buffier breasts. Blackish *nigrescens,* formerly called "Dusky Seaside Sparrow," was found only near Titusville, FL, and became extinct in June 1987.

VOICE: Song like Red-winged Blackbird but buzzier.

RANGE: Fairly common in grassy tidal marshes; accidental inland. Rare in ME; very rare to the Maritime Provinces.

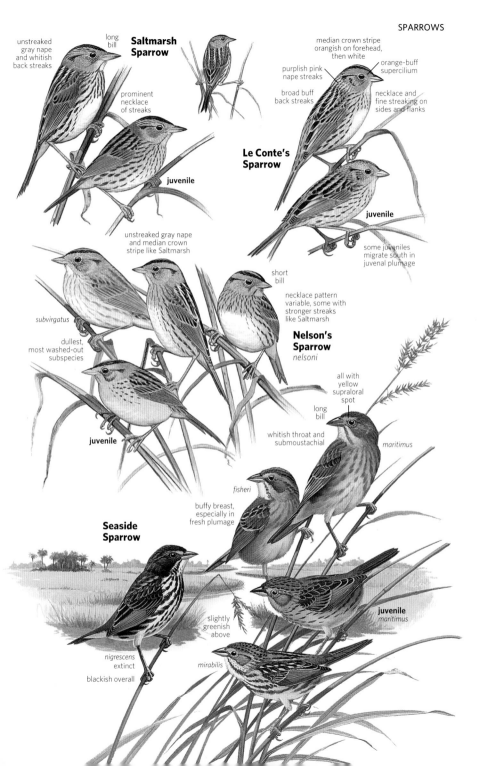

Saltmarsh Sparrow

unstreaked gray nape and whitish back streaks

long bill

prominent necklace of streaks

juvenile

unstreaked gray nape and median crown stripe like Saltmarsh

median crown stripe orangish on forehead, then white

purplish pink nape streaks

orange-buff supercilium

broad buff back streaks

necklace and fine streaking on sides and flanks

Le Conte's Sparrow

juvenile

some juveniles migrate south in juvenal plumage

short bill

necklace pattern variable, some with stronger streaks like Saltmarsh

Nelson's Sparrow
nelsoni

subvirgatus

dullest, most washed-out subspecies

all with yellow supraloral spot

long bill

whitish throat and submoustachial

maritimus

juvenile

fisheri

buffy breast, especially in fresh plumage

Seaside Sparrow

juvenile
maritimus

slightly greenish above

nigrescens extinct

blackish overall

mirabilis

GENUS *MELOSPIZA*
Members of the genus *Melospiza* are found in brushy habitats. Their relatively long, rounded tails are pumped in flight.

Lincoln's Sparrow *Melospiza lincolnii* L 5¾" (15 cm)
Buffy wash and fine streaks on breast and sides, contrasting with whitish, unstreaked belly. Note broad gray eyebrow, whitish chin and eye ring. Briefly held **juvenal** plumage is paler overall than juvenile Swamp Sparrow. Distinguished from juvenile Song Sparrow by shorter tail, slimmer bill, and thinner malar stripe, often broken. Often raises slight crest when disturbed.
VOICE: Two call notes: a flat *tschup,* repeated in a series as an alarm call; and a sharp, buzzy *zeee.* Rich, loud song, a rapid, bubbling trill.
RANGE: Found in brushy bogs and mountain meadows; in winter prefers thickets. Uncommon east of Mississippi River.

Song Sparrow *Melospiza melodia* L 4¾-6¾" (13-17 cm)
All subspecies have long, rounded tail, pumped in flight. All show broad grayish eyebrow and broad, dark malar stripe bordering whitish throat. Highly variable. Upperparts are usually streaked. Underparts whitish, with streaking on sides and breast that often converges in a central spot. **Juvenile** is buffier overall, with finer streaking. The numerous subspecies vary geographically in size, bill shape, overall coloration, and streaking. Brownish eastern birds, now generally treated as one subspecies, nominate *melodia,* breed west to northeastern BC; winter west to west TX. Large AK subspecies, the largest resident on the Aleutians, reach an extreme in the gray-brown *maxima;* paler subspecies such as *fallax* inhabit southwestern deserts; *morphna* represents the darker, redder subspecies of the Pacific Northwest; *heermanni* is one of the blackish-streaked CA subspecies.
VOICE: Typical song has three or four short clear notes followed by a buzzy *tow-wee,* then a trill. Distinctive call note is a nasal, hollow *chimp.*
RANGE: Generally common. Found in brushy areas, especially dense streamside thickets.

Swamp Sparrow *Melospiza georgiana* L 5¾" (15 cm)
Gray face; rich rufous upperparts and wings; variable black streaks on back; white throat. **Breeding adult** has variable reddish crown, gray breast, and whitish belly. **Winter adult** is buffier overall; crown is streaked, shows gray central stripe; sides are rich buff. Briefly held **juvenal** plumage is usually even buffier; darker overall than juvenile Lincoln's or Song Sparrow; wings and tail redder. **Immature** resembles winter adult. Resident birds on mid-Atlantic coast, *nigrescens,* have a broader band of black across forehead, at least in breeding plumage; also have broader black streaks on back and longer bills.
VOICE: Typical song is a slow, musical trill, all on one pitch. Two call notes: a prolonged *zeee,* softer than Lincoln's Sparrow, and an Eastern Phoebe–like chip.
RANGE: Nests in dense, tall vegetation in marshes and bogs. Winters in marshes and brushy fields. Fairly common in western breeding range, but otherwise generally rare in the West.

often appears slightly crested when agitated

broad gray supercilium

well-outlined buffy submoustachial stripe

Lincoln's Sparrow

finer streaking than Song against buffy breast

juvenile *heermanni*

buffy wash with blurry streaks on breast and sides

similar to adult but overall, pattern a little more muted and more streaked

juvenile

compare carefully to Lincoln's Sparrow

northwestern subspecies are dark and ruddy

thicker streaks on breast and sides than Lincoln's

whitish submoustachial

California subspecies are very dark

morphna

Alaskan subspecies, especially those on Aleutians, are very large

maxima

heermanni

all *Melospiza* have long, rounded tails

fallax

smallest and palest subspecies

Song Sparrow

many subspecies have blurry streaks below often with central spot on breast

variable reddish crown

breeding

ruddy rufous wings in all plumages

grayish breast

melodia

dark back streaks

Swamp Sparrow

juvenile

winter adult

immature

dull, blurry streaks confined to sides

See subspecies map, page 556

Fox Sparrow *Passerella iliaca* L 7" (18 cm)

A large and highly variable species. Most subspecies have reddish rump and tail; reddish in wings; underparts heavily marked with triangular spots merging into a larger spot on central breast. Base of bill is yellowish orange in winter, darker and more grayish in summer. The many named subspecies are divided into four subspecies groups; may represent distinct species. The brightest, *iliaca*, and slightly duller *zaboria*, found west of Hudson Bay (**"Red"** group), breed in the far north, from Seward Peninsula, AK, to NL; winter mostly in southeastern U.S. Western mountain subspecies have gray head and back; range from small-billed Rockies *schistacea* (**"Slate-colored"** group) to large-billed CA *stephensi* (**"Thick-billed"** group). Breeding birds of the Canadian Rockies, *altivagens,* like *schistacea* generally, but are somewhat more reddish. Birds from the White Mountains of eastern CA and NV and east across the mountains of NV *(canescens)* are genetically intermediate between "Slate-colored" and "Thick-billed" groups, but at least in the White Mountains, most chip like "Thick-billed." Dark coastal subspecies (**"Sooty"** group), with browner rumps and tails, vary from sooty *fuliginosa* of the Pacific Northwest and gradually lighten to palest *unalaschcensis* of southwestern AK. Rather retiring but responds to pishing and comes up on a visible perch and starts chipping. When feeding, scratches with both feet together, like a towhee.

VOICE: Songs are sweet, melodic in northern "Red" group; include harsher trills in other subspecies. Thick-billed Pacific subspecies give a sharp *chink* call, like California Towhee; others give a *tschup* note, like Lincoln's Sparrow but louder, though in "Slate-colored" call is a little different, more of a *tewk,* and slightly downslurred.

RANGE: Uncommon to common; found in undergrowth in coniferous or deciduous woodlands and chaparral. Within "Sooty" group, paler northernmost subspecies migrate the farthest south. The "Slate-colored" group migrates primarily southwest to CA. Some from the "Red" group winter in the West, and "Sooty" is accidental in the East.

GENUS *ZONOTRICHIA*

The genus *Zonotrichia* includes some of the largest sparrows, which form flocks in brushy areas during the nonbreeding season. They are usually cooperative for viewing.

Harris's Sparrow *Zonotrichia querula* L 7½" (19 cm)

A large sparrow with at least some blackish on crown and face, a pink bill, and a dark postocular mark in rear of face. In **breeding** plumage, crown and throat are black and face is silvery gray, except for the postocular spot. **Winter adult's** crown is blackish; cheeks buffy; throat may be all-black or show white flecks or partial white band. **Immature** resembles winter adult but shows less black; white throat is bordered by dark malar stripe.

VOICE: Song is a series of long, clear, quavering whistles, often beginning with two notes on one pitch followed by two notes on another pitch. Calls include a loud *wink* and a drawn-out *tseep.*

RANGE: Fairly common but local. Nests in stunted boreal forest; winters in open woodlands and brushlands. Rare to casual in migration and in winter in rest of North America outside mapped range; more in the West.

Fox Sparrow

small bill

large bill

"Slate-colored"
schistacea

gray back

in both groups, rufous tail contrasts with gray rump and back

"Thick-billed"
stephensi

"Red"
iliaca

rufous in cheek

reddish streaks on gray back

bright rufous tail

"Sooty"
unalaschcensis

very reddish *iliaca* from east of Hudson Bay illustrated here; *zaboria* from west of Hudson Bay is just slightly duller; both winter in Southeast

thick, dark streaks coalescing on breast

"Sooty" subspecies overall dark brown with little contrast between tail and back; more southerly breeding subspecies darker than those from south-coastal and southwestern Alaska

rufous streaking on underparts

"Sooty"
fuliginosa

black crown and bib

large pink bill

breeding

dark postocular spot in all plumages

winter adults

Harris's Sparrow

dark malar stripe

dark chest patch

brownish flank streaking

immature

White-throated Sparrow *Zonotrichia albicollis*

L 6¾" (17 cm) Conspicuous and strongly outlined white throat; mostly dark bill; dark crown stripes and eye line. Broad eyebrow is yellow in front of eye; remainder is either white or tan. Upperparts rusty brown; underparts grayish, sometimes with diffuse streaking. **Juvenile's** eyebrow and throat are grayish, breast and sides heavily streaked.
VOICE: Song is a thin whistle, generally two single notes followed by three triple notes: *pure sweet Canada Canada Canada,* often heard in winter. Calls include a sharp *pink* and a drawn-out, lisping *tseep.*
RANGE: Common in woodland undergrowth, brush, and gardens. Generally rare in the West, south of breeding range.

See subspecies map, page 556

White-crowned Sparrow *Zonotrichia leucophrys*

L 7" (18 cm) Black-and-white striped crown; pink, orange, or yellowish bill; whitish throat; underparts mostly gray. **Juvenile's** head is brown and buff, underparts streaked. **Immature** has tan and rufous-brown (chocolate brown in *oriantha*) head stripes; compare with immature Golden-crowned Sparrow. Nominate *leucophrys* (mainly found in the eastern Canadian tundra) and *oriantha* (High Sierra, southern Cascades to Mount St. Helens, and Rockies) have a black supraloral area and large, dark pink bill; supraloral and bill are a little darker in *oriantha,* and underparts are a slightly paler; *gambelii* (from AK to Hudson Bay) has whitish supraloral and a smaller, orange-yellow bill; in coastal *nuttalli* (not shown) and *pugetensis,* breast and back are browner, bill dull yellow, supraloral pale. Some, perhaps most, *nuttalli* maintain an immature-like plumage into second summer.
VOICE: Songs for all subspecies are often heard in winter; three of the subspecies (*oriantha, pugetensis,* and *nuttalli*) give one or more thin, whistled notes followed by a sweet twittering trill; resident coastal *nuttalli* is particularly geographically variable in song dialects, even from immediately adjacent areas; *leucophrys* and *gambelii* give a more mournful song with no trill at the end. Calls include a loud *pink,* sharper and more downslurred in *oriantha,* and sharp *tseep.*
RANGE: Generally common in woodlands, grasslands, and roadside hedges. Subspecies *oriantha* winters in Mexico, irregularly along border in southeastern AZ. A few *pugetensis* winter in Central Valley, CA; casual to coastal San Diego County; accidental western NV.

Golden-crowned Sparrow *Zonotrichia atricapilla*

L 7" (18 cm) Yellow patch tops black crown; back brownish, streaked with dark brown; breast, sides, and flanks grayish brown. Bill dusky above, pale below. Yellow is less distinct on **immature's** brown crown. Briefly held **juvenal** plumage has dark streaks on breast and sides. **Winter adults** are duller overall; amount of black on crown varies.
VOICE: Song is a series of three or more plaintive, whistled notes: *oh dear me.* Calls include a soft *tseep* and a flat *tsick.*
RANGE: Fairly common in stunted boreal bogs and in open areas near tree line, especially in willows. Winters in dense woodlands, tangles, and brush; casual in East. Rare well east of coastal region in migration and winter; small numbers in Reno, NV, region. Casual farther south but widespread records in East.

yellow supraloral spot

dark bill

tan and dark brown head stripes

tan-striped morph

White-throated Sparrow

white-striped morph

juvenile

white throat

richly colored back

brown and light tan head stripes

like *leucophrys*, but lateral crown stripe more extensive and darker

lower white stripe normally cut off just in front of eye

bill darker red

adult like *leucophrys*, but slightly more blackish in supraloral, usually dark reddish bill, and slightly paler gray below

immature
leucophyrys

pinkish bill

immature
oriantha

adult
oriantha

adult
leucophrys

breeds in mountains of West; most winter in Mexico

yellow bill

White-crowned Sparrow

like *oriantha*, but note head pattern and bill color

juveniles of all subspecies streaked below

head stripes duller

immature
pugetensis

supraloral area whitish

yellow-orange bill

juvenile
gambelii

blackish brown streaks down back with pale brown edges

wintering White-crowned over much of West

immature
gambelii

many have dark malar streak

adult
pugetensis

buffy wash on sides and flanks

adult
gambelii

Golden-crowned Sparrow

dull yellow forehead; head pattern otherwise muted

yellow crown with broad black eyebrow

winter adult crown pattern much more muted

juvenile

mostly dark bill

immature

winter adult

breeding

GENUS *JUNCO*

Rather confiding, all have long tails with white outer tail feathers. Often the most numerous visitor to feeding stations in winter, especially from northern regions and in the mountains.

See subspecies map, page 556

Dark-eyed Junco *Junco hyemalis* L 6¼" (16 cm)

Variable. Most subspecies with gray or brown head and breast. Note white outer tail feathers in flight. **Juveniles** of all subspecies are streaked. **Male** of the widespread **"Slate-colored Junco"** group has a dark gray hood; upperparts gray or with varying brown at center of back; **female** brownish gray overall. "Slate-colored" winters mostly in eastern North America; uncommon in the West. Male **"Oregon Junco"** of the West has slaty to blackish hood, rufous-brown to buffy brown back and sides; females have duller hood color. In both sexes, lower border of hood more convex than "Slate-colored." "Oregon" group has eight subspecies (two resident in Baja California Norte); more southerly subspecies are paler. "Oregon" types winter mainly in the West; some to central Great Plains; very rare during winter in the East. **"Pink-sided Junco,"** *mearnsi* (part of "Oregon" group)—breeding in central Rockies and wintering from central Great Plains to the foothills of the Southwest and northern Mexico, rarely to southern CA—has very broad, bright pinkish cinnamon sides that sometimes meet across the breast, blue-gray hood, blackish lores. **"White-winged Junco"** subspecies, *aikeni*—breeding in the Black Hills area and wintering largely in the Front Range south to north-central NM, rarely on western Great Plains and casually to Southwest and CA—is mostly pale gray above, usually with two thin, white wing bars; also larger, with bigger bill, more white on tail. In **"Gray-headed Junco"** of the southern Rockies, pale gray hood is barely darker than underparts; back is rufous. "Gray-headed" winters on western Great Plains and in foothills of the Southwest and northern Mexico, rarely to CA; accidental in Midwest. In much of mountainous AZ and NM, largely resident "Gray-headed," *dorsalis,* has even paler throat and large, bicolored bill, black above and bluish below. Intergrades between some subspecies are frequent.
VOICE: Song is a musical trill on one pitch, often heard in winter. Varied calls include a sharp *dit;* in flight, a rapid twittering. Some songs and calls of "Gray-headed" *dorsalis* are more suggestive of Yellow-eyed Junco.
RANGE: Breeds in coniferous or mixed woodlands. In migration and winter, found in a wide variety of habitats, usually in flocks, which in Southwest and western Great Plains contain multiple subspecies.

Yellow-eyed Junco *Junco phaeonotus* L 6¼" (16 cm)

Bright yellow eyes, set off by black lores. Pale gray above, with a bright rufous back and rufous-edged greater wing coverts and tertials; underparts paler gray. **Juveniles** similar to juveniles of the two "Gray-headed" subspecies of Dark-eyed Junco; eye is brown, becoming pale before changing to yellow of adult; look for rufous on wings.
VOICE: Song is a variable series of clear, thin whistles and trills. Calls include a high, thin *seep,* similar to Chipping Sparrow.
RANGE: Resident in coniferous and pine-oak slopes, generally above 6,000 feet. Some move lower in winter. Casual to west TX.

grayish hood

buffy rufous sides

"Slate-colored"
hyemalis

some have faint white tips to wing coverts, similar to "White-winged"

♀

"Oregon"
♀ *shufeldti*

♂

females browner above

blackish hood

"Oregon"
♂ *shufeldti*

juvenile

all juvenile juncos are streaked

larger and paler gray than "Slate-colored"

"White-winged"
aikeni

♂

faint white tips to wing coverts create wing bars

Dark-eyed Junco

"Slate-colored" ♂

bluish gray hood and dark lores

pale gray head and underparts

dark lores

pale bill

dark lores

rufous back

all subspecies have white outer tail feathers

♂

"Pink-sided"
mearnsi

broad pinkish buff sides and flanks

pinkish buff meets across lower breast on some

mostly dark upper mandible

"Gray-headed"

caniceps

paler throat than *caniceps*

dorsalis

sometimes called "Red-backed Junco"

Yellow-eyed Junco
palliatus

striking yellow eye

rufous back

bicolored bill

pale throat

rufous greater coverts

juvenile

Yellow-breasted Bunting *Emberiza aureola* L 6" (15 cm)

Breeding **male** rufous-brown above, bright yellow below; white wing patch. East Asian *ornata* has black on forehead and base of breast band. **Female** and immatures have a pale median crown stripe; outlined auricular with pale spot; yellowish underparts with sparse streaking.
VOICE: Call is a *tzip,* similar to Little Bunting.
RANGE: Declining, Asian species; casual to AK, mostly on western Aleutians, but at least one record for St. Lawrence Island, AK.

Gray Bunting *Emberiza variabilis* L 6¾" (17 cm)

A large, heavy-billed bunting; shows no white in tail. Breeding **male** is gray overall, prominently streaked with blackish above. Adult **female** is brown; chestnut rump is conspicuous in flight. **Immature male** resembles adult female above but is mostly gray below with some gray on the head; immature plumage is largely held through first spring.
VOICE: Call is a sharp *zhii.*
RANGE: Asian species, casual spring vagrant on western Aleutians.

Reed Bunting *Emberiza schoeniclus* L 6" (15 cm)

Note solid chestnut lesser wing coverts (often hidden); heavy, gray bill with curved culmen; cinnamon wing bars; dark lateral crown stripes, paler median crown stripe. **Breeding male** has black head and throat, broad white submoustachial stripe, white nape; upperparts streaked black and rust. Often flicks its tail, showing white outer tail feathers.
VOICE: Call is a *seeoo,* falling in pitch; flight note, a hoarse *brzee.*
RANGE: Eurasian species, casual vagrant on westernmost Aleutians in late spring; one fall record on St. Lawrence Island, AK. All records are of pale East Asian *pyrrhulina.*

Pallas's Bunting *Emberiza pallasi* L 5" (13 cm)

Smaller than Reed; smaller bill with straighter culmen; grayish lesser wing coverts; shorter tail. **Female** has more indistinct eyebrow and lateral crown stripes than female Reed; lacks median crown stripe.
VOICE: Call, a *cheeep,* recalls Eurasian Tree Sparrow, very unlike Reed.
RANGE: Asian species. Casual to St. Lawrence Island, AK; accidental to Buldir Island and Point Barrow, AK.

Rustic Bunting *Emberiza rustica* L 5¾" (15 cm)

A slight crest, whitish nape spot, prominent pale line extending back from eye. **Breeding male** unmistakable. **Female** and fall and winter males have duller brownish head pattern, pale spot at rear of ear patch.
VOICE: Call note is a hard, sharp *jit* or *tsip.* Song, a soft, bubbling warble.
RANGE: Eurasian species; uncommon in spring on western and central Aleutians, rare in fall; very rare on other Bering Sea islands; casual to Pacific region in fall and winter; accidental SK (winter).

Little Bunting *Emberiza pusilla* L 5" (13 cm)

Small with small triangular bill, creamy white eye ring, chestnut ear patch. In **breeding** plumage, shows chestnut crown stripe bordered by black stripes, pattern muted in winter birds.
VOICE: Call note is a sharp *tsick.*
RANGE: Eurasian species, very rare in fall to St. Lawrence Island, AK. Casual on Pribilofs and western Aleutians and in CA.

pale median crown stripe and pale spot in rear of auricular

light yellow underparts

distinctive head pattern with gray on sides of nape

mostly pale bill in all plumages

♀

Gray Bunting

large size

black face; east Asian *ornata* also with black on sides of breast and forehead

white in outer tail feathers, as with most *Emberiza* buntings ♀

dark rufous-brown upperparts and white shoulder patch

rufous-brown rump, often best seen in flight

bright yellow underparts ♂

Yellow-breasted Bunting *ornata*

lacks white in tail in all plumages, unusual in *Emberiza* buntings

overall dark gray with black back streaks ♂

paler brown median crown stripe

curved culmen

dark bill

contrasting rufous lesser coverts, often hidden

striking head pattern

extensive rufous on wing

even by first spring plumage, intermediate between male and female

immature ♂

breeding ♂

♀

smaller and smaller billed than Reed with straight, not curved, culmen

pale lower mandible

no median crown stripe ♀

pale wing bars

fall ♂

Reed Bunting *pyrrhulina*

striking head pattern

breeding ♂

Pallas's Bunting *polaris*

black on head veiled on adult male Reed and Pallas's in fresh fall plumage

gray (adult male) to grayish lesser wing coverts diagnostic from Reed, but often hidden

straight culmen

immatures often have breast streaking in fall

juvenile

pale spot at rear of crown and in ear coverts in all plumages

Rustic Bunting

chestnut markings below

often raises slight crest

black-and-white head pattern

♀

bright chestnut breast band and streaks down sides and flanks

breeding ♂

chestnut median crown stripe and auricular with broad dark lateral crown stripes

prominent whitish eye ring

fall birds, especially immatures, are duller

small size

immature

straight culmen

Little Bunting

breeding ♂

In North America, this diverse family now includes *Piranga* tanagers formerly with Thraupidae, the tanagers. Also included are various seedeaters including Northern Cardinal, certain grosbeaks, the *Passerina* and other buntings, and Dickcissel. SPECIES: 48 WORLD, 18 N.A.

Hepatic Tanager *Piranga flava* L 8" (20 cm)

Large grayish cheek patch and gray wash on flanks set off brighter throat, breast, and cap in both sexes; dark bill with gray base small "tooth" hard to see. In a few females, the lemon yellow is replaced by a more orangish color. **Adult male** plumage is acquired by second fall; dull red plumage retained year-round. Juvenile resembles yellow-and-gray **female** but is heavily streaked overall. Birds from eastern part of U.S. range *(dextra)* are somewhat more richly colored than those from farther west *(hepatica)*.

VOICE: Song suggests Black-headed Grosbeak. Call is a single low, sharp *chuck,* unlike all other *Piranga* tanagers.

RANGE: Inhabits mixed mountain forests. Breeds on mountains of Southwest; rare to eastern CA and CO. Very rare migrant in lowlands and in winter in southeastern AZ and southern CA. Casual to southern NV and north, central, and south TX; accidental to WY, LA, and IL.

Summer Tanager *Piranga rubra* L 7¾" (20 cm)

Adult male is rosy red year-round. **First-spring male** usually has red head. Some **females,** especially of eastern *rubra,* show overall reddish wash; most have a mustard tone, lack olive of female Scarlet Tanager; bill larger. Western birds *(cooperi)* are larger, longer billed, and paler; females generally grayer above.

VOICE: Song is robinlike; call is a staccato *ki-ti-tuck.*

RANGE: Common in pine-oak woods in the East, cottonwood groves in the West. Eastern *rubra* occurs rarely but regularly in West. Rare in winter in coastal CA.

Scarlet Tanager *Piranga olivacea* L 7" (18 cm)

Breeding male bright red and black. In late summer, becomes splotchy green-and-red as he molts to yellow-green winter plumage. **Female** has uniformly olive head, back, and rump; whitish wing linings; bill smaller and stubbier than Summer Tanager. **First-spring male** resembles adult male, but note brownish primaries and secondaries. Some immatures show faint wing bars.

VOICE: Robinlike song (hoarser than Summer Tanager) of raspy notes, *querit queer query querit queer,* is heard in deciduous forests. Call is a hoarse *chip-burr;* sometimes just first part is given.

RANGE: Found in deciduous forests. Winters in South America; accidental in U.S. in winter. Very rare vagrant in the West, most in late fall.

color brightest on
forehead and throat
in all plumages

"tooth"
on upper
mandible

dull red
plumage

dark
bill

grayish
cheek

**Hepatic
Tanager**
hepatica

red head and
patches elsewhere

greenish primaries
and secondaries

♀

grayish
wash

adult ♂

1st spring ♂

**Summer
Tanager**
rubra

large
bill

overall
rosy red

some females
with patchy dull red
throughout

red morph ♀

overall ochre
plumage

♀

adult ♂

small
bill

♀

1st spring ♂

wings
contrast
darker to
olive-green
back

**Scarlet
Tanager**

black
contrasts
with greenish
remiges

brownish
primaries

shorter tail
than Summer

1st fall ♂

**breeding
adult ♂**

all adult males
have solid black wings
and tail

**fall adult
♂**

Western Tanager *Piranga ludoviciana* L 7¼" *(18 cm)*

Conspicuous wing bars, often paler and thinner in **female** (especially an immature female) and can appear virtually absent when worn, upper bar yellow in **male.** Adult male's red head becomes yellow-ish and finely streaked in **winter** plumage. Many adult males still in breeding plumage migrate south in July and Aug. Other adult males molt and migrate south in Sept. in fresh winter plumage. Female's grayish back and scapulars ("saddle") contrast with greenish yellow nape and rump, a diagnostic character from the smaller and smaller-billed Scarlet Tanager (page 488). Some females are duller below, grayer above.

VOICE: Song is like Scarlet Tanager. Call is a *pit-er-ick;* also a whistled flight call.

RANGE: Breeds in coniferous forests. Rare in winter north along coastal slope to central CA. Uncommon migrant on western Great Plains. Rare on western Gulf Coast in migration, casual in the East, mostly in late fall and in winter at feeders.

Flame-colored Tanager *Piranga bidentata* L 7¼" *(18 cm)*

Has gray bill with visible "teeth"; blackish rear border to ear patch; streaked back; white wing bars and tertial tips; whitish tail corners. Hybrids with Western Tanager are somewhat regularly noted in southeastern AZ. **Male** of nominate west Mexican subspecies, *biden-tata,* is flaming orange, eastern *sanguinolenta* male redder. **Female** and immatures are colored like female Western Tanager. **First-spring males** have brighter yellow head; some spotting.

VOICE: Song similar to Western and Scarlet Tanagers; call is a low-pitched *prreck,* also like Western but huskier.

RANGE: Resident from western Panama to northern Mexico; very rare to mountains of southeastern AZ (nearly annual in recent years) in spring and summer; casual to west and south TX.

Crimson-collared Grosbeak *Rhodothraupis celaeno*

L 8½" *(22 cm)* Stubby, mostly black bill; long tail; black on head vari-able. **Adult male's** collar and much of underparts an intense shade of red; upperparts darker. **Adult female** is olive above; has thin yel-lowish wing bars; yellow-green rear collar; yellowish olive underparts. Immatures show less black than female; throat, which is more sooty black, blends more with chest. Male shows some red and black patches by first spring. Often skulks on or near ground; often raises rear crown feathers. Prefers to eat green leaves but will also take fruit.

VOICE: Song is a variable warble; call is a penetrating, rising and falling *seeiyu.*

RANGE: Found in thickets in woodland and second growth. Endemic to northeastern Mexico; casual to south TX, mainly in winter; multiples in some winters.

gray "saddle"

white wing
bars and
tertial
edges

gray
morph ♀

1st fall ♂

larger
bill than
Scarlet

♀

some
very dull
below

winter
adult ♂

reddish
head

**Western
Tanager**

black "saddle"
and yellow
rump

breeding
adult ♂

1st spring ♂

head and
below bright
yellow with a
tinge of orange

**Flame-colored
Tanager**
bidentata

faint dark
facial mask

streaked
back

white tertial
tips

dark bill
with slightly
protruding
"teeth"

♀

faint facial
mask

deep orange
coloration

adult ♂

yellow-olive
coloration on
female and
immature male

adult ♀

on adult female,
black hood distinctly
defined on breast

black head

thick
stubby
bill with
curved
culmen

intense deep
red on nape
and below

adult ♂

**Crimson-collared
Grosbeak**

Northern Cardinal *Cardinalis cardinalis* L 8¾" *(22 cm)*

Conspicuous crest; cone-shaped reddish bill. **Male** is red overall, with black face. **Female** is buffy brown or buffy olive, tinged with red on wings, crest, and tail. **Juvenile** browner overall, dusky bill; juvenile female lacks red tones. Bill shape helps distinguish female and juveniles from similar Pyrrhuloxia. Geographically variable, especially in color of males. Males from the Southwest *(superbus)* are particularly bright red; also have a longer crest and less black around bill.

VOICE: Song is a loud, liquid whistling with many variations, including *cue cue cue* and *cheer cheer cheer* and *purty purty purty.* Both sexes sing almost year-round. Common call is a sharp chip.

RANGE: Abundant throughout the East, inhabits woodland edges, swamps, streamside thickets, and suburban gardens, also the Sonoran Desert and riparian areas of the Southwest. Nonmigratory, but has expanded its range in East northward during the 20th century. In West, formerly a few *(superbus)* to Colorado River in southeastern CA, where they likely bred. Casual to Inyo County, CA, and Las Vegas, NV. Also a small introduced population of eastern birds in Los Angeles, CA; local escapes seen elsewhere.

Pyrrhuloxia *Cardinalis sinuatus* L 8¾" *(22 cm)*

Thick, strongly curved, pale bill helps distinguish this species from female and juvenile Northern Cardinal. **Male** is gray overall, with red on face, crest, wings, tail, and underparts. **Female** shows little or no red.

VOICE: Song is a liquid whistle, thinner and shorter than Northern Cardinal; call is a sharper, more metallic *chink.*

RANGE: Fairly common in thorny brush, mesquite thickets, desert, woodland edges, and ranchlands. More migratory than the resident range would suggest, and small flocks present in winter from locations where they do not breed. Casual to southern CA and central Great Plains and on oil rigs in the Gulf of Mexico; accidental to OR and ON.

Dickcissel *Spiza americana* L 6¼" *(16 cm)*

Yellowish eyebrow, thick bill, and chestnut wing coverts are distinctive. **Breeding male** has black bib under white chin, bright yellow breast. **Female** lacks black bib, but has some yellow on breast; chestnut wing patch muted. **Winter adult male's** bib is less distinct. **Immatures** are duller overall than adults, breast and flanks lightly streaked; female may show almost no yellow or chestnut. Compare with shorter-winged female House Sparrow (page 528).

VOICE: Common call, often given in flight, is a distinctive electric-buzzer *bzrrrrt.* Song is a variable *dick dick dickcissel.*

RANGE: Breeds in open weedy meadows, grainfields, and prairies. Abundant and gregarious, especially in migration, but numbers and distribution vary locally from year to year outside core breeding range. Irregular east of the Appalachians; occasional breeding is reported outside mapped range. Uncommon breeder in limited western range. Rare migrant to both coasts; more common in the East, where a few winter, often at feeders with House Sparrows. Casual in winter to CA. Accidental to southeastern AK.

long crest

black surrounds red bill

straight culmen

♂

dark bill

Northern Cardinal
cardinalis

longer crest than eastern birds

less black around red bill than eastern birds

southwestern
♂ *superbus*

male brighter red than eastern birds

juvenile ♂

♀

overall buffy brown and dull red

Pyrrhuloxia
fulvescens

yellowish bill with strongly curved culmen

grayish overall with pale buffy breast

♀

♂

overall gray with patchy bright red

breeding ♀

winter adult ♂

Dickcissel

black bib and yellow breast

chestnut lesser and median coverts

longish wings

pale supercilium

large bill

immature ♂

breeding ♂

broad, pale yellow submoustachial

immature ♀

chestnut median coverts

Rose-breasted Grosbeak *Pheucticus ludovicianus*

L 8" *(20 cm)* Large size; very large, triangular bill; upper mandible paler than Black-headed Grosbeak. **Breeding male** has rose red breast, white underparts, white wing bars, white rump. Rose red wing linings show in flight. Brown-tipped **winter** plumage is acquired before fall migration. **Female's** streaked plumage and yellow wing linings resemble female Black-headed, but underparts are more heavily and extensively streaked. Compare also with smaller, immature Purple Finch (page 518). Similar **first-fall male** is buffier above, with buffy wash across breast; often has a few red feathers on breast; red wing linings distinctive.

VOICE: Rich, warbled songs of Rose-breasted and Black-headed are nearly identical, but Rose-breasted's call, a sharp *eek,* is squeakier.

RANGE: Common in wooded habitat along watercourses. Rare throughout West in migration. Very rare in winter on CA coast.

Black-headed Grosbeak *Pheucticus melanocephalus*

L 8¼" *(21cm)* Large, with a very large, triangular bill, upper mandible darker than Rose-breasted Grosbeak. **Male** has burnt orange underparts, all-black head. In flight, both sexes show yellow wing linings. **Female** plumage is generally buffier above and below than female Rose-breasted, with less streaking below. **First-fall male** Black-headed is rich buff or butterscotch below, with little or no streaking.

VOICE: Songs of Black-headed and Rose-breasted are nearly identical, but Black-headed's *ik* call is lower pitched. Juvenile's begging call, a two-part, downslurred *swee-o,* is a characteristic sound in mid- to late summer, even from southbound migrants.

RANGE: Common in open woodlands and forest edges. Rare in winter on CA coast. Casual during migration and winter to the Midwest and East, often at feeders. Also to AK. Hybridizes occasionally with Rose-breasted in range of overlap on the Great Plains.

Yellow Grosbeak *Pheucticus chrysopeplus* L 9¼" *(24 cm)*

Male distinguished by large size, massive bill, yellow plumage. **Females** similar to male but duller; crown streaked; immature male has yellower head than female, like adult male by second fall.

VOICE: Call and song are like Black-headed Grosbeak.

RANGE: Found from western Mexico to southern Guatemala, breeding north to northern Sonora. Casual vagrant to southeastern AZ in late spring to early summer, chiefly in open woodlands and river courses of low mountains. Records away from AZ more problematic: One adult male with a deformed bill from Albuquerque in winter was accepted. A badly abraded bird with a deformed bill from Inyo County, CA, was not accepted; nor was a winter bird from IA.

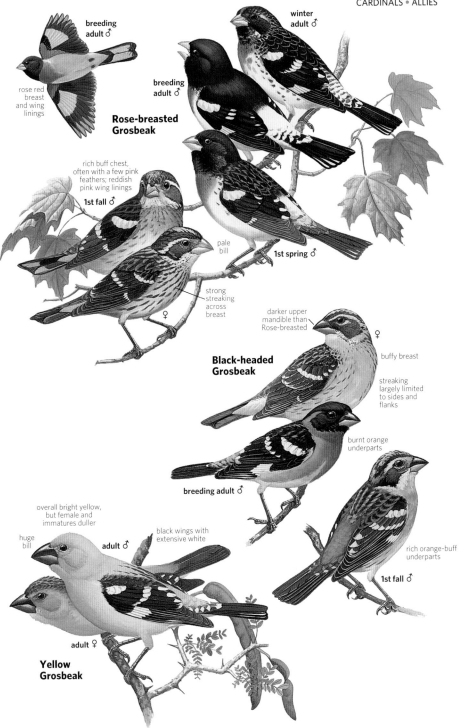

breeding
adult ♂

winter
adult ♂

rose red
breast
and wing
linings

breeding
adult ♂

**Rose-breasted
Grosbeak**

rich buff chest,
often with a few pink
feathers; reddish
pink wing linings

1st fall ♂

pale
bill

1st spring ♂

strong
streaking
across
breast

♀

darker upper
mandible than
Rose-breasted

♀

**Black-headed
Grosbeak**

buffy breast

streaking
largely limited
to sides and
flanks

burnt orange
underparts

breeding adult ♂

overall bright yellow,
but female and
immatures duller

black wings with
extensive white

huge
bill

adult ♂

rich orange-buff
underparts

1st fall ♂

adult ♀

**Yellow
Grosbeak**

Blue Bunting *Cyanocompsa parellina* L 5½" (14 cm)

Smaller than Blue Grosbeak; lacks wing bars. Found in brushy fields and woodland edges. **Adult male** is blackish blue overall year-round, looks blackish in poor light; paler blue on crown, cheeks, shoulder, and rump. Immature male very similar, but with brownish cast to wings. The blue of male Indigo Bunting (page 498) is paler, and Indigo lacks contrasting paler blue highlights. **Female** is a uniform cinnamon-brown and lacks the blurry streaking characteristic of all female and immature Indigos. For both sexes, note thick, strongly curved bill and rounded, notched tail. TX records likely represent *beneplacita* from northeastern Mexico of the eastern Middle American *parellina* group. Females of *indigotica* subspecies, from western and southwestern Mexico (found north to Sinaloa), are paler and grayer; they may well differ vocally as well, raising the issue of whether they represent a separate species (more study needed).

VOICE: Chip notes of *parellina* group of subspecies are strongly suggestive of Hooded Warbler. Song, seldom heard in TX, is a short warble with a longer introductory note.

RANGE: Tropical species, found on both slopes of Mexico south to Nicaragua. Very rare and irregular winter visitor to south TX; accidental LA. Found in woodland thickets. Many claims pertain to Indigo Buntings, which do winter in south TX in small numbers.

Blue Grosbeak *Passerina caerulea* L 6¾" (17 cm)

Wide chestnut wing bars, large heavy bill, and larger overall size distinguish **male** from male Indigo Bunting (page 498). **Females** of these two species also similar; compare bill shape, wing bars, and overall size. Juvenile resembles female; in first fall, some **immatures** are richer brown than female. **First-spring male** shows variable blue on head, occasionally lacking; resembles adult male by second winter. Frequently twitches and spreads tail when perched.

VOICE: Listen for distinctive call, a loud, explosive *chink*. Song is a series of rich, rising and falling warbles.

RANGE: Fairly common; found in low, overgrown fields, streamsides, woodland edges, and brushy roadsides. Uncommon to rare migrant (mainly fall) north to New England and the Maritime Provinces, also coast of central and northern CA. Casual to OR and BC, accidental southeastern AK. Accidental in winter in CA and AZ.

Painted Bunting *Passerina ciris* L 5½" (14 cm)

Adult male's gaudy colors are retained year-round. **Female** is bright green above, paler yellow-green below. **Juvenile** is much drabber; look for telltale hints of green above, yellow below. Fall molt in eastern nominate subspecies takes place on breeding grounds; more western *pallidior* molts on winter grounds. First-winter male resembles adult female; by spring, may show tinge of blue on head, red on breast.

VOICE: Song is a rapid series of varied phrases, thinner and sweeter than Indigo Bunting. Call is a loud, rich chip.

RANGE: Locally common in low thickets, streamside brush, and woodland borders. Casual north on Atlantic coast to NY; in Midwest north to northern ON. Accidental to Akimiski Island, James Bay. Rare fall (Aug.) migrant to southeastern AZ. Very rare, mainly fall, to CA. Otherwise casual in West. Some may be escapes. Declining in Southeast.

overall a rich
cinnamon-
brown

large
dark bill

♀

**Blue
Bunting**

adult ♂

deep, dark blue
overall with paler
blue highlights

♀

thick
bill

**Blue
Grosbeak**
caerulea

deep buffy
color with
rusty buff
wing bars

chestnut
wing bars

immature

breeding
adult ♂

variable
amount of
blue on head

1st spring ♂

frequently
twitches tail

unmistakable

bright green
upperparts similar
overall in coloration
to larger female
Scarlet Tanager

adult ♂

**Painted
Bunting**

yellow
eye ring

♀

overall plain and grayish,
usually with some greenish
on lower back

juvenile

Varied Bunting *Passerina versicolor* L 5½" (14 cm)

Breeding male's plumage is colorful in good light; otherwise appears black. In **winter,** colors are edged with brown. **Female** is plain gray-brown or buffy brown above, slightly paler below; resembles female Indigo Bunting but lacks streaks and has a plainer wing; note also that Varied Bunting's culmen is slightly more curved. First-spring male resembles female. Adult males from southeastern NM and TX *(versicolor)* have a duller, less contrasting but more extensive reddish nape; also have reddish tinge to throat and pale blue rumps. Adult males from southwestern NM and southeastern AZ *(pulchra)* have a smaller but brighter and more contrasting patch of red on nape; also lack reddish tinge to throat, are darker bellied, and have on average a more purplish blue rump. Some treat *"dickeyae"* as a valid subspecies, but the differences from *pulchra* are slight and variable; if recognized, *dickeyae* would comprise the western portion of the U.S. range of Varied Bunting.

VOICE: Song is similar to Painted Bunting. Call is similar to Indigo Bunting, but slightly smoother.

RANGE: Locally common in thorny thickets in washes and canyons. Accidental to southeastern CA and southern ON.

Indigo Bunting *Passerina cyanea* L 5½" (14 cm)

Breeding male deep blue. Smaller than Blue Grosbeak (page 496); bill much smaller; lacks wing bars. In **winter** plumage, blue is obscured by brown and buff edges. **Female** is brownish, always with diffuse streaking on breast and flanks. Young birds resemble female.

VOICE: Song is a series of varied phrases, usually paired. Calls include a sharp *pit* or *spitch;* flight call is a dry buzz.

RANGE: Common in woodland clearings and borders. Rare but regular to Atlantic Canada and throughout the West; casual to AK and in winter in Pacific states.

Lazuli Bunting *Passerina amoena* L 5½" (14 cm)

Adult male bright turquoise above and on throat; cinnamon across breast; thick white upper wing bars. **Female** is grayish brown above, rump grayish blue; whitish underparts with buffy wash across breast. **Juveniles** resemble female but have distinct fine streaks across breast; immature male is mostly blue by first spring. Winter adult male's blue color is obscured by brown and buff edges. **Winter females** and immatures more richly and extensively colored below, more like female Indigo Bunting, but note absence of streaks below, which characterize *all* female and immature male Indigo Buntings. Occasionally hybridizes with Indigo; hybrids tend to look largely like Indigo, though blue is paler and belly is white.

VOICE: Song is a series of varied phrases, sometimes paired; faster and less strident than Indigo Bunting.

RANGE: Found in open deciduous or mixed woodlands and chaparral, especially in brushy areas near water. Casual to East, NT, and southeastern AK.

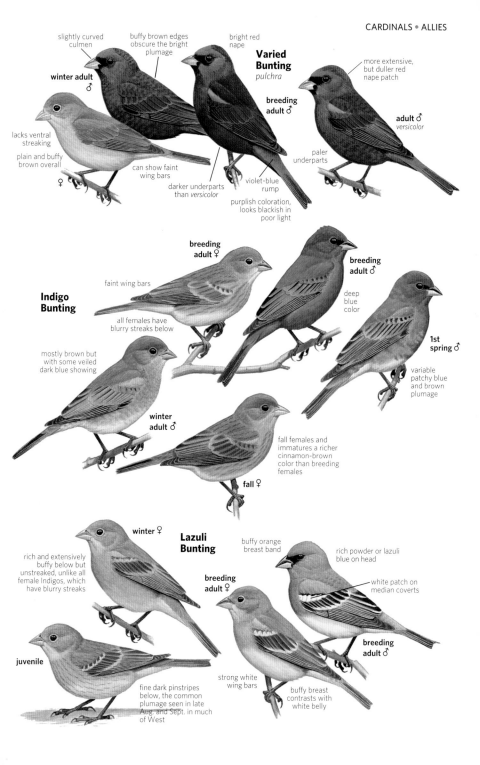

slightly curved culmen

buffy brown edges obscure the bright plumage

bright red nape

Varied Bunting
pulchra

more extensive, but duller red nape patch

winter adult ♂

breeding adult ♂

adult ♂
versicolor

lacks ventral streaking

plain and buffy brown overall

paler underparts

♀

can show faint wing bars

darker underparts than *versicolor*

violet-blue rump

purplish coloration, looks blackish in poor light

breeding adult ♀

breeding adult ♂

faint wing bars

deep blue color

Indigo Bunting

all females have blurry streaks below

1st spring ♂

mostly brown but with some veiled dark blue showing

variable patchy blue and brown plumage

winter adult ♂

fall females and immatures a richer cinnamon-brown color than breeding females

fall ♀

winter ♀

Lazuli Bunting

buffy orange breast band

rich powder or lazuli blue on head

rich and extensively buffy below but unstreaked, unlike all female Indigos, which have blurry streaks

breeding adult ♀

white patch on median coverts

breeding adult ♂

juvenile

fine dark pinstripes below, the common plumage seen in late Aug. and Sept. in much of West

strong white wing bars

buffy breast contrasts with white belly

BLACKBIRDS Family Icteridae

Strong, direct flight and pointed bills mark this diverse group that includes blackbirds, grackles, cowbirds, and orioles, among others. SPECIES: 104 WORLD; 25 N.A.

See subspecies map, page 557

Eastern Meadowlark *Sturnella magna* L 9½" (24 cm)

Black V-shaped breast band on yellow underparts is characteristic of both meadowlark species after post-juvenal molt. In fresh **fall** plumage, birds are more richly colored overall, with partly veiled breast band and rich buffy flanks. On Eastern females, yellow does not reach submoustachial area, and barely does so on males. In widespread northern nominate subspecies *magna*, dark centers are visible on central tail feathers, uppertail coverts, secondary coverts, and tertials. Southeastern *argutula* is smaller and darker, especially those from FL. Southwestern *lilianae* is pale, like Western Meadowlark, but note more extensively white tail. South TX birds, *hoopesi*, are intermediate in color. Birds from Cuba (*hippocrepis; not shown*) are more like Western in plumage, including more extensive yellow into face; calls are unlike Eastern or Western and song is closer to Western; likely represent a different species.

VOICE: Song is a clear, whistled *see-you see-yeeer;* distinctive call is a high, buzzy *drzzt,* given in a rapid series in flight.

RANGE: Generally fairly common in fields and meadows; has declined in the East in recent decades. Eastern *magna* is casual to eastern CO (most results from northeast); *lilianae* is casual to AZ west to Colorado River.

Western Meadowlark *Sturnella neglecta* L 9½" (24 cm)

Plumages similar to those of Eastern Meadowlark, but in **spring** and summer yellow extends well into the submoustachial area, especially in males; yellow often veiled in **fall.** Black breast band is thinner on Western and yellow below is a little paler. Lack of dark centers to feathers of upperparts helps to separate from the more easterly subspecies of Eastern, in areas where ranges overlap. Also, in fresh fall and winter plumage, upperparts, flanks, and undertail region are much paler. Distinguished from pale *lilianae* subspecies of Eastern by mottled cheeks, more mottled postocular and lateral crown stripes, and less white in tail. Northwestern *confluenta* is darker above and can show dark feather centers like Eastern.

VOICE: Song is a series of bubbling, flutelike notes of variable length, usually accelerating toward the end. Sharp *chuck* note; rattled flight call similar to Eastern, but lower pitched; also gives a whistled *wheet.*

RANGE: Western is gregarious in winter; large flocks often gather along roadsides, while Eastern usually prefers taller cover. Overall Western is more migratory than Eastern. In south TX, large flocks are found on roadsides throughout; Easterns occur mainly closer to Gulf Coast in taller cover. Western is casual to East Coast, where most records are of singing birds in late spring and early summer. Casual also to AK. Eastern and Western Meadowlarks hybridize in Midwest, where Western is uncommon and local.

darker above than *magna*

paler than *magna*

more extensive white in tail

strongly contrasting black lateral and postocular lines

spring *argutula*

spring *hoopesi*

southwestern spring *lilianae*

pale cheek in all plumages

coloration like Western except for head

spring *magna*

Eastern Meadowlark

whitish submoustachial

stiff, fluttery flight

spring *magna*

extensive white in tail

more richly colored above than Western, with stronger head pattern

short, broad triangular wings

fall *magna*

juvenile *magna*

strong side and flank streaks against rich buffy ground color when in fresh fall and winter plumage

Western Meadowlark *neglecta*

yellow extends into submoustachial, unlike Eastern

more limited white in tail than Eastern, especially *lilianae*, forming a broad triangular dark wedge

breast band of Western averages thinner and yellow on underparts averages paler than Eastern (except *lilianae*)

spring

lateral and postocular lines paler than Eastern

mottled cheeks

spring *confluenta*

fall

grayer above than *magna* Eastern

northwestern birds are somewhat darker

juvenile

flanks more spotted, with a whiter background, than Eastern *magna*

Bobolink   *Dolichonyx oryzivorus*   L 7" *(18 cm)*

Breeding male entirely black below; hindneck is buff, fading to whitish by midsummer; scapulars and rump white. Male in **spring** migration shows pale edgings. **Breeding female** is buffy overall, with dark streaks on back, rump, and sides; head is striped with dark brown. Juvenile resembles female, but lacks streaking below; has indistinct spotting on throat and upper breast. All **fall** birds resemble female, but are rich yellow-buff below — especially, on average, the immatures. In all plumages, note sharply pointed tail feathers.

VOICE: Male's loud, bubbling *bob-o-link* song, often given in flight, is heard in spring and summer. Flight call heard year-round and carrying a great distance is a repeated, whistled *ink*. Also gives a blackbird-like *chuck* call.

RANGE: Nests primarily in hayfields, weedy meadows. Most birds migrate east of the Great Plains. Rare to very rare migrant in West away from breeding grounds; on West Coast most are in fall. Accidental to AK, northwestern Canada. Winters in South America.

Red-winged Blackbird   *Agelaius phoeniceus*   L 8¾" *(22 cm)*

Glossy black **male** has red shoulder patches broadly tipped with buffy yellow. In perched birds, red patch may not be visible; only the buffy or whitish border shows. **Females** are dark brown above, heavily streaked below; sometimes show a red tinge on wing coverts or pinkish wash on chin and throat. **First-year male** like very dark female with reddish shoulder patch. Males in subspecies of Central Valley, CA, and central coast region, known as **"Bicolored Blackbird,"** nearly or totally lack the buffy band behind red shoulder patch. Females have darker bellies, more like female Tricolored; but note chestnut-buff edging on feathers of upperparts, except when worn away; more rounded wings; stouter bill. The subspecies *aciculatus* of Kern Basin in south-central CA has a bill like Tricolored.

VOICE: Song is a liquid, gurgling *konk-la-reee,* ending in a trill. Most common call is a *chack* note. Also frequents feeding stations.

RANGE: Abundant, often found in immense flocks in winter. Generally nests in thick vegetation of freshwater marshes, sloughs, and dry fields; forages in surrounding fields, orchards, and woodlands.

Tricolored Blackbird   *Agelaius tricolor*   L 8¾" *(22 cm)*

More pointed wings and bill than Red-winged Blackbird. Glossy black (slightly grayish sheen) **male** has dark red shoulder patches, often hidden, broadly tipped with white; tips are buffy white in fresh fall plumage. **Females** usually lack any red on shoulder and never show pinkish on throat; plumage is sooty brown and streaked overall; darker than female Red-winged Blackbird, particularly on belly; note more pointed wings and bill. In fresh fall plumage, all Tricolored Blackbirds have grayish buff edging on feathers of upperparts, unlike chestnut-buff of Red-winged. Distinction between females is more difficult when feathers are worn.

VOICE: Song is a harsh, braying *on-ke-kaaangh;* lacks Red-winged's liquid tones. Call lower pitched than Red-winged.

RANGE: Gregarious; found year-round in large flocks (often segregated by sex) in open country, dairy farms; nests in large colonies in marshes. Numbers have declined drastically in recent decades, due largely to habitat degradation.

Bobolink

buffy
nape

black
face and
underparts

buffy edges
to feathers when fresh

**early
spring** ♂

breeding
♂

white
rump

dark pink
bill

rich
yellow-buff
overall

strong head
pattern

strong back
streaks

breeding ♀

spiky tail
tips

very long
primary
projection

fall

pale
supercilium

many with
pinkish throat

adult ♀

immature ♀

strongly
streaked below

**Red-winged
Blackbird**

dark belly on
female Bicolored, like
female Tricolored

thick
bill

adult ♂

red shoulders
most visible
when singing

1st year ♂

**"Bicolored
Blackbird"**
♀

**"Bicolored
Blackbird"
adult**
♂

yellowish border to
red patch largely or
completely absent in
Bicolored

males

when feeding on
ground, tail cocked up like
Brown-headed Cowbird

red shoulder
often completely
hidden, showing
only white bar

glossy black
plumage with
gray sheen

dark red shoulder
with broad white
(spring) or buffy
(fall) border

**Tricolored
Blackbird**

♀

long,
thin
bill

dark
belly

Yellow-headed Blackbird *Xanthocephalus xanthocephalus*
L 9½" (24 cm) **Adult male's** yellow head and breast and white wing patch contrast sharply with black body. **Adult female** is dusky brown, lacks wing patch; eyebrow, lower cheek, and throat are yellow or buffy yellow; belly streaked with white. **Juvenile** is dark brown with buffy edgings on back and wing; head mostly tawny. **Immature male** resembles smaller female but is darker with blackish lores and more extensive yellow on head; primary coverts tipped with white; acquires adult plumage by following fall.
VOICE: Song begins with a harsh, rasping note, ends with a long, descending buzz. Call note is a distinctive rich *croak.*
RANGE: Prefers freshwater marshes or reedy lakes; often seen foraging in nearby farmlands and livestock pens. Locally common throughout most of range; uncommon and very local in the Midwest. Rare fall and winter visitor to the East Coast often in mixed-species blackbird flocks. Casual in spring and fall as far north as southern AK.

Rusty Blackbird *Euphagus carolinus* L 9" (23 cm)
Adults and fall immatures have yellow eyes. Fall adults and immatures are broadly tipped with rust; tertials and wing coverts edged with rust. **Fall female** has broad, buffy eyebrow, outlined at bottom by broad dark lores; buffy underparts, gray rump. **Fall male** is darker, especially the adult; eyebrow fainter. The rusty feather tips wear off by spring, producing the dark **breeding** plumage. Juveniles have dark eyes. On all, note the long and very slender, spikelike bill.
VOICE: Call is a low *tschak;* song, a high, squeaky *koo-a-lee.* Female sings as well, although her song is not as well described.
RANGE: Overall fairly common to uncommon and declining in wet woodlands and swamps; nests in shrubs or conifers near water. Gregarious in fall and winter. A late fall migrant; not until very late Sept. or early Oct. in the northern tier of states; mid- to late Oct., even early Nov., elsewhere; earlier reports are suspect. Very rare in West in late fall and winter; fall migrants often solitary and found around aquatic habitats. Rare wintering Rustys in West often join Brewer's Blackbirds. Also casual to Bering Sea islands.

Brewer's Blackbird *Euphagus cyanocephalus* L 9" (23 cm)
Male has yellow eyes; **female's** are usually brown. Male is black year-round, with purplish gloss on head and neck, greenish gloss on body and wings. **Immature males** show variable buffy feather edgings, but never on tertials or wing coverts, as in Rusty Blackbird; note also the shorter, thicker bill. Female and juveniles are gray-brown.
VOICE: Typical call is a low *check;* song, a wheezy *que-ee* or *k-seee.*
RANGE: Common in open habitats including city parks. Often forages in parking lots. Gregarious. Very local in Southeast, mostly at livestock pens. Casual in winter to East Coast and north and west to YT and AK.

BLACKBIRDS

deep yellow
throat and breast

spring adult ♂

deep yellow
head

immature ♂

♀

juvenile

**Yellow-headed
Blackbird**

white wing patch on
primary coverts

spring
adult ♂

head pattern distinctive
with rusty crown and
prominent pale supercilium

pale eye

**Rusty
Blackbird**

fall ♀

contrasting
gray rump

breeding ♀

long,
thin bill

rusty tips to wing
coverts and tertials

fall ♂

breeding ♂

**Brewer's
Blackbird**

buffy tipping to
head, back, and
much of upperparts

wings
uniformly dark

♂

bill slightly
thicker
than Rusty

dark eye

♀

immature ♂

colored plumage
with more gloss
than Rusty, visible
in good light when
close

See subspecies map, page 557

Common Grackle *Quiscalus quiscula* L 12½" (32 cm)

Long, keel-shaped tail; pale yellow eyes. Plumage appears all-black at a distance. In good light, **males** show glossy purplish blue head, neck, and breast. Females are smaller and duller than males. **Juveniles** are sooty brown, with brown eyes. Widespread subspecies *versicolor,* **"Bronzed Grackle,"** occurs in most of New England and west of the Appalachians; it has a bronze back, blue head, and purple tail. Smaller **"Purple Grackle,"** *quiscula* of the Southeast, has a narrow bill, purple head, bottle green back, and blue tail. An intergrade population from the mid-Atlantic *(stonei)* shows variable head color and purplish back with iridescent bands of variable color.

VOICE: Song is a short, creaky *koguba-leek;* call note, a loud, deep *chuck.*
RANGE: Abundant and gregarious, roaming in mixed flocks in open fields, marshes, parks, and suburban areas. Casual to very rare to Pacific states and AK.

Boat-tailed Grackle *Quiscalus major*

♂L 16½" (42 cm) ♀L 14½" (37 cm) Large grackle with a very long, keel-shaped tail; smaller overall size, duller eye color, and more rounded crown than Great-tailed Grackle. **Adult male** is iridescent blue-black. **Adult female** is tawny brown with darker wings and tail. **First-fall male** is black but lacks iridescence; **juvenile** shows a hint of spotting or streaking on breast. Immatures resemble respective adults by mid-fall. Male and female eye color is mostly brown in nominate subspecies of coastal TX and LA and *westoni* of FL; *alabamensis* of coastal MS to northwestern FL and the largest subspecies, *torreyi,* on the Atlantic coast, have a yellow iris.

VOICE: Calls include a quiet *chuck* and a variety of rough squeaks, rattles, and other chatter. Most common song is a series of harsh *jeeb* notes.
RANGE: Common, seldom strays beyond coastal saltwater marshes except in FL, where inhabits inland lakes and streams. Nests in small colonies. Range is slowly expanding northward on the Atlantic coast.

Great-tailed Grackle *Quiscalus mexicanus*

♂L 18" (46 cm) ♀L 15" (38 cm) A large grackle with very long, keel-shaped tail and golden yellow eyes. **Adult male** is iridescent black with purple sheen on head, back, and underparts. **Adult female's** upperparts are brown; underparts cinnamon-buff on breast to grayish brown on belly; shows less iridescence than male. **Juveniles** resemble adult female but are even less glossy and show some streaking on underparts. Immature males are duller, with shorter tails and darker eyes than adults by mid-fall. Females west of central AZ are smaller and paler below than subspecies to the east. In narrow zone of range overlap, distinguished from Boat-tailed by bright yellow eyes, larger size, and flatter crown.

VOICE: Songs include clear whistles and loud *clack* notes; show marked geographical variation between eastern *prosopidicola* and western *nelsoni.* Call note is a low *chut;* makes often a louder *clack.*
RANGE: Common, especially in open flatlands with scattered groves of trees and in marshes and wetlands. Casual far north of breeding range to BC; accidental to NS; rapidly expanding north and west in western U.S.

Common Grackle

purplish gloss to head

"Purple Grackle"
quiscula

bluish head contrasts with bronze-green back and underparts

sooty brown overall

keel-shaped tail

overall blackish

♂

"Bronzed Grackle"
versicolor

juvenile

bill appears thicker based than Great-tailed

Boat-tailed Grackle
major

distinctive rounded head

juvenile

1st fall ♂

cinnamon-buff underparts

♀

western Gulf Coast adult ♂

where ranges overlap, note dark eye of *major* Boat-tailed, unlike yellow eye of all Great-tailed

blue-green gloss overall

long, keel-shaped tail

juvenile

streaked underparts

buffy supercilium

Great-tailed Grackle

♀

cinnamon-buff underparts

long, rather thin bill

♂

purple gloss with no head and body contrast as in Common Grackle

western ♀

nelsoni is smaller and paler than other subspecies

very long, keel-shaped tail

Shiny Cowbird *Molothrus bonariensis* 7½" (19 cm)

Sleeker, with longer tail, flatter head, and longer, more pointed bill than Brown-headed Cowbird. **Male** blackish with blue or purple gloss on head, breast, and back. **Female** and juveniles resemble female Brown-headed except for shape, darker color, more prominent eyebrow, and slimmer all-black bill.

VOICE: Song, whistled notes followed by trills. Male's high-pitched sweet flight call is not like other cowbirds. Other call a soft *chup*; females give a chatter.

RANGE: Mainly South American species, which spread through the West Indies, arriving in south FL in 1985. Uncommon and local in coastal south FL. Elsewhere mostly noted in spring: very rare to casual in Southeast; accidental to OK, ME, and Maritime Provinces. Has declined in U.S. somewhat over the last two decades.

Brown-headed Cowbird *Molothrus ater* L 7½" (19 cm)

Male's brown head contrasts with metallic green-black body. **Female** is gray-brown above, paler below; overall darkest in eastern *ater*. **Juvenile** is paler above, more heavily streaked below; pale edgings give its back a scaled look; juvenile *obscurus* is paler. Young males molting to adult plumage in late summer are a patchwork of buff, brown, and black. Southwestern *obscurus*, "Dwarf Cowbird," is distinctly smaller than eastern nominate *ater*; Rockies and Great Basin *artemisiae* is largest. Feeds with tail cocked up. Gregarious; often mixes with other blackbirds and starlings during nonbreeding season. During breeding season, courtship and parasitic activities occur primarily in the morning, feeding in the afternoon. All cowbirds lay their eggs in nests of other species; a single female Brown-headed will travel up to four miles through woodland to lay up to several dozen eggs in a season. People who feed birds in spring at the edges or openings of large woodland tracks facilitate cowbird's brood parasitism.

VOICE: Male's song is a squeaky gurgling. Calls include a harsh rattle and squeaky whistles.

RANGE: Common; found in woodlands, farmlands, suburbs. Rare visitor to AK, where recorded in fall northwest to St. Lawrence Island.

Bronzed Cowbird *Molothrus aeneus* L 8¾" (22 cm)

Red eyes distinctive at close range. Bill larger than Brown-headed Cowbird. **Adult male** is black with bronze gloss; wings and tail blue-black; thick ruff on nape and back gives a hunchbacked look. **Adult female** of the TX subspecies, *aeneus*, is duller than the male; **juveniles** are dark brown. In southwestern *loyei*, females and juveniles are gray.

VOICE: Call is a harsh, guttural *chuck*. Song is wheezy and buzzy, often delivered by displaying male in spectacular "helicopter" fluttery flight over an often indifferent-looking female. The collective whistles of roosting males in winter can suggest European Starlings. Females give a rattle.

RANGE: Locally common in open country, brushy areas, and wooded mountain canyons; forages in flocks. Nominate *aeneus* found in TX west to Pecos River; both subspecies occur in west TX where species is scarce. Western *loyei* very local in winter. Uncommon and local in southeastern CA (mainly along Colorado River); casual elsewhere in southern CA and north to UT and CO.

Shiny Cowbird

pale supercilium

♀

black spikelike bill

♂

black with blue or purple gloss overall

brown head contrasts with black body

♂

short tail

Brown-headed Cowbird

immature ♂ in molt

young males seen in this transitional plumage in late summer and early fall can be confusing

streaked below and scaly above, can be confused with female Red-winged Blackbird

juvenile

Yellow Warbler

♀

conical bill, smaller than Bronzed Cowbird, distinctive for all Brown-headed Cowbirds

parasitic female Brown-headed Cowbirds lay eggs in nests of other species, so young raised by other parents

Brown-headed Cowbirds often feed with tail cocked up

♀

red eye

thick bill

ruff often raised, giving thick-necked look

highly iridescent plumage

grayish plumage, unlike female *aeneus*

red eye

Bronzed Cowbird
loyei

♂

♀ *loyei*

blackish plumage

less scaly above with larger bill than juvenile Brown-headed

juvenile

Texas
♀ *aeneus*

Orchard Oriole *Icterus spurius* L 7¼" *(18 cm)*

Adult male is chestnut overall, with black hood. During winter, plumage veiled by buff tips. **Female** is olive above, yellowish below. Immature male resembles female; acquires black bib and, sometimes, traces of chestnut during first winter. Smaller size, lack of orange tones or whitish belly, and thinner, more curved bill distinguish female and immature male from Baltimore and Bullock's Orioles (page 514). Compare especially with *nelsoni* subspecies of Hooded Oriole. A subspecies breeding in eastern Mexico from southern Tamaulipas to Veracruz, *fuertesi,* has been documented on three occasions in Cameron County, TX. Adult males have an ochre, rather than chestnut, coloration. Some have treated *fuertesi* as a separate species.

VOICE: Calls include a sharp *chuck.* Song is a loud, rapid burst of whistled notes, downslurred at the end.

RANGE: Locally common in suburban shade trees and orchards. An early fall migrant, especially adult males, with numbers moving south in late July and Aug. Rare to AZ, CA, the Maritime Provinces. Casual to OR. Accidental to WA, BC, and southeastern AK.

See subspecies map, page 557

Hooded Oriole *Icterus cucullatus* L 8" *(20 cm)*

Bill long and slightly curved. **Breeding male** is orange or orange-yellow; note black patch on throat. Western birds, *nelsoni,* breeding east locally to southeastern NM and El Paso region, are yellower. The rather uncommon *sennetti* from south TX is orange; nominate *cucullatus,* an uncommon and declining breeder along Rio Grande (from about Langtry to Big Bend) is also orange to a deep reddish orange about the head. Both TX subspecies have more extensive black on face. All **winter adult males** have buffy brown tips on back, forming a barred pattern; compare with Streak-backed Oriole. Hooded **female** and immature male lack pale belly of Bullock's Oriole (page 514); bill is more curved. Compare *nelsoni* also with female and immature male Orchard Orioles, which are smaller and purer lemon yellow below, with smaller bill, but beware of recently fledged juvenile Hooded Orioles in Aug. to early Sept. with short (but thick-based) bill and seemingly smaller size; they even give a type of *chuck* call, similar to Orchard. Immature male Hooded acquires black patch on throat during winter.

VOICE: Calls include a distinctive whistled, rising *wheet;* song is a series of whistles, trills, and rattles.

RANGE: Common in varied habitats, especially near palms. Rare in winter in coastal southern CA and south TX. Casual to Pacific Northwest. Accidental to YT, KY, and ON. Breeding has expanded northward on West Coast.

Streak-backed Oriole *Icterus pustulatus* L 8¼" *(21 cm)*

Distinguished from winter Hooded Oriole by broken streaks (characteristic of *microstictus* from west Mexico) on upper back; deeper orange head; and much thicker-based, straighter bill. **Female** is duller than **male.** Immature male resembles adult female.

VOICE: *Wheet* call is softer than Hooded Oriole and does not rise in pitch. Chatter calls resemble Baltimore Oriole.

RANGE: Tropical species, casual mostly in fall and winter in southeastern AZ (has nested), southern CA; accidental to OR, NM, CO, east TX, and WI.

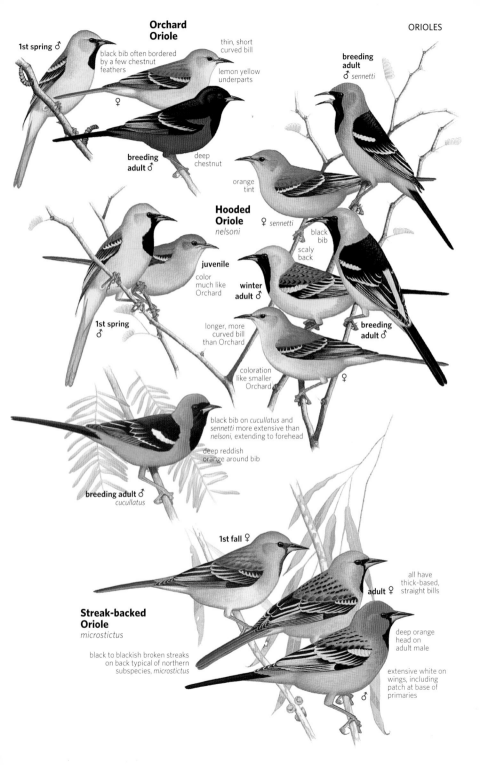

Orchard Oriole

1st spring ♂

black bib often bordered by a few chestnut feathers

thin, short curved bill

lemon yellow underparts

♀

breeding adult ♂

deep chestnut

breeding adult ♂ *sennetti*

orange tint

♀ *sennetti*

Hooded Oriole *nelsoni*

juvenile

color much like Orchard

black bib

scaly back

winter adult ♂

1st spring ♂

longer, more curved bill than Orchard

breeding adult ♂

coloration like smaller Orchard

♀

black bib on *cucullatus* and *sennetti* more extensive than *nelsoni*, extending to forehead

deep reddish orange around bib

breeding adult ♂ *cucullatus*

1st fall ♀

adult ♀

all have thick-based, straight bills

Streak-backed Oriole *microstictus*

deep orange head on adult male

black to blackish broken streaks on back typical of northern subspecies, *microstictus*

extensive white on wings, including patch at base of primaries

♂

Black-vented Oriole *Icterus wagleri* L 8¾" (22 cm)
Long, narrow, mostly black bill is bluish gray at base and slightly decurved at tip; exceedingly long, graduated tail held together in a point. **Adult** has solid black head, back, undertail coverts, tail, and wings, except for yellow-orange shoulders; the border between breast and belly is chestnut. The orange coloration with a peach or persimmon wash differs from other North American orioles. First-winter and **first-spring** birds have variable amount of black on lores, chin, and back. Young juvenile lacks black.
VOICE: Call is a nasal *nyeh,* often repeated.
RANGE: Resident from southern Sonora, western Chihuahua and southern Nuevo León to central Nicaragua; casual to south TX; accidental to west TX (Big Bend) and southeastern AZ.

Altamira Oriole *Icterus gularis* L 10" (25 cm)
Distinguished from Hooded Oriole (page 510) by much larger size and stockier body shape, much thicker-based, mostly blackish bill, and, in **adult,** by orange shoulder patch. Lower wing bar whitish. Males have an all-black tail; females often with some olive. **Immatures** are duller than adults, lack yellow shoulder patch, and have an olive back. Juvenile lacks black bib; like adult by second fall. Occasionally hybridizes with Audubon's Oriole in south TX.
VOICE: Calls include a low, raspy *ike ike ike;* song is a series of clear, varied whistles.
RANGE: Found in southernmost TX in tall trees and willows. Its huge hanging nest is a distinctive sight.

Audubon's Oriole *Icterus graduacauda* L 9½" (24 cm)
Male has greenish yellow back. Female is slightly duller, showing more of a greenish back, head averages a duller black. Juvenile has extensive black on head by fall. Both sexes have an all-black tail. Rather secretive, tending to feed low in understory; often seen foraging on ground.
VOICE: Song is a series of soft, tentative, three-note warbles. Both sexes sing. Call is a nasal *nyyyee* and a high-frequency buzz.
RANGE: Tropical species, resident but uncommon in south TX in woodlands and brushlands. Accidental southern IN.

Spot-breasted Oriole *Icterus pectoralis* L 9½" (24 cm)
Adults have an orange or yellow-orange patch on shoulders; black lores and throat; dark spots on upper breast; extensive white on wings. **Juveniles** are yellower overall and lack black on lores and throat; **immatures** may lack breast spots. FL introduction appears to be one of the brighter, more southerly subspecies, either *guttulatus* or *espinachi.*
VOICE: Male's song, heard throughout the year, is a long, loud series of melodic whistles. Female sings a less complex song. Call is a nasal *nyeh,* a sharp *whip;* also a nasal chatter call.
RANGE: Middle American species, introduced and now established in southeastern FL and first found nesting in 1949. Prefers suburban gardens. The never large FL population has declined over past three decades.

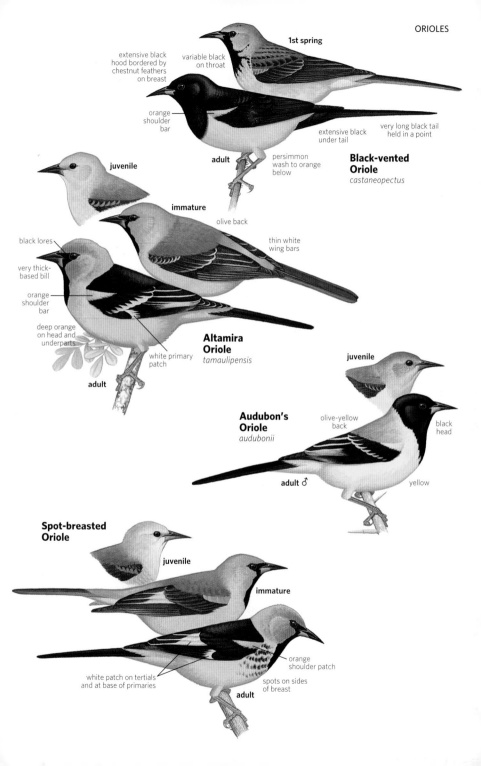

Black-vented Oriole
castaneopectus

1st spring

extensive black hood bordered by chestnut feathers on breast

variable black on throat

orange shoulder bar

adult

persimmon wash to orange below

extensive black under tail

very long black tail held in a point

juvenile

immature

olive back

thin white wing bars

black lores

very thick-based bill

orange shoulder bar

deep orange on head and underparts

white primary patch

Altamira Oriole
tamaulipensis

adult

Audubon's Oriole
audubonii

juvenile

olive-yellow back

black head

yellow

adult ♂

Spot-breasted Oriole

juvenile

immature

white patch on tertials and at base of primaries

orange shoulder patch

spots on sides of breast

adult

Baltimore Oriole *Icterus galbula* L 8¼" *(21 cm)*

Adult male has black hood and back, bright orange rump and under-parts; large orange patches on tail. **Adult females** are brownish olive above and orange below, with varying amounts of black on head and throat; those with maximum black (shown) resemble first-spring males. Extent and intensity of color on underparts of **fall immatures** is highly variable; dullest birds easily confused with Bullock's, but note more distinctly contrasting wing bars, palish lores, no eye line or yellowish supercilium, more contrasty (less blended) auriculars, and a yellowish, not grayish, rump.

VOICE: Common call is a rich *hew-li;* also gives a series of rattles different in quality (dryer) and more stuttery than Bullock's series of *cheh* notes. Song is a musical, irregular sequence of *hew-li* and other notes.

RANGE: Common breeder in deciduous woodland over much of the East. Some winter at feeders in the South. Rare to very rare to West; rare to Newfoundland. Accidental to YT.

Bullock's Oriole *Icterus bullockii* L 8¼" *(21 cm)*

Formerly considered same species as Baltimore Oriole; some interbreeding on Great Plains. **Adult male** has less black on head: crown, eye line, throat patch; note bold white patch on wing, entirely orange outer tail feathers. **Females** and **immatures** have yellow throat and breast, unlike Baltimore's extensive orange; note Bullock's dark eye line, weakly defined yellowish supercilium, and less contrasting, plainer black wing bars. Most birds show dark "teeth" intruding into white of median covert bar. By **first spring,** males have black lores, chin.

VOICE: Song is a mix of whistles and harsher notes; call is a harsh *cheh* or series of same and a whistled *pheew.*

RANGE: Breeds where shade trees grow. Small numbers winter in coastal CA, casual elsewhere. Casual vagrant to the East, where many reports are of dull, immature Baltimores. Also casual to AK, including St. Lawrence Island (several fall records).

Scott's Oriole *Icterus parisorum* L 9" *(23 cm)*

Adult male's black hood extends to back and breast; rump, wing patch, and underparts bright lemon yellow. Adult female is olive and streaked above, dull greenish yellow below; throat shows variable amount of black. Immature male's head is mostly black by first spring. **Female** and immature larger, grayer, and more streaked above; overall slightly more olive, less yellow; straight bill (slightly curved in Hooded Oriole, page 510).

VOICE: Common call note is a harsh *shack;* also a scolding *chah-chah.* song is a mixture of rich, whistled phrases, reminiscent of Western Meadowlark. Females sing a weaker song.

RANGE: Found in arid and semiarid habitats. Casual to Channel Islands and northern CA; accidental to OR and WA. Casual east to MN, WI, LA, KY, GA, and NC. A few winter in southern CA, including along coast.

all with thick
wing bars and
orangish below

fall immatures

yellow-
ochre
rump

Baltimore Oriole

maximum black
spring adult ♀

black hood

fall immature ♂
strongly contrasting
wing bars

**breeding
adult ♂**

**1st
spring ♀**

**1st spring
♂**

black lores
and chin

Bullock's Oriole

orange supercilium
and black eye line

pale yellow
head and
grayish back

gray
rump

large white
wing patch

dusky
eye line

yellowish
throat
and
breast
♀

immature ♀

extensive
white
belly

dark "teeth"
extend into white
wing bars

**breeding
adult ♂**

**Scott's
Oriole**

**1st
spring ♂**

adult ♂

long,
straight
bill

stippled
dusky black
above

yellow
shoulder

immature ♀

olive-yellow
below, duller
than female
Hooded

bright
lemon
yellow
underparts

FRINGILLINE AND CARDUELINE FINCHES • ALLIES Family Fringillidae

Seedeaters with undulating flight. Many nest in the North; in fall, flocks of "winter finches" may roam south. SPECIES: 209 WORLD; 23 N.A.

See subspecies map, page 557

Gray-crowned Rosy-Finch *Leucosticte tephrocotis*

5½-8¼" (14-21 cm) Dark brown, with gray on head; pink on wings and underparts; underwing silvery. Female shows less pink; some worn one-year-old females of *tephrocotis* group lack gray head bands and resemble female Brown-capped Rosy-Finch. **Juveniles** are grayish. All have yellow bill in **winter,** black by spring. All rosy-finches highly gregarious, except when breeding. Two groups of subspecies differ significantly from each other in head pattern. Differences within each group are slight, other than size in some cases. Gray-headed *littoralis* group comprises three subspecies: resident *griseonucha* from Aleutians and Kodiak, AK; resident *umbrina* (shown) from Pribilofs, St. Matthew, and Hall Islands; and migratory continental *littoralis,* **"Hepburn's Rosy-Finch."** All have gray faces, with gray extending well below the eye; *griseonucha* and *littoralis* are by far the largest subspecies of all rosy-finches. The three other more easterly continental subspecies—widespread and highly migratory nominate *tephrocotis; wallowa* from northeastern OR; and *dawsoni* from the Sierra Nevada and White Mountains — closely resemble one another. All have narrower gray head bands with no gray below the eye.

VOICE: Call, a high, chirping *chew,* is often given in courtship flight and in unison when flocks take flight, which is frequent.

RANGE: Descends from higher elevations in winter. Both nominate *tephrocotis* and *littoralis* are highly migratory and winter throughout mapped range. Casual east to Great Plains. Accidental to ON, OH, QC, and ME (both *littoralis* and *tephrocotis* recorded).

Brown-capped Rosy-Finch *Leucosticte australis*

L 6" (15 cm) Plumages, behavior, and voice like Gray-crowned. Lacks gray head band of other North American rosy-finches. **Male** rich brown; darker crown; extensive pink on underparts. **Female** much drabber; some young female Gray-crowneds can be very similar.

VOICE: Call similar to Gray-crowned Rosy-Finch.

RANGE: The least migratory rosy-finch. Migrates south to Sandia Mountains, NM, but unrecorded away from Rocky Mountain region.

Black Rosy-Finch *Leucosticte atrata* L 6" (15 cm)

Plumages, behavior, and voice like Gray-crowned Rosy-Finch. Darkest rosy-finch. **Male** is blackish; in fresh plumage, scaled with silver-gray; has gray head band; shows extensive pink. **Female** is blackish gray with little pink. Told from Gray-crowned by darker coloration and absence of brownish tones above.

VOICE: Call similar to Gray-crowned Rosy-Finch.

RANGE: Casual to eastern CA, northern AZ, and western NE.

extensive gray face

"Hepburn's Rosy-Finch" winter ♂
littoralis

yellow bill on all fall and winter rosy-finches, becoming black by spring

gray head band

winter ♂
tephrocotis

Pribilof, St. Matthew, and Hall Islands *umbrina* and Aleutian and Commander Islands (Russian Far East) *griseonucha* are much larger than *littoralis*

extensive gray face, like *littoralis*

Pribilofs winter ♂
umbrina

Gray-crowned Rosy-Finch

brownish edges above, compare to female Black Rosy-Finch

like male but with less pink

dark sooty overall with little pink

winter ♀
tephrocotis

juvenile
tephrocotis group

Brown-capped unrecorded away from Rockies region

some worn 1st-spring and summer females from *tephrocotis* group of subspecies look very similar

rich brown back and breast with no gray head band

dark sooty cap

Brown-capped Rosy-Finch

extensive rosy-pink on wings and below

winter ♀

breeding ♂

silver-gray edges above on fresh plumaged females and males

Black Rosy-Finch

stunning — black overall with contrasting rose and silver-gray head band

dark sooty gray overall with reduced pink

winter ♀

breeding ♂

Finches

Purple Finch *Carpodacus purpureus* L 6" (15 cm)

Not purple, but rose red over most of **adult male's** body, brightest on head and rump. Back is streaked; tail notched. Pacific coast subspecies, *californicus,* is buffier below and more diffusely streaked than the widespread *purpureus,* especially in female types. **Adult female** and immature male (plumage held for over one year) are heavily streaked below; also note dark auricular and whitish eyebrow.

VOICE: Calls include a musical *chur-lee* and in flight, a sharp *pit,* a bit sharper in *californicus.* Song is a rich warbling, longer and more variable in *purpureus;* shorter than Cassin's.

RANGE: Fairly common; found in coniferous or mixed woodland borders, suburbs, parks; in the Pacific states, inhabits coniferous forests, oak canyons, and lower mountain slopes. Nominate subspecies, *purpureus,* breeds across boreal forest of Canada west to northern BC. Rare in migration and winter on western Great Plains south to NM. Casual farther west, including AK with multiple records for St. Lawrence Island. Pacific *californicus* breeds north to southwestern BC. Casual to rare in migration to western Great Basin, Mojave Desert, and CO deserts.

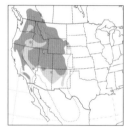

Cassin's Finch *Carpodacus cassinii* L 6¼" (16 cm)

Adult male has contrasting crimson cap. Throat and breast paler than Purple Finch; streaks on sides and brown malar stripe more distinct. Tail strongly notched. Undertail coverts always distinctly streaked, unlike many Purples. **Adult female** and immature male (plumage held for over one year) otherwise closely resemble Purple. Cassin's facial pattern is slightly less distinct; culmen is straighter and longer; has longer primary projection.

VOICE: Gives a dry *tee-dee-yip* call. Lively song, a variable warbling, longer and more complex than Purple Finch, especially *californicus.*

RANGE: Fairly common in upper mountain forests, evergreen woodlands. Casual in fall and winter east of range to western Great Plains; also to West Coast. Accidental ON.

House Finch *Carpodacus mexicanus* L 6" (15 cm)

Male has brown cap; front of head, bib, and rump are typically red but can vary to orange or occasionally yellow. Bib is clearly set off from streaked underparts. Tail is squarish. **Adult female** and juveniles are streaked with brown overall; lack distinct ear patch and eyebrow of Purple and Cassin's Finches. Young males like adult by first fall.

VOICE: Lively, high-pitched song consists chiefly of varied three-note phrases; includes strident notes, unlike Purple Finch's song; usually ends with a nasal *wheer.* Calls include a whistled *wheat.*

RANGE: Common; found in semiarid lowlands and slopes up to about 6,000 feet. Introduced in the East in the 1940s; especially numerous in towns. Casual to southeastern AK.

Common Rosefinch *Carpodacus erythrinus* L 5¾" (15 cm)

Strongly curved culmen. Lacks distinct eyebrow. Adult male has red head, breast, and rump. Female and immatures diffusely streaked above and below except on pale throat; note long primary projection.

VOICE: Call is a soft, nasal *djuee.*

RANGE: Eurasian species; very rare migrant, chiefly in spring, on western Aleutians and other western AK islands. Accidental CA.

adult ♂
purpureus

overall a deep
rose-red color

white belly and
flanks with no
brownish, as in
californicus

dark auricular
and distinct white
facial streaks

distinct
dark
streaks
above

short thick bill
with slightly
curved culmen

**Purple
Finch**

Pacific region
adult ♂
californicus

distinct
and rather
broad streaks
below

unlike *purpureus*, has
a brownish tinge to
back, flanks, and belly

pale facial stripes
less obvious

♀ *purpureus*

contrasting
crimson-red
crown

wide brown
malar streak

adult ♂

pale pink throat
and breast with
white belly

**Cassin's
Finch**

browner,
less contrasty
streaking below
against a buffier
background color

Pacific region ♀
californicus

less obvious streaking
above against a more
olive background color

pale eyebrow, but face
pattern less distinct
than Purple

larger bill
than Purple with
straight culmen

long primary
projection, longer
than Purple

males can be
yellow to orange

cinnamon tinge
to auriculars

adult ♀

notched tail as
in Purple

House Finch
frontalis

broad
reddish
eyebrow

red throat
and breast

all plumages have
streaked undertail
coverts, unlike most
Purples

variant ♂

brownish
streaks on
flanks

typical ♂

immature male
plumage like female
until about one year
of age

known widely
in Old World as
Scarlet Rosefinch

bright red head and rump
and extensive red below
often with buff tinge

**Common
Rosefinch**
grebnitskii

curved
culmen

muted
brown back

more
square-ended
tail

very
indistinct
face pattern

adult ♂

indistinct
face pattern

♀

♀

long primary
projection

immature male like
female until just
over a year of age

olive-tinged
above with
pale wing bars

Red Crossbill *Loxia curvirostra* L 5½-7¾" (14-20 cm)

Bill with crossed tips identifies both crossbill species. Red Crossbill lacks the bold white bars of White-winged Crossbill. Most **males** are reddish overall, brightest on crown and rump, but may be pale rose or scarlet or largely yellow; always have red or yellow on throat. Most **females** are yellowish olive; may show patches of red; throat is always gray, except in a small northern subspecies where yellow extends to center, but not sides, of throat. **Juveniles** are boldly streaked; a few juveniles and a very few adult males show white wing bars, the upper bar thinner than the lower. Immatures are like the respective adult but juvenile wing is retained. All birds except adult males have olive edges on wings. Subspecies vary widely in size, including bill size; extremes are shown here. All have their "home range." Distinct differences in vocalizations have led some authorities to believe that there may be a half dozen or more cryptic separate species in the Red Crossbill complex. All have large heads and short, notched tails.

VOICE: Calls, given chiefly in flight, vary among subspecies. Song begins with several two-note phrases followed by a warbled trill.

RANGE: Fairly common, inhabits coniferous woods. May nest at any time of year, especially in southern range. Highly irregular in their wanderings, dependent upon cone crops. Any subspecies may turn up almost anywhere. Irruptive migrant. Has bred in East as far south as GA.

White-winged Crossbill *Loxia leucoptera* L 6½" (17 cm)

All ages have black wings with white tips on the tertials; two bold, broad white wing bars. Upper wing bar is often hidden by scapulars. **Adult male** is bright pink overall, paler in winter. **Immature male** is largely yellow, with patches of red or pink. **Adult female** is yellowish olive overall. **Juvenile** is heavily streaked; wing bars thinner than adults.

VOICE: Distinctive flight call is a rapid series of harsh *chet* notes. Variable song, often delivered in display flight, combines harsh rattles and musical warbles.

RANGE: Inhabits coniferous woods. Highly irregular wanderings, dependent upon spruce cone crops. Irruptive migrant. Casual to southern OR, northeastern NV, southern UT, northern NM, northern TX, and NC.

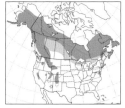

See subspecies map, page 557

Pine Grosbeak *Pinicola enucleator* L 9" (23 cm)

Large and plump, with long tail. Bill is dark, stubby, strongly curved. Two white wing bars, sometimes tinged with pink in adult male. **Male's** gray plumage is tipped with red on head, back, and underparts; pinker in fresh fall plumage. **Female** and immatures are grayer overall; head, rump, and underparts variably yellow or reddish; some females and immature males are **russet.**

VOICE: Typical flight call is a whistled *pui pui pui;* alarm call, a musical *chee-vli.* Location call shows considerable geographic variation. Song is a rather short, musical warble.

RANGE: Uncommon; inhabits open coniferous woods. In winter, found also in deciduous woods, orchards, and suburban shade trees. Usually unwary and approachable. Irruptive winter migrant in the East. In West, *montana* is irruptive, casually reaching CA, northern AZ, and western Great Plains. Asian *kamtschatkensis* is casual to western Aleutians (Attu and Shemya Islands) in spring.

Red Crossbill

variant ♂

juvenile

some with faint whitish wing bars

female typically has olive body with darker wings and gray throat

northern ♀ *minor*

overall size and bill size, especially, varies significantly and geographically

typical ♀

southwestern ♂ *stricklandi*

overall reddish coloration

typical ♂

juvenile

White-winged Crossbill
leucoptera

immature ♂

thinner crossed bill than Red Crossbill

♀

pink body a little more reddish pink in summer

white tertial tips

winter adult ♂

black wings with thick white wing bars

yellow-olive head and back; body otherwise gray

stubby black bill with curved culmen

♀

white wing bars

adult ♂

deep pinkish red

russet variant

Pine Grosbeak
leucura

American Goldfinch *Spinus tristis* L 5" (13 cm)

Breeding adult male is bright yellow with black cap; black wings have white bars, yellow shoulder patch; uppertail and undertail coverts white, with black-and-white tail. **Female** is duller overall, olive above; lacks black cap and yellow shoulder patch. White undertail coverts distinguish female from most Lesser Goldfinches, which are smaller. **Winter adults** and immatures are either brownish or grayish above; bill darker than when breeding; male may show some black on forehead. **Juvenal** plumage, held into Nov., has cinnamon-buff wing markings and rump.

VOICE: Song is a lively series of trills, twitters, and *swee* notes. Distinctive flight call, *per-chik-o-ree.*

RANGE: Common and gregarious; found in weedy fields, open second-growth woodlands, and roadsides, especially in thistles and sunflowers. Casual north to southeastern and south-coastal AK and YT.

Lesser Goldfinch *Spinus psaltria* L 4½" (11 cm)

Smaller than American Goldfinch. All birds have a white wing patch at base of primaries. Entire crown black on **adult male;** back varies from black in eastern part of range to greenish in western birds. Most **adult females** are dull yellow below; except for a few extremely pale birds, they lack the white undertail coverts typical of American. **Immature male** lacks full black cap. Juveniles resemble adult female.

VOICE: Call is a plaintive, kittenlike *tee-yee.* Song is somewhat similar to American Goldfinch. Frequently imitates other species.

RANGE: Common in dry, brushy fields, woodland borders, and gardens. Range expanding northward. Casual in central BC, MT, and Great Plains; accidental farther east and to YT.

Lawrence's Goldfinch *Spinus lawrencei* L 4¾" (12 cm)

Wings extensively yellow; upperparts grayish in breeding plumage, acquired by wear; large yellow patch on breast. **Male** has black face and yellowish tinge on back. **Winter** birds are browner above, duller below. **Juvenile** is faintly streaked, unlike other goldfinches.

VOICE: Call is a bell-like *tink-ul.* Mixes *tink* notes into jumbled, melodious song. Lawrence's and Lesser Goldfinches often mimic other species' songs.

RANGE: Fairly common in spring and early summer; may sometimes flock with other goldfinches, but generally prefers drier interior foothills and mountain valleys; also western fringe of desert near watercourses. Erratic but usually uncommon at other seasons. Irregular fall movements to southwestern U.S. In some winters, locally numerous in southeastern AZ with fewer in southern NM; absent in other winters. Casual as far east as west TX and north to NV. Accidental to western CO.

FINCHES

American Goldfinch
tristis

breeding ♀

black wings and tail; white under tail

breeding ♂

black forehead and forecrown

bright yellow body

yellow shoulder; brownish back

winter adult ♂

whitish undertail coverts

prominent wing bars

winter ♀

juvenile

smaller than American

Lesser Goldfinch
hesperophila

♀

pale yellow underparts

black-backed adult ♂
psaltria

pale ♀

blackish cap contrasts with green back

green-backed adult ♂

immature ♂

white patch at base of primaries

black forehead, face, and chin

brownish wash on back in fresh fall plumage in both sexes, wears to more grayish by spring

winter ♀

winter ♂

Lawrence's Goldfinch

yellow breast

prominent yellow on wings in all plumages

distinctive white oval tail spots like many *Setophaga* wood-warblers

breeding ♂

juvenile

Common Redpoll *Acanthis flammea* L 5¼" (13 cm)

Red or orange-red cap or "poll," black chin. Closely resembles Hoary Redpoll, but usually has distinct streaks on flanks, rump, undertail coverts; bill slightly larger. **Male** usually has bright rosy breast and sides. Both sexes paler, buffier in winter. **Juveniles** lack red cap until late summer molt; males acquire pinkish breast by end of second summer. Greenland-breeding *rostratas* is larger (10 percent larger on average) and bigger billed than *flammea,* found elsewhere in North America; *rostratus* has heavier and more extensive streaking below and is browner (darker) in coloration, which obscures paler marking found on faces in *flammea*. Extent of interbreeding with Hoary is unknown.

VOICE: When perched, gives a *swee-ee-eet* call; flight call is a dry rattling. Song combines trills and twittering.

RANGE: Fairly common; breeds in subarctic forests and tundra scrub. Unwary. Forms large winter flocks; frequents brushy, weedy areas, also catkin-bearing trees like alder and birch. Rather numerous to border states in some winters. Casual to northern CA, NV, northern NM, TX.

Hoary Redpoll *Acanthis hornemanni* L 5½" (14 cm)

Closely resembles Common but usually frostier, paler overall, has slightly smaller bill. Minimal or no streaking below and on rump. **Male's** breast usually paler, pinker than Common; color restricted to breast.

VOICE: Calls and song similar to Common.

RANGE: Fairly common; nests on or near the ground above Arctic tree line. The subspecies *hornemanni,* of Canadian Arctic islands and Greenland and rarely to casually farther south in East, once from Fairbanks, AK (Mar. 1964), is larger and paler than more widespread *exilipes*. Rare sightings, especially of *exilipes,* almost always with Commons, occur south of Canada in winter to MT and northern WY; casual to WA, northeastern OR; accidental UT.

Pine Siskin *Spinus pinus* L 5" (13 cm)

Prominent streaking; yellow at base of tail and in flight feathers conspicuous in flight; bill thinner than other finches. Some are washed overall with pale yellow. **Juvenile's** overall yellow tint is lost by late summer.

VOICE: Calls include a rising *tee-ee* and, in flight, a harsh, descending *chee*. Song is similar to American Goldfinch but much huskier.

RANGE: Gregarious; flocks with goldfinches in winter. Found in coniferous, mixed woods in summer; forests, shrubs, fields in winter. Erratic in winter. Casual to Bering Sea islands and throughout the Aleutians.

Evening Grosbeak *Coccothraustes vespertinus* L 8" (20 cm)

Stocky, noisy finch. Big bill pale yellow or greenish by spring, whitish by fall; prominent white inner wing patch. Yellow forehead, eyebrow on **adult male;** dark brown and yellow body. Grayish tan **female** has thin, dark malar stripe, white-tipped tail; second wing patch, on primaries, conspicuous in flight. **Juvenile** has brown bill; female resembles adult female; male yellower overall, wing and tail like adult male.

VOICE: Loud, strident call is a *clee-ip* or *peeer.*

RANGE: Breeds in mixed woods; in the West, mainly in mountains. In winter frequents woodlots, shade trees, and feeders; numbers and range limits vary greatly, but substantial overall decline during the past two decades, especially in the East. Casual to southeastern AK.

juvenile

breeding ♀

breeding ♂

extensive
pinkish red
breast

**Common
Redpoll**
flammea

FINCHES

larger and darker
than *flammea* with
more extensive
streaking below

winter ♀
rostrata

winter ♀

streaked
undertail
coverts

breeds in Iceland,
Greenland and Baffin
Island; a few in winter
in East in redpoll
invasion years

faint flank
streaks

winter ♂

winter ♀
exilipes

slightly
smaller bill
than Common

**Hoary
Redpoll**

paler upperparts
than Common

very
pale pink

larger and
overall paler
than *exilipes*

winter ♂
hornemanni

pale rump

faint flank
streaks

winter ♂
exilipes

thin, sharply
pointed bill

pale wing bars

Pine Siskin
pinus

juvenile

prominent yellow
wing stripe and
yellow patches at
base of tail

streaked
underparts

yellow forehead
and eyebrow

white patch on
inner wing

**Evening
Grosbeak**
vespertinus

large
bill

breeding ♂

♂

♀

white
primary
patch

juvenile ♂

short tail with
white tail spots

female with gray on head
and back; buffy below

breeding ♀

Oriental Greenfinch *Chloris sinica* L 6" *(15 cm)*
Adult male has greenish face and rump, dark grayish olive nape and crown, bright yellow wing patch and undertail coverts. **Adult female** is paler, with a brownish head. **Juvenile** has same yellow areas as adults but is streaked overall.
VOICE: Call is a nasal *dzweeee.*
RANGE: Asian species. Casual migrant, mainly in spring, on outer Aleutians; one Pribilofs record. A record in winter from Arcata, CA, was judged to be of questionable origin.

Common Chaffinch *Fringilla coelebs* L 6" *(15 cm)*
Has white patches on lesser coverts, base of primaries, wing bar; outer tail feathers white. **Male's** crown and nape are blue-gray; shows pinkish below, pinkish brown above. **Female's** head is mostly gray with brown lateral stripes.
VOICE: Call is a metallic *pink-pink;* also a *hweet.*
RANGE: Palearctic species; casual to northeastern North America; reports elsewhere possibly escaped cage birds.

Brambling *Fringilla montifringilla* L 6¼" *(16 cm)*
Adult male has tawny orange shoulders, spotted flanks; head and back fringed with buff in fresh fall plumage that wears down to black by spring. **Female** and juvenile have mottled crown, gray face, striped nape.
VOICE: Flight call is a nasal *check-check-check;* also gives a nasal *zwee.* Song consists of just a short buzzy note.
RANGE: Eurasian species; fairly common but irregular migrant on Aleutians; rare on Pribilofs and St. Lawrence Island, AK; casual in fall and winter in Canada and northern U.S.; recorded south to central CA. Has nested on Attu Island (1996).

Hawfinch *Coccothraustes coccothraustes* L 7" *(18 cm)*
Stocky; yellowish brown above; pinkish brown below; has black throat and lores; shows conspicuous white band on extended wing. Big bill is blue-black in spring, yellowish in fall. Note club-shaped inner primaries, visible at close range. Female resembles **male,** but is duller; has grayish secondaries and inner primaries. Walks with parrotlike waddle.
VOICE: Call is a loud, explosive *ptik.*
RANGE: Eurasian species. Rare spring stray on western Aleutians; casual on other islands in Bering Sea and from Nome and vicinity.

Eurasian Bullfinch *Pyrrhula pyrrhula* L 6½" *(17 cm)*
Cheeks, breast, and belly intense reddish pink in **male,** brown in **female.** Black cap and face, gray back, prominent whitish bar on wing, distinct white rump. In profile, top of head and bill form unbroken curve. Juvenile resembles female, but with brown cap. All records are of northeast Asian *cassinii,* which was described in 1869 from a Jan. specimen taken in Nulato, along the Yukon River, central AK.
VOICE: Call is a soft, piping *pheew.*
RANGE: Eurasian species. Casual migrant on Aleutians; casual in winter on AK mainland, where recorded south to Petersburg, southeastern AK.

Oriental Greenfinch
kawarahiba

dark grayish olive nape and crown

adult ♂

extensive yellow on wings in all plumages

female duller than male

adult ♀

juvenile streaked below

juvenile

Common Chaffinch

brown lateral crown stripes

blue-gray crown and nape

♀

♂

pinkish below

extensive white on wings in all plumages

Brambling

glossy black head and back

orange breast and shoulders

pale-fringed feathers with blackish bases

breeding ♂

white rump in all plumages

gray face bordered by black

fall ♂

pale nape spot

♀

breeding ♂

thick blue-black bill is yellowish in fall and winter

Hawfinch
japonicus

club-shaped inner primaries visible at close range

pale wing patch

short tail

Eurasian Bullfinch
cassinii

bright reddish pink color characteristic of northeast Asian *cassinii*

♂

gray nape and back

white rump in all plumages

pale brownish cheeks and underparts

♀

OLD WORLD SPARROWS Family Passeridae
Old World family. Gregarious; two species have become established in North America.
SPECIES: 39 WORLD; 2 N.A.

House Sparrow *Passer domesticus* L 6¼" (16 cm)
Breeding male has gray crown, chestnut nape, black bib, black bill. Fresh **fall** plumage edged with gray, obscuring markings; bill becomes brownish. **Female** and juvenile identified by streaked back, buffy eye stripe.
VOICE: Calls include a sweet *cheelip* and monotonous chirps.
RANGE: Common and aggressive, omnipresent in populated areas. Gregarious in winter. Also known as English Sparrow. Casual to YT and southeastern and western AK; appearance in western AK likely results from introductions to Russian Far East.

Eurasian Tree Sparrow *Passer montanus* L 6" (15 cm)
Gregarious all year. Brown crown, black ear patch distinguish **adult**. House Sparrow has gray crown, more extensively black throat. **Juvenile** has dark mottling on crown, dark gray throat and ear patch.
VOICE: Similar to House Sparrow but harder.
RANGE: Old World species, introduced and locally common in parks and farmlands within mapped range. Accidental to IN, WI, KY, MB, ON.

WEAVERS Family Ploceidae
Large, primarily African family. Breeding males are often highly colored. Known for elaborate woven nests. Orange Bishop not accepted by ABA. SPECIES: 108 WORLD; 1 N.A.

Orange Bishop *Euplectes franciscanus* L 4" (10 cm)
Breeding male is bright orange-red with black cap, breast, and belly; long tail coverts obscure tail. **Females,** immatures, and winter birds are streaked above. Compare especially with Grasshopper Sparrow (page 474); note Orange Bishop's thicker, pinkish bill; short, blunt tail, often flicked open. Also known as Northern Red Bishop.
VOICE: Complex song is high and buzzy. Calls include a sharp *tsip* and a mechanical *tsik tsik tsk.*
RANGE: Native to sub-Saharan Africa; widely introduced. Established in Los Angeles, CA (1980s), and Phoenix, AZ (1998), areas, where it favors weedy areas, especially river bottoms.

ESTRILDID FINCHES Family Estrildidae
Large, Old World family found from Africa to Australia and South Pacific islands. Most are small, with pointed tails. Related to weavers. Nutmeg Mannikin not accepted by American Birding Association. SPECIES: 140 WORLD; 1 N.A.

Nutmeg Mannikin *Lonchura punctulata* L 4½" (11 cm)
Small, with heavy bill, pointed tail. **Adult** rich reddish brown above, scaled below; has thick black bill. Also known as Scaly-breasted Munia. **Juveniles** are tan; bill slate gray.
VOICE: Song, *tiks* and whistles, is nearly inaudible; call, a loud *kibee.*
RANGE: Favors weedy areas. Widespread in Southeast Asia; rapidly spreading in greater Los Angeles area (north to Santa Barbara, CA).

House Sparrow
domesticus

gray crown

blackish bill

chestnut-brown nape

white wing bar

head pattern more subdued

bill paler

fall ♂

buffy eyebrow

black bib

pale bill

♀

plain underparts; dark streaked back

breeding ♂

juvenile

chestnut-brown cap and bold blackish spot on white cheeks

whitish collar

displaying ♂

Eurasian Tree Sparrow
montanus

adult

Orange Bishop
franciscanus

unmistakable, but other species of African bishops are similarly colored

breeding ♂

chestnut-brown head and back

thick, blackish bill

adult

Nutmeg Mannikin
punctulata

intricately scaled breast and belly

thick bill

overall tannish brown

juvenile

dark eye stands out on blank face

thick, pinkish bill

streaked back

♀

short, blunt tail often flicked open

pointed tail

ACCIDENTALS • EXTINCT SPECIES

These 92 species have been recorded for North America, but for nearly all there are fewer than three records in the past two decades or five records in the last hundred years. Four species that have gone extinct in the past two centuries are also included.

adult
anser

Graylag Goose *Anser anser*

L 29-32" (74-81 cm) WS 59-66" (150-168 cm) Palearctic species that has become widely domesticated (page 48). The nominate European subspecies is a common summer resident on Iceland; recently recorded from Greenland. From 24 Apr. to 2 May 2005, one landed and remained on a ship some 120 miles southeast of St. John's, Newfoundland. More equivocal on origin is record from Wallingford, CT, Feb.–March 2009. Largest and bulkiest of gray geese with heavy head, neck, and bill. Pink legs and orange (in nominate *anser*) bill. Head and neck gray, uniform with rest of body, unlike other gray geese, which have a darker head and neck. In flight, shows striking pale gray forewing and contrasting pale gray underwing coverts.

adult

Lesser White-fronted Goose *Anser erythropus*

L 22-26" (55-66 cm) Palearctic species. Closely resembles Greater White-fronted Goose (page 14) and plumages similar, but smaller and more stocky, with shorter neck and stubbier bill. The yellow orbital ring is conspicuous. Darker neck than Greater White-fronted (except for the *elgasi* and *flavirostris* subspecies of Greater). Wings extend beyond the tail when folded. Specimen record from Attu Island, AK, on 5 June 1994. Population is declining, especially from the western Palearctic.

adult
♂

Labrador Duck *Camptorhynchus labradorius*

L 22½" (57 cm) **EX** Extinct. An endemic North American species. Note the unique bill shape that broadens toward tip. Much of **adult male**'s head, neck, and chest white; remainder of body plumage blackish. Adult females, immatures, and eclipse males more grayish brown overall, the throat being whiter than the head. Never common, it was best known from the winter grounds on the mid-Atlantic coast, especially the southern shore of Long Island, NY, where the last definite record (specimen) was obtained in 1875. Breeding grounds unknown, perhaps Labrador, perhaps farther north. Accidental inland in spring from Montreal, QC (1862).

adult

Light-mantled Albatross *Phoebetria palpebrata*

L 31-35" (79-89 cm) WS 72-86" (183-218 cm) A circumpolar species of the southern oceans that breeds on subantarctic islands. A very graceful flyer. **Adult** has dark head with prominent white eye crescents and strikingly pale mantle and body. Long, dark, wedge-shaped tail is distinctive. At close range, note bluish line of skin on the lower mandible (sulcus). Juvenile is similar to adult but browner overall, with less prominent eye crescents and gray sulcus. One individual was well documented at Cordell Bank, off northern CA, on 17 July 1994.

Wandering Albatross *Diomedea exulans*

L 42-53" (107-135 cm) WS 100-138" (254-351 cm) Circumpolar in southern oceans, breeding on subantarctic islands. A polytypic species (five to seven subspecies), some authorities recognize up to five species. Huge size and immense wingspan (reaches over 11 feet). Massive pinkish bill and white underwing with narrow dark trailing edge and primary tips. **Adult** has extensively white back and wings. Juvenile is dark chocolate brown with conspicuous white face. Maturation takes up to 15 years. Becomes white first on the mantle, body, and head, eventually spreading to upperwing coverts. Two records: one onshore record at Sea Ranch, Sonoma County, CA, 11 to 12 July 1967; another at Perpetua Bank, OR, 13 Sept. 2008, thought to be *antipodensis*. Five European records.

adult
♀

newelli

Townsend's Shearwater *Puffinus auricularis*

L 13" (33 cm) WS 33" (83 cm) Two subspecies: nominate *auricularis* breeds on the Revillagigedo Islands off western Mexico; *newelli* with whiter under tail and longer tail breeds on HI. One record of *newelli;* turned into a rehab center after it had been attracted to a night construction worker's head lamp at Del Mar, CA, on 1 Aug. 2007. Similar to Manx Shearwater (page 92) but more sharply black and white with more dark on the longest undertail coverts and a longer tail.

Black-bellied Storm-Petrel *Fregetta tropica*

L 8" (20 cm) WS 18" (46 cm) A widespread southern ocean species, one was well photographed off Manteo, NC, on 31 May 2004; another was photographed off Hatteras on 16 July 2006 and yet another was photographed there on 14 Aug. 2010. Black-and-white coloration distinctive on all individuals, but diagnostic black line up through white belly (separating it from another southern oceans species, White-bellied Storm-Petrel, *F. grallaria*) can be hard to see. Note long legs and feet project past tail. Foraging behavior distinctive: splashes breast into water and then springs forward pushing off with long legs.

Ringed Storm-Petrel *Oceanodroma hornbyi*

L 8¼-9" (21-23 cm) South American species of the Humboldt Current from Chile to southern Ecuador; casual to Colombia. Nesting grounds unknown, but possibly in the central Andes. Recent well-documented record off San Miguel Island, CA, on 2 Aug. 2005. Note large size and striking plumage pattern, including blackish cap and breast band; tail is deeply forked.

Swinhoe's Storm-Petrel *Oceanodroma monorhis*

L 8" (20 cm) WS 18" (46 cm) Breeds close to northeast Asia in shallower waters off Russian Far East, Korea, Japan, and China. Winters in northern Indian Ocean. A few may breed in eastern North Atlantic (especially Selvagem Grande) where first discovered in 1983. Four records off NC, one on 2 June 2008, particularly well documented. One possible, with marginal photographic documentation, off Kodiak Island, AK, 5 Aug. 2003. Only all-dark storm-petrel in North Atlantic with stout bill, moderate pale bar across the upperwing, and forked tail. Somewhat broad wing, like Band-rumped (page 96), and wingflaps rather shallow with lots of gliding. White base to primary shafts visible at close range.

Tristram's Storm-Petrel *Oceanodroma tristrami*

L 10" (25 cm) WS 22" (56 cm) Breeds on Leeward Hawaiian Islands and Volcano and southern Izu Islands off southern mainland Japan. One certain record, one photographed and measured on Southeast Farallon Island off central CA, 22 Apr. 2006. Larger and grayer than Black Storm-Petrel (page 98) with paler carpal area and more deeply forked tail.

Great Frigatebird *Fregata minor*

L 37" (95 cm) WS 85" (216 cm) Extensive breeding range in Indian and Pacific Oceans. Closely resembles Magnificent Frigatebird (page 102). Adult male distinguished from Magnificent by russet bar on upper wing coverts and pink feet and often by whitish scallops on axillars. Note **adult female**'s dark head with pale gray throat, rounder (less tapered) black belly patch, and red orbital ring. When fresh, juvenile has rusty wash to head and chest, and pink feet. Specimen from Perry, OK, 3 Nov. 1975. Two photographed records of adults off CA: adult male in Monterey Bay, 13 Oct. 1979, and adult female Southeast Farallon Island, 14 Mar. 1992.

Lesser Frigatebird *Fregata ariel*

L 30" (76 cm) WS 73" (185 cm) Widespread in southwestern and central Pacific and Indian Ocean; a few colonies in South Atlantic. Our smallest frigatebird; in all plumages a white spur extends from the flanks into the axillaries. Juvenile has pale, rusty head. Recorded four times from North America: an **adult male** photographed at Deer Isle, ME, 3 July 1960; an adult female (photographed) found moribund then discarded near Basin, WY, 11 July 2003; an adult male photographed at Lake Erie Metropark, Wayne County, MI, 18 Sept. 2005; and an immature female photographed at Lanphere Dunes, Arcata, CA, 15 July 2007.

Nazca Booby *Sula granti* L 32" (81 cm) WS 62" (158 cm)

Recently split from the Masked Booby (page 102), this eastern tropical Pacific endemic ranges north to Mexico. One debatable (on origin) record: an immature landed on a ship off northern Baja California and rode to San Diego, 29 May 2001. Other records of immatures off southern and central CA remain problematic. Similar to Masked Booby in all plumages; subtle shape differences include shorter and thinner bill, shorter legs, and longer wings and tail. Note **adult's** orange-pink bill and more orange (not yellow) iris. Juvenile averages paler than Masked, and pale collar is less marked or absent.

Yellow Bittern *Ixobrychus sinensis*

L 15" (38 cm) WS 21" (53 cm) Widespread Asian species. One specimen record from Attu Island, AK, 17 to 22 May 1989. In **adults** (sexes similar), head and neck are buffy, cap and tail are black, and neck is streaked. Juvenile is more streaked overall. In flight, note black primary coverts and flight feathers. An Asian congener, Schrenck's Bittern *(Ixobrychus eurhythmus),* though scarce, is highly migratory and could occur in North America. It is slightly larger than Yellow Bittern, more cinnamon dorsally, and has a slaty gray trailing edge to its wing.

Bare-throated Tiger-Heron *Tigrisoma mexicanum*

L 30" (76 cm) Tropical species. Found in wetlands from southern Sonora and southern Tamaulipas south to northwestern Colombia. A second-year bird was at Bentsen-Rio Grande State Park, TX, 21 Dec. 2009 to 20 Jan. 2010. Large and American Bittern–like wing shape in flight (page 112). Largest tiger-heron, with long neck and bill. Adults with black cap, gray face, bare yellow throat, and finely barred neck. Juvenile boldly barred and spotted, including remiges, with cinnamon-buff and brown.

subadult

Gray Heron *Ardea cinerea*

L 33-40" (84-102 cm) WS 61-69" (155-175 cm) Widespread Old World species; one found on Newfoundland coast in Oct. 1999, subsequently died in a rehabilitation center. There is a sight record from St. Paul Island, AK, 1 Aug. 1999, and one was photographed there 1 to 2 Oct. 2007; another was photographed on Shemya Island, Aleutians, 29 Apr. to 2 May 2010. Similar to Great Blue Heron (page 112), but smaller, with shorter legs and neck. In all plumages lacks rufous thighs of Great Blue; in flight, leading edge of wing shows prominent white area, rather than rufous.

adult
cinerea

Intermediate Egret *Mesophoyx intermedia* *L 27" (69 cm)*

Widespread Old World species found in Africa and from India to Australia. Breeds north in Asia to Japan. One found dead (Asian nominate race) on Buldir Island, western Aleutians, AK, 30 May 2006, had likely arrived and died some days earlier. Similarly shaped and colored like the larger Great Egret (page 114), but with shorter bill with a distinct dark tip.

breeding
adult
intermedia

Chinese Egret *Egretta eulophotes* *L 27" (65 cm) WS 41" (104 cm)*

Threatened Asian species, breeding on islands off Korea, China, and perhaps the Russian Far East; winters in the Philippines and Borneo, some on coastal mainland of Southeast Asia. One specimen record from Agattu Island, AK, 16 June 1974. Note shorter legs than Little Egret (page 114). In **breeding** plumage has shaggy crest, turquoise lores, entirely orange-yellow bill, and black legs with yellow feet. Nonbreeding birds lack crest; legs and feet are yellowish green and bill mostly dark.

breeding
adult

Western Reef-Heron *Egretta gularis*

L 23½" (60 cm) WS 37½" (95 cm) Old World species; casual to West Indies. Six records of **dark morphs**, but perhaps involving only two individuals: Nantucket Island, MA, 26 Apr. to 13 Sept. 1983; Stephenville Crossing, western Newfoundland, 14 June to 6 Sept. 2005; Cape Breton Island, NS, 26 June to 1 Aug. 2006; the ME-NH border, 9 Aug. to 20 Sept. 2006; South Amboy, NJ, 30 June and 14 July 2007, and Brooklyn, NY, 8 July 2007. All of those sightings from 2005 to 2007 could have involved the same wandering bird. Structurally resembles closely related Little Egret (page 114), but has slightly thicker neck; thicker-based bill is longer and a little more curved. Two color morphs. Much more numerous dark morph is slaty gray overall with white chin and throat; lores and bill dusky yellow; legs black and feet yellow. White morph resembles Little Egret, but note slight structural differences; immatures often have scattered dark feathers.

dark-
morph
adult
gularis

breeding
adult ♂

adult
nigra

light-
morph
adult
naso

1st
summer ♂

olivascens

Chinese Pond-Heron *Ardeola bacchus*

L 18" (46 cm) WS 34" (86 cm) Migratory East Asian species that occurs rarely, but with increasing frequency, to Japan and Korea. A breeding-plumaged adult was on St. Paul Island, AK, 4 to 9 Aug. 1996; another was collected on Attu Island, Aleutians, on 20 May 2010. Members of this genus are short and stocky and most have entirely white wings, rump, and tail (especially visible in flight); yellow legs and feet. **Breeding male** has bright chestnut head, neck, and upper breast, slaty lower breast, and blue-based bill. Breeding female lacks slaty lower breast. Immatures and nonbreeding adult much duller with streaked neck and not separable from several other congeners from south and Southeast Asia.

Crane Hawk *Geranospiza caerulescens*

L 18-21" (46-53 cm) WS 36-41" (91-104 cm) Neotropical species from northeastern and northwestern Mexico to South America. One wintered at Santa Ana National Wildlife Refuge, south TX, 20 Dec. 1987 to 9 Apr. 1988. Distinctive, long profile with small head, long orange legs, and long banded tail; iris reddish. Northeastern subspecies (*nigra*) is darkest. In flight, note white crescent at base of primaries. Juvenile has some whitish in face and under tail, whitish barring below, and duller soft parts.

Collared Forest-Falcon *Micrastur semitorquatus*

L 20" (52 cm) WS 31" (79 cm) Neotropical species found from northeastern and northwestern Mexico to South America. Recorded once in south TX at Bentsen-Rio Grande Valley State Park, 22 Jan. to 24 Feb. 1994 (light-morph adult). Distinctive structure: very short, rounded wings (wing tips barely reach base of tail), long graduated tail, long legs. Three color morphs: most numerous **light morph** is black above, white below; note black crescent on white cheek, thin white bars on tail; juvenile similar but browner above, barred and more buffy below. In buff morph, white areas replaced with buff; dark morph is rare. Usually seen within forest canopy; often located by loud calls.

Red-footed Falcon *Falco vespertinus*

L 11" (27 cm) WS 29" (73 cm) A medium-size falcon that breeds in eastern Europe and western Asia and winters in south and southwestern Africa. Regular, especially in spring, to northwestern Europe; casual to Iceland. One **first-summer male** was present at Martha's Vineyard, MA, 8 to 24 Aug. 2004. Adult male is slaty gray overall with rufous thighs and under tail. Adult female has buffy crown and underparts, dark moustache, pale sides of neck; adults have red legs and feet. Juvenile similar but browner and more streaked, duller legs and feet. Often hovers.

Paint-billed Crake *Neocrex erythrops* *L 7¼-8" (18-20 cm)*

Found from eastern Panama through South America and Galápagos Islands. Specimens from Brazos County, TX, 17 Feb. 1972 (*erythrops*) and near Richmond, VA, 15 Dec. 1978 (*olivascens*). Olive-brown above with gray forecrown and face; throat whitish; sides, flanks, and undertail coverts barred; red legs and yellow-green bill with bright orange base.

Spotted Rail *Pardirallus maculatus* L 10-11¼" *(25-28 cm)*

Resident on Cuba and Isle of Pines, and Hispaniola *(maculatus)*, and from central Mexico to South America and the Galápagos Islands (both *insolitus*). Specimens *(insolitus)* from Beaver County, PA, 12 Nov. 1976, and from Brown County, TX, 9 Aug. 1977. Rather large, blackish rail. **Adult** is spotted with white on head and upperparts; remainder of underparts banded with white; long, slender greenish yellow bill has red spot at base; iris, legs, and feet red. Juveniles are polymorphic, have duller legs and bill, and brown iris. Dark morph is plain dark brown above, sooty gray below; pale morph has grayish brown throat and breast with fine white bars on breast; barred morph has gray throat with white spots, breast and belly barred with white.

adult
insolitus

Sungrebe *Heliornis fulica* L 11" *(28 cm)*

Tropical species found from central Tamaulipas, Mexico, to northern Argentina. One of three finfoot species in the world (family Heliornithidae). A female was present at Bosque del Apache National Wildlife Refuge, NM, 13 and 18 Nov. 2008. Swims and slowly rocks neck back and forth like a Common Moorhen, but does not dive. Black areas on crown, sides of face and neck, white supercilium and neck. Auricular white (male) to tawny buff **(female).** Bill rather long and pointed, somewhat stout at base. Slight crest to rear crown.

♀

Double-striped Thick-knee *Burhinus bistriatus*

L 16½" *(42 cm)* Resident from northeastern Mexico to Brazil; recent nesting record from Great Inagua, Bahamas. A specimen record on 5 Dec. 1961 from the King Ranch, Kleberg County, TX. A bird at Yuma, AZ, was transported. Crepuscular and terrestrial, with ploverlike gait of runs and abrupt stops. Adults with dark lateral crown stripe and bold white supercilium; dark-tipped yellow bill and yellow legs. Juvenile slightly duller. In flight, upper wing two-tone; prominent broken white bar on primaries.

bistriatus

Greater Sand-Plover *Charadrius leschenaultii* L 8½" *(32 cm)*

Old World species. One wintered at Bolinas Lagoon, CA, 29 Jan. to 8 Apr. 2001 (photos, measured in hand); another was photographed at Huguenot Memorial Park, Duval County, FL, 14 to 26 May 2009. Overall closely resembles Lesser Sand-Plover (page 166), but bill distinctly longer; legs longer and not as dark, more greenish yellow. In flight, wing stripe broader and feet project beyond tail. In breeding plumage, colored breast band is not as dark or extensive.

winter

Collared Plover *Charadrius collaris* L 5" *(14 cm)*

Resident from northern Mexico to South America. Only record north of Mexico at Uvalde, TX, 9 to 11 May 1992. Small with disproportionately long legs, small thin black bill, pinkish legs; lacks white collar around nape. Adult with dark forecrown, often with rusty border; auriculars, nape, and sides of breast often with rusty fringes, especially in males; narrow but complete black breast band, often with distinct rusty fringes at sides, especially in male. Juvenile has incomplete breast band and pale rusty edges above.

adult

breeding
adult
ostralegus

adult
himantopus

adults

adult

Eurasian Oystercatcher *Haematopus ostralegus*
L 16½" (42 cm) Palearctic species, breeding west to Iceland; rare migrant to Greenland (over 30 records). Two spring records from Newfoundland: Fox Island, Tors Cove, 22 to 25 May 1994; and Eastport, 3 Apr. to 2 May 1999. Similar to American Oystercatcher (page 172), but has black back and red iris. In flight, white wing bar is bolder and more extensive, and white extends up back. Nonbreeding birds have a white bar across throat.

Black-winged Stilt *Himantopus himantopus* L 13" (33 cm)
Widespread in Old World. Two photographed spring records for AK from western Aleutians: Nizki Island, 24 May to 3 June 1983; and two at Shemya Island, 1 to 9 June 2003. One specimen record from St. George Island, Pribilofs, AK, 15 May 2003. Similar to Black-necked Stilt (page 172), but paler on rear neck and lacks white spot above eye. **Adults** vary from having entirely white head and neck to being darker; adult males have glossy black backs, females browner.

Eskimo Curlew *Numenius borealis* L 14" (36 cm) **E**
Formerly common, now probably extinct. Only known nesting area was Anderson River region, NT (nests found 1862 to 1866); possibly bred farther west. Wintered mainly on the Pampas of Argentina. Migrated up through Great Plains in spring; to northeast Arctic Canada and then over the Atlantic to South America in fall. The last certain record was of an adult female shot by a "sportsman" on Barbados on 4 Sept. 1963 (specimen eventually secured; at Academy of Natural Sciences, Philadelphia). All sightings since then not adequately documented. Likely extinction was due to unregulated market hunting, especially prevalent on central Great Plains in the two decades following the U.S. Civil War. Rare by 1900, thought possibly extinct by 1940; but a few persisted, as up to two were well photographed in Mar. and Apr. from 1959 to 1962 at Galveston Island, TX, the last verified North American records. Resembles a small Whimbrel (page 182), but upperparts darker, bill less curved; wing linings pale cinnamon. Calls are poorly known; one call reportedly a rippling *tr-tr-tr* and a soft whistle.

Slender-billed Curlew *Numenius tenuirostris* L 15" (39 cm) **E**
Critically endangered, possibly extinct. Only known nests found near Tara, north of Omsk, southwestern Siberia, Russia, early in 20th century. Wintered in western Mediterranean region, where several recorded from one Moroccan location until 1995. In North America a specimen from Crescent Beach, ON, from "about 1925." Size of Whimbrel but patterned like Eurasian Curlew (page 184), but with slender bill and black heart-shaped spots, not chevrons, on sides and flanks.

Solitary Snipe *Gallinago solitaria* L 12" (30 cm)
Asian species. Found in the Himalaya and northeast Asia, as close to AK as Kamchatka and western Chukota, Russian Far East. Some populations migratory, others altitudinal migrants. One certain record, a specimen taken at Attu Island, 24 May 2010. Another likely record, supported by marginal photos, at St. Paul Island, Pribilofs, 10 Sept. 2008. Large, dark and stocky, single eye stripe and very long bill. Overall gingery brown, including nearly solid area on breast sides.

Eurasian Woodcock *Scolopax rusticola* L 13" (33 cm)

Widespread Old World species. Casual to North America, where most records are old and from the Northeast; older records are from Newfoundland (1862), QC (twice in 1862), PA (1886, 1890), NJ (1859), and AL (1889). The last record, and the only one accepted from the 20th century, was one at Goshen, NJ, 2 to 9 Jan. 1956, although one from OH (specimen lost) in 1935 may have been this species. All dated records fall between early Nov. and early Mar. Distinctly larger than similar American Woodcock (page 204); also duller and heavily barred below.

Oriental Pratincole *Glareola maldivarum*

L 9" (23 cm) WS 23½-25½" (60-65 cm) Asian species. Winters south to Australia. Recorded twice in AK: a specimen from Attu Island, 19 to 20 May 1985; one at Gambell, St. Lawrence Island, 5 June 1986. Short-tailed pratincole with no white trailing edge to wing and chestnut underwings. Collared Pratincole *(G. pratincola),* breeding in the western Palearctic, has occurred once on Barbados and could occur in eastern North America. Collared has longer tail than Oriental and has a white trailing edge to wing.

breeding
adult

Swallow-tailed Gull *Creagrus furcatus*

L 23" (58 cm) WS 52" (132 cm) Breeds on Galápagos Islands and on Isla Malpelo, Colombia. Otherwise pelagic, ranging south to central Chile, and north casually to all Costa Rica and Nicaragua. Two CA records, one at Pacific Grove and Moss Landing, 6 to 8 June 1985; another on 3 Mar. 1996, 15 miles west of Southeast Farallon Island. All plumages unmistakable. Much larger than Sabine's Gull (page 212) but with similar wing pattern in all plumages. Note very long drooped bill. **Adults** in breeding plumage have a slaty gray hood with scarlet eye ring.

breeding
adult

Gray-hooded Gull *Chroicocephalus cirrocephalus*

L 16" (41 cm) WS 43" (109 cm) A native to Africa and South America. One adult was photographed at Apalachicola, FL, 26 Dec. 1998. **Breeding adult** has pale gray hood with darker border, long dark red bill, and long red legs; wing pattern distinctive. In winter loses hood and has dark ear spot and smudge around eye and dark tip to bill. First-year has similar outer wing pattern to adult, but a diagonal brown bar across the inner wing and a dark secondary bar, and a dark tail band. The pinkish yellow bill has a dark tip.

breeding
adult
cirrocephalus

Whiskered Tern *Chlidonias hybrida*

L 9½-10" (24-25 cm) WS 26½-28½" (67-72 cm) Widespread Old World species. Two North American records, both adults. In 1993 one at Cape May, NJ, 12 to 15 July, later moved to DE shore, 19 July to 24 Aug. Another at Cape May, 8 to 12 Aug. 1998. Like congeners, secures food by picking it off surface. **Adults** have short stout dark red bill and medium length red legs; dark gray underparts set off contrasting white cheeks and under tail. In winter, head and underparts white with thin black postocular patch, blackish bill and legs; compare head pattern to "ear muff" effect of White-winged Tern (page 238).

breeding
adult
hybridus

adult

Great Auk *Pinguinus impennis* L 30" *(81 cm)* **EX**

Extinct. North Atlantic species known in North America from three nesting colonies on islands off QC and Newfoundland, the largest on Funk Island off Newfoundland. Wintered within breeding range and south to MA, casually to SC. Extirpated from Funk Island about 1800; last definite record was two clubbed on Eldey Stack, Iceland, on 3 June 1844. Flightless. Resembled a large Razorbill with similarly shaped bill. Distinct, white circular patch in lores. Winter plumage imperfectly known.

adult ♂

Scaly-naped Pigeon *Patagioenas squamosa* L 14" *(36 cm)*

Resident throughout most of the West Indies. Two old specimen records from Key West, FL: 24 Aug. 1898 and 6 May 1929. A large dark pigeon; in good light, head and upper breast are dark maroon; feathers on sides of neck more reddish, tipped black, forming diagonal lines giving scaled appearance; dark red bill with yellow tip; orange-red iris; orange orbital ring.

turtur

European Turtle-Dove *Streptopelia turtur* L 10" *(26 cm)*

Widespread in Eurasia and Africa. Records from south FL (1990), St.-Pierre, St.-Pierre and Miquelon (2001), and MA (2001). This species strays annually to Iceland. It is known to ride ships. Scapulars and wing coverts have black centers with bold orange-brown edges. Blue-gray panel in center of wing, black-and-white neck patch, and orange eye. Compare carefully to Oriental Turtle-Dove (page 268).

adult ♂

Passenger Pigeon *Ectopistes migratorius*

L 15¾" *(40 cm)* **EX** Extinct. Believed to have been the most abundant bird species in North America—in migration, flocks said to have numbered in the millions. Formerly found in eastern North America, casually in the West. By 1870s relegated to scattered breeding locations. Last records were a specimen from OH in 1900 and a reliable sight record in MO in 1902. Last individual died in captivity in a Cincinnati zoo on 1 Sept. 1914. Destruction of old-growth deciduous forests and overhunting, especially in breeding colonies, led to its demise. Resembled a large Mourning Dove. **Adult male** bluish gray above and pinkish below; female browner above and paler below; juvenile similar to female.

adult

Carolina Parakeet *Conuropsis carolinensis*

L 13½" *(34 cm)* **EX** Extinct. Only native breeding North American psittacid. Formerly resident in Southeast, especially along rivers. Last certain records were from FL and KS in 1904; a reliable sight record from MO in 1905 and perhaps another in 1912. Last captive died on 21 Feb. 1918. **Adult** had green body with yellow patches on shoulder, thighs, and vent, a yellow head, and reddish orange face. Immature entirely green, except for orange patch on forehead.

Oriental Scops-Owl *Otus sunia* L 7½" (19 cm) WS 21" (53 cm)

A small, nocturnal, insectivorous owl of East Asia. Multiple subspecies, northern ones are migratory. Two records (*japonicus*) of rufous morphs from Aleutian Islands, AK: a dried wing found on Buldir Island, 5 June 1977, and one found alive on Amchitka Island, 20 June 1979, subsequently died (specimen). Three color morphs: gray-brown, reddish gray, and **rufous**. Fine dark streaks on head; breast streaked vertically and horizontally with thin crossbars. Short ear tufts. Northern subspecies may be specifically distinct, calls differ.

rufous morph
japonicus

Mottled Owl *Ciccaba virgata* L 14" (36 cm) WS 33" (84 cm)

A medium-size and very vocal nocturnal owl found in a variety of woodland habitats from northwestern and northeastern Mexico to South America. A road-killed specimen was salvaged in front of Bentsen-Rio Grande Valley State Park, south TX, on 23 Feb. 1983. Also, a controversial record from Weslaco, TX, 5-11 July 2006. Note round head with no ear tufts and streaked underparts; brown facial disk with bold white eyebrows and whiskers is distinctive. Larger Barred Owl (page 284) has prominent barring across upper breast and paler facial disk.

Brown Hawk-Owl *Ninox scutulata* L 12¼" (31 cm)

East Asian species ranging north to Ussuriland, Korea, and Japan. Eleven subspecies recognized; darker northern breeding *japonica* is migratory. One record from St. Paul Island, Pribilofs, AK, 27 Aug. to 3 Sept. 2007; another was found dead (photographed) on Kiska island, Aleutians, 1 Aug. 2008. Overall slate brown coloration with a round head and yellow eyes, and a long banded tail. Also widely known by an alternative English name, Brown Boobook.

japonica

Stygian Owl *Asio stygius* L 17" (43 cm) WS 42" (107 cm)

A medium-size, forest-dwelling nocturnal owl occurring from northern Mexico to South America; also Cuba, Hispaniola, and Gonâve Island, West Indies. Two winter records of birds found roosting and photographed at Bentsen-Rio Grande Valley State Park, TX: 9 Dec. 1994 and 26 Dec. 1996. Deep chocolate brown overall with close-set ear tufts and blackish facial disk with contrasting white forehead; underparts show distinct dark streaks and crossbars. Compare to Long-eared Owl (page 282), which is browner and has a rufous facial disk.

Gray Nightjar *Caprimulgus indicus* L 11-12¾" (28-32 cm)

Asian species, formerly known as Jungle Nightjar. One desiccated specimen (*jotaka*) salvaged on Buldir Island, AK, on 31 May 1977. Overall color is grayish brown, patterned with black, buff, and grayish white. Note long wing-tip projection. Adult male has large white subterminal patch on inner primaries and white tips to all but central pair of tail feathers; on female, wing patch and tail tips more buffy. The two more northerly and migratory subspecies (*jotaka* and *hazarae*) have different vocalizations and are treated as a distinct species by most authors now, *C. jotaka*.

robustus

♂ *jotaka*

Antillean Palm-Swift *Tachornis phoenicobia* L 4¼" (11 cm)

A tiny Caribbean swift that is resident in the Greater Antilles (except Puerto Rico). Two were present and photographed at Key West, FL, 7 July to 13 Aug. 1972. Distinctive are dark cap, dark sides, and thin line across breast contrasting with white throat, belly, and rump; tail has shallow fork. Batlike flight with rapid wingbeats and short glides and twists; generally flies low, among trees and palms.

Xantus's Hummingbird *Hylocharis xantusii* L 3½" (9 cm)

Endemic to southern Baja California, Mexico. Accidental vagrant to southern CA (Anza-Borrego State Park, 27 Dec. 1986, and Ventura, 30 Jan. to 27 Mar. 1988) and southwestern BC (Gibsons, 16 Nov. 1997 to 21 Sept. 1998). Plumages and calls similar to White-eared Hummingbird (page 300), but note buff on underparts and rufous in tail; **male** has black forehead and ear patches.

Cinnamon Hummingbird *Amazilia rutila* L 4-4½" (10-12 cm)

Resident in lowlands from Sinaloa and Yucatán Peninsula, Mexico, south to Costa Rica. Two photo records from Southwest: 21 to 23 July 1992 at Patagonia, AZ; 18 to 21 Sept. 1993 at Santa Teresa, NM. Adults have a cinnamon tail and underparts and a black-tipped red bill. Immature is similar but upperparts edged cinnamon when fresh, and upper mandible is mostly dark.

Bumblebee Hummingbird *Atthis heloisa* L 2¾-3" (7-8 cm)

Endemic to montane forests of Mexico north of the Isthmus of Tehuantepec. Two specimens taken on 2 July 1896 in Ramsey Canyon, AZ. That two would be taken on the same date from one locality with no records since seems unlikely, but the specimens are extant and the records have not yet been refuted. The question is whether the collector was actually in the present-day Ramsey Canyon or was farther south in Mexico. A tiny hummingbird with a short bill and a short, rounded or double-rounded, rufous-based tail with white tips. **Adult males** have an elongated magenta-rose gorget. Females and immatures closely resemble female-type Calliope Hummingbird (page 308), which have darker tails that fall shorter than or equal to wing tips; on Bumblebee, tail extends beyond wing tips.

Amazon Kingfisher *Chloroceryle amazona* L 11¼" (29 cm)

Tropical species found from Tamaulipas and southern Sinaloa, Mexico, to Argentina and Uruguay. One substantiated record, a **female** at Laredo, TX, 24 Jan. to 3 Feb. 2010. Almost as large as Belted Kingfisher, but colored like Green Kingfisher (page 310). From Green, much larger with more massive bill, lack of white on wings, and a tufted crest. Male with cinnamon band, female with single solid green band (Green has two green bands). Calls include low, slightly raspy *check*.

Eurasian Hoopoe *Upupa epops* L 10½" (27 cm)
Widespread Old World species. One North American record, a
specimen *(saturata)* from Old Chevak, Yukon-Kuskokwim Delta,
AK, 2 to 3 Sept. 1975. Unmistakable: pinkish brown coloration; long
crest (sometimes raised); long, thin, slightly decurved bill. Striking
black-and-white wing pattern in flight; wingbeats slow and floppy.
The two subspecies from equatorial Africa and farther south and
from Madagascar are treated as separate species by some authors.

adult
saturata

Eurasian Wryneck *Jynx torquilla* L 6" (17 cm)
Widespread in Old World. Two fall records from AK: a specimen
(chinensis) from Cape Prince of Wales, 8 Sept. 1945, and one photographed at Gambell, St. Lawrence Island, 2 to 5 Sept. 2003. A
dead bird found in southern IN was believed to have been artificially
transported. Patterned in browns and grays, wrynecks are quite
unlike a woodpecker, except for the sharply pointed bill. Note dark
mask, dark vertical band on sides of back, and the long and sparsely
barred tail. Often forages on ground, but also perches on branches in
a somewhat horizontal posture.

adult

Greenish Elaenia *Myiopagis viridicata* L 5½" (14 cm)
Resident in Mexico from southern Durango and southern Tamaulipas,
south to northern Argentina. Recorded once in North America on upper
TX coast at High Island, 20 to 23 May 1984. Overall greenish above with
a contrasting grayish head and dark eye stripe with distinct, pale supercilium. Primaries edged with olive; bright yellowish on secondaries;
short primary projection. Grayish throat and olive breast contrast with
yellow belly. Distinctive call note, a high, thin, and descending *seei-seeur*.

White-crested Elaenia *Elaenia albiceps* L 6" (15 cm)
South American species. Six subspecies arranged into three groups.
Southern and most migratory subspecies group, *chilensis* (only subspecies), breeds in central and southern Chile and extreme southwestern Argentina. Winters north to northern Peru, smaller numbers
elsewhere, casually to Falklands. One *(chilensis)* well substantiated
(photo and sound recordings) at South Padre Island, Cameron
County, TX, 9 to 10 Feb. 2008. Overall grayish olive above, paler
below with bold wing markings and broad white area on crown. Call
a *feeoo* with downward inflection. Another photographed *elaenia* from
Santa Rosa Island, northwestern FL, 28 Apr. 1984, thought to be a
Caribbean Elaenia *(E. martinica)*, was perhaps this species.

chilensis

Social Flycatcher *Myiozetetes similis* L 6¾-7¼" (17-18 cm)
Common from northeastern and northwestern Mexico to northeastern
Argentina. Only one fully documented certain record at Bentsen-Rio
Grande Valley State Park, TX, 7 to 14 Jan. 2005; one at Anzalduas
County Park, TX, 17 Mar. to 5 Apr. 1990 is controversial. Suggests a
diminutive Great Kiskadee (page 346), but black bill is much smaller.
Adult also lacks rufous in the wings and tail, which eliminates Middle
American races of Great Kiskadee, although juvenile Social Flycatcher
does show rufous edges. The reddish orange central crown patch is
concealed by dark gray. Distinctive call is a loud *che cheechee cheechee
cheechee;* also a harsh *cree-yooo.*

adult
primulus

aurantioatrocristatus

immature ♂
uropygialis

adult ♂

adult ♀

Crowned Slaty Flycatcher

Empidonomus aurantioatrocristatus L 7" *(18 cm)* Southern subspecies, *pallidiventris*, is an austral migrant breeding from northern Bolivia and the interior of Brazil south to central Argentina. Migrates north, some to Amazonia. One record, an adult male specimen (likely *pallidiventris*) taken from near Johnsons Bayou, Cameron Parish, LA, 3 June 2008; an additional record from Panama (Dec. 2007). Dark crown with semi-concealed yellow center, distinct pale supercilum, broad dark eye stripe.

Gray-collared Becard *Pachyramphus major* L 5¾" *(15 cm)*

Tropical species found from eastern Sonora and central Nuevo León, Mexico, south to northern Nicaragua. Distinctive pale west Mexican subspecies, *uropygialis* (found south to Oaxaca), has bred as close to AZ as Yécora and Sahuaripa. One record, an **immature male** photographed at Cave Creek Canyon, AZ, 5 June 2009. Note short white supraloral stripe and graduated tail with black subterminal marks and white (adult male) to cinnamon (female) tip. Adult male *uropygialis* with black crown and back, pale gray hind collar and pale underparts; female with cinnamon crown bordered black and rufous markings on wings; underparts and hind collar pale lemon. Immature male intermediate with cinnamon on back and rump, but white markings on wings and tips to outer tail feathers. In more easterly nominate race, female has black crown and pale areas are more buffy cinnamon.

Masked Tityra *Tityra semifasciata* L 9" *(23 cm)*

Common from northwestern and northeastern Mexico to Brazil. One record from south TX at Bentsen-Rio Grande Valley State Park, 17 Feb. to 10 Mar. 1990. Large and chunky; **males** are pale gray above and whitish below with contrasting black on face, most of wings, and thick subterminal tail band. Bare skin on face and base of thick bill is pinkish red. Female is similar, but is darker and duller. Distinctive call is a double, nasal grunt, *zzzr-zzzrt.*

Yucatan Vireo *Vireo magister* L 6" *(15 cm)*

Resident on Yucatán Peninsula and its offshore islands; also on Grand Cayman and islands off Honduras. One record from Bolivar Peninsula, TX, 28 Apr. to 27 May 1984. Overall brownish above with dark eye line, but no dark lateral crown stripe as in Red-eyed Vireo (page 354). Dull whitish below with grayish brown side and flanks, short primary projection, and large, heavy bill. Call a nasal *benk,* often in a series.

Cuban Martin *Progne crytoleuca* L 7½" *(19 cm)*

Breeds in Cuba and Isle of Pines; wintering grounds unknown; presumably South America. Only acceptable record is specimen taken on 9 May 1895 at Key West, FL. Adult male like Purple Martin (page 368), but has relatively longer and more deeply forked tail; in hand, note concealed white feathers on belly. **Female** resembles female Purple, but with unmarked white belly and undertail coverts; lacks grayish collar. Separation from Caribbean (*P. dominicensis*) and Sinaloa Martins (*P. sinaloae),* with which it sometimes is treated as conspecific, is difficult; Cuban is darker overall with dark shaft smudges on the undertail coverts and usually with some shaft streaking on the breast and sides.

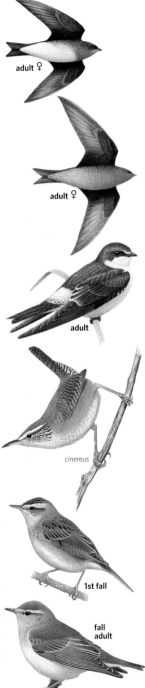

Gray-breasted Martin *Progne chalybea* L 6¾" (17 cm)
Breeds in Mexico from southern Sinaloa and southern Tamaulipas,
south to Argentina. Withdraws from northeastern portion of range
in winter. Two old specimen records for south TX: Rio Grande City,
25 Apr. 1880, and Hidalgo County, 18 May 1889; all other reports
unsubstantiated. Similar to female Purple Martin (page 368), but
smaller with a less deeply forked tail; also browner on forehead, a
less well-defined collar, and paler underparts.

adult ♀

Southern Martin *Progne elegans* L 7" (18 cm)
An austral migrant from South America. One specimen record from
Key West, FL, 14 Aug. 1890. Adult male resembles Purple Martin
(page 368), but smaller with slightly longer and more forked tail, and
in hand, lacks concealed white patch on sides and flanks. **Female** is
darker below than female Purple Martin.

adult ♀

Mangrove Swallow *Tachycineta albilinea* L 5¼" (13 cm)
A small swallow found from coastal slopes of central Sonora and
southern Tamaulipas, Mexico, south through Panama; isolated population in Peru. An adult was photographed at the Viera Wetlands in
Brevard County, FL, 18 to 25 Nov. 2000. **Adults** have an iridescent
greenish crown and back; auriculars and lores black; narrow white
line usually meets across forehead; partial white collar; and distinct
white rump. Juvenile brownish above. Seldom found far from water.

adult

Sinaloa Wren *Thryothorus sinaloa* L 5¼" (13 cm)
West Mexican endemic resident from northern Sonora to western
Oaxaca. Two records in southeastern AZ of duller northern subspecies, *cinereus*: the first one at Patagonia from 25 Aug. 2008 through fall
2009, built two nests; the other at Huachuca Canyon, Fort Huachuca,
14 to 18 Apr. 2009. Shaped like a Carolina Wren (page 384), but coloration much drabber. Note streaked face and barred undertail (also
known as Bar-vented Wren). Song consists of loud rich phrases, often
with rapid repetition of notes; wide variety of call notes. Skulking.

cinereus

Sedge Warbler *Acrocephalus schoenobaenus* L 4¾" (12 cm)
A common *Acrocephalus* breeding from northwestern Europe to western Siberia and northwestern China; winters sub-Saharan Africa. One
immature photographed at Gambell, St. Lawrence Island, AK, 30 Sept.
2007. Overall buffy brown with a wedge-shaped tail and long primary
projection. Distinctly patterned with long bold supercilium and dark
lateral and complete eye stripe with paler median crown area, and
a faintly streaked back. Immature faintly streaked across the breast.

1st fall

Wood Warbler *Phylloscopus sibilatrix* L 5" (13 cm)
Western Palearctic breeder; winters in tropical Africa. Two fall records
for western AK: one collected on Shemya Island, western Aleutians, 9
Oct. 1978; one photographed on St. Paul Island, 7 Oct. 2004. There
are at least two records for Japan. A colorful *Phylloscopus* with yellow
throat and upper breast contrasting sharply with white lower breast
and belly; yellow supercilium is well defined by a complete dark eye
line; dark-centered tertials have sharply defined pale edges. Note very
long primary projection, which contributes to short-tailed appearance.

fall
adult

Pallas's Leaf-Warbler *Phylloscopus proregulus* L 3½" (9 cm)

A tiny kinglet-size Old World Warbler breeding from southwestern Siberia east to Ammurland, Ussuriland, and Sakhalin Island, Russian Far East. Winters primarily in southeastern China, northern Indochina. Frequent vagrant to northwestern Europe. One photographed at Gambell, St. Lawrence Island, AK, 25 to 26 Sept. 2006. Distinctive; deep yellow median crown stripe, supercilium; sides of crown quite dark. Very small bill; distinct wing bars, tertial tips, yellow rump, often visible as species frequently hovers. Most now regard species as monotypic. Call is a soft, nasal rising *chuee.*

fall immature

Lesser Whitethroat *Sylvia curruca* L 5½" (14 cm)

Old World species. Only record was one photographed at Gambell, St. Lawrence Island, AK, 8 to 9 Sept. 2002. Distinctive, with a rather long tail, outer rectrices tipped with white; gray crown with a dark mask, a warm brown back and wings, and whitish underparts with a tan wash on the sides and flanks. Multiple subspecies groups are treated by Old World authorities as up to four species. Call is a hard *tik,* often repeated.

1st fall
blythi

Mugimaki Flycatcher *Ficedula mugimaki* L 5¼" (13 cm)

Highly migratory species of East Asia. Only record is from Shemya Island, AK, 24 May 1985, supported only by very marginal photos. Note long wings. Adult male is striking, with blackish head and upper parts, white wing patch, short but broad downcurving white supercilium, and extensively orange underparts. Female is brownish above, burnt orange on throat and breast; has two thin pale wing bars. **First-year male** closer to female, but with partial, broad and downcurving supercilium. In Asian range feeds from mid- to upper canopy.

1st year ♂

Spotted Flycatcher *Muscicapa striata* L 6" (15 cm)

Widespread breeder in the Palearctic, east to about Lake Baikal; winters in Africa, south of the Sahara. One photographed at Gambell, St. Lawrence Island, AK, 14 Sept. 2002. Overall grayish brown above, with fine streaking on crown and forecrown, and indistinct whitish eye ring; lacks distinct malar and submoustachial markings; below indistinctly streaked, not spotted, across throat and breast. Juvenile is spotted with buff above and has dark mottling below.

adult

Rufous-tailed Robin *Luscinia sibilans* L 5¼" (13 cm)

A small East Asian chat, breeding in northeastern Asia as close to AK as Kamchatka; winters from southeast China to Indochina, a few to Thailand. Three AK records: two from Attu Island, 4 June 2000 (supported by marginal photos) and 4 June 2008 (specimen of a female); another photographed on St. Paul Island, Pribilofs, 8 June 2008. Superficially resembles a small nominate *guttatus* Hermit Thrush (page 398), but longer legged. Brown upperparts with contrasting rufous upper tail coverts and tail; pale below with white throat and dark scaling across breast and down sides, perhaps fainter in immatures, especially when worn. Also note pale eye ring and supraloral line and slight whitish insertion on sides of neck. Quite secretive (skulks in wet forested gullies in Asia); often shivers tail, but less so and not for sustained periods as in Siberian Blue Robin. Call is a low *tuc-tuc.*

Siberian Blue Robin *Luscinia cyane* L 5½" (14 cm)

Highly migratory Asian species. One certain North American specimen record from Attu Island, AK, 21 May 1985; an additional spring
sighting of an adult male from YT is disputed. Adult male is deep blue
above, clear white below. **Adult female** brownish above with faint buffy
eye ring; buffy wash across breast with faint stippling, and most have
some bluish on tail (lacking on immature females). Immature male
has some blue on scapulars, wings, and tail. Frequently vibrates tail.

adult ♀

Brown-backed Solitaire *Myadestes occidentalis*

L 8¼" (21 cm) Found in mountains from northern Sonora, Mexico, as
close as Sierra Huachinera, some 80 miles south of AZ, and southern
Nuevo León to central Honduras. A singing bird was in Miller Canyon,
Huachuca Mountains, 16 July 2009, then present 18 July to 1 Aug. in
nearby Ramsey Canyon. A worn bird from Madera Canyon, Santa
Rita Mountains, AZ, 4 to 7 Oct. 1996, now accepted too. Shape and
coloration like Townsend's Solitaire (page 396), but upperparts
brown, wing plainer, and with white eye crescents. Amazing song
with initial hesitant call notes, accelerating into a long rocking gargle
of flute-like notes.

occidentalis

Orange-billed Nightingale-Thrush

Catharus aurantiirostris L 6½" (17 cm) Widespread in Neotropics.
Two migration records from south TX: one photographed in hand,
8 Apr. 1996, at Laguna Atascosa National Wildlife Refuge; and a
specimen from Edinburg on 28 May 2004. Remarkable was a singing
bird on territory along Iron Creek, Spearfish Canyon, Black Hills, SD,
10 July to 19 Aug. 2010. Orange-brown above, pale gray and whitish
below with distinctive bright orange bill, legs, and orbital ring. In flight,
lacks pale underwing bar of northern breeding *Catharus* thrushes.

Black-headed Nightingale-Thrush *Catharus mexicanus*

L 6½" (17 cm) Found from northeastern Mexico to western Panama.
Only record was one at Pharr, in south TX, 28 May to 29 Oct. 2004.
Distinctive, with blackish crown and face, whitish throat and belly;
otherwise gray below. Bright orange orbital ring, bill, and legs.

Eurasian Blackbird *Turdus merula* L 10½" (27 cm)

Palearctic species. A **male** was found dead on 16 Nov. 1994 at
Bonavista, Newfoundland. Other records, mainly from Ontario, are
of uncertain origin. A dozen Greenland records. Male is all black;
orange-yellow orbital eye ring and bill (duller on immatures). Female
is browner with pale throat and dark-streaked chest; soft parts duller.

♂ *merula*

Song Thrush *Turdus philomelos* L 8-9¼" (20-23 cm)

Old World species found from Europe and Scandinavia to about Lake
Baikal, Russia. Northern and eastern populations migratory. Annual in
very small numbers to Iceland, chiefly in fall; one Greenland specimen.
One record from St.-Fulgence, eastern QC, 11 to 17 Nov. 2006. Larger
than all *Catharus* thrushes. Plumage vaguely suggestive of Swainson's
Thrush (page 398), but much more heavily and extensively marked
below with arrow-shaped spots; auricular is strongly patterned. Call is
a sharp *tick*, unlike any *Catharus* thrush.

fall immature

Red-legged Thrush   *Turdus plumbeus* L 10½" (27 cm)

West Indian species found on northern Bahamas, Cuba, Cayman Brac, Hispaniola, Puerto Rico, and Dominica; formerly (until late 19th century) Honduras's Swan Islands. One FL record: one photographed at Maritime Hammock Sanctuary, Melbourne Beach, 31 May 2010. Six subspecies; FL record of nominate *plumbeus* with white chin and black throat; overall coloration dark lead gray except for prominent white tail tips, red legs, and red orbital ring. Prefers woodlands and thickets.

plumbeus

Citrine Wagtail   *Motacilla citreola* L 6½" (17 cm)

Palearctic species that winters farther north than either Western or Eastern Yellow Wagtail. Single record at Starkville, MS, 31 Jan. to 1 Feb. 1992. All plumages have gray back and bold, well-defined white wing bars. Breeding adult male has bright yellow head and underparts and a dark nape. Adult female and **winter adult male** have yellow-centered, grayish brown auriculars completely surrounded by yellow. Immatures lack all yellow, but have a similar face pattern. Call is loud buzzy *tsweep,* like Eastern Yellow Wagtail.

winter adult ♂

Tree Pipit   *Anthus trivialis* L 6" (15 cm)

Palearctic breeder, wintering mainly in Africa south of the Sahara and in India. Three records for western AK: a specimen from Cape Prince of Wales, 23 June 1972, two photographed at Gambell, St. Lawrence Island, 6 June 1995 and 21 and 27 Sept. 2002. Resembles Olive-backed Pipit (page 414), but browner above with distinct back streaks; face pattern more blended, no dark cheek spot; fine flank streaks; call similar.

Gray Silky-flycatcher   *Ptilogonys cinereus* L 7½" (20 cm)

Resident (some seasonal movement) from northern Mexico to Guatemala; largely montane. Two accepted TX records: Laguna Atascosa National Wildlife Refuge, 31 Oct. to 11 Nov. 1985, and El Paso, 12 Jan. to 5 Mar. 1995. Four southern CA records have been questioned on origin. **Adult males** are crested and gray with bright yellow undertail coverts; note white eye ring and white base of tail. Females and juveniles are similar, but duller.

adult ♂

Bachman's Warbler   *Vermivora bachmanii* L 4¾" (12 cm) **E**

Probably extinct; the last definite record was in 1962 near Charleston, SC. Once bred in canebrakes and wet woodlands; was known very locally in the southeastern U.S., from southeastern MO and Logan County in southern KY east to SC, but was probably never numerous. Most migrants were taken at Key West, FL, and at Mandeville, LA. Casual to NC and VA. Wintered in Cuba and on Isle of Pines. Bill is very thin, long, somewhat downcurved; undertail coverts white in both sexes. **Male** has yellow forehead, chin, and shoulders; black crown and bib. Immature male has less black on crown and throat, less yellow on shoulders, and more white on lower belly. **Female** drabber, crown gray, throat and breast gray or yellow. Distinctive song, typically a rapid series of buzzes on one pitch; similar to Blue-winged Warbler's alternative song.

adult ♀

adult ♂

Worthen's Sparrow *Spizella wortheni* L 5½" (14 cm)
Critically endangered. Only extant populations are in northeastern
Mexico in Coahuila and Nuevo León. Only U.S. record at Silver City,
NM, 16 June 1884, the type specimen; probably part of a small resi-
dent population, subsequently extirpated. Resembles *arenacea* Field
Sparrow (page 466), but crown solidly rufous, rump grayish, legs and
feet dark; vocalizations differ.

Pine Bunting *Emberiza leucocephalos* L 6½" (17 cm)
Primarily an eastern Palearctic species. Two fall records from Attu
Island, AK, both males: one photographed, 18 to 19 Nov. 1995; and
a specimen (nominate *leucocephalos*), 6 Oct. 1993. Breeding male is
rusty overall, with chestnut head, and bold white eyebrow and cheek
patch; in winter, duller head lacks white crown patch. Female is duller
still with a weak malar, but has a rusty eyebrow and at least a small
white spot at rear of cheek.

fall adult ♂
leucocephalus

Yellow-browed Bunting *Emberiza chrysophrys* L 6" (15 cm)
Breeds southeastern Siberia, winters in central and eastern China.
One photographed at Gambell, St. Lawrence Island, AK, 15 Sept.
2007. Distinctive with broad well-defined yellowish supercilium; pale
spot in rear of ear coverts. The white outer tail feathers are characteristic of most other *Emberiza*, but not the vaguely similar, smaller, and
shorter tailed Savannah Sparrow (page 472). Often appears slightly
crested. Call is a short *ziit*.

fall

Yellow-throated Bunting *Emberiza elegans* L 6" (15 cm)
East Asian species. One record of a male photographed on Attu Island,
AK, 25 May 1998. Note prominent crest. Yellow-and-black head pattern of **adult male** is striking. Female and immature male are similar,
but duller, with brownish auriculars.

adult ♂

Tawny-shouldered Blackbird *Agelaius humeralis*
L 8" (20 cm) Resident on Cuba and in Haiti. Only U.S. record was two
secured (specimens) at the Key West Lighthouse, FL, 27 Feb. 1936.
Smaller and slimmer than Red-winged Blackbird (page 502) and has
a slim, pointed bill; lesser coverts tawny, not red, and rear border has
a narrow blended edge. More arboreal than Red-winged and buzzy,
muffled song is more drawn out; calls differ somewhat too.

adult ♂

Eurasian Siskin *Spinus spinus* L 4¾" (12 cm)
Palearctic species. Two records of males from Attu Island, AK: a
specimen, 21 to 22 May 1993, and a sight record, 4 June 1978. About
six records from northeastern North America, but the origin of these
has been questioned; a male photographed at St.-Pierre and Miquelon
on 23 June 1983 is perhaps the most compelling. Unrecorded from
Greenland. **Male** is distinctive with black forecrown and chin, olive
above, and extensively yellow below. **Female** is much duller, the yellow
restricted to sides of breast, and a wash of yellow on face, eyebrow,
and rump; juvenile duller still. Some Pine Siskins (page 524) are very
similar, but wing coverts of Eurasian average darker.

♀

♂

SUBSPECIES MAPS

To complement the illustrations and species accounts where subspecies are detailed, we are including maps for 37 species that detail the breeding and in some cases winter and migration ranges of subspecies. We believe that this information is much more easily conveyed via a map than through words. This list is not exhaustive; note that for an additional 59 species we have included brief subspecies annotations on the maps within the main text. The subspecies chosen here are mostly field identifiable, at least to subspecies group. Where groups are identified, we have mapped them using a thicker "divider line" than used for the individual subspecies.

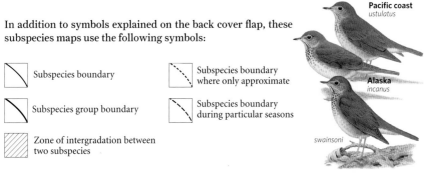

In addition to symbols explained on the back cover flap, these subspecies maps use the following symbols:

Subspecies boundary

Subspecies boundary where only approximate

Subspecies group boundary

Subspecies boundary during particular seasons

Zone of intergradation between two subspecies

Pacific coast
ustulatus

Alaska
incanus

swainsoni

Swainson's Thrush

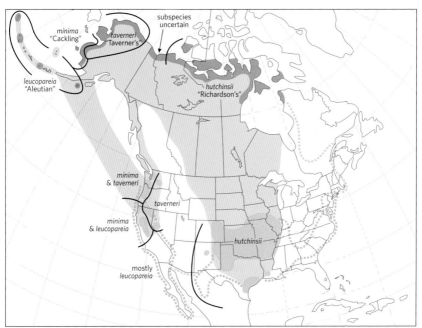

Cackling Goose, *Branta hutchinsii* (page 20)

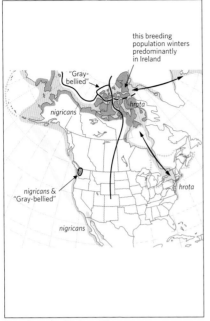

Brant, *Branta bernicla* (page 18)

Common Eider, *Somateria mollissima* (page 38)

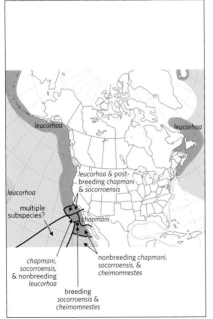

Leach's Storm-Petrel, *Oceanodroma leucorhoa* (page 96)

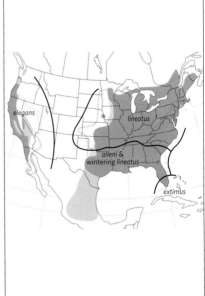

Red-shouldered Hawk, *Buteo lineatus* (page 138)

Red-tailed Hawk, *Buteo jamaicensis* (page 144)

Merlin, *Falco columbarius* (page 148)

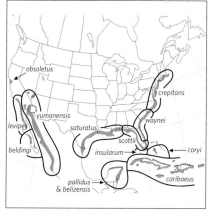

Clapper Rail, *Rallus longirostris* (page 158)

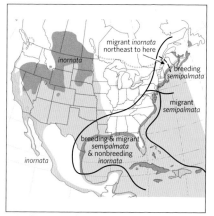

Willet, *Tringa semipalmata* (page 174)

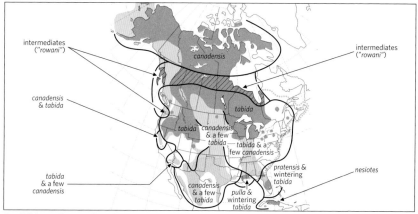

Sandhill Crane, *Grus canadensis* (page 162)

Dunlin, *Calidris alpina* (page 198)

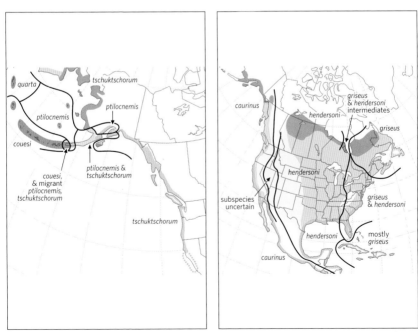

Rock Sandpiper, *Calidris ptilocnemis* (page 188)

Short-billed Dowitcher, *Limnodromus griseus* (page 202)

Common Nighthawk, *Chordeiles minor* (page 292)

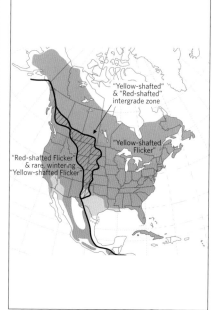

Northern Flicker, *Colaptes auratus* (page 322)

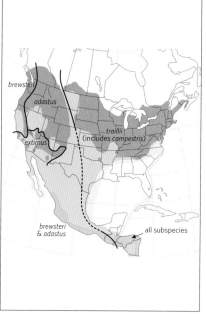

Willow Flycatcher, *Empidonax traillii* (page 330)

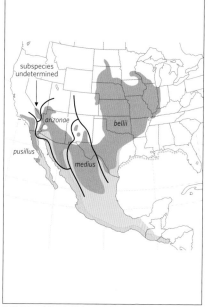

Bell's Vireo, *Vireo bellii* (page 352)

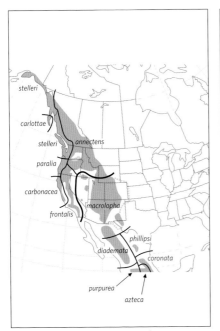

Steller's Jay, *Cyanocitta stelleri* (page 358)

Gray Jay, *Perisoreus canadensis* (page 356)

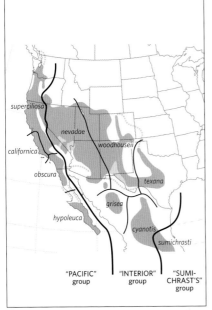

Western Scrub-Jay, *Aphelocoma californica* (page 360)

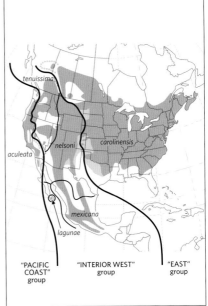

White-breasted Nuthatch, *Sitta carolinensis* (page 380)

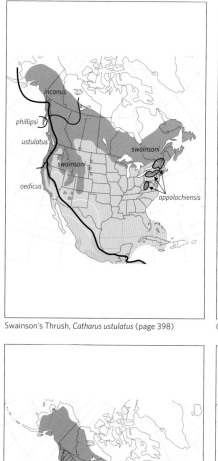

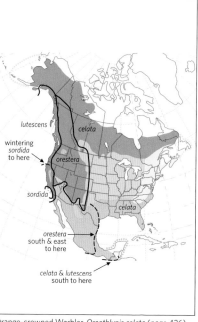

Swainson's Thrush, *Catharus ustulatus* (page 398)

Orange-crowned Warbler, *Oreothlypis celata* (page 426)

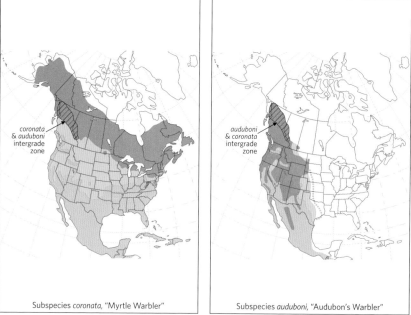

Subspecies *coronata*, "Myrtle Warbler"

Yellow-rumped Warbler, *Setophaga coronata* (page 434)

Subspecies *auduboni*, "Audubon's Warbler"

Yellow-rumped Warbler, *Setophaga coronata* (page 434)

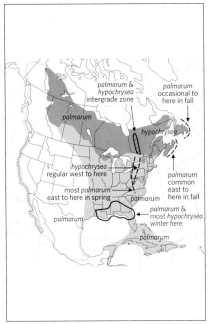

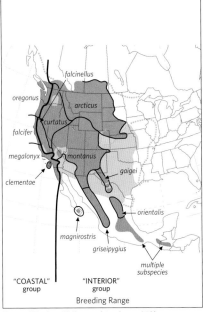

Palm Warbler, *Setophaga palmarum* (page 438)

Spotted Towhee, *Pipilo maculatus* (page 460)

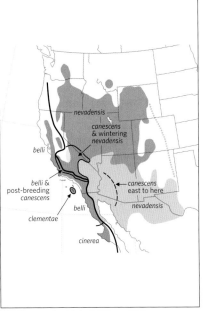

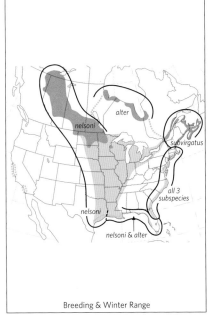

Sage Sparrow, *Amphispiza belli* (page 470)

Nelson's Sparrow, *Ammodramus nelsoni* (page 476)

Fox Sparrow, *Passerella iliaca* (page 480)

Fox Sparrow, *Passerella iliaca* (page 480)

White-crowned Sparrow, *Zonotrichia leucophrys* (page 482)

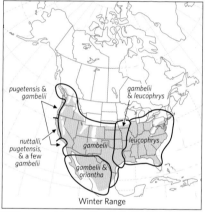

White-crowned Sparrow, *Zonotrichia leucophrys* (page 482)

Dark-eyed Junco, *Junco hyemalis* (page 484)

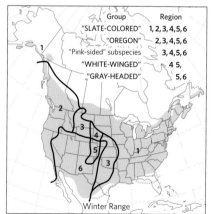

Dark-eyed Junco, *Junco hyemalis* (page 484)

Seaside Sparrow, *Ammodramus maritimus* (page 476)

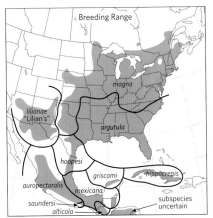

Eastern Meadowlark, *Sturnella magna* (page 500)

Common Grackle, *Quiscalus quiscula* (page 506)

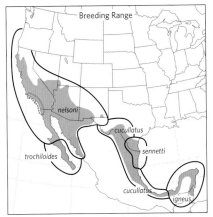

Hooded Oriole, *Icterus cucullatus* (page 510)

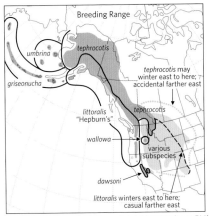

Gray-crowned Rosy-Finch, *Leucosticte tephrocotis* (page 516)

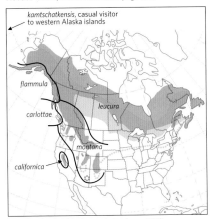

Pine Grosbeak, *Pinicola enucleator* (page 520)

APPENDIX

Light-mantled Albatross

● AOU and ABA Checklist Differences

As noted in the introduction of this book, the AOU and ABA checklist committees serve different purposes. But both produce species lists, and for the areas in common the lists are nearly identical. Updates since the publication of the fifth edition of this guide are as follows.

The ABA now accepts records of **Light-mantled Albatross** *Phoebetria palpebrata* and **European Turtle-Dove** *Streptopelia turtur.* The ABA no longer accepts the FL record of **Caribbean Elaenia** *Elaenia martinica.*

This leaves **Azure Gallinule** *Porphyrula flavirostris* as the sole difference. The record involved a specimen from Suffolk County, NY, on 14 Dec. 1986 and was initially accepted by both committees, but subsequent information revealed that it may have escaped from a local aviculturist. The ABA subsequently removed the species, but the AOU retained it on a 4-4 vote. (Two-thirds agreement is required for a motion to pass; in this case the motion was to remove.)

● Greenland

Greenland is the largest island in the world, most of it lying north of the Arctic Circle, and forms the most northeastern part of North America. More than 230 species have been recorded there, and nearly all are substantiated by specimens. The definitive ornithological reference is *An Annotated Checklist to the Birds of Greenland* (1994) by David Boertmann (*Bioscience* 38). Boertmann divides Greenland into four regions (North, West, Northeast, and Southeast) and details the bird distribution for each. Greenland's avifauna is a mix of both Palearctic and Nearctic species, many of which are strays respectively from Europe (including Iceland) and mainland North America. The breeding avifauna includes Pink-footed, Greater White-fronted (the endemic breeding subspecies *flavirostris*), and Barnacle Geese, White-tailed Eagle, Fieldfare, Redwing, White Wagtail (nominate *alba*), and Meadow Pipit. Three subspecies of Rock Ptarmigan are found in Greenland, two of which (*saturata* and *capta*) are endemic. Two subspecies of Dunlins breed in Greenland, one of which (*arctica*) is an endemic breeder in the northeast; the other (*schinzii*) breeds in southern Greenland as well as northwestern Europe. Merlin specimens from Greenland are of *subaesalon*, an endemic breeder on Iceland; also recorded is the more widespread *aesalon*, breeding mainly in northern Europe. Red Crossbills collected are of nominate *curvirostra* from the Palearctic. These, along with the following list of species *not* recorded in our area of coverage, should alert observers to the potential visitors to northeastern North America. For polytypic species, the trinomial subspecies name is given, if known:

Eurasian Spoonbill *Platalea leucorodia leucorodia.* One Oct. record (1909) for the West. **Ruddy Shelduck** *Tadorna ferruginea.* Four collected in the summer of 1892, an invasion year for the species in northwestern Europe (see also page 48). **Common Scoter** *Melanitta nigra*, recently split from Black Scoter (page 42). Casual to Greenland from northwestern Europe. **Water Rail** *Rallus aquaticus hibernans.* Four records. The subspecies *hibernans* is endemic to Iceland. **Spotted Crake** *Porzana porzana.* Eleven records, nearly all in fall, from western

Redwing

White Wagtail
alba

Greenland. **Oriental Plover** *Charadrius veredus.* One May record (1948) from western Greenland. A remarkable record, as it breeds on the steppes of East Asia and winters in Australia. **Rook** *Corvus frugilegus frugilegus.* One Mar. record (1901) for the Southeast. **Carrion Crow** *Corvus corone cornix.* Two spring records (1897, 1907) were of the "Hooded Crow." **Meadow Pipit** *Anthus pratensis pratensis.* Scarce breeder in eastern Greenland. **White's Thrush** *Zoothera aurea aurea.* One Oct. record (1954) for the Northeast. It is here treated as a separate species from the Scaly Thrush, *Z. dauma,* following other authorities. **Blackcap** *Sylvia atricapilla atricapilla.* One Nov. record (1916) for the Southeast. **Lesser Redpoll** *Acanthis cabaret.* One Sept. record (1933) for the Southeast.

Dunlin
schinzii

● Bermuda

Bermuda is a series of some 300 small islands, most of which are uninhabited. It is best known as the only breeding site of the Bermuda Petrel (also called the Cahow), which was discovered early in the 17th century and then thought to be extinct until it was rediscovered in 1951. The White-tailed Tropicbird reaches its northernmost breeding range here. The only endemic breeding land bird is a nonmigratory subspecies of White-eyed Vireo *(bermudianus).* Established exotics not found in North America include European Goldfinch, Common Waxbill, and Orange-cheeked Waxbill. Great Kiskadee (page 346) is a well-established, introduced species in Bermuda. Bermuda is well known for its migrants and vagrants, which make up most of Bermuda's extensive species list—in excess of 360 species. Surprising northern species that have occurred include Northern Hawk Owl, Snowy Owl, Bohemian Waxwing, White-winged Crossbill, and Pine Grosbeak. The single Snowy Owl (1987) took to preying upon the endangered Bermuda Petrels and had to be collected. (Sometimes conservationists have to make hard choices!) Most of the vagrants come from North America (including the West) or Europe, but a few (Large-billed Tern and Fork-tailed Flycatcher) are from South America. Red-necked Stint and, even more surprisingly, Dark-sided Flycatcher (a late Sept. specimen of nominate *sibirica*) are from Asia. The most recent birding references for Bermuda are *A Guide to the Birds of Bermuda* (1991) by Eric Amos and *A Birdwatching Guide to Bermuda* (2002) by Andrew Dobson.

Bermuda Petrel

White-tailed Tropicbird

The following are the species recorded from Bermuda but not from mainland North America:

West Indian Whistling-Duck *Dendrocygna arborea.* One record (1907). This species is resident on some of the Bahamian islands. (A recent record from VA is of uncertain origin.) **Ferruginous Duck** *Aythya nyroca.* A winter sight record (1987). **Striated Heron** *Butorides striata.* One record (1985) of a long-staying bird. **Booted Eagle** *Hieraaetus pennatus.* A Sept. sight record (1989). **White Tern** *Gygis alba.* A remarkable Dec. record (1972). Photographic evidence indicates that it was not the expected nominate subspecies from the south Atlantic but, rather, one of the Pacific subspecies!

ART CREDITS

The following artists contributed the illustrations for the 6th edition: Jonathan Alderfer, David Beadle, Peter Burke, Marc R. Hanson, Cynthia J. House, H. Jon Janosik, Donald L. Malick, Killian Mullarney, Michael O'Brien, John P. O'Neill, Kent Pendleton, Diane Pierce, John C. Pitcher, H. Douglas Pratt, David Quinn, Chuck Ripper, N. John Schmitt, Thomas R. Schultz, Daniel S. Smith, Patricia A. Topper, and Sophie Webb.

Front cover—P. A. Topper; **Visual index**—various artists credited below; **Title page**—J. Alderfer; **6**—T. R. Schultz; except Wilson's Storm Petrel by J. Alderfer; **7**—D. L. Malick; except Cerulean Warbler by H. D. Pratt; **8**—C. J. House; except Short-billed Dowitcher by J. Alderfer; and Alder and Willow Flycatchers by D. Beadle; **9**—N. J. Schmitt; except Western Tanager by P. Burke; and Hammond's Flycatcher by D. Beadle; **10**—J. Alderfer; **11**—J. Alderfer; except American Black Duck by C. J. House; Common Tern by T. R. Schultz; and *Myiarchus* tails by P. Burke; **12**—J. Alderfer; except Greater and Lesser Yellowlegs by J. C. Pitcher; Eurasian Hobby by N. J. Schmitt; and Eurasian Hoopoe by D. Quinn; **13**—D. L. Malick; except Eurasian Collared-Dove by J. Alderfer and N. J. Schmitt; **15**—K. Mullarney; except Greater White-fronted Goose by T. R. Schultz; **17**—C. J. House; except Pink-footed Goose by K. Mullarney; **19**—C. J. House; except Brant by J. Alderfer; **21**—C. J. House; except Cackling Goose (*taverneri*, *hutchinsii*, and flight) and Canada Goose (*parvipes*) by N. J. Schmitt; **23**—C. J. House; except Tundra and Trumpeter Swan (heads) by K. Mullarney; and Whooper Swan by N. J. Schmitt; **25**—C. J. House; except Muscovy Duck (flight) by N. J. Schmitt; **27**—C. J. House; except Mottled Duck (*fulgivula*) by T. R. Schultz; and American Black Duck (head) by N. J. Schmitt; **29**—C. J. House; except Eastern Spot-billed Duck by J. Alderfer; Green-winged Teal (female) by N. J. Schmitt; and Baikal Teal by K. Mullarney; **31**—C. J. House; **33**—C. J. House; except Cinnamon Teal (female) by N. J. Schmitt; and Garganey by K. Mullarney; **35**—C. J. House; **39**—C. J. House; except Common Eider (heads and flight) by J. Alderfer; **41**—C. J. House; except Harlequin Duck (adult males) by J. Alderfer; **43**—C. J. House; except Common Scoter and White-winged Scoter (*fusca*) by K. Mullarney; and White-winged Scoter (*stejnegeri*) by J. Alderfer; **45**—C. J. House; except goldeneye hybrid by J. Alderfer; **47**—C. J. House; except Common Merganser (*merganser*) by J. Alderfer; **49**—C. J. House; except Egyptian Goose by N. J. Schmitt; **50**—C. J. House; except Muscovy Duck by N. J. Schmitt; **51**—C. J. House; except Garganey and Baikal Teal by K. Mullarney; **52**—C. J. House; except Common Eider by J. Alderfer; **53**—C. J. House; **55**—K. Pendleton; **57**—K. Pendleton; except Northern Bobwhite (*floridanus* and *taylori*) by N. J. Schmitt; **59**—N. J. Schmitt; except Gray Partridge and Chukar (standing) by K. Pendleton; **61–65**—K. Pendleton; **67**—K. Pendleton; except Gunnison Sage-Grouse, Sharp-tailed Grouse (displaying male), and Greater Sage-Grouse (displaying male) by J. Alderfer; **69–71**—D. Quinn; **73**—J. Alderfer; **75**—H. J. Janosik; except Clark's Grebe (winter adult) and Western Grebe (winter adult) by J. Alderfer; **77–79**—J. Alderfer; **81**—J. Alderfer; except Great-winged Petrel by D. Beadle; **83**—J. Alderfer; **85**—M. O'Brien; **87**—J. Alderfer; except Herald Petrel by M. O'Brien; **89**—J. Alderfer; except Flesh-footed Shearwater, Short-tailed Shearwater (with dark underwing), and Sooty Shearwater (flight) by M. R. Hanson; **91**—J. Alderfer; **93**—M. R. Hanson; except Manx Shearwater (small flight and swimming) and Black-vented Shearwater by J. Alderfer; **95–99**—J. Alderfer; **101**—H. J. Janosik; except White-tailed Tropicbird (*dorotheae*) by J. Alderfer; **103**—J. Alderfer; except Magnificent Frigatebird by H. J. Janosik; **105–109**—J. Alderfer; **107**—J. Alderfer; **111**—H. J. Janosik; except Brown Pelican (*californicus*) by J. Alderfer; **113**—P. Burke; except Great Blue Heron by D. Pierce; **115**—D. Pierce; except Cattle Egret (flight

and *coromandus*) and Snowy Egret (flight) by T. R. Schultz; and Little Egret and Great Egret (*modesta*) by D. Quinn; **117**—D. Pierce; except Little Blue Heron (flight) and Reddish Egret by T. R. Schultz; **119**—P. Burke; except Green Heron by D. Pierce; **121**—P. Burke; except White and Scarlet Ibises by D. Pierce; **123**—D. Pierce; except American Flamingo (flight) by J. Alderfer; **125**—D. L. Malick; **127**—K. Pendleton; except Osprey by D. L. Malick; and Hook-billed Kite (flying adult female, flying pale juvenile, and perched juvenile) by N. J. Schmitt; **129**—D. L. Malick; **131**—D. L. Malick; except Bald Eagle (3rd year) by N. J. Schmitt; **133**—N. J. Schmitt; **135**—D. L. Malick; except three flying juveniles by N. J. Schmitt; **137**—N. J. Schmitt; except Common Black-Hawk (perched juvenile) and Zone-tailed Hawk (perched) by D. L. Malick; **139**—N. J. Schmitt; **141**—N. J. Schmitt; except Short-tailed Hawk (perched) by D. L. Malick; **143**—D. L. Malick; except White-tailed Hawk (flight) and Ferugginous Hawk (perched juvenile and flying dark-morph adult) by N. J. Schmitt; **145**—N. J. Schmitt; except Rough-legged Hawk by D. L. Malick; **147**—D. L. Malick; except Eurasian Hobby and Aplomado Falcon (flight) by N. J. Schmitt; **149**—D. L. Malick; except American Kestrel (dorsal flight), Eurasian Kestrel (flight), and Merlin (*suckleyi* and flight) by N. J. Schmitt; **151**—D. L. Malick; except all flight by K. Pendleton; **152**—K. Pendleton; except Hook-billed Kite by N. J. Schmitt; **153**—N. J. Schmitt; **154**—N. J. Schmitt; except Ferruginous Hawk, Rough-legged Hawk, and Swainson's Hawk by K. Pendleton; **155**—K. Pendleton; **157**—M. R. Hanson; **159**—T. R. Schultz; **161**—M. R. Hanson; except Purple Swamphen and Eurasian Moorhen by D. Quinn; **163**—D. Pierce; **165**—J. Alderfer; **167**—K. Mullarney; except European Golden-Plover by J. Alderfer; and Wilson's Plover by J. C. Pitcher; **169**—J. C. Pitcher; except all flight figures by D. S. Smith; **171**—K. Mullarney; except Mountain Plover (dorsal flight) and Eurasian Dotterel (flight) by D. S. Smith; and Northern Jacana by H. J. Janosik; **173**—H. J. Janosik; except American Oystercatcher (*frazari*) by T. R. Schultz; **175**—C. J. Pitcher; except Willet by M. O'Brien; and both flying yellowlegs by D. S. Smith; **177**—J. C. Pitcher; except Marsh Sandpiper and Common Redshank by D. Quinn; and Common Greenshank (flight) and Spotted Redshank (flight) by D. S. Smith; **179–181**—J. C. Pitcher; **183**—K. Mullarney; except Upland Sandpiper (flight) and Whimbrel by D. S. Smith; and Little Curlew by N. J. Schmitt; **185**—N. J. Schmitt; except all flight figures and Long-billed Curlew by D. S. Smith; **187**—J. Alderfer; except flight figures of Bar-tailed and Marbled Godwits by D. S. Smith; **189**—J. C. Pitcher; except Black Turnstone (flight) by J. Alderfer; and Rock and Purple Sandpipers (flight) by D. S. Smith; **191**—T. R. Schultz; except Surfbird by J. C. Pitcher; and Red Knot (flight) and Sanderling (flight) by D. S. Smith; **193**—J. C. Pitcher; except Semipalmated Sandpiper (breeding female) by T. R. Schultz; **195**—J. C. Pitcher; except Temminck's Stint (flight) by D. S. Smith; **197**—J. C. Pitcher; except White-rumped and Baird's Sandpipers (flight) by D. S. Smith; **199**—T. R. Schultz; except all flight figures by D. S. Smith; and Stilt Sandpiper (standing figures) by J. Alderfer; **201**—K. Mullarney; except all flight figures by D. S. Smith; **203–205**—J. Alderfer; **207**—K. Mullarney; **208**—D. S. Smith; except Little Ringed Plover by K. Mullarney; **209**—D. S. Smith; **210**—D. S. Smith; except Common Redshank by D. Quinn; and all phalaropes by K. Mullarney; **211**—D. S. Smith; **213–249**—T. R. Schultz; **251**—T. R. Schultz; except Dovekie (breeding adults and swimming winter) by C. Ripper; and Dovekie (flying winter) by J. Alderfer; **253**—C. Ripper; except Thick-billed Murre and Common Murre (flight) by J. Alderfer; **255**—C. Ripper; except Black Guillemot (all *mandtii*) by J. Alderfer; **257**—C. Ripper; except Long-billed Murrelet by J. Alderfer; **259**—C. Ripper; **261**—J. Alderfer; **263**—J. Alderfer; except Atlantic Puffin by C. Ripper; **265**—H. D. Pratt; **267**—N. J. Schmitt and J. Alderfer; except Zenaida

560

Dove (female) by J. Alderfer; **269**—D. L. Malick; except Spotted Dove and Oriental Turtle-Dove by N. J. Schmitt and J. Alderfer; **271**—D. L. Malick; except Budgerigar and Rosy-faced Lovebird by N. J. Schmitt; **273-277**—N. J. Schmitt; **279**—H. D. Pratt; **281**—D. Quinn; except Groove-billed and Smooth-billed Anis by H. D. Pratt; **283-291**—D. L. Malick; **293**—N. J. Schmitt; except Common Nighthawk by T. R. Schultz; and Common Pauraque by C. Ripper; **295**—N. J. Schmitt; except perched Mexican Whip-poor-will by T. R. Schultz; spread tails of Eastern Whip-poor-will, Buff-collared Nightjar, and Common Poorwill by C. Ripper; and spread tail of Mexican Whip-poor-will by S. Webb; **297**—N. J. Schmitt; **299**—N. J. Schmitt; except Elegant Trogon and Eared Quetzal by J. P. O'Neill; **301**—H. D. Pratt; except Green Violetear and Green-breasted Mango by S. Webb; **303**—H. D. Pratt; except Lucifer Hummingbird by S. Webb; **305-309**—H. D. Pratt; **311**—D. L. Malick; except Green Kingfisher (flight) by N. J. Schmitt; **313-319**—D. L. Malick; **321**—D. L. Malick; except Great Spotted Woodpecker by D. Quinn; **323**—D. L. Malick; except American Three-toed Woodpecker (bacatus) by T. R. Schultz; **325**—D. L. Malick; **327-335**—D. Beadle; **337**—H. D. Pratt; **339**—P. Burke; **341**—P. Burke; except Sulphur-bellied Flycatcher by T. R. Schultz; **343**—J. Alderfer; **345**—H. D. Pratt; except Gray Kingbird by D. Beadle; and Fork-tailed Flycatcher (immature) by N. J. Schmitt; **347**—D. Beadle; except Rose-throated Becard by J. Alderfer; and Great Kiskadee by H. D. Pratt; **349**—H. D. Pratt; except Brown Shrike by D. Quinn; and flight figures by N. J. Schmitt; **351**—H. D. Pratt; except Thick-billed Vireo by D. Beadle; **353**—H. D. Pratt; except Gray Vireo by P. Burke; **355**—H. D. Pratt; except Hutton's Vireo and Ruby-crowned Kinglet by N. J. Schmitt; **357**—D. Beadle; except Clark's Nutcracker and Gray Jay by H. D. Pratt; **359**—H. D. Pratt; **361**—N. J. Schmitt; except Western Scrub-Jay (texana) by T. R. Schultz; and Florida Scrub-Jay and Mexican Jay (juvenile) by H. D. Pratt; **363**—H. D. Pratt; except Tamaulipas and Northwestern Crows by N. J. Schmitt; **365**—N. J. Schmitt; except Chihuahuan Raven (flight) and Common Raven (flight and calling bird) by J. Alderfer; **367**—D. Beadle; **369**—H. D. Pratt; except Sky Lark and Brown-chested Martin by D. Beadle; and Purple Martin (western female) by T. R. Schultz; **371**—H. D. Pratt; except Common House-Martin by D. Quinn; **373**—H. D. Pratt; except Cave Swallow (pelodoma flight) by D. Beadle; **375-377**—M. O'Brien; **379**—M. O'Brien; except Verdin and Bushtit by J. P. O'Neill; **381**—H. D. Pratt; **383**—H. D. Pratt; except Brown Creeper by T. R. Schultz; and Cactus Wren (nest) by N. J. Schmitt; **385**—H. D. Pratt; except Winter Wren and Bewick's Wren (calaphonus) by N. J. Schmitt; and Pacific Wren (pacificus) by D. Beadle; **387**—N. J. Schmitt; except American Dipper by H. D. Pratt; **389**—H. D. Pratt; except Wrentit by M. O'Brien; **391-395**—D. Quinn; **397**—H. D. Pratt; **399**—T. R. Schultz; **401**—D. Quinn; except Wood Thrush by T. R. Schultz; **403**—H. D. Pratt; except White-throated Thrush by P. Burke; **405**—H. D. Pratt; except Blue Mockingbird by J. Alderfer; and Northern Mockingbird (flight) by N. J. Schmitt; **407**—H. D. Pratt; except Sage and California Thrashers by N. J. Schmitt; **409**—N. J. Schmitt; **411**—H. D. Pratt; except Common Myna by J. Alderfer; **413**—H. D. Pratt; except Siberian Accentor by D. Quinn; **415**—D. Quinn; **417**—H. D. Pratt; **419-421**—D. Pierce; **423**—D. Pierce; except Snow Bunting (1st winter female flight) by J. Alderfer; **425**—H. D. Pratt; **427**—H. D. Pratt; except Nashville Warbler by T. R. Schultz; **429**—H. D. Pratt; except Virginia's Warbler by T. R. Schultz; and Crescent-chested Warbler by D. Beadle; **431**—H. D. Pratt; except Yellow Warbler by T. R. Schultz; **433**—H. D. Pratt; **435**—H. D. Pratt; except Yellow-rumped Warbler (fall female coronata) by T. R. Schultz; **437**—H. D. Pratt; **439**—H. D. Pratt; except Palm Warbler by T. R. Schultz; **441**—H. D. Pratt; **443**—H. D. Pratt; except Bay-breasted Warbler (fall male) by D. Beadle; **445**—H. D. Pratt; except

Black-and-white Warbler by T. R. Schultz; **447**—H. D. Pratt; except Louisiana and Northern Waterthrushes by T. R. Schultz; **449**—T. R. Schultz; **451**—H. D. Pratt; except Wilson's Warbler by T. R. Schultz; **453**—P. Burke; except Common Yellowthroat and Yellow-breasted Chat (female auricollis) by T. R. Schultz; **455**—H. D. Pratt; except Red-faced Warbler by D. Beadle; and Golden-crowned Warbler and Rufous-capped Warbler by P. Burke; **457**—H. D. Pratt; except Western Spindalis by T. R. Schultz; **459**—P. Burke; except White-collared Seedeater (female and 1st-winter male) and Black-faced Grassquit by H. D. Pratt; **461-463**—P. Burke; **465**—T. R. Schultz; **467**—D. Pierce; except Rufous-crowned Sparrow (juvenile) by T. R. Schultz; **469**—T. R. Schultz; **471**—D. Pierce; except Sage Sparrow (canescens) by D. Beadle; **473**—D. Pierce; except Vesper Sparrow by D. Beadle; and Savannah Sparrow by T. R. Schultz; **475**—T. R. Schultz; except Grasshopper Sparrow by T. R. Schultz, and Orange Bishop by N. J. Schmitt; **477**—N. J. Schmitt; except Seaside Sparrow by D. Pierce; **479**—D. Pierce; except Lincoln's Sparrow and Song Sparrow (juvenile) by T. R. Schultz; **481**—D. Pierce; except Fox Sparrow (iliaca) by T. R. Schultz; **483**—D. Pierce; except White-crowned Sparrow by T. R. Schultz; **485**—D. Pierce; except Dark-eyed Junco (montanus and mearnsi) by D. Beadle; and Dark-eyed Junco (flight) by N. J. Schmitt; **487**—D. Quinn; **489-491**—P. Burke; **493**—D. Pierce; except Northern Cardinal (superbus) by D. Beadle; **495**—D. Pierce; **497**—D. Pierce; except Painted Bunting by T. R. Schultz; **499-501**—T. R. Schultz; **503**—H. D. Pratt; except Bobolink by T. R. Schultz; and Tricolored Blackbird (males) by N. J. Schmitt; **505-507**—H. D. Pratt; **509**—D. Beadle; except Brown-headed Cowbird by N. J. Schmitt; Shiny Cowbird by P Burke; and Bronzed Cowbird (female aeneus) by H. D. Pratt; **511-515**—P. Burke; **517**—N. J. Schmitt; except Gray-crowned Rosy-Finch (littoralis and male tephrocotis) by D. Beadle; and Gray-crowned Rosy-Finch (umbrina) and Brown-capped Rosy-Finch (male) by D. Pierce; **519**—T. R. Schultz; **521-523**—D. Pierce; **525**—D. Pierce; except Common Redpoll (rostratus) by N. J. Schmitt; **527**—D. Pierce; except Common Chaffinch by D. Quinn; **529**—N. J. Schmitt; **530**—J. Alderfer; except Lesser White-fronted Goose by D. Quinn; **531**—J. Alderfer; **532**—J. Alderfer; except Great and Lesser Frigatebirds by T. R. Schultz; and Yellow Bittern by D. Quinn; **533**—D. Quinn; except Bare-throated Tiger-Heron and Gray Heron by J. Alderfer; **534**—T. R. Schultz; except Chinese Pond-Heron by D. Quinn; and Paint-billed Crake by D. Beadle; **535**—D. Beadle; except Sungrebe by J. Alderfer; and Greater Sand-Plover and Collared Plover by D. Quinn; **536**—D. Quinn; except Eskimo Curlew by D. S. Smith; **537**—D. Quinn; except Swallow-tailed Gull by T. R. Schultz; **538**—J. Alderfer; except Scaly-naped Pigeon by D. Beadle; and European Turtle-Dove by D. Quinn; **539**—T. R. Schultz; except Brown Hawk-Owl by D. Quinn; and Gray Nightjar by D. Beadle; **540**—D. Beadle; except Xantus's Hummingbird by S. Webb; and Amazon Kingfisher by J. Alderfer; **541**—D. Quinn; except Greenish Eleania by D. Beadle; White-crested Elaenia by N. J. Schmitt; and Social Flycatcher by J. Alderfer; **542**—T. R. Schultz; except Masked Tityra by J. Alderfer; and Yucatan Vireo and Cuban Martin by D. Beadle; **543**—D. Beadle; except Mangrove Swallow by J. Alderfer; Sinaloa Wren by N. J. Schmitt; and Sedge and Wood Warblers by D. Quinn; **544**—D. Quinn; **545**—D. Quinn; except Brown-backed Solitaire by T. R. Schultz; and Orange-billed and Black-headed Nightingale-Thrushes by D. Beadle; **546**—D. Quinn; except Red-legged Thrush by T. R. Schultz; Gray Silky-flycatcher by J. Alderfer; and Bachman's Warbler by H. D. Pratt; **547**—D. Quinn; except Worthen's Sparrow and Tawny-shouldered Blackbird by D. Beadle; **548**—T. R. Schultz; **558**—J. Alderfer; except Redwing by D. Quinn; **559**—T. R. Schultz; except Bermuda Petrel by M. O'Brien; and White-tailed Tropicbird by H. J. Janosik.

ABOUT THE AUTHORS

Jon L. Dunn has served as chief consultant for five editions of the *National Geographic Field Guide to the Birds of North America* and co-authored the sixth. Dunn has served as a member of the California Bird Records Committee for 24 years and is currently on the American Ornithologists' Union Committee on Classification and Nomenclature of North and Middle American Birds and the American Birding Association Checklist Committee, where he has served as chair.

Jonathan Alderfer, artist and editor, has contributed extensively to several editions of the *National Geographic Field Guide to the Birds of North America* and co-authored the sixth edition. He has served on the Maryland/District of Columbia Bird Records Committee and as associate editor of the American Birding Association's magazine *Birding*.

ACKNOWLEDGMENTS

The authors wish to thank the following individuals and institutions for their valuable assistance in the preparation of the sixth edition of this guide:

Several individuals deserve special recognition for their multiple contributions to this guide: Kimball L. Garrett, Los Angeles County Museum of Natural History, Los Angeles, CA, and Daniel D. Gibson, University of Alaska Museum, Fairbanks, AK, answered numerous questions and assisted the authors and artists with issues that arose throughout the production of this edition. David Willard, Field Museum of Natural History, Chicago, IL, made available the collection at the Field Museum for Thomas R. Schultz to assist in his painting 101 new illustrations. Alan Wormington read the entire text and made numerous suggestions. David Quady reviewed and offered numerous suggestions for all of the owl accounts. Larry Sansone provided the artists with numerous digital images to assist with their illustrations. We also wish to thank the following individuals and institutions who also assisted in the preparation of this guide: John Anderton; Tony Bennett; Edward S. Brinkley; Allen Chartier; Britt Griswold; Ed Harper; Paul Hess; Bruce Mactavish; Guy McCaskie; Bill Pranty; J. V. Remsen, Jr., Museum of Natural Science, Louisiana State University, Baton Rouge, LA; Michael L. P. Retter; Noel F. R. Snyder, Philip Unitt, San Diego Natural History Museum, San Diego, CA; Jack Withrow, University of Alaska Museum, Fairbanks, AK.

Paul Lehman, the chief map consultant for the sixth edition's range maps, wishes to thank the following regional, state, and provincial consultants for their expert advice:

Christian Artuso; Giff Beaton; Edward Brinkley; Richard Cannings; Russell Cannings; Allen Chartier; Alan Contreras; Ricky Davis; Richard Erickson; Rachel Farrell; Doug Faulkner; Shawneen Finnegan; Ted Floyd; Kimball Garrett; Mary Gustafson; Matt Heindel; Steve Heinl; Paul Hess; Tom Hince; Bill Howe; Rich Hoyer;

Dan Kassebaum; Richard Knapton; Dave Krueper; Tony Leukering; Mark Lockwood; Derek Lovitch; Bob Luterbach; David Mackay; Jeff Marks; Steve McConnell; Eric Mills; Steve Mirick; Steve Mlodinow; Ted Murin; Kenny Nichols; Ron Pittaway; Bill Pranty; Nick Pulcinella; Michael L. P. Retter; Steve Rottenborn; Scott Schuette; Larry Semo; Bill Sheehan; John Sterling; Mark Stevenson; Mike Todd; Bill Tweit; Philip Unitt; Sandy Williams; Alan Wormington.

Many individuals were extremely helpful in preparing previous editions of this guide. Although much has changed since the first edition was published in 1983, the previous editions are the foundation on which the current edition is based. Erik A. T. Blom drafted the maps and was co-chief consultant with Jon L. Dunn for the first and second editions. Claudia P. Wilds was the chief consultant through the early stages of the first edition. The individuals who assisted in the preparation of previous editions are listed on the acknowledgments pages of those editions—their many and varied contributions continue to be greatly appreciated.

The following institutions assisted in many ways with the preparation of previous editions of this guide. Especially important was the loan of specimens in their care to the artists who created the illustrations:

Denver Museum of Natural History, Denver, CO; Field Museum of Natural History, Chicago, IL; Museum of Natural Science, Louisiana State University, Baton Rouge, LA; Museum of Vertebrate Zoology, University of California, Berkeley, CA; National Museum of Natural History, Smithsonian Institution, Washington, DC; Natural History Museum of Los Angeles County, Los Angeles, CA; Natural History Museum, Tring, UK; Patuxent Wildlife Research Center, Laurel, MD; Royal Ontario Museum, Toronto, Canada; San Diego Natural History Museum, San Diego, CA; Santa Barbara Museum of Natural History, Santa Barbara, CA; Western Foundation of Vertebrate Zoology, Camarillo, CA; World Museum, Liverpool, UK.

For this second printing we are especially indebted to Kenneth P. Able and Daniel D. Gibson for their helpful comments.

INDEX

The page number for the main entry for each species is listed in **boldface** type and refers to text page opposite the illustration.

**NATIONAL GEOGRAPHIC FIELD GUIDE
TO THE BIRDS OF NORTH AMERICA**
Jon L. Dunn and Jonathan Alderfer

Published by the National Geographic Society
John M. Fahey, Jr., *Chairman of the Board and Chief Executive Officer*
Timothy T. Kelly, *President*
Declan Moore, *Executive Vice President; President, Publishing*
Melina Gerosa Bellows, *Executive Vice President, Chief Creative Officer, Books, Kids, and Family*

Prepared by the Book Division
Barbara Brownell Grogan, *Vice President and Editor in Chief*
Jonathan Halling, *Design Director, Books and Children's Publishing*
Marianne R. Koszorus, *Director of Design*
Susan Tyler Hitchcock, *Senior Editor*
Carl Mehler, *Director of Maps*
R. Gary Colbert, *Production Director*
Jennifer A. Thornton, *Managing Editor*
Meredith C. Wilcox, *Administrative Director, Illustrations*

Staff for This Book
Bridget A. English, *Editor*
Jennifer Seidel, *Text Editor*
Chalkley Calderwood, *Art Director*
Linda Makarov, *Designer*
Paul Lehman, *Chief Map Researcher and Editor*
Sven M. Dolling, *Map Production Manager*
Michael McNey, *Assistant, Map Production*
Gregory Ugiansky, *Assistant, Map Production*
Judith Klein, *Production Editor*
Mike Horenstein, *Production Manager*
Robert Waymouth, *Illustrations Specialist*
Jodie Morris, *Design Assistant*

Manufacturing and Quality Management
Christopher A. Liedel, *Chief Financial Officer*
Phillip L. Schlosser, *Senior Vice President*
Chris Brown, *Technical Director*
Robert L. Barr, *Manager*
Rahsaan Jackson, *Imaging Technician*

Since 1888, the National Geographic Society has funded more than 13,000 research, exploration, and preservation projects around the world. National Geographic Partners distributes a portion of the funds it receives from your purchase to National Geographic Society to support programs including the conservation of animals and their habitats.

National Geographic Partners
1145 17th Street NW
Washington, DC 20036-4688 USA

Get closer to National Geographic explorers and photographers, and connect with our global community. Join us today at nationalgeographic.com/join

For information about special discounts for bulk purchases, please contact National Geographic Books Special Sales: specialsales@natgeo.com

For rights or permissions inquiries, please contact National Geographic Books Subsidiary Rights: bookrights@natgeo.com

Copyright © 2011 National Geographic Society. All rights reserved. Reproduction of the whole or any part of the contents without written permission from the publisher is prohibited.

NATIONAL GEOGRAPHIC and Yellow Border Design are trademarks of the National Geographic Society, used under license.

ISBN: 978-1-4262-0828-7
ISBN: 978-1-4262-2048-7 (special sale ed.)

The Library of Congress has cataloged the 4th edition as follows:

Field guide to the birds of North America
    Includes index.
    1. Birds—North America—Identification I. National Geographic Society (U.S.) II. Title: Birds of North America.
QL681.F53   1987
598.297   86-33249

Printed in China
19/RRDH/2

BIRDING WITH THE BEST

From gorgeous photography to practical and easy-to-use guides to kids-focused books, National Geographic has it all.

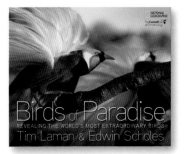

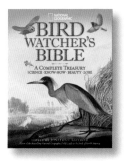

DOWNLOAD THE APP!
National Geographic Birds: Field Guide to North America
for iPhone®, iPod touch®, and iPad®

iPad, iPhone, and iPod touch are trademarks of Apple Inc.,
registered in the U.S. and other countries.
App Store is a service mark of Apple Inc.

 Like us on Facebook.com: Nat Geo Books

Follow us on Twitter.com: ©NatGeoBooks

NATIONAL GEOGRAPHIC

Copyright © 2013 National Geographic Society

AVAILABLE WHEREVER BOOKS ARE SOLD
nationalgeographic.com/books